AF183327

Soil Management

Soil Management

Edited by **Lester Bane**

Callisto Reference
New York

Published by Callisto Reference,
106 Park Avenue, Suite 200,
New York, NY 10016, USA
www.callistoreference.com

Soil Management
Edited by Lester Bane

© 2016 Callisto Reference

International Standard Book Number: 978-1-63239-771-3 (Hardback)

This book contains information obtained from authentic and highly regarded sources. Copyright for all individual chapters remain with the respective authors as indicated. All chapters are published with permission under the Creative Commons Attribution License or equivalent. A wide variety of references are listed. Permission and sources are indicated; for detailed attributions, please refer to the permissions page and list of contributors. Reasonable efforts have been made to publish reliable data and information, but the authors, editors and publisher cannot assume any responsibility for the validity of all materials or the consequences of their use.

The publisher's policy is to use permanent paper from mills that operate a sustainable forestry policy. Furthermore, the publisher ensures that the text paper and cover boards used have met acceptable environmental accreditation standards.

Trademark Notice: Registered trademark of products or corporate names are used only for explanation and identification without intent to infringe.

Printed in China.

Contents

Preface

It is often said that books are a boon to mankind. They document every progress and pass on the knowledge from one generation to the other. They play a crucial role in our lives. Thus I was both excited and nervous while editing this book. I was pleased by the thought of being able to make a mark but I was also nervous to do it right because the future of students depends upon it. Hence, I took a few months to research further into the discipline, revise my knowledge and also explore some more aspects. Post this process, I began with the editing of this book.

Soil is the basic requirement for agriculture as it stores nutrients and water. Soil management is concerned with the protection of soil and its enhancement. It includes topics such as soil chemistry, soil ecology, soil microbiology, soil fertility, soil improvement studies, etc. This book is a compilation of chapters that discuss the most vital concepts and emerging trends in the field of soil management. From theories to research to practical applications, case studies related to all contemporary topics of relevance to this field have been included in this book. It will serve as a vital tool for all researching and studying this field as it offers new insights into soil management.

I thank my publisher with all my heart for considering me worthy of this unparalleled opportunity and for showing unwavering faith in my skills. I would also like to thank the editorial team who worked closely with me at every step and contributed immensely towards the successful completion of this book. Last but not the least, I wish to thank my friends and colleagues for their support.

Editor

1

Surface Hardness as an Indicator of Soil Strength of Agricultural Soils

Gaius D. Eudoxie*, Dennison Phillips, Raymond Springer

Department of Food Production, University of the West Indies, Saint Augustine, Trinidad and Tobago.
Email: *gaius.eudoxie@sta.uwi.edu

ABSTRACT

Soil strength is an important quality of agricultural soils prone to traffic. Surface hardness (SH) measured by the clegg impact tester (CIT) was evaluated as an indicator for assessing soil strength. Proctor tests were performed on a diverse range of soils to examine the relationships between bulk density (BD), penetration resistance (PR), SH and water content. All three indices showed typical response curves with increasing water content, with notable differences among the soils. Maximum dry bulk density (MDBD), peak penetration resistance (PPR) and peak surface hardness (PSH) showed values of 1.98 $Mg \cdot m^{-3}$, 8.2 MPa and 248 Cmax for Piarco, River Estate and Piarco respectively. Corresponding critical moisture contents (CMC) were much greater for MDBD compared to PPR and PSH. SH showed a significant positive correlation with PR, but not BD. Further divulgence into the relationships between SH and other soil properties as well as crop response will facilitate greater use of the CIT.

Keywords: Clegg Impact Tester; Surface Hardness; Penetration Resistance; Moisture-Density Relationship

1. Introduction

Physical properties of agricultural soils play a crucial role in productivity and determine to some extent strategies to improve soil quality. In the Caribbean the evaluation of soil physical properties has been limited to textural distribution [1], which provides only partial information for decision making. The nature and influence of physical parameters such as density, Atterberg limits, and soil strength indices has received limited attention [2,3], whilst they dictate the timing and influence of physical soil manipulation. Further, the relationships between soil strength and other soil parameters are not fully understood [4].

Reference [5] indicated that for accurate assessment of changes in the soil fabric due to compaction, measurement of BD is not adequate. PR is frequently used in the assessment of surface strength in association with bulk density. Reference [6] indicated that PR is regarded as a useful measure of soil impedance and root growth. Although PR determination is relatively rapid compared to BD, both applications are limited at extremes of moisture content [7] where high and low soil strength interferes with point measurements, and bulk sampling respectively. Reference [2] illustrated the bell shaped response for these two parameters with increasing water content. Another limitation of PR is the relatively great spatial variation associated with a point measurement, rather than a bulk soil measurement. Reference [8] stated that to strengthen the value of point measurements of PR further studies to develop relations between PR and non-point measurements of shear strength are needed. Reference [9] highlighted the interrelatedness of BD and PR in assessing soil strength and further stated that using only one of these variables may lead to misleading results. Although these two parameters may provide adequate information for evaluating soil compaction and strength, newer technologies such as the clegg impact tester (CIT) may prove useful across a wider moisture content range.

SH measured by the CIT has been defined as the ability of a surface to absorb the impact energy created by any object striking that surface. Softer surfaces will absorb a larger percentage of the energy generated upon impact than a harder surface [10]. The CIT measures Cmax which is a measure of the strength of soil layers. This is achieved by measuring the dynamic rebound of the soil owing to a standard weight falling from a constant height. Reference [11] describes the CIT as a modified Proctor hammer with an attached accelerometer. They indicated that the unit operates on the principle that the rate at which the hammer rebounds (soil resistance) is related to soil strength, BD or hardness. The harder the surface, the higher will be the rebound and so will be the

*Corresponding author.

Cmax. Agronomy research using the CIT has been restricted to evaluating turf root zone firmness on player performance, with decreases in maximum deceleration as a result of increases in moisture content [12]. No attempt has been made to correlate surface hardness with other soil strength indices.

SH may provide quantitative data on soil strength that has not been previously documented. It was hypothesized that this in-situ test can be used as a single test to ascertain the overall strength of agricultural soils at a range of soil moisture content.

2. Materials and Methods

Six agricultural soils (**Table 1**) were selected to represent a diversity of soil physical properties. The soils were collected (0 - 20 cm), air-dried and ground to pass a 2 mm sieve, prior to characterization. Standard physical and chemical analyses were conducted. Particle size distribution was performed using the hydrometer method [13]. pH was determined in a soil to water extract of 1:2.5 [14]. Organic matter (OM) content of the soils was assessed by Walkley and Black wet oxidation method [15], except for Chagaramus which was done by the loss on ignition method due to the high content of organic matter. BD was determined using the core method [16] and particle density (PD) via the pyconometer method [17]. Atterberg limits were assessed according to [18].

Soil compaction and strength indicators were assessed on prepared samples (<4.75 mm) using the standard Proctor test explained in detail by [19]. Compaction was done in three layers at different water contents (ranging from 5% - 65% by wt) using 25 blows from a standard Proctor hammer (mass—2.5 kg; drop height—305 mm) in a cylindrical mould of diameter 102.05 mm. Moisture contents for compacting each soil were chosen based on the consistency limits. The soils were left to equilibrate over night prior to compaction. Compacted soils were

Table 1. Properties of selected soils.

Soil	Sand	Clay	PL	LL	PI	**OM**	BD	PD	pH
	←		%			→	← Mg·m^{-3} →		
River Estate	38.7	34.8	29	38	9	1.26	1.46	2.61	5.78
Cunupia	34.1	44.4	34	49	15	1.04	1.34	2.6	4.49
Piarco	71.6	12.8	ND[a]	ND	ND	0.86	1.52	2.67	4.41
Chagaramus	19.3	43.3	39	66	17	25.1[b]	0.45	1.84	4.04
Talparo	16.3	74.7	35	60	25	0.68	1.26	2.52	4.52
Princes Town	20.9	64.6	41	79	38	2.09	1.22	2.36	7.13

[a]Not determined due to low clay content; [b]Determined by loss on ignition (LOI) according to (Nelson and Sommers, 1996); PL is plastic limit, LL is liquid limit, PI is plasticity index, OM is organic matter, BD is bulk density and PD is particle density.

oven dried at 105˚C for 24 hrs and weighed to determine dry bulk density. For soils containing expanding clay minerals, volume measurements were made directly on the soil to cater for shrinkage during drying. Additionally, replicate cores were used to determine PR and SH, using a spring type cone penetrometer (Eijkelkamp; 06.01.SA) and the CIT (Simon Deakin Instruments; CIST/882) respectively. Five readings with a 1 cm^2 cone, (cone angle of 60˚), inserted to 2 cm were averaged to obtain the mean surface PR, whilst the reading corresponding to the forth drop of the hammer (mass—2.25 kg; drop height—457 mm) was recorded as the Cmax [20]. Moisture contents of the compacted soils were determined gravimetrically.

3. Results and Discussion

3.1. Soil Physical Properties

Table 1 shows textural diversity among the selected soils, with clay content ranging from 12.8% for Piarco to 74.7% for Talparo. This variation had dual importance; firstly, the range is representative of the textural classes of local agricultural soils and secondly, the validation of the CIT in measuring and monitoring soil strength required a range of soil properties. All soils were moderately acid except for Princes Town which was slightly alkaline. OM contents were typical for mineral soils (<5%), however, Chagaramus had an elevated OM content of 25.1%. Importantly, this soil has not been surveyed and this report represents the first detailed analytical description of this soil. The high carbon content is associated with peat accumulation. The area in which this soil occurs is currently under agricultural intensification, highlighting the need for characterization for optimal use. BD values were consistent with that expected from both mineral and organic soils [21], decreasing with both clay and OM. A strong and significant negative correlation was seen between BD and OM ($r^2 = -0.961$, $p = 0.002$), indicating the dominance of OM on BD.

Reference [2] showed a similar relationship for some wetland soils in Trinidad. Reference [22] indicated that OM imparted better structural arrangement to soils which lowered BD. Excluding the organic soil, clay content was the dominant factor influencing BD. A similar trend was observed for PD, values decreasing with increasing clay and OM contents, however, the relationships were not significant. The low PD for Chagaramus is associated with the high OM. The specific gravity of OM is much lower than for crystalline minerals [21].

The Atterberg limits varied among the soils. Values for plastic limit (PL) ranged from 29% - 41% with Piarco being undetermined due to lack of cohesiveness. For both limits moisture contents increased with increasing content of clay and expanding minerals (**Table 1**), particularly evident with the greater PI for Princes Town versus Talparo, although the latter has a greater clay content. Reference [23] showed that Princes Town had a much greater proportion of montmorillonite and vermiculite than Talparo, which was dominated by illite and kaolinite. They additionally showed that the Cunupia series had a similar clay distribution to Talparo. **Table 1** shows that the PL for Talparo and Cunupia was similar, supporting this conclusion. It can be inferred that clay mineralogy imparts a strong influence on the Atterberg limits. Although Chagaramus had a clay content of 43.3% it showed a low PL and liquid limit (LL) of 11% and 28% respectively. During preparation of the sample, especially for determination of LL, the soil showed signs of aggregate integrity, which could have resulted in reduced specific surface for water absorption. Reference [3] showed increased aggregate strength under wetting for tropical soils with higher levels of OM. Although the limit values were low, Chagaramus still fell into the category of a highly plastic soil [24]. This presents challenges for use as this soil is mostly water saturated in its natural state. Talparo and Princes Town showed the highest PI, which implies that they are most prone to stress related deformation [2]. The moisture contents identified are typical during the rainy season and slumping is a common phenomenon on these soils, where proper drainage has not been implemented. The monitoring of soil moisture contents through simple devices like the TDR meter would greatly improve decision making with respect to applied stresses on these soils.

3.2. Soil Compactability and Strength

Traditionally, BD and PR have been used to assess soil compaction and strength, respectively. Many studies, both local [2] and international [25] have shown these indices to be dually important, especially in relation to crop response. However, the literature is devoid of any information on the use of the CIT in assessing soil strength conditions for agricultural and resource management purposes. A major objective of this study was to determine the feasibility of the CIT to assess both soil compaction and strength.

BD values varied across soils showing the expected trends with increasing soil water content (**Figure 1**). Reference [26] explained this behaviour where BD increased until peak critical values were attained and then decreased with further increases in soil water content. Variation occurred among the soils in both the shape and position of the relationship. Piarco showed responses in the lower moisture range, with a moisture content >15% resulting in reduced BD. The BD values for Piarco and Chagaramus was always higher and lower respectively, than the other soils regardless of moisture content. This finding is in agreement with other works on similar soils [26]. These authors observed that the clay soils had greater aggregate stability than the sandy soils. They also inferred that lower BDs are associated with a higher degree of aggregation in soils with high clay and OM contents. They explained that the relative amount of organic material acts to dilute the soil mineral matter effectively lowering density and reducing aggregate slaking and buffering the effects of compaction. Minimal variation of BD with water content was observed for Chagaramus.

MDBD decreased with increasing clay, expanding minerals and OM content, whilst corresponding critical moisture contents (CMC) increased (**Table 2**). The percentage increase in BD when compared to in-situ values, varied from 11.6% in River Estate to 166% for the peaty clay. The greater compactability of peat [27] allowed for a significantly greater amount of material to be used during compaction. In this study a compactive effort of 549 kJ·m^{-3}, equivalent to 25 compaction blows with the standard Proctor hammer [28] resulted in true changes in BD at moisture contents well below the plastic limit. The data indicates that all soils are prone to compactive levels

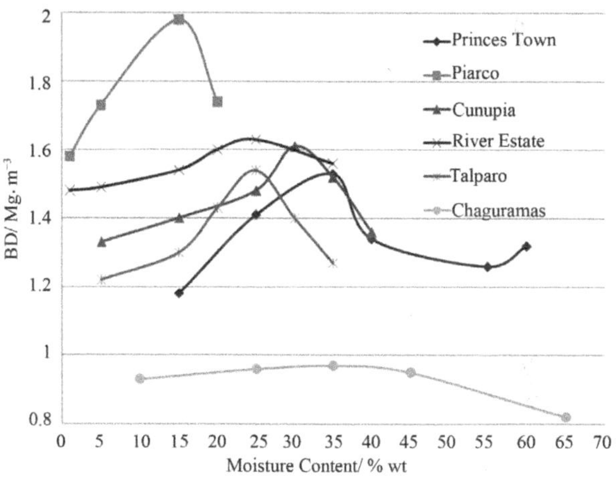

Figure 1. Dry bulk density and moisture relationships for six soils tested at high compaction (25 Proctor blows).

Table 2. Maximum dry bulk density (MDBD), penetration resistance (PPR), peak surface hardens (PSH) and the corresponding moisture contents.

Soil	MDBD	PPR	PSH
	Mg m^{-3}	MPa	C_{max}
River Estate	1.63 (24)[a]	8.2 (4.6)	206 (4.8)
Cunupia	1.61 (30)	5.95 (13)	195 (5)
Piarco	1.98 (15)	2.4 (5)	248 (5.2)
Chagaramus	0.97 (34)	6 (25)	164 (25)
Talparo	1.55 (25)	7.6 (14)	232 (13.8)
Princes Town	1.53 (34.5)	4.1 (26)	208 (15)

[a]Values in parentheses are critical moisture contents. Coefficients of variation values ranged from 2% - 8%.

that may limit plant development. It is clear that monitoring tools are necessary to prevent soil degradation via compaction.

PR showed a similar trend to BD with increasing water content (**Figure 2**). However, clear differences were observed among the soils. Piarco showed a notably lower response to water content compared to the other soils, whilst Chagaramus which had low BDs showed PR values similar to and even greater than some clay soils. The difference in response of Piarco soil to compaction and soil strength as measured by BD and PR respectively supports statements made by [29] that either parameter alone cannot be used to assess changes in the soil fabric due to compaction. Peak PR and corresponding CMC increased with increasing clay content (**Table 2**). Values ranged from 2.4 - 8.2 MPa and were generally in the range reported by [2]. Princes Town showed an unusually low value, considering that it is a cohesive soil. The preponderance of smectitic mineral in Princes Town [23] seemingly influenced PR, through the large water holding capacity. Although the Chagaramus peaty clay has not been mineralogically characterized, its parent material originates from the Northern Range, alluding to a dominance of 1:1 clay minerals. The influence of mineralogy on PR as assessed via the cone index requires further investigation. Reference [30] stated that OM increases soil strength and reduces the effects of compaction. This was evident in Chagaramus, which showed the 3rd highest PPR. A strong relationship exists among OM content, applied stress and soil strength. Reported PPR values were mostly greater than the ranges reported in the literature for limiting root growth [31]. Additionally, these peak values occurred below the PL, at moisture contents typically experienced during crop cultivation. An appropriate moisture content to prevent physical limitations should be between the CMC for BD and PR.

Cmax which has typically been used to determine SH of playing surfaces [10] and strength of engineering materials [11] showed the typical response curve with increasing water content. SH initially increased with increasing water content for all soils except Cunupia and Princes Town, and then decreased steadily after peak hardness was achieved (**Figure 3**).

Reference [11] using a similar compactive effort showed a similar relationship between Cmax and moisture content for a marl soil. Peak SH ranged from 164 - 248 Cmax with corresponding moisture contents of 5% - 25% wt. PSH and associated CMC seemed related to OM content. The former variable showed lower values for higher OM levels, with an inverse effect seen for the latter. Reference [10] alluded to the increases in SH with reductions in soil moisture at high compaction. Reference [32] showed no significant differences between sand and sand plus peat treatments for BD but significant differences were observed for PSH, with the peat treatment showing lower values. CMC for PSH was similar or slightly lower among the soils compared to PPR, but were much lower compared to PDBD. Reference [10] linked changes in soil BD to increase surface hardness. However, correlations between these indices showed a strong positive and significant ($p < 0.001$) relationship between SH and PR ($r^2 = 0.796$), indicating that in this study BD was not an indicator of surface conditions and reversibly, SH was not able to assess bulk soil condition. Reference [25] also found poor relations between cone index and BD. Although SH may substitute for PR, no data exist to facilitate correlation to crop response, in-situ values under traffic or other soil properties. However, the CIT is practical where the cone penetrometer in limited, such as in sandy, dry or wet soils, or where surface layers interferes with cone insertion. Reference [25] further indicated that in high strength soils (>6 MPa) the ability to maintain uniform insertion speed and force using a hand held penetrometer is greatly diminished.

Authors have shown moisture content to be the most dominant factor affecting soil strength indices [25] and have developed predictive empirical relationships. Regression equations for the influence of water content on all three indices were significant (**Table 3**), with r^2 values of 19.5%, 34.6% and 47.4% for BD, PR and SH respectively. The data implies that moisture content explained the greater part of the variability associated with

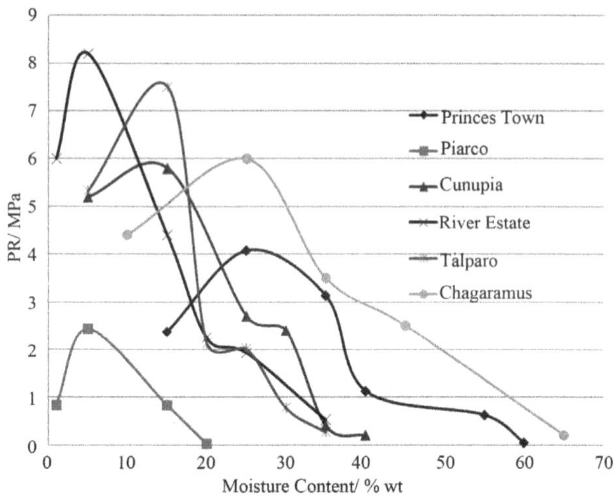

Figure 2. Penetration resistance and moisture relationships for six soils tested at high compaction (25 Proctor blows).

Table 3. Regression statistics for BD, PR and SH with MC.

Variable	p value	Equation	SE Coefficient	r^2
BD	0.010	1.59 - 0.00698 MC	0.00255	19.5
PR	<0.001	4.80 - 0.082 MC	0.02023	34.6
SH	<0.001	195 - 2.67 MC	0.5055	47.4

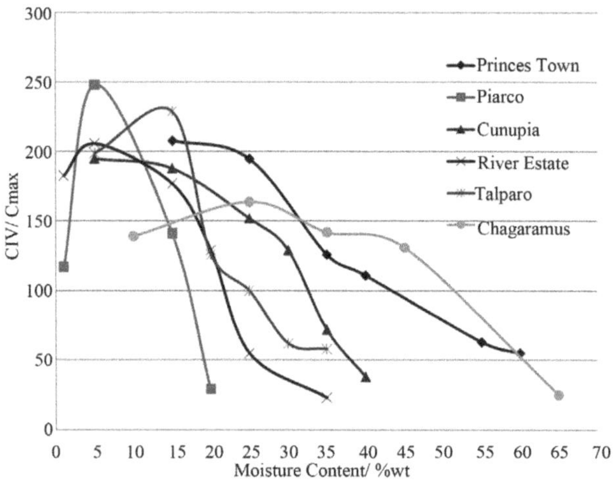

Figure 3. Surface hardness and moisture relationships for six soils tested at high compaction (25 Proctor blows).

SH. The implication of this resides in the use of moisture content as a determinant of soil strength and compaction. When used with SH may prove to be a strong indicator of soil strength.

4. Conclusion

The CIT was shown to be useful in assessing soil strength but not compaction on a range of agricultural soils. Similar responses to increasing water content after compaction was observed for SH, compared to BD and PR. Peak values and their associated CMC varied among soils and indices, with clay mineralogy and OM content having the most influential role. SH showed greater responsiveness to moisture content. Further correlation of this instrument to growth indices would provide needed information to improve its applicability to agricultural soil management.

REFERENCES

[1] N. Ahmad, "Soils of the Caribbean," Ian Randle Publishers, Kingston, 2011.

[2] E. I. Ekwue, R. Stone and S. Ramphalie, "Engineering Properties of Some Wetland Soils in Trinidad," *Transactions of the ASABE*, Vol. 18, No. 1, 2002, pp. 37-45.

[3] M. N. Wuddivira, R. J. Stone and E. I. Ekwue, "Clay, Organic Matter and Wetting Effects on Splash Detachment and Aggregate Breakdown under Intense Rainfall," *Soil Science Society of America Journal*, Vol. 73, No. 1, 2009, pp. 226-232. doi:10.2136/sssaj2008.0053

[4] W. J. Busscher, P. J. Bauer, C. R. Camp and R. E. Sojka, "Correction of Cone Index for Soil Water Content Differences in a Coastal Plain Soil," *Soil Tillage Research*, Vol. 43, No. 3-4, 1997, pp. 205-217. doi:10.1016/S0167-1987(97)00015-9

[5] J. Lipiec and R. Hatano, "Quantification of Compaction Effects on Soil Physical Properties and Crop Growth," *Geoderma*, Vol. 116, No. 1-2, 2003, pp. 107-136. doi:10.1016/S0016-7061(03)00097-1

[6] A. G. Bengough and C. E. Mullins, "Mechanical Impedance to Root Growth: A Review of Experimental Techniques and Root Growth Responses," *Journal of Soil Science*, Vol. 41, No. 3, 1990, pp. 341-358. doi:10.1111/j.1365-2389.1990.tb00070.x

[7] W. R. Whalley, J. Lipiec, W. Stepniewski and F. Tardieu, "Control and Measurement of the Physical Environment in Root Growth Experiments," In: A. L. Smit, A. G. Bengough, C. Engels, M. van Noordwijk, S. Pellerin and S. C. van de Geijn, Eds., *Root Methods A Handbook*, Springer-Verlag, Berlin, 2000, pp. 76-112.

[8] J. Glinski and J. Lipiec, "Soil Physical Conditions and Plant Roots," CRC Press, Boca Raton, 1990.

[9] D. J. Campbell and J. K. Henshall, "Bulk Density," In: K. A. Smith and C. E. Mullins, Eds., *Soil Analysis, Physical Methods*, Marcel Dekker, New York, 1991, pp. 329-366.

[10] J. T. Brosnan, A. S. McNitt and T. J. Serensits, "Effects of Varying Surface Characteristics on the Hardness and Traction of Baseball Field Playing Surfaces," *International Turf Society Research Journal*, Vol. 11, 2009, pp. 1-13.

[11] O. S. B. Al-Amoudi, I. M. Asi, H. I. A. Wahhab and Z. A. Khan, "Clegg Hammer-California Bearing Ratio Correlations," *Journal of Materials in Civil Engineering*, Vol. 14, No. 6, 2002, pp. 512-523. doi:10.1061/(ASCE)0899-1561(2002)14:6(512)

[12] J. N. Rogers III and D. V. Waddington, "Impact Absorption Characteristics on Turf and Soil Surfaces," *Agronomy Journal*, Vol. 84, No. 2, 1992, pp. 203-209. doi:10.2134/agronj1992.00021962008400020016x

[13] G. W. Gee and D. Or, "Particle Size Analysis," In: J. H. Dane and G. C. Topp, Eds., *Methods of Soil Analysis, Part* (4), *Physical Methods*, 3rd Edition, SSSA, Madison, 2002, pp. 255-294.

[14] G. W. Thomas, "Soil pH and Soil Acidity," In: D. L. Sparks, *et al.*, Eds., *Methods of Soil Analysis, Part* (3), *Chemical Methods*, SSSA, Madison, 1996, pp. 475-490.

[15] D. W. Nelson and L. E. Sommers, "Total Carbon, Organic Carbon and Organic Matter," In: D. L. Sparks, *et al.*, Eds., *Methods of Soil Analysis, Part* (3), *Chemical Methods*, SSSA, Madison, 1996, pp. 995-996.

[16] R. B. Grossman and T. G. Reinsch, "Bulk Density and Linear Extensibility," In: J. H. Dane and G. C. Topp, Eds., *Methods of Soil Analysis*, Part (4), *Physical Methods*, 3rd Edition, SSSA, Madison, 2002, pp. 201-228.

[17] A. L. Flint and L. E. Flint, "Particle Density," In: J. H. Dane and G. C. Topp, Eds., *Methods of Soil Analysis*, Part (4), *Physical Methods*, 3rd Edition, SSSA, Madison, 2002, pp. 229-241.

[18] R. A. Mc Bride, "Atterberg Limits," In: J. H. Dane and G. C. Topp, Eds., *Methods of Soil Analysis, Part* (4), *Physical Methods*, 3rd Edition, SSSA, Madison, 2002, pp. 389-399.

[19] R. J. Stone, E. I. Ekwue and R. O. Clarke, "Engineering Properties of Sewage Sludge in Trinidad," *Journal of Agricultural Engineering Research*, Vol. 70, No. 2, 1998, pp. 221-230. doi:10.1006/jaer.1998.0266

[20] F. B. Lopez and L. E. Chinnery, "Surface Hardness Char-

acteristics of Cricket Pitches as Determined with the Clegg Hammer," *International Turf Society Research Journal*, Vol. 11, 2009, pp. 38-40.

[21] N. C. Brady and R. R. Weil, "The Nature and Properties of Soils," 14th Editon, Prentice Hall, London, 2008

[22] B. D. Kay and D. A. Angers, "Soil Structure," In: M. E. Sumner, Ed., *Handbook of Soil Science*, CRC Press, New York, 1999, pp. A-229-A-269.

[23] N. Ahmad and C. E. Davis, "The Effects of Drying on Release of Native and Added Potassium on Six West Indian Soils with Contrasting Mineralogy," *Soil Science*, Vol. 112, No. 2, 1971, pp. 100-106. doi:10.1097/00010694-197108000-00003

[24] A. R. Jumikis, "Soil Mechanics," Robrt E Kriger Co., Malabar, 1984.

[25] R. E. Sojka, W. J. Busscher and G. A. Lehrsch, "In Situ Strength, Bulk Density and Water Content Relationships of a Durinodic Xeris Haplocalcid Soil," *Soil Science*, Vol. 166, No. 8, 2001, pp. 520-529. doi:10.1097/00010694-200108000-00003

[26] E. I. Ekwue and R. J. Stone, "Effect of Peat on the Compactibility of Some Trinidadian Soils," *Journal of Agricultural Engineering Research*, Vol. 57, No. 2, 1994, pp. 129-136. doi:10.1006/jaer.1994.1012

[27] R. J. Stone and E. I. Ekwue, "Soil Compressibility as Influenced by Sewage Sludge Incorporation," *Journal of Agricultural Engineering Research*, Vol. 64, No. 3, 1996, pp. 227-235. doi:10.1006/jaer.1996.0063

[28] M. Diaz-Zorita, J. H. Grove and E. Perfect, "Laboratory Compaction of Soils Using a Small Mold Procedure," *Soil Science Society of America Journal*, Vol. 65, No. 6, 2001, pp. 1593-1598. doi:10.2136/sssaj2001.1593

[29] D. L. McQueen and T. G. Shepherd, "Physical Changes and Compaction Sensitivity of a Fine-Textured, Poorly Drained Soil (Typic Endoaquept) under Varying Durations of Cropping Manawatu Regions, New Zealand," *Soil Tillage Research*, Vol. 63, No. 3-4, 2002, pp. 93-107. doi:10.1016/S0167-1987(01)00231-8

[30] E. I. Ekwue, "Organic Matter Effects on Soil Strength Properties," *Soil Tillage Research*, Vol. 16, No. 3, 1990, pp. 289-297. doi:10.1016/0167-1987(90)90102-J

[31] J. J. H. Van den Akker and B. Soane, "Compaction," In: L. Rattan, Ed., *Encyclopaedia of Soil Science*, 2nd Edition, CRC Press, Boca Raton, 2005, pp. 285-293. doi:10.1021/ja0410542

[32] C. Waltz, S. Burnett, V. Quisenberry and B. McCarty, "Soil Amendments Affect Compaction, Soil Strength," *Golf Course Management*, Vol. 68, No. 11, 2000, pp. 49-55.

2

Nematodes for Soil Quality Monitoring: Results from the RMQS BioDiv Programme

Cécile Villenave[1], Anne Jimenez[1], Muriel Guernion[2], Guénola Pérès[2], Daniel Cluzeau[2], Thierry Mateille[3], Bernard Martiny[3], Mireille Fargette[3], Johannes Tavoillot[3]

[1]ELISOL Environnement and IRD UMR 210 ECO & SOLS, Montpellier, France; [2]UMR CNRS 6553 EcoBio, University of Rennes 1, Rennes, France; [3]IRD UMR 1062 CBGP, Montferrier sur Lez, Monteferrier, France.

Email: cecile.villenave@elisol-environnement.fr, anne.jimenez@elisol-environnement.fr, muriel.guernion@laposte.net, guenola.peres@univ-rennes1.fr, daniel.cluzeau@univ-rennes1.fr, thierry.mateille@ird.fr, bernard.martiny@ird.fr, mireille.fargette@ird.fr, Johannes.Tavoillot@ird.fr

ABSTRACT

A French programme, "Réseau de mesure de la qualité des sols: biodiversité des organismes" (RMQS BioDiv) was developed in Brittany (27,000 km^2 in the western part of France) as an initial assessment of soil biodiversity on a regional scale in relation to land use and pedoclimatic parameters. The nematode community assemblages were compared among the land use categories. Crops were characterised by a high abundance of bacterial-feeders, particularly opportunistic bacterial-feeders belonging to Rhabditidae. Meadows presented a higher total abundance of nematodes than did crops (20.6 ind·g^{-1} dry soil vs. 13.1 ind·g^{-1} dry soil), and they were mainly linked to the great abundance of plant-parasitic nematodes, particularly *Meloidogyne*, but with a very high heterogeneity between sampled plots. The nematodes were most abundant in forests (23.7 ind·g^{-1} dry soil) and presented the most structured community (SI = 82.2 in forests vs. 58.6 and 55.5 in crops and meadows, respectively). Forests had also the higher fungal component (fungal-feeders and facultative plant-feeders belonging to the Tylenchidae) leading to a significant higher part of the fungal decomposition pathway in forests than in crops. The ability of different taxonomic levels of nematode identification to discriminate among different cropping systems (*i.e.*, continuous cropping system, crop with meadow in the rotation, meadow with crop in the rotation and permanent meadow) was also tested. The family level (48 families identified in these samples) was more efficient than the other taxonomic levels (86 taxa, 17 functional guilds and 6 trophic groups): best statistical significant discrimination for time spent in identification. The relation between the nematode ecological indices, the abundance of nematode trophic groups and the crop management practices were studied. The effects of fertilization, ploughing frequency, use of pesticides and management systems on ecological indices, particularly on the Maturity Indices, were observed.

Keywords: Agricultural Practices; Bioindicators; Crop; Forest; Meadow; Soil Biodiversity Monitoring

1. Introduction

The Convention on Biological Diversity (CBD) recognized the importance of biodiversity for ecosystem functioning and for the provisioning of soil services to humankind ([1]; www.cbd.int). Soil biota is thought to harbour a large portion of the biodiversity of the world, but there is still limited knowledge regarding this domain. The European Union (EU) has underscored that an interest in the biological compartment of soil and that the role of biodiversity in the maintenance of ecological functions (*i.e.*, life support functions) in the soil requires more attention. Therefore, the EU has mandated biological research on soil management policies, including characterrizing species and the biological functions of some soil organisms [2]. In Europe, indicators related to soil biodiversity are still rarely measured [3]; according to the report of the European project ENVASSO [4], only a few countries have monitoring sites for soil fauna, such as Germany, the Netherlands, Ireland, Austria, Latvia, Estonia and France [3,5-7]. Several international syntheses have reported on soil biota as bioindicators of human activities [8-12].

To fulfill the need expressed by the European Union to study soil biodiversity (species and function) in relation to pedoclimatic parameters and land use (mainly agricultural practices), a French programme, RMQS BioDiv, was developed at a wide regional scale (Brittany) from 2006 to 2009. It studied most soil biological guilds (total macrofauna, earthworms, mites, collembola, nematodes, microorganisms) [13] in locations where physical and chemical soil properties and agricultural practices have already been monitored [14,15].

The objectives of this study focusing on soil nematodes were to 1) investigate the relations between nematodes and land uses, 2) prioritize the different levels of taxonomic determination based on their ability to distinguish between situations, 3) investigate the relations between nematodes and agricultural management practices, and 4) incorporate nematodes as part of the soil quality descriptors in ecosystem service management for French and European policies.

2. Materials and Methods

2.1. Sampling Site Network

The sampling site network is located in Brittany in the western part of France (**Figure 1**), between 2° to 5° W longitude and 47° to 49° N latitude. The climate is temperate oceanic, with the mean annual air temperature ranging from 7.6°C to 15.9°C and the mean annual rainfall ranging from 1400 mm in the west to 700 mm in the eastern part of the region. Cambisol and Luvisol are the major recorded soils. The study site network, comprising 109 monitoring sites, was based on a 16 × 16 km systematic grid covering the entire area (27,000 km²) and was also used by the RMQS, which monitored the physical and chemical soil parameters and agricultural practices [15].

To link the biological data to the agricultural, physical and chemical data, the RMQS BioDiv sampling area was located 5 meters from the RMQS area for each sampling site. Within the network, the primary land uses were meadows and crops (90%) and forests (10%). Because of its variability in geology, climate and land-use, the Brittany region offers a large gradient of pedoclimatic land use and management. The biological samples were collected in 2006 and 2007 springs. The "baselines" obtained for nematological indicators are consequently valid for spring conditions. This study was conducted in real farmer fields and forests; it is not an experiment conducted in experimental station. Then, it is possible that the land use reflect differences in soil type. As a matter of fact, farmers can select different kind of areas of land for cropping, pasture or forestry on the basis of soil characteristics, such as the ability to cultivate the land (stone content, slope) or the organic content (carbon, nitrogen…) or the physical properties (soil structure, bulk density); however this will not be analysed in this article.

2.2. Nematode Sampling

For each site sampled on the BioDiv grid, a single sample was composited from 32 samples collected from the surface soil layer (0 - 15 cm). In forest soils, it was chosen to remove the litter before soil sampling in order to compare forest soil without humus with cultivated soil

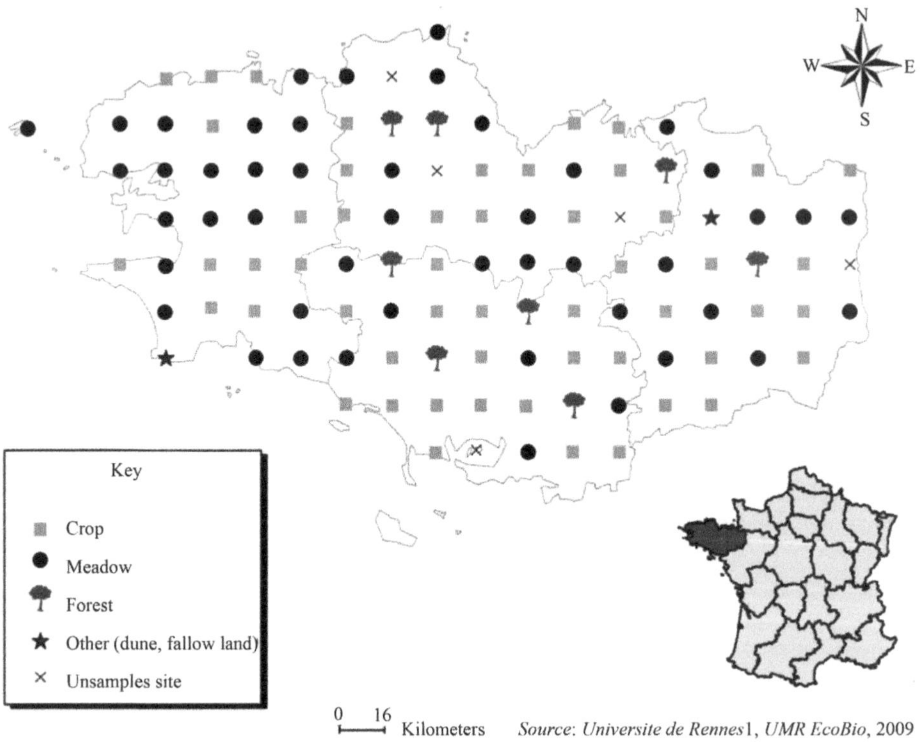

Figure 1. Location and land use of RMQS BioDiv sites (n = 109) in the Brittany region of France.

even if, doing so, important information on the ecosystem functioning in forests was lost. The soil was sieved through 6 mm mesh to remove large stones and plant debris. The soil samples (approximately 3 kg fresh weight) were then homogenised, shipped and stored at 4˚C until being used.

For each fresh sample, the nematodes were extracted from approximately 300 g wet soil by elutriation, followed by an active pass through a cotton wool filter for 48 hours using the Seinhorst method [16]; they were then counted using a binocular microscope. The composition of the soil nematofauna was determined after fixing in a formaldehyde-glycerol mixture and transferring to mass slides [17]. On average, 200 nematodes per mass slide were identified to the family or genus level at 400× magnification. For some morphotypes, it was impossible to find adults, allowing only identification to genus or family level using several identification books [18-20].

The nematode taxa were assigned to trophic groups modified from [21]: bacterial-feeders (Ba), fungal feeders (Fu), facultative plant-feeders (Fpf), obligate plant-feeders (Pf), omnivores (Om) and carnivores (Ca). The nematodes were also allocated to coloniser-persister (cp) classes following [22]. The coloniser-persister scale ranged from 1 to 5 and could vary within a trophic group; thus, bacterial-feeders with a cp class of 1 were placed in the functional guild Ba_1, and fungal-feeders with a cp class of 4 were placed in the functional guild Fu_4.

Eight nematode ecological indices were calculated after Bongers [22], Ferris *et al.* [23] and Yeates [24].

Maturity Index ($MI = \Sigma vi\ pi$), where vi is the coloniser-persister value assigned to taxon i (from cp 1 to cp 5) and pi is the frequency of taxon i; the MI is calculated for all of the free-living nematodes; lower values of this Index indicate disturbed and nutrient enriched environments whereas higher values indicate more stable conditions. The BaMI, FuMI and PPI are the same indices calculated for the bacterial-feeding nematodes, fungal-feeding nematodes and plant-feeding nematodes, respecttively.

Enrichment Index ($EI = 100 \times [e/(e + b)]$), Structure Index ($SI = 100 \times [s/(b + s)]$) and Channel Index ($CI = 100(0.8\ Fu_2)/(3.2\ Ba_1 + 0.8\ Fu_2)$), where e is the abundance of nematodes in the basal component weighted by their ke values; b is the abundance of nematodes in the basal component weighted by their kb values; s is the abundance of nematodes in the structural component weighted by their ks values; ke is the weighting assigned to the guilds Ba_1 and Fu_2 (enrichment component); kb is the weighting assigned to the guilds Ba_2 and Fu_2 (basal component); and ks is the weighting assigned to the guilds $Ba_{3\text{-}5}$, $Fu_{3\text{-}5}$, $Om_{4\text{-}5}$ and $Pr_{2\text{-}5}$ (structural component).

The EI is meant to assess the food web response to the availability of resources [23]. The SI indicates whether the soil community is basal (typical for disturbed systems) or structured (typical of more stable systems) [23]. The CI indicates the putative predominant decomposition pathways, *i.e.*, it is meant to quantify the relative importance of the fungal-fed and the bacterial-fed trophic channels of the soil decomposer food web [23].

The Nematode Channel Ratio (NCR) was calculated after [24]: $NCR = B/(B + F)$, where B and F are, respecttively, the relative contributions of the bacterial- and fungal-feeding nematodes to the total nematode abundance.

2.3. Explanatory Variables

The explanatory data were from the RMQS programme and were collected by the Orleans INRA [14]. These data included the physicochemical soil parameters, land use and agricultural practices. Relation between nematological data and five physico-chemical soil parameters were studied. Total organic carbon (TOC) and total nitrogen (N) contents were measured by dry combustion (NF ISO 10694 and 13878); C ranged from 6.0 to 50.5 with a mean of 25.8 $g \cdot kg^{-1}$; total N ranged from 0.7 to 4.5 $g \cdot kg^{-1}$ with a mean of 2.3 $g \cdot kg^{-1}$. C/N ranged from 8.2 to 22.6 with a mean of 11.0. pH_{water} ranged from 4.0 to 8.6 with a mean of 5.9. Assimilable P (P_2O_5 Olsen method NF ISO 11263) ranged from 0.00 to 0.32 with a mean of 0.10 $g \cdot kg^{-1}$. Other soil physico-chemical characteristics are given in **Table 1**. To assess the relations between the biological data and the agricultural practices, nine agricultural management variables were grouped: land use (crop field, permanent meadow, meadow with crops in rotation, crops with meadow in rotation, or forest), management system (intensive, integrated, sustainable systems, or organic farming), use of fertilisers (no fertilisers, mineral fertilisers, organic fertilisers, or organic + mineral fertilisers), ploughing frequency (high, low), time since last tillage (<5 months, >5 months), and pesticide use (yes, no).

2.4. Statistical Analysis

Spearman correlation tests were made between nematological parameters and physico-chemical parameters (quantitative explanatory variables). The global nematode parameters (densities of trophic groups and nematode ecological indices) were compared among the three land uses sampled in Brittany: crops, meadows and forests. The relations between the nematode parameters and agricultural practices were also tested. The tests used were non-parametric Kruskal-Wallis test (5% signifycance level) followed by Bonferroni post-hoc multiple comparisons to test for differences between the means.

Table 1. Main physico-chemical characteristics of the 108 sampled locations.

	Minimum	Mean	Standard deviation	Maximum
Clay (%)	5.9	17.89	5.25	49.7
Silt (%)	14.00	51.2	13.42	73.8
Sand (%)	9.8	30.9	14.91	80.1
C ($g \cdot kg^{-1}$)	5.97	25.44	10.19	50.5
Total N ($g \cdot kg^{-1}$)	0.72	2.345	0.8	4.5
C/N	8.2	10.91	1.83	20.68
pH_{water}	4.00	5.85	0.66	8.2
Polsen ($g \cdot kg^{-1}$)	0.00	0.11	0.06	0.319
ECC ($cmol^{+} \cdot kg^{-1}$)	2.5	7.22	2.44	13,00
Exchangeable Ca ($cmol^{+} \cdot kg^{-1}$)	0.01	5.61	2.6	11.3
Exchangeable Mg ($cmol^{+} \cdot kg^{-1}$)	0.09	0.81	0.78	7.52
Exchangeable K ($cmol^{+} \cdot kg^{-1}$)	0.08	0.39	0.24	1.34
Water content (%)	12.16	23.58	5.26	42.39

Mann-Whitney tests were also used when comparing only two modalities of a variable. The analyses were performed using XLStat 2008 (Addinsoft). The multivariate patterns in the nematode communities were investigated on transformed abundances from 4 datasets at different taxonomic levels: the highest level of identification (86 taxa), family level (48 families), functional guilds (17) and trophic groups (6). A permutational multivariate ANOVA (PERMANOVA) was performed to determine the statistical significance of the effects of treatments and sampling time on the nematode community structure. The PERMANOVA tests were based on 9999 restricted permutations of the data. Pairwise comparisons were performed within treatments and sampling times to test for significant ($p < 0.05$) nematode community patterns. Non-metric multi-dimensional scaling (nMDS) was performed on the data from two different datasets (86 taxa, 17 functional guilds) to illustrate the differences in plot separation when using different identification levels [25]. MDS relies on a two-dimensional map in which the degree of similarity among the observations is indicated by the proximity of their representative points (the closer, the more similar). The axes of the map hold no specific value and can be rotated or mirrored without influencing the relative distances between the observations. PERMANOVA and MDS were performed using PRIMER v6 with PERMANOVA+ [26].

3. Results

3.1. Relation between Nematode Parameters and Five Main Physico-Chemical Parameters

Positive correlations were measured between total nematode abundance and soil C and N contents (**Table 2**). Opportunistic bacterial-feeders were correlated to available P. Others bacterial-feeders were correlated with soil C. Fungal-feeders were correlated with soil C, total N and available P. Plant-parasitic nematodes were correlated with C and N contents. MI was positively correlated Polsen. BaMI was positively correlated with C/N and pH and negatively correlated with available P. FuMI was positively correlated with C/N, available P and pH. NCR was positively correlated with C, N and available P. EI was positively correlated with N, Polsen and pH. CI was positively correlated with available P. PPI and SI were not correlated with these physico-chemical parameters.

3.2. Regional Average of Nematode Parameters in Crops, Meadows and Forests

Nematodes were more abundant in meadows and forests than in crops (**Table 3**). The opportunistic bacterial-feeding nematodes (Ba_1) and the other bacterial-feeders (Ba_{234}) showed different land use distributions; the Ba_1 were significantly more abundant in crops than in meadows or forests. Land use had no influence on the density of the Ba234. Fungal-feeders (Fu) and facultative plant-feeders (RHF) were more numerous in forests than in other land uses. Omnivores (Om) were more abundant in forests than in crops, with intermediate density in meadows. Carnivores (Ca) were more abundant in crops than in meadows. The highest abundances of plant parasitic nematodes (Pf) were found in meadows. These differences in nematode trophic group densities observed among the land uses resulted in completely different community structures (**Figure 2**). Microbivorous (Ba and

Table 2. Relations between the abundances of the nematode trophic groups and the nematode ecological indices and the five physico-chemical soil parameters using Spearman correlations and significant difference among means for the Brittany region. Values indicated in the table are the p-value of the significant correlations.

Nematode trophic groups		Carbon	Total Nitrogen	C/N	P_{olsen}	pH_{water}
(ind·g^{-1} dry soil)		(g·kg^{-1})	(g·kg^{-1})		(g·kg^{-1})	
Opportunistic bacterial-feeders	Ba_1	ns	ns	ns	<0.0001	ns
Others bacterial-feeders	Ba_{234}	0.046	ns	ns	ns	ns
Fungal-feeders	Fu	0.031	0.011	ns	0.006	ns
Omnivores	Om	ns	ns	ns	ns	ns
Carnivores	Ca	ns	ns	ns	ns	ns
Plant-parasitic nematodes	Pf	0.003	0.0001	ns	ns	ns
Facultative plant feeders	RHF	ns	ns	ns	ns	ns
Total		0.006	0.005	ns	ns	ns
Nematode ecological indices						
Maturity Index	MI	ns	ns	ns	0.0005	ns
Maturity Index on bacterial-feeders	BaMI	ns	ns	0.011	0.0001(-)[1]	0.0002
Maturity Index on fungal-feeders	FuMI	ns	ns	0.018	0.001	0.021
Nematode Channel Ratio	NCR	0.015	0.002	ns	<0.0001	ns
Plant Parasitic Index	PPI	ns	ns	ns	ns	ns
Enrichment Index	EI	ns	0.011	ns	0.0005	0.009
Structure Index	SI	ns	ns	ns	ns	ns
Channel Index	CI	ns	ns	ns	<0.0001	ns

[1](-) indicate a negative correlation.

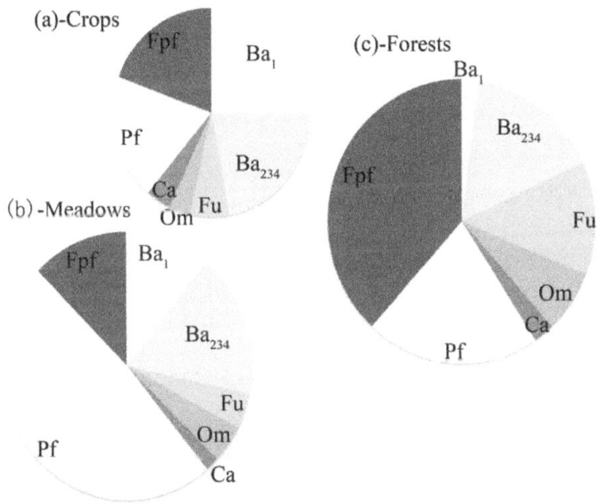

Figure 2. The relative abundances of the nematode trophic groups in crops, meadows and forests. The pie-charts are proportional to the nematode densities: (a) Crops; (b) Meadows; (c) Forests.

Fu) nematodes dominated in crops, whereas strict and facultative fungal-feeders (Fu and RHF) prevailed in forests. Plant-feeders (Pf) were the most abundant trophic group in meadows.

The nematode ecological indices also differed significantly among the three land uses (**Table 4**). The MI, FuMI, SI and CI were significantly higher in forests than in crops or meadows. The BaMI increased from crops to meadows and from meadow to forests. The EI and NCR were significantly higher in crops than in forests. Moreover, the EI was higher in crops than in meadows, whereas the NCR did not differ significantly between those two land-uses. Finally, the PPI increased from forests to crops and from crops to meadows.

3.3. Regional Average of Nematode Taxa Composition in Crops, Meadows and Forests

The densities of the dominant taxa found in the 109 samples from Brittany are presented in **Table 4**. Among bacterial-feeding nematodes (Ba), *Alaimus* (cp 4), *Wilsonema* (cp 2), *Prismatolaimus* (cp 3) and *Metateratocephalus* (cp 3) were mostly found in forests. In contrast, *Eucephalobus* (cp 2) were more abundant in cropped areas. The densities of Rhabditidae (cp 1), and Panagrolaimidae (cp 1), were 10 times higher in crops than in forests. Tylolaimorphorus (cp 4) and Tylencholaimus (cp 4) were more abundant in forests than in cultivated areas.

Among plant-feeders (Pf and RHF), the main difference between the crops and meadows concerned the high abundance of the genus *Meloidogyne* (cp 3) in meadows. The high variability of its density in meadows (CV = 149%) clearly indicates heterogeneous population levels among plots. However, M. naasi occurred abundantly in

Table 3. Means (M), standard deviation (SD), minimum (min) and maximum (max) values of the abundance (ind 100 g^{-1} dry soil) of the nematode trophic groups and the nematode ecological indices recorded by the RMQS Biodiv programme (n = 109) for the three main land uses: crops (n = 52), meadows (n = 47) and forests (n = 8).

		Crops			Meadows			Forests		
		min - max	M ± SD		min - max	M ± SD		min - max	M ± SD	
Nematodes trophic groups										
Opportunistic bacterial-feeders	Ba_1	32 - 1511	326.1 ± 270.2	c*	9 - 994	224.3 ± 204.1	b	0 - 123	50.3 ± 52.8	a***
Others bacterial-feeders	Ba_{234}	46 - 1399	291.5 ± 211.2	a	93 - 1175	360.5 ± 261.3	a	40 - 668	381 ± 203.6	a^ns
Fungal-feeders	Fu	6 - 262	84.5 ± 61.6	a	0 - 570	88.5 ± 91.7	a	34 - 609	305.2 ± 202.9	b**
Omnivores	Om	0 - 313	45.9 ± 58.6	a	0 - 522	96 ± 112.7	ab	1 - 342	169.4 ± 106	b**
Carnivores	Ca	3 - 207	53.7 ± 45.8	b	0 - 178	35 ± 40.4	a	0 - 198	60 ± 71.5	ab*
Plant-parasitic nematodes	Pf	20 - 983	254.6 ± 214.4	a	0 - 3922	999.1 ± 1046.7	b	12 - 3018	497.3 ± 979.7	a***
Facultative plant-feeders	RHF	11 - 817	255.7 ± 195.4	a	0 - 1211	257.7 ± 305.3	a	78 - 1698	907.1 ± 559.8	b**
Total		384 - 3548	1312 ± 589.7	a	370 - 5271	2061.1 ± 1289.4	b	173 - 5794	2370.3 ± 1551.3	b*
Nematode ecological indices										
Maturity Index	MI	1.4 - 3	1.98 ± 0.33	a	1.4 - 3	2.18 ± 0.44	a	2.3 - 3.4	2.99 ± 0.38	b***
Maturity Index on bacterial-feeders	BaMI	1.2 - 2.1	1.56 ± 0.22	a	1.2 - 2.1	1.69 ± 0.22	b	1.9 - 2.5	2.36 ± 0.18	c***
Maturity Index on fungal-feeders	FuMI	2 - 2.8	2.08 ± 0.16	a	2 - 3.7	2.17 ± 0.34	a	2.5 - 3.6	2.97 ± 0.34	b***
Nematode Channel Ratio	NCR	56 - 99	86.9 ± 8.5	b	66 - 100	86.3 ± 8.2	b	43 - 88	59.1 ± 14.9	a***
Plant parasitic Index	PPI	2.1 - 3	2.48 ± 0.23	b	2 - 3	2.65 ± 0.26	c	2 - 2.4	2.1 ± 0.13	a***
Enrichment Index	EI	46 - 96	76.4 ± 11.5	b	13 - 95	64.4 ± 18.6	a	6 - 72	40.9 ± 23.8	a***
Structure Index	SI	16 - 91	58.6 ± 17.5	a	0 - 91	55.5 ± 22.9	a	49 - 94	82.2 ± 14.7	b**
Channel Index	CI	1 - 24	7.4 ± 5.7	a	0 - 69	11.1 ± 11.5	a	11 - 100	51.7 ± 33.6	b***

*Varying letters (a, b, c) indicate significant differences between means (Kruskal-Wallis test followed by Bonferoni correction, *$p < 0.05$; **$p < 0.01$; ***$p < 0.01$).

several meadow samples. *Paratylenchus* species (cp 2) were found in some forest plots in extremely high densities (maximum 24 ind·g^{-1} dry soil). Meadows and forests contained more Hoplolaimidae than did crops. Among omnivorous nematodes (Om), Belondiridae (cp 5) and Qudsianematidae (cp 4) were mainly found in forests whereas Thornematidae (cp 5) and Nordiidae (cp 4) were more common in meadows. The carnivorous (Ca) genus *Clarkus* (cp 4) was mostly found in cropped areas.

3.4. Discrimination of the Cropping Modalities by the Taxonomic Rank

Four taxonomic levels of identification (genus, family, functional guilds, trophic groups) were used with PERMANOVA analysis to discriminate among five modalities of land-use (**Table 5**): meadows with crops in rotation (Mc), permanent meadows (M), crops with meadows in rotation (Cm), continuous crops (C), and forests (F). None of the four taxonomic levels distinguished between the Mc and M modalities, or the C and Cm modalities (except the functional-guild level for C/Cm, with a low significance, p = 0.054). Both functional-guild and trophic-group levels did not distinguish between the Mc and Cm modalities. Otherwise, all of the taxonomic levels concerned were able to distinguish between the other land-use modalities. Two MDS analyses were performed,

Table 4. Means (M), standard deviation (SD), minimum (min) and maximum (max) values of the abundance (ind 100 g^{-1} dry soil) of the nematode taxa recorded by the RMQS Biodiv programme (n=109) for the three main land uses: crops (n = 52), meadows (n = 47) and forest (n = 8).

Superfamilies/ families/subfamilies	Genera	Crops		Meadows		Forests	
		min - max	M ± SD	min - max	M ± SD	min - max	M ± SD
Plant and root hair feeders							
Belonolaiminae	*Morulaimus*	0 - 0	0 ± 0	0 - 0	0 ± 0	0 - 18	2.3 ± 6
Criconematidae	*Criconema* + *Hemicriconemoides*	0 - 0	0 ± 0	0 - 0	0 ± 0	0 - 62	7.7 ± 20.4
Heteroderidae	*Heterodera*	0 - 66	2.7 ± 12.8	0 - 697	49.3 ± 130.7	0 - 0	0 ± 0
Hoplolaimidae	*Helicotylenchus*	0 - 93	3.4 ± 13.2	0 - 2116	98.6 ± 323.9	0 - 539	92.2 ± 181
Hoplolaimidae	*Rotylenchus*	0 - 267	12 ± 39.6	0 - 43	1.8 ± 6.6	0 - 411	52.1 ± 135.7
Longidoridae	*Longidorus*	0 - 235	11.1 ± 40	0 - 214	13.7 ± 43.7	0 - 27	3.4 ± 8.9
Longidoridae	*Xiphinema*	0 - 0	0 ± 0	0 - 286	6.3 ± 40.5	0 - 0	0 ± 0
Meloidogynae	*Meloidogyne*	0 - 415	33.1 ± 75.4	0 - 3559	599.5 ± 893.2	0 - 0	0 ± 0
Paratylenchidae	*Paratylenchus* + *Gracilacus*	0 - 271	24.7 ± 55.3	0 - 665	88.6 ± 146.5	0 - 2425	329.9 ± 792.8
Pratylenchidae	*Pratylenchus*	0 - 933	92.4 ± 154.8	0 - 890	70.3 ± 134.6	0 - 27	5 ± 9.4
Telotylenchidae	*Tylenchorhynchus* + *Amplimerlinius**	0 - 579	56.5 ± 95.5	0 - 557	69.8 ± 115.5	0 - 37	4.7 ± 12.4
Trichodoridae	*Paratrichodorus*	0 - 25	0.7 ± 3.7	0 - 0	0 ± 0	0 - 0	0 ± 0
Tylenchidae	*Boleodorus* + *Coslenchus* + *Filenchus* + unidentified genera	11 - 817	255.7 ± 195.4	0 - 1211	257.7 ± 305.3	78 - 1698	907.1 ± 559.8
Omnivores							
Belondiridae	*Dorylaimellus* + other genera	0 - 0	0 ± 0	0 - 81	3.3 ± 12.9	0 - 119	24.7 ± 40.1
Dorylaimoidea	unidentified genera	0 - 21	2.4 ± 5.3	0 - 47	5.6 ± 9.8	0 - 27	5.3 ± 9.6
Nordiidae	*Pugentus*	0 - 67	4.8 ± 10.8	0 - 186	22.5 ± 36.2	0 - 135	16.8 ± 44.6
Nordiidae	unidentified genera	0 - 0	0 ± 0	0 - 36	1.4 ± 6	0 - 27	3.3 ± 8.8
Qudsianematidae	*Eucumenicus*	0 - 83	6.2 ± 17.9	0 - 110	6.5 ± 20.1	0 - 18	4.2 ± 7.2
Qudsianematidae	*Eudorylaimus*	0 - 86	5.4 ± 13.5	0 - 55	5.9 ± 13	0 - 294	51.7 ± 96.3
Qudsianematidae	unidentified genera	0 - 249	11.8 ± 34.7	0 - 71	6.3 ± 12.1	0 - 225	56.5 ± 77.1
Thornenematidae	*Mesodorylaimus*	0 - 101	6.1 ± 6.2	0 - 241	24.8 ± 48.6	0 - 13	1.7 ± 4.4
Thornenematidae	*Prodorylaimus*	0 - 108	7.7 ± 17.4	0 - 296	17 ± 47.1	0 - 27	5.3 ± 9.6
Thornenematidae	unidentified genera	0 - 30	1.4 ± 4.9	0 - 36	2.5 ± 7.3	0 - 0	0 ± 0
Carnivores							
Anatonchidae	*Michonchus*	0 - 10	0.5 ± 1.9	0 - 28	0.7 ± 4.1	0 - 27	3.4 ± 8.9
Aphelenchoididae	*Seinura*	0 - 50	2.2 ± 7.7	0 - 36	1.4 ± 5.5	0 - 27	3.4 ± 8.9
Aporcelaimidae	*Aporcelaimus* + other genera	0 - 83	17.8 ± 20	0 - 107	14.1 ± 19.7	0 - 54	12.8 ± 20.1
Discolaimidae	*Discolaimus*	0 - 0	0 ± 0	0 - 26	1.1 ± 4.7	0 - 0	0 ± 0
Mononchidae	*Clarkus*	0 - 124	18 ± 25.1	0 - 40	4 ± 9	0 - 1	0.2 ± 0.4
Mononchidae	*Coomansus*	0 - 9	0.2 ± 1.3	0 - 43	0.9 ± 6.1	0 - 0	0 ± 0

Continued

Mononchidae	*Mononchus*	0 - 27	2.2 ± 6.1	0 - 66	3.2 ± 10.3	0 - 27	3.3 ± 8.8
Mononchidae	*Mylonchulus*	0 - 10	1 ± 2.7	0 - 14	0.5 ± 2.4	0 - 27	3.4 ± 8.9
Mononchidae	*Prionchulus*	0 - 45	1.4 ± 6.6	0 - 14	0.5 ± 2.3	0 - 13	2.4 ± 4.5
Nygolaimidae	*Paravulvus*	0 - 86	8.6 ± 17.7	0 - 77	5.9 ± 16.8	0 - 183	30.8 ± 59.1
Tripylidae	*Tripyla*	0 - 16	1 ± 3.5	0 - 57	3.1 ± 9.9	0 - 15	3.7 ± 6.1
Bacterial-feeders							
Achromodoridae	*Achromodora*	0 - 33	1.1 ± 5	0 - 14	0.7 ± 2.9	0 - 0	0 ± 0
Alaimidae	*Alaimus*	0 - 100	15.9 ± 21.9	0 - 98	9.6 ± 21	0 - 135	61 ± 37.9
Alaimidae	*Amphidelus*	0 - 19	0.7 ± 2.9	0 - 29	0.6 ± 4	0 - 0	0 ± 0
Bastianiidae	*Bastiania*	0 - 18	0.4 ± 2.5	0 - 13	0.3 ± 1.9	0 - 13	1.7 ± 4.4
Cephalobidae	*Acrobeles*	0 - 271	5.3 ± 37.5	0 - 0	0 ± 0	0 - 18	2.3 ± 6
Cephalobidae	*Acrobeloides*	0 - 400	35.7 ± 66.9	m - 126	30.1 ± 29.1	18 - 189	67.3 ± 60.3
Cephalobidae	*Acrobeloides* + *Cephalobus*	0 - 34	1.9 ± 6.9	0 - 26	1.6 ± 5.1	0 - 46	5.8 ± 15.3
Cephalobidae	*Cephalobus*	0 - 200	33.7 ± 38	0 - 400	46.8 ± 82	0 - 135	22.2 ± 43
Cephalobidae	*Cervidellus*	0 - 17	1 ± 3.1	0 - 40	2 ± 7	0 - 12	1.5 ± 4
Cephalobidae	*Chiloplacus*	0 - 114	15.2 ± 25.9	0 - 80	12.7 ± 20.2	0 - 6	1.1 ± 2.1
Cephalobidae	*Eucephalobus*	4 - 400	118 ± 79.8	10 - 948	m ± 189	0 - 135	24.1 ± 43.8
Cephalobidae	*Heterocephalobus*	0 - 20	2.6 ± 5.4	0 - 46	4.4 ± 10.3	0 - 0	0 ± 0
Cephalobidae	*Stegelata* + *Acrobelophis*	0 - 24	2.3 ± 5.6	0 - 38	3.2 ± 8.4	0 - 0	0 ± 0
Cephalobidae	*Zeldia*	0 - 38	0.9 ± 5.4	0 - 0	0 ± 0	0 - 0	0 ± 0
Diplogasteridae	unidentified genera	0 - 13	0.5 ± 2.1	0 - 28	1 ± 4.5	0 - 0	0 ± 0
Diplopeltidae	*Cylindrolaimus*	0 - 107	2.1 ± 14.8	0 - 0	0 ± 0	0 - 0	0 ± 0
Leptolaimidae	*Chronogaster* + *Leptolaimus*	0 - 47	1.7 ± 7.5	0 - 11	0.2 ± 1.6	0 - 6	0.8 ± 2
Monhysteridae	*Monhystera* + others	0 - 27	1.7 ± 5.2	0 - 33	4.5 ± 8.6	0 - 0	0 ± 0
Neodiplogasteridae	unidentified genera	0 - 131	5.7 ± 22.1	0 - 64	5.1 ± 13.5	0 - 0	0 ± 0
Odontopharyngidae	*Odontopharynx*	0 - 0	0 ± 0	0 - 0	0 ± 0	0 - 27	3.4 ± 8.9
Osstellidae	*Drilocephalobus*	0 - 14	0.5 ± 2.6	0 - 0	0 ± 0	0 - 0	0 ± 0
Panagrolaimidae	*Panagrolaimus* + others	0 - 77	10.9 ± 18	0 - 233	23.3 ± 39.4	0 - 13	1.7 ± 4.4
Plectidae	*Anaplectus*	0 - 55	5.1 ± 10.7	0 - 175	13.2 ± 28.6	0 - 27	5.8 ± 9.2
Plectidae	*Plectus*	0 - 207	37.6 ± 49.6	0 - 261	42.9 ± 53.4	8 - 198	69.1 ± 62.2
Plectidae	*Plectus* + *Anaplectus*[1]	0 - 67	3.3 ± 10	0 - 29	4.7 ± 7.9	0 - 3	0.3 ± 0.9
Plectidae	*Wilsonema*	0 - 0	0 ± 0	0 - 0	0 ± 0	0 - 102	20.1 ± 34.2
Prismatolaimidae	*Prismatolaimus*	0 - 22	2.2 ± 5.4	0 - 50	3.7 ± 10.2	0 - 87	40.2 ± 33.4
Rhabditidae	unidentified genera	0 - 866	232.9 ± 197.7	0 - 754	157 ± 158.7	0 - 66	24.4 ± 28.4
Rhabditidae	*Dauer*	0 - 601	74.5 ± 104.6	0 - 227	35 ± 50.8	0 - 62	20.8 ± 20.6
Rhabdolaimidae	*Rhabdolaimus*	0 - 0	0 ± 0	0 - 14	0.3 ± 2	0 - 6	0.8 ± 2
Teratocephalidae	*Metateratocephalus*	0 - 0	0 ± 0	0 - 4	0.1 ± 0.5	0 - 154	49.1 ± 55.6

Continued

Teratocephalidae	*Teratocephalus*	0 - 0	0 ± 0	0 - 11	0.4 ± 1.8	0 - 37	4.7 ± 12.4
Fungal-feeders							
Anguinidae	*Ditylenchus*	0 - 42	3.8 ± 8.8	0 - 71	4.1 ± 11.3	0 - 37	10.5 ± 12.3
Aphelenchidae	*Aphelenchus*	0 - 140	26.5 ± 27.7	0 - 285	33 ± 47.1	0 - 27	5.2 ± 9.3
Aphelenchoididae	*Aphelenchoides*	0 - 262	47.2 ± 49.4	0 - 143	35 ± 32.5	0 - 169	59.3 ± 64
Diphterophoridae	*Diphterophora*	0 - 62	3.8 ± 10.2	0 - 107	7.6 ± 19.7	0 - 0	0 ± 0
Diphterophoridae	*Tylolaimorphus*	0 - 0	0 ± 0	0 - 0	0 ± 0	0 - 335	115.7 ± 132.2
Leptonchidae	*Tylencholaimus* + other genera	0 - 23	0.9 ± 3.6	0 - 110	7.2 ± 20.2	0 - 278	111.1 ± 117.3
Others[2]			31		11.1		4

[1]Separation between the two genus was not possible; [2]Unidentified nematodes which were groups with nematodes having the most similar morphological structures to calculate trophic groups abundances and indices.

Table 5. Results of the pairwise test of the PERMANOVA comparing the structure of the nematode communities based on 4 taxonomic levels of identification: p values of the Mann-Whitney or Kruskal-Wallis and Bonferroni tests.

Land-use modalities		86 genera		48 families		17 functional guilds		7 trophic groups	
Meadow (with crop in rotation)/	Crop	0.001	***	0.001	***	0.001	***	0.002	**
Meadow (with crop in rotation)/	Permanent meadow	0.657	ns	0.593	ns	0.214	ns	0.115	ns
Meadow (with crop in rotation)/	Crop (with meadow in rotation)	0.002	**	0.003	**	0.082	ns	0.051	ns
Meadow (with crop in rotation)/	Forests	0.001	***	0.001	***	0.001	***	0.001	***
Crop/	Permanent meadow	0.001	***	0.001	***	0.001	***	0.001	***
Crop/	Crop (with meadow in rotation)	0.054	ns	0.024	*	0.246	ns	0.325	ns
Crop/	Forests	0.001	***	0.001	***	0.001	***	0.001	***
Permanent meadow/	Crop (with meadow in rotation)	0.001	***	0.001	***	0.002	**	0.001	***
Permanent meadow/	Forests	0.001	***	0.001	***	0.001	***	0.001	***
Crop (with meadow in rotation)/	Forests	0.001	***	0.001	***	0.001	***	0.001	***

The symbol *Indicates that Mann-Whitney or Kruskal-Wallis and Bonferoni tests are significant with $p < 0.05$; **Significant at $p < 0.01$; ***Significant at $p < 0.01$.

one at the genus levels (**Figure 3(a)**) and the other at the functional-guild level (**Figure 3(b)**). The forest, crop and meadow samples were significantly separated from each other at the genus level, whereas the functional-guild level was not discriminative. Neither of the two taxonomic ranks distinguished the effect of rotations (C vs. Cm and M vs. Mc).

3.5. Relations between Nematodes Communities, Land Use and Technical Cropping Practices

The nematode ecological indices, such as the MI, BaMI, NCR and PPI, significantly responded to land use and technical cropping practices; 33 significant differences were obtained between nematode ecological indices and the crop managements (**Table 6**). In contrast, the total abundance of nematodes and the abundance of the different trophic groups were discriminated to a lesser extent by more or less technical cropping practices (**Table 6**); only 18 significant differences were recorded between

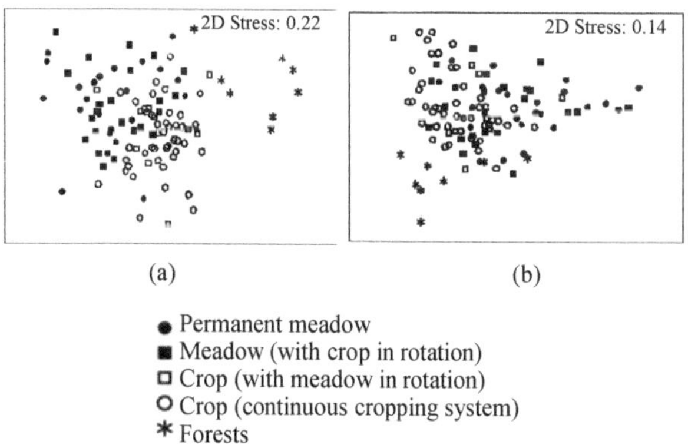

Figure 3. Multidimensional scaling on (a) nematode taxa (86) and (b) nematode functional guilds (17) for the 109 sites.

nematode abundances and the crop managements.

The abundances of plant-parasitic nematodes (Pf) and of the opportunistic bacterial-feeders (Ba_1) were significantly different between crops and meadows (with or without crops in rotation). Consequently, the EI and PPI

Table 6. Relations between the abundances of the nematode trophic groups and the nematode ecological indices and the crop management modalities studied using non-parametric Kruskal-Wallis and Mann-Whitney tests.

Nematode trophic groups		Land uses[1]	Managment system[2]	Organic amendment[3]	Ploughing frequency[4]	Time since last tillage[5]	Use of pesticides[6]
Opportunistic bacterial-feeders	Ba_1	***	ns	**	ns	*	ns
Others bacterial-feeders	Ba_{234}	ns	ns	ns	**	*	ns
Fungal-feeders	Fu	ns	ns	ns	**		ns
Omnivores	Om	ns	**	ns	ns	**	ns
Carnivores	Ca	*	ns	ns	ns		**
Plant-parasitic nematodes	Pf	***	**	ns	**	***	ns
Facultative plant feeders	RHF	ns	ns	ns			ns
Total		**	***	ns	**	**	ns
Nematode ecological indices							
Maturity Index	MI	***	**	***	***	***	*
Maturity Index on bacterial-feeders	BaMI	**	*(8%)	***	***	***	*
Maturity Index on fungal-feeders	FuMI	ns	**	***	ns	ns	*
Nematode Channel Ratio	NCR	ns	ns	***	**		***
Plant parasitic Index	PPI	***	ns	*(5%)	***	**	*
Enrichment Index	EI	**	ns	***	**	***	ns
Structure Index	SI	ns	ns	*(6%)	ns	ns	ns
Channel Index	CI	*	ns	***	**	**	***

The symbol *Indicate that Mann-Whitney or Kruskal-Wallis and Bonferoni tests are significant with $p < 0.05$; **Significant at $p < 0.01$; ***Significant at $p < 0.01$; ns means not significant difference; crop, crop with meadow in rotation, [1]Meadow with crop in rotation, permanent meadow; [2]Intensive, moderate, sustainable, organic; [3]None, slurry, manure, both slurry and manure; [4]Low, high; [5]<5 months, >5 months; [6]Yes, no.

significantly differed by land-use. Crops with meadow in the rotation exhibit intermediate values for these parameters. Total density was significantly higher in permanent meadows than in continuous crops.

The total abundance as well as the omnivorous and plant-feeder abundances were significantly higher under sustainable crop systems vs. intensive crop systems (respectively from 28.4, 13.5, 1.7 $ind \cdot g^{-1}$ under sustainable vs. 15.0, 4.7, 6.0 $ind \cdot g^{-1}$ under intensive) but did not distinguish among the other systems. Three of the indices were higher in sustainable systems than in intensive systems (**Figure 4(a)**): MI (2.5 vs. 2.0), BaMI (1.8 vs. 1.6) and FuMI (2.3 vs. 2.1). Organic management did not lead to significantly different nematode community parameters when compared to intensive, moderate or sustainable managements. Differences were observed between plots that were not fertilised and fertilised plots, regardless of the type of amendment. The Ba_1 were less abundant in no fertilised plots (0.7 $ind \cdot g^{-1}$ vs. 2.9 $ind \cdot g^{-1}$ in average). The MI, BaMI, FuMI, CI were higher without fertilisation (respectively 2.4 vs. 2.1 in average; 1.8 vs. 1.6 in average and 2.2 vs. 2.1 in average), whereas the NCR and EI were lower without fertilisation (respectively 82.4 vs. 86.9 in average and 54.4 vs. 71.5 in average) (**Figure 4(b)**).

Although the nematode community did not show differences among tillage practices, deep plough, pseudo-ploughing, superficial ploughing or no ploughing (results not shown), the frequency of the ploughing significantly impacted the nematodes, as did the time since last tillage. When time since last tillage was less than five months, several parameters were reduced in comparison with plots where time since last tillage was greater than five months (**Figure 4(d)**): MI (1.96 vs. 2.23), BaMI (1.55 vs. 1.72), PPI (2.66 vs. 2.50), Ba_{234} (2.9 vs. 3.7 $ind \cdot g^{-1}$), Om (0.5 vs. 1.0 $ind \cdot g^{-1}$), Pf (3.9 vs. 9.1 $ind \cdot g^{-1}$) and total abundance (14.0 vs. 20.3 $ind \cdot g^{-1}$); in contrast, the EI (76.1 vs. 63.8) and Ba_1 (3.3 vs. 2.2 $ind \cdot g^{-1}$) increased; Ca were not affected (0.4 vs. 0.5 $ind \cdot g^{-1}$). Low ploughing frequency led to higher abundances of nematodes than with high ploughing frequency (23.4 vs. 13.0 $ind \cdot g^{-1}$) and Ba_{234} (4.1 vs. 2.8 $ind \cdot g^{-1}$), Fu (1.1 vs. 0.6 $ind \cdot g^{-1}$), PF (11.8 vs. 3.0 $ind \cdot g^{-1}$) leading to higher MI (2.24 vs 1.99), BaMI (1.73 vs. 1.56), PPI (2.68 vs. 2.50), CI (13.1 vs. 6.9) and lower EI (61.1 vs. 76.1) (**Figure 4(c)**). The use of pesticides led to fewer changes in the

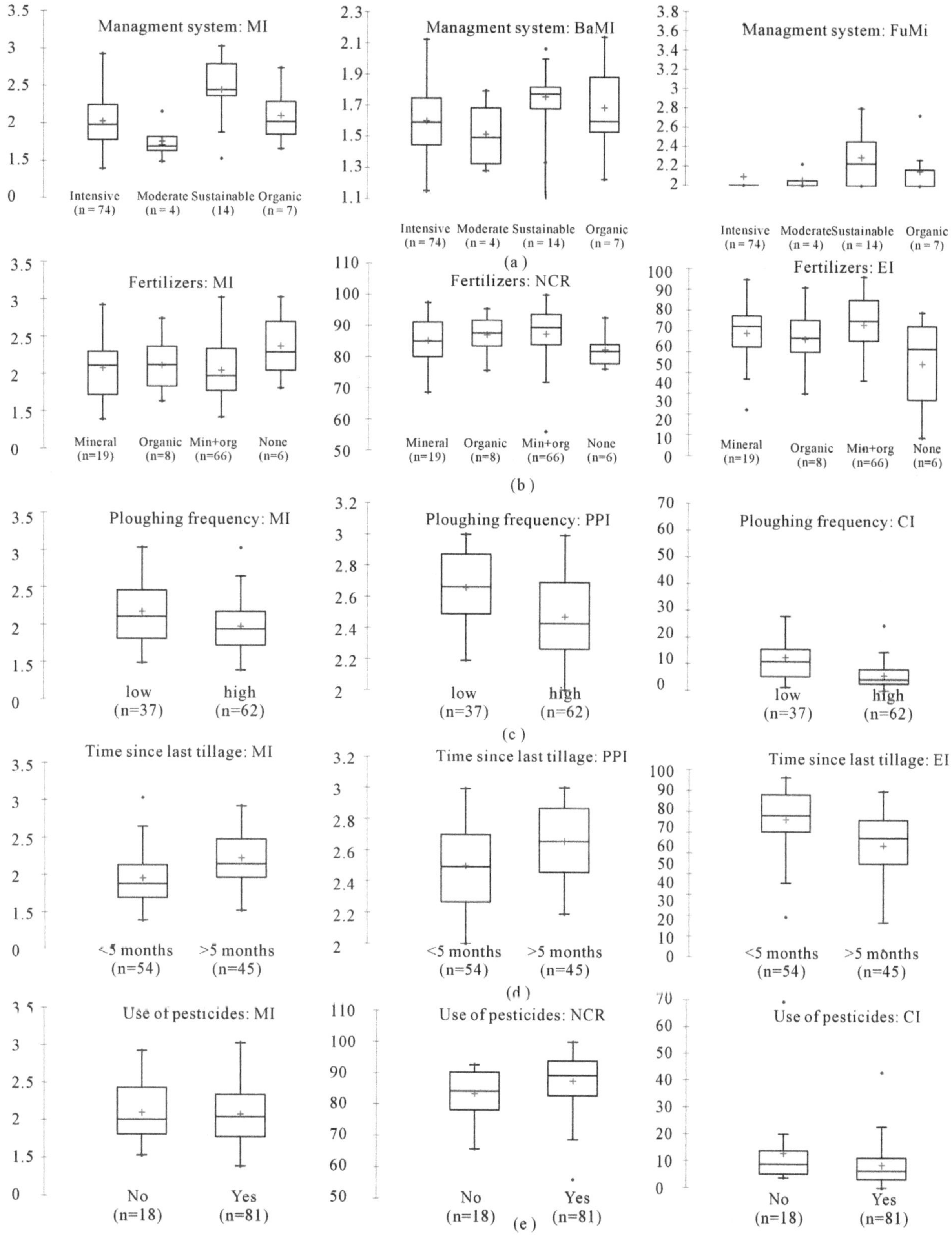

Figure 4. Box plot of the nematode parameters in relation to agricultural practices: (a) Management system and the nematode ecological indices MI, BaMI and FuMI; (b) Fertilisers and the nematode ecological indices MI, NCR and EI; (c) Ploughing frequency and the nematode ecological indices MI, PPI and CI; (d) Time since last tillage and the nematode ecological indices MI, PPI and EI; (e) Use of pesticides and the nematode ecological indices MI, NCR and CI (n = 99; forest sites excluded from these analyses).

nematode communities, but when pesticides were not used, the decomposition pathway was more bacterial than fungal (NCR increased and CI decreased) (**Figure 4(e)**).

4. Discussion

The study site network comprises 109 monitoring sites located on a 16 × 16 km systematic grid covering a very large area of 27,000 km^2. The sites covered contrasting pedoclimatic conditions and land uses from natural forests to permanent meadows to intensive cropping.

4.1. Nematode Abundances and Soil Organic Matter Content

Soil characteristics are the first elements that structure the nematode community even if the relation between nematodes and soil physico-chemical characteristics is rarely studied [27,28]. We analysed the relation between the nematode community characteristics and five important soil parameters (C content, total N content, C/N, available P, pH), whatever the land uses and agricultural practices. The total nematode abundance as well as the plant-parasitic nematode, the bacterial-feeder and the fungal-feeder abundances were positively correlated with the soil C content: the more organic resources are available, the more nematodes are abundant [28-30]. Moreover opportunistic bacterial-feeder abundance was highly correlated with available P, which is in agreement with the assumption that these nematodes multiply under rich conditions [9,22]. Bacterial-feeding nematodes are, with protozoa, the main grazers of soil bacteria. Interactions between bacteria and nematodes have important repercussions on soil functioning increasing the nutrient availability [29,31].

Nematode abundances, whatever the trophic groups, not correlated with the soil pH. The Nematode Channel Ratio (NCR) increased with increasing soil C, N and available P contents, indicating that, in soil with higher fertility in the Britanny area, the decomposition pathway was more bacterial than fungal. Enrichment Index and Maturity indices based on bacterial-feeders and fungal-feeders (BaMI and FuMI) were higher in soil high available P confirming the relation of these indices with the high nutrient status of soils. The structure Index was not significantly correlated to soil chemical fertility indicating that disturbances other than reduction of soil organic matter content can lead to short and simple soil foodwebs (*i.e.* low SI).

4.2. Regional Average of Nematode Parameters and Taxa Composition in Crops, Meadows and Forests

Using this unique dataset, it was possible to clearly distinguish the nematode communities of the three main land uses. As found by van Ekeren *et al.* [32] in the Netherlands, and Hanel [33,34] in the Czech Republic, the nematode abundance was greater in grasslands than in crops; meadows were characterised by values ranged from 3.7 to 52.7 with an average of 20.6 ind·g^{-1}, and croplands were characterised by values ranged from 3.8 to 35.5 with an average of 13.2 ind·g^{-1}. The differences in densities between the crops and meadows were mainly linked to the higher abundances of plant-parasitic nematodes in the meadows, as also seen in van Ekeren *et al.* [32]. The crop areas were dominated by bacterial-feeders, particularly opportunistic bacterial-feeders, indicating greater nutrient enriched conditions in crops than in meadows [35,36]. Even if it was not precisely quantified in this study, it can be said that usually, crops receive more fertilizers (mineral) than meadows in Brittany, which explain the greatest abundances of opportunistic bacterial-feeders in crops. The fungal-feeders, omnivores and facultative plant-feeders were present in similar densities in these two agricultural uses of lands.

Forests were characterised by total densities of nematodes ranging from 1.7 to 57.9 with an average of 23.7 ind·g^{-1} and by the dominance of facultative plant-feeders, also called myco-phytophages (9.1 ind·g^{-1} soil). High variations in the abundances of plant-parasitic nematode observed in the forest areas (ranging from 0.1 to 30.3 in·g^{-1} soil) were probably due to the different types of forest sampled, from old spruce or oak forests with low phytoparasitic nematode densities to young copse forests with high densities of this trophic group [37].

The analysis of decomposition pathways by nematodes shows differences among land uses. The very low CI in crops and meadows (7.4 and 11.1 on average) indicated a dominant bacterial-decomposition pathway, more enhanced than that found by [38] (18 and 24 respectively in crop-fields and meadows). The CI in forests was higher than in crops and meadows; extremely variable values (11 to 100, average = 51.7) may reflect the different types of forests, e.g., the coniferous forests exhibited higher CI than the deciduous forests [38]. The NCRs in crops and meadows (86 on average) were comparable to those found by van Eekeren *et al.* [32]. The EI was on average 76.4 (min: 46, max: 96) in crops, which is a high value at a regional scale. The high EI in meadows (greater than 64 on average; min: 13, max: 95) could be because of the large use of organic amendments in Brittany, which can lead to water pollution [39]. The maturity indices (MI, BaMI and FuMI) followed the same trends, indicating that the agricultural fields are more disturbed and show mostly very enriched conditions than the forests, with the crops being even more enriched (EI, FuMI) than the meadows [24]. The SI in forests was characterised by values ranging from 49 to 94, with an average of 82, which was significantly higher than in the meadows and crops (average of 56 and 59, respectively), indicating longer and more complex micro-food webs in forests than in agricultural lands. However, very high variability in the SI was measured in agricultural lands, from 16 to 91 in crops and 0 to 91 in meadows, reflecting the effects of the agricultural practices.

Cephalobidae were the predominant bacterial-feeders in Brittany meadows, as also reported by [28] from various experiments. Rhabditidae outnumbered Cephalobidae in crops, a situation that was only reported in conventionally managed arable land by [24]. Plectidae were more represented in forests, particularly *Wilsonema*, which were found only in these ecosystems. *Prismatolaimus*, and Teratocephalidae, were also quite exclusively present in the forests. Hanel [33] found the same pattern for these bacterial-feeders, and he also found more *Tylencholaimus* in forests than in crops and meadows. However, he did not find higher densities of *Tylolaimophorus* in forests. The predators *Clarkus* and *Mylonchulus* have previously been found not to be affected by agricultural practices such as tillage [40], and *Clarkus* were even more abundant in crops in Brittany than in the other land uses. *Clarkus* and *Aporcelaimellus* were also found to be abundant predators in agricultural fields by Vestergard [41]. Plant-feeding nematode densities are very heterogeneous across all land uses, varying from 0 to 39.2 ind·g^{-1}. The nematode communities were dominated by *Tylenchorhynchus* and *Pratylenchus* in the cropped fields, by *Meloidogyne* in the meadows and by *Paratylenchus* in the forests. The greater abundances of the plant-parasitic nematodes, especially *Meloidogyne*, that were detected in the perennial sites than in the annual sites may be linked in part to the vegetation state on the sampling date; in March and April, the meadows have developed a higher root biomass than have the annual crops that have just begun their vegetative cycle.

4.3. The Taxonomic Rank of Determination as a Tool to Discriminate Modalities

The objective of this study was not to determine the nematode diversity at species level; it would have required a much more intensive effort to identify the 21,820 observed nematodes. Such a nematode species inventory is absent for France, and it should be of great interest in the future to have a complete study of the biodiversity of these organisms and, particularly, its loss [4]. However, for the ecological analysis and evaluation of the soil ecosystem services supported by the micro-regulators [11,42,43], and for the commercial use of nematodes as bioindicators [44], a time/money compatible taxonomical resolution is required. Total nematode abundance is know, since long time, to be well correlated to herbage production in pasture [45]. However, when studying other land uses, it is not always such a good parameter to evaluate soil quality as it includes organisms with very different roles in agrosystem functioning, such as parasites of plants and micro-regulators [46]. Trophic groups are the first pertinent level of nematode grouping [21], followed by functional guilds (trophic group × cp class), families, whose genera and species usually belong to the same functional guild, and finally, the highest level of identification, species or genera [23,24].

In this study, four different taxonomic levels (genus: 86; families: 48; functional guilds: 17 and trophic groups: 6) were used to distinguish among five land uses. The distinction between permanent meadows, continuous crops and forests was highly significant for all taxonomic levels. The distinction between permanent meadows and meadows with crops in the rotation was never possible for any taxonomic level. Continuous crops and crops with meadows in rotation were significantly different only for the family level. Finally, meadows with crops in rotation and crops with meadows in rotation were significantly different for both genus and family levels. From these results, it appeared that analysis at the family level (and genus for families that include species with different trophic behaviours) was the most efficient for soil quality monitoring programmes using nematodes as descriptors.

4.4. Relations between Biological Data and Agricultural Practices

The nematodes parameters (the densities of the trophic groups and the nematode ecological indices) have been partly confronted to explanatory variables such as the agricultural management practices.

The management modalities considered here influence the abundance of soil nematodes significantly. The lowest abundances were found under both intensive and organic management, whereas the highest was found under sustainable management. Several studies have shown that the abundance of soil nematodes is increased under organic agriculture compared to intensive agriculture [28,47], however, such results were not found in French Brittany. Sustainable strategies led to higher MIs (including those of bacterial- and fungal-feeders), indicating a less disturbed situation with significantly more persistent microbivores than opportunist microbivores. For some indices, no differences were found between organic, moderate and conventional management systems, as was expected [48,49], possibly because of the low numbers of plots with organic (=7) and moderate (n = 4) management systems. Tillage is known to greatly influence the soil biological activities and nematode structures [30,35].

No differences were found in the nematode patterns or indices between the upper or deep tillage practices; however, the ploughing frequency and time since last tillage led to disturbances in the nematode communities, as shown by the reduction in the Maturity Index, the increase in the Enrichment Index and the decrease in the BaMI. The decomposition pathways were more bacterial-based in areas with high ploughing frequencies and time since last tillage reduced (higher CI, lower NCR).

According to the literature, mineral and organic amendments can cause increases in microbivorous nematodes [23,29,50]. In this study, very few plots received no fertiliser (neither organic nor mineral), but they exhibited higher Maturity Indices and lower NCR than those fertilised with organic and mineral amendments in combination, indicating a decrease of the opportunistic Rhabditidae and an enhancement of the fungal-feeders in these communities. The only significant, measured effects of pesticides on the nematode parameters were on NCR and CI, suggesting the involvement of pesticides in the fungal decomposition pathway.

This study shows the relevance of nematode descriptors, especially the ecological indices, to distinguish among most agricultural practices. These parameters could be used as bioindicators of land use practices.

5. Conclusion

The RMQS BioDiv programme provided the opportunity to develop a database with relevant frame of references for nematodes assemblages. This dataset could contribute to the development of biological indicators at the European scale. This initial inventory will be used over time to assess the dynamics of the biological status of soils from land use, agricultural practices, contamination and climate changes.

6. Acknowledgements

We thank all of the farmers who kindly allowed us to access to their land and all of the contributors who collected soil data, especially Vincent MERCIER. This programme was part of "Bioindicator-Phase1" projects funded by ADEME (French Environment and Energy Management Agency).

REFERENCES

[1] UNCED, "United Nations Conference on Environment and Development Agenda 21, Rio de Janeiro," 1992.

[2] European Commission, "Communication from the Commission to the Council, the European Parliament, the European Economic and Social Committee and the Committee of the Regions: Thematic Strategy for Soil Protection," Vol. 231, 2006. http://ec.europa.eu/environment/soil/

[3] X. Morvan, N. P. A. Saby, D. Arrouays, C. Le Bas, R. J. A. Jones, F. G. A. Verheijen, P. H. Bellamy, M. Stephens and M. G. Kibblewhite, "Soil Monitoring in Europe: A Review of Existing Systems and Requirements for Harmonisation," *Sciences of Total Environment*, Vol. 391, No. 1, 2008, pp. 1-12.

[4] A. Bispo, G. Pérès, D. Cluzeau, U. Graefe, J. Römbke, M. Rutgers, M. Fuchs, J. P. Sousa, R. Schulte, M. Dombos, B. Simon, A. Gal, J. Cortet, R. Chaussod, K. Ritz, R. E. Creamer, A. Winding, M. English, J. Boixadera and J. Rubio, "ENVASSO (Environmental Assessment of Soil for Monitoring) WP 5—Decline in Soil Biodiversity," 2007.

[5] S. Krück, M. Joschko, R. Schultz-Sternberg, B. Kroschewski and J. Tessmann, "A Classification Scheme for Earthworm Populations (Lumbricidae) in Cultivated Agricultural Soils in Brandenburg, Germany," *Journal of Plant Nutrition and Soil Science*, Vol. 169, No. 5, 2006, pp. 651-660.

[6] D. Cluzeau, G. Pérès, M. Guernion, R. Chaussod, J. Cortet, M. Fargette, F. Martin-Laurent, T. Mateille, C. Pernin, J.-F. Ponge, N. Ruiz-Camacho, C. Villenave, L. Rougé, V. Mercier, A. Bellido, M. Cannavacciuolo, D. Piron, D. Arrouays, L. Boulonne, C. Jolivet, P. Lavelle, E. Velasquez, O. Plantard, C. Walter, B. Foucaud-Lemercier, S. Tico, J.-L. Giteau and A. Bispo, "Integrating Biodiversity in Soil Quality Monitoring Networks: Example of the Pilot Programme at the Regional Scale, the 'RMQS BioDiv' (Intégration de la Biodiversité des Sols Dans les réSeaux de Surveillance de la Qualité des Sols: Exemple du Programme-Pilote à L'éChelle Régionale, le RMQS BioDiv)," *Study and Soil Management* (*Etude et Gestion des Sols*), Vol. 16, No. 3-4, 2009, pp. 187-201.

[7] M. Rutgers, A. J. Schouten, J. Bloem, Eekeren van N. J. M, R. G. M. de Goede, G. A. J. M. Jagers op Akkerhuis, A. Van der Wal, C. Mulder, L. Brussaard and A. M. Breure, "Biological Measurements in a Nationwide Soil Monitoring Network," *European Journal of Soil Science*, Vol. 60, No. 5, 2009, pp. 820-832.

[8] M. G. Paoletti, "Using Bioindicators Based on Biodiversity to Assess Landscape Sustainability," *Agriculture, Ecosystems & Environment*, Vol. 74, No. 1-3, 1999, pp. 1-18.

[9] T. Bongers and H. Ferris, "Nematode Community Structure as a Bioindicator in Environmental Monitoring," *Trends in Ecology and Evolution*, Vol. 14, No. 6, 1999, pp. 224-228. doi:10.1016/S0169-5347(98)01583-3

[10] J. Cortet, A. G. De Vauflery, N. P. Balaguer, L. Gomot, C. Texier and D. Cluzeau, "The Use of Invertebrate Soil Fauna in Monitoring Pollutant Effects," *European Journal of Soil Biology*, Vol. 35, No. 3, 1999, pp. 115-134. doi:10.1016/S1164-5563(00)00116-3

[11] L. Brussaard, "Soil Biodiversity for Agricultural Sustainability," *Agriculture, Ecosystems and Environment*, Vol. 121, No. 3, 2007, pp. 233-244.

[12] A. Bispo, D. Cluzeau, R. E. Creamer, U. Graefe, P. H. Krogh, J. P. Sousa, G. Peres, M. Rutgers, A. Winding and J. Römbke, "Indicators for Monitoring Soil Biodiversity," *Integrated Environment Assessment Management*, Vol. 5, No. 4, 2009, pp. 717-719.

[13] D. Cluzeau, M. Guernion, R. Chaussod, F. Martin-Lau-

rent, C. Villenave, J. Cortet, N. Ruiz-Camacho, C. Pernin, T. Mateille, L. Philippot, A. Bellido, L. Rougé, D. Arrouays, A. Bispo and G. Pérès, "Integration of Biodiversity in Soil Quality Monitoring: Baselines for Microbial and Soil Fauna Parameters for Different Land-Use Types," *European Journal of Soil Biology*, Vol. 49, 2012, pp. 63-72.

[14] D. Arrouays, C. Jolivet, L. Boulonne, G. Bodineau, N. P. A. Saby and E. Grolleau, "A New Initiative in France: A Multi-Institutional Soil Quality Monitoring Network," *Comptes Rendus de l'Academie d'Agriculture de France*, Vol. 88, No. 2, 2002, pp. 93-103.

[15] N. P. A. Saby, J. Thioulouse, C. Jolivet, C. Ratié, L. Boulonne, A. Bispo and D. Arrouays, "Multivariate Analysis of the Spatial Patterns of 8 Trace Elements Using the French Soil Monitoring Network Data," *Sciences of Total Environment*, Vol. 407, No. 21, 2009, pp. 5644-5652.

[16] J. W. Seinhorst, "Modifications of the Elutriation Method for Extracting Nematodes from Soil," *Nematologica*, Vol. 8, No. 2, 1962, pp. 117-128.

[17] ISO 23611-4, "Soil Quality—Sampling of Soil Invertebrates. Part 4: Sampling, Extraction and Determination of Soil-Inhabiting Nematodes," Geneva, 2007.

[18] I. Andrassy, "Klasse Nematoda (Ordnungen Monhysterida, Desmoscolecida, Araeolaimida, Chromadorida, Rhabditida)," Akademie-Verlag, Berlin, 1984.

[19] T. Bongers, "Nematodes from Netherlands (De Nematoden van Nederland)," Koninklijke Nederlandse Natuurhistorische Vereniging, Utrecht, 1994.

[20] M. R. Siddiqi, "Tylenchida. Parasites of Plants and Insects," CABI, 2000. doi:10.1079/9780851992020.0000

[21] G. W. Yeates, T. Bongers, R. G. M. de Goede, D. W. Freckman and S. S. Georgieva, "Feeding Habits in Soil Nematode Families and Genera—An Outline for Soil Ecologists," *Journal of Nematology*, Vol. 25, No. 3, 1993, pp. 315-331.

[22] T. Bongers, "The Maturity Index: An Ecological Measure of Environmental Disturbance Based on Nematode Species Composition," *Oecologia*, Vol. 83, No. 1, 1990, pp. 14-19.

[23] H. Ferris, T. Bongers and R. de Goede, "A Framework for Soil Food Web Diagnostics: Extension of the Nematode Faunal Analysis Concept," *Applied Soil Ecology*, Vol. 18, No. 1, 2001, pp. 13-29. doi:10.1016/S0929-1393(01)00152-4

[24] G. W. Yeates, "Nematodes as Soil Indicators: Functional and Biodiversity Aspects," *Biology and Fertility of Soils*, Vol. 37, No. 4, 2003, pp. 199-210.

[25] J. B. Kruskal and M. Wish, "Multidimensional Scaling. on Quantitative Applications in the Social Sciences," Sage Publications, London, 1978.

[26] M. J. Anderson, R. N. Gorley and K. R. Clarke, "Permanova+ for Primer. Guide to Software and Statistical Methods," PRIMER-E, Ltd., Plymouth, 2008.

[27] C. Villenave and P. Cadet, "Influence of Soil Organic Matter and Ion Concentration on Some Senegalese Plant-Parasitic Nematodes," *European Journal of Soil Biology*, Vol. 35, No. 4, 1999, pp. 189-197. doi:10.1016/S1164-5563(10)70005-4

[28] G. W. Yeates, R. D. Bardgett, R. Cook, P. J. Hobbs, P. J. Bowling and J. F. Potter, "Faunal and Microbial Diversity in Three Welsh Grassland Soils under Conventional and Organic Management Regimes," *Journal of Applied Ecology*, Vol. 34, No. 2, 1997, pp. 453-470. doi:10.2307/2404890

[29] C. Villenave, K. Ekschmitt, S. Nazaret and T. Bongers, "Interactions between Nematodes and Microbial Communities in a Tropical Soil Following Manipulation of the Soil Food Web," *Soil Biology & Biochemistry*, No. 36, 2004, pp. 2033-2043.

[30] C. Villenave, A. O. Ba and B. Rabary, "Analysis of Soil Biological Functionning by the Study of Nematofauna: Direct Seeding Versus Tillage in the Highlands near Antsirabe (Madagascar) (Analyse du Fonctionnement Biologique du sol Par L'étude de la Nematofaune: Semis Direct Versus Labour sur les Hautes Terres près d'Antsirabé (Madagascar))," *Study and Soil Management*, Vol. 16, No. 3-4, 2009, pp. 369-378.

[31] U. Irshad, C. Villenave, A. Brauman and C. Plassard, "Grazing by Nematodes on Rhizosphere Bacteria Enhances Nitrate and Phosphorus Availability to Pinus Pinaster Seedlings," *Soil Biology & Biochemistry*, Vol. 43, No. 10, 2011, pp. 2121-2126.

[32] N. van Eekeren, L. Bommele, J. Bloem, T. Schouten, M. Rutgers, R. G. M. deGoede, D. Reheul and L. Brussaard, "Soil Biological Quality after 36 Years of Ley-Arable Cropping, Permanent Grassland and Permanent Arable Cropping," *Applied Soil Ecology*, Vol. 40, No. 3, 2008, pp. 432-446. doi:10.1016/j.apsoil.2008.06.010

[33] L. Hanel, "Secondary Successional Stages of Soil Nematodes in Cambisols of South Bohemia," *Nematologica*, Vol. 41, No. 1-5, 1995, pp. 197-218.

[34] L. Hanel, "Recovery of Soil Nematode Populations from Cropping Stress by Natural Secondary Succession to Meadow Land," *Applied Soil Ecology*, Vol. 22, No. 3, 2003, pp. 255-270. doi:10.1016/S0929-1393(02)00152-X

[35] H. Okada and H. Haradaba, "Effects of Tillage and Fertilizer on Nematode Communities in a Japanese Soybean Field," *Applied Soil Ecology*, Vol. 35, No. 3, 2007, pp. 582-598.

[36] W. Liang, Y. Lou, L. Q. S. Zhong, X. Zhang and J. Wang, "Nematode Faunal Response to Long-Term Application of Nitrogen Fertilizer and Organic Manure in Northeast China," *Soil Biology & Biochemistry*, Vol. 41, No. 5, 2009, pp. 883-890.

[37] L. Hanel, "Soil Nematodes in Five Spruce Forests of the Beskydy Montains, Czech Republic," *Fundamental and Applied Nematology*, Vol. 19, No. 1, 1996, pp. 15-24.

[38] L. Ruess, "Nematode Soil Faunal Analysis of Decomposition Pathways in Different Ecosystems," *Nematology*, Vol. 5, No. 2, 2003, pp. 179-181. doi:10.1163/156854103767139662

[39] J. Molénat, P. Durand, C. Gascuel-Odoux, P. Davy and G. Gruau, "Mechanisms of Nitrate Transfer from Soil to Stream in an Agricultural Watershed of French Brittany," *Water Air Soil Pollution*, Vol. 133, No. 1-4, 2002, pp. 161-183.

[40] D. W. Freckman and C. H. Ettema, "Assessing Nematode Communities in Agroecosystems of Varying Human Intervention," *Agriculture, Ecosystems & Environment*, Vol. 45, No. 3-4, 1993, pp. 239-261.

[41] M. Vestergard, "Nematode Assemblages in the Rhizosphere of Spring Barley (*Hordeum Vulgare* L.) Depended on Fertilisation and Plant Growth Phase," *Pedobiologia*, Vol. 48, No. 3, 2004, pp. 257-265. doi:10.1016/j.pedobi.2004.01.003

[42] G. A. Adams and D. H. Wall, "Biodiversity above and below the Surface of Soils and Sediments: Linkages and Implications for Global Change," *Bioscience*, Vol. 50, No. 12, 2000, pp. 1043-1048. doi:10.1641/0006-3568(2000)050[1043:BAABTS]2.0.CO;2

[43] P. Lavelle, T. Decaëns, A. Aubert, S. Barot, M. Blouin, F. Bureau, P. Margerie, P. Mora and J.-P. Rossi, "Soil Invertebrates and Ecosystem Services," *European Journal of Soil Biology*, Vol. 42, Suppl. 1, 2006, pp. S3-S15.

[44] M. W. Trett, B. C. Urbano, S. J. Forster and S. P. Trett, "Commercial Aspects of the Use of Nematodes as Bioindicators," In: M. J. Wilson and T. Kakouli-Duarte, (Eds), *Nematodes as Environmental Indicators*, CABI, Wallingford, 2009, pp. 1-44.

[45] G. W. Yeates, "Soil Nematodes in Terrestrial Ecosystems," *Journal of Nematology*, Vol. 11, No. 3, 1979, pp. 213-229.

[46] T. Bongers, H. van der Meulen and G. Korthals, "Inverse Relationship between the Nematode Maturity Index and Plant Parasitic Index under Enriched Nutrient Conditions," *Applied Soil Ecology*, Vol. 6, No. 2, 1997, pp. 195-199.

[47] P. Coll, E. Le Cadre, E. Blanchart, P. Hinsinger and C. Villenave, "Organic Viticulture and Soil Quality: A Long-Term Study in Southern France," *Applied Soil Ecology*, Vol. 50, No. 1, 2011, pp. 37-44.

[48] D. A. Neher, "Soil Community Composition and Ecosystem Processes—Comparing Agricultural Ecosystems with Natural Ecosystems," *Agroforestry Systems*, Vol. 45, No. 1-3, 1999, pp. 159-185. doi:10.1023/A:1006299100678

[49] C. Mulder, D. De-Zwart, H. J. Van Wijnen, A. J. Schouten and A. M. Breure, "Observational and Simulated Evidence of Ecological Shifts within the Soil Nematode Community of Agroecosystems under Conventional and Organic Farming," *Functional Ecology*, Vol. 17, No. 4, 2003, pp. 516-525. doi:10.1046/j.1365-2435.2003.00755.x

[50] C. Villenave, S. Saj, A.-L. Pablo, S. Sall, D. Djigal, J.-L. Chotte and M. Bonzi, "Influence of Long-Term Organic and Mineral Fertilization on Soil Nematofauna when Growing Sorghum Bicolor in Burkina Faso," *Biology and Fertility of Soils*, Vol. 46, No. 7, 2010, pp. 659-670.

3

Characteristics of Soil Organic Carbon, Total Nitrogen, and C/N Ratio in Chinese Apple Orchards

Shunfeng Ge, Haigang Xu, Mengmeng Ji, Yuanmao Jiang*

State Key Laboratory of Crop Biology, College of Horticulture Science and Engineering, Shandong Agricultural University, Taian, China.
Email: *ymjiang@sdau.edu.cn

ABSTRACT

Soil organic carbon and nitrogen are used as indexes of soil quality assessment and sustainable land use management. At the same time, soil C/N ratio is a sensitive indicator of soil quality and for assessing the carbon and nitrogen nutrition balance of soils. We studied the characteristics of soil organic carbon and total nitrogen by investigating a large number of apple orchards in major apple production areas in China. High apple orchard soil organic carbon content was observed in the provinces of Heilongjiang, Xinjiang, and Yunnan, whereas low content was found in the provinces of Shandong, Henan, Hebei, and Shaanxi, with the values ranging between 6.44 and 7.76 $g \cdot kg^{-1}$. Similar to soil organic carbon, soil total nitrogen content also exhibited obvious differences in the 12 major apple producing provinces. Shandong apple orchard soil had the highest total nitrogen content (1.26 $g \cdot kg^{-1}$), followed by Beijing (1.23 $g \cdot kg^{-1}$). No significant difference was noted between these two regions, but their total nitrogen content was significantly higher than the other nine provinces, excluding Yunnan. The soil total nitrogen content for Xinjiang, Heilongjiang, Hebei, Henan, and Gansu was between 0.87 and 1.03 $g \cdot kg^{-1}$, which was significantly lower than that in Shandong and Beijing, but significantly higher than that in Liaoning, Shanxi, and Shaanxi. Six provinces exhibited apple orchard soil C/N ratio higher than 10, including Heilongjiang (15.42), Xinjiang (13.38), Ningxia (14.45), Liaoning (12.24), Yunnan (11.03), and Gansu (10.63). The soil C/N ratio was below 10 in the remaining six provinces, in which the highest was found in Shaanxi (9.47), followed by Beijing (8.98), Henan (7.99), and Shanxi (7.62), and the lowest was found in Hebei (6.80) and Shandong (6.05). Therefore, the improvement of soil organic carbon should be given more attention to increase the steady growth of soil C/N ratio.

Keywords: Chinese Apple Orchard; Soil Organic Carbon; Total Nitrogen; C/N Ratio

1. Introduction

China currently has the largest apple cultivation area and production in the world, with 1.99×10^6 ha and 35×10^6 t in the 2011/2012 season [1]. Fruit production has gradually become an important pillar industry to increase the income of farmers. In addition, an increasing amount of nitrogen (N) fertilizers are used to improve yield with the corresponding intensive production of fruits. At present, N utilization has already reached 400 $kg \cdot hm^{-2}$ to 600 $kg \cdot hm^{-2}$ in Chinese apple orchards, which is four to five times more than that in other countries, and is showing an annual increasing trend [2,3]. Excessive use of N fertilizers not only cuts down utilization rate and decreases agricultural production benefits, but also leads to negative consequences, such as aggravated physiological fruit disease and environmental pollution [4-6].

Soil organic carbon and nitrogen are the main nutrition used for vegetation growth, and are also used as indexes of soil quality assessment and sustainable land use management [7-9]. Soil organic carbon and nitrogen not only can reflect the soil fertility level, but can also explain the regional ecological system evolution. The relationship between them can be represented as soil C/N ratio, a sensitive indicator of soil quality and for assessing carbon and nitrogen nutrition cycling of soils [10]. Many researches have shown that soil organic carbon and nitrogen content are not only affected by climate, altitude, terrain, but also land use management. At present, the superior apple production areas are concentrated in Bohai coast area (Shandong, Hebei, Liaoning, Beijing, Tianjin),

*Corresponding author.

Northwest Loess altiplano (Shaanxi, Gansu, Shanxi, Ningxia, Qinghai), Yellow River original (Henan, Jiangsu, Anhui), and Southwest Cool Highlands area [11]. The geographic latitude and longitude, soil type, management level, and fertilization conditions varied between the four superior apple production areas; thus, apple orchard soil carbon and nitrogen characteristics are different among these regions. This study aims to conduct a survey of Chinese apple orchard soils to establish the soil carbon and nitrogen status.

2. Materials and Methods

2.1. Sampling Sites

Soils from the superior apple production areas in China were sampled, including the Bohai coastal area, Northwest Loess altiplano, Yellow River original, and Southwest Cool Highlands area (**Figure 1**). Sampling sites covered 8583 apple orchards in 36 counties of 12 provinces, including Shandong province (Qixia, Penglai, Longkou, Zhaoyuan, Muping, Laishan, Fushan, Jiaonan, Rushan, Yiyuan), Shanxi province (Yuncheng and Taigu), Shaanxi province (Baishui, Luochuan, Qian), Gansu province (Tianshui, Pingliang, Qingyang), Heilongjiang province (Mudanjiang), Liaoning province (Suizhong, Xingcheng), Beijing (Changping), Hebei province (Shijiazhuang, Baoding, Changli), Henan province (Sanmenxia, Shangqiu), Ningxia Autonomous Region (Yongning, Zhongning, Qingtongxia), Yunnan province (Xishan, Zhaoyang, Malong), and Xinjiang Uygur Autonomous Region (Tekesi, Gongliu, Qiaolaketiereke), respectively.

2.2. Soil Sampling and Determination

Composite soil samples were taken from 0 cm to 30 cm depth (more than 20 random points) in each apple orchard in mid-august 2012. The moist soils were sieved (2 mm) after removing the visible plant debris, including roots. Soil organic carbon was determined by wet digestion according to the Turin method, and soil total nitrogen was determined by the Kjeldahl method.

2.3. Data Analyses

All data were analyzed by ANOVA using SPSS (Standard released version 11.5; SPSS Inc., IL, USA).

Figure 1. Geographical location of the sampling apple orchards.

3. Results and Discussion

3.1. Soil Organic Carbon Content

Soil organic carbon not only affects the soil fertility, but also has influence on releasing or holding CO_2 from the atmosphere through various channels, thereby possibly affecting the atmosphere—soil carbon balance [8,12,13]. **Table 1** shows obvious difference in soil organic carbon content between the 12 major apple production areas. The highest soil organic carbon content (15.57 g·kg^{-1}) was found in Heilongjiang province, followed by Xinjiang (14.88 g·kg^{-1}). No significant difference was found between these two areas, but their soil organic carbon content was significantly higher than that in the other nine areas (except Yunnan province). Soil organic carbon content in Yunnan and Beijing apple orchards was higher, with the values of 12.79 and 11.04 g·kg^{-1}, respectively. Soil organic carbon content in Liaoning (9.55 g·kg^{-1}) and Gansu provinces (9.25 g·kg^{-1}) were lower than that of Yunnan and Beijing, but no significant difference (P = 0.05) was found between these four areas. Soil organic carbon content in Ningxia, Shandong, Henan, Hebei, and Shaanxi province ranged from 6.44 g·kg^{-1} to 7.76 g·kg^{-1}, which were significantly lower than that in Heilongjiang, Xinjiang, Yunnan, and Beijing. The lowest apple orchard soil organic carbon content was found in Shanxi province (5.41 g·kg^{-1}, which only accounts for 34.75% that of Heilongjiang province). Some management measures, such as fertilization and irrigation in the apple production system, can significantly affect soil organic carbon content. Higher soil organic carbon content was found in Heilongjiang, Xinjiang, and Yunnan apple orchards, which may be attributed to the following two reasons: First, the soil type and climate in these areas are different with other areas. Second, much more organic fertilizer was used in the apple production. However, soil organic carbon content is relatively low in Shandong, Henan, Hebei, and Shaanxi apple orchards, which is ascribed to a change in fertilizer use strategy with greater use of inorganic fertilizers and less organic fertilizers in these areas. Many studies have shown that, organic combined with inorganic fertilizer is more beneficial to the improvement of the soil organic carbon content compared with the single application of inorganic chemical fertilizer [14,15]. In short, Chinese apple orchard soil organic carbon content is very low, with an average of 9.52 g·kg^{-1}.

Table 1. Soil organic carbon contents of apple orchards in different areas in China.

Area	Maximum (g·kg^{-1})	Minimum (g·kg^{-1})	Mean (g·kg^{-1})	SD
Heilongjiang	21.53	11.42	15.57 a	4.35
Liaoning	14.65	6.67	9.55 bc	2.17
Beijing	15.43	8.38	11.04 b	2.25
Shanxi	10.22	3.87	5.41 d	3.67
Hebei	9.45	4.81	6.53 c	1.83
Shannxi	9.29	4.38	6.44 c	2.17
Shandong	12.53	6.03	7.62 c	2.73
Henan	10.45	5.87	7.43 c	3.11
Gansu	13.22	6.83	9.25 bc	2.93
Ningxia	10.73	5.32	7.76 c	1.86
Xinjiang	22.63	9.74	14.88 a	4.36
Yunnan	16.74	9.89	12.79 ab	2.71

Note: Data in the same column with different letter indicate significant difference (P < 0.05), the same as follows.

3.2. Soil Total Nitrogen Content

Soil total nitrogen is used as an important index for soil fertility evaluation, and reflects the soil nitrogen status [9]. Similar to soil organic carbon, soil total nitrogen content also exhibited obvious differences in the 12 major apple production areas (**Table 2**). The highest soil total nitrogen content (1.26 g·kg^{-1}) was found in Shandong province, followed by Beijing (1.23 g·kg^{-1}). No significant difference was found between these two areas, but their soil total nitrogen content was significantly higher than the other nine areas, which is ascribed to the high nitrogen fertilizer application rate in these two areas. Soil total nitrogen content in Yunnan apple orchards is higher, with a value of 1.16 g·kg^{-1}. Soil total nitrogen content in Xinjiang, Heilongjiang, Hebei, Henan, and Gansu province ranged from 0.87 g·kg^{-1} to 1.03 g·kg^{-1}, which were significantly lower than that in Shandong and Beijing, but significantly higher than that in Liaoning, Shanxi, and Shaanxi. Although nitrogen fertilizer inputs is also at a higher level in Shaanxi province, the total nitrogen content is low because of the poor nitrogen retention ability of the sandy soil [16,17].

Table 2. Soil total nitrogen contents of apple orchards in different areas in China.

Area	Maximum (g·kg^{-1})	Minimum (g·kg^{-1})	Mean (g·kg^{-1})	SD
Heilongjiang	1.23	0.92	1.01 b	0.05
Liaoning	0.92	0.57	0.78 c	0.04
Beijing	1.43	1.08	1.23 a	0.02
Shanxi	1.02	0.57	0.71 c	0.04
Hebei	1.25	0.73	0.96 b	0.06
Shannxi	0.89	0.59	0.68 c	0.03
Shandong	1.56	0.83	1.26 a	0.06
Henan	1.35	0.74	0.93 b	0.04
Gansu	1.20	0.71	0.87 b	0.03
Ningxia	0.85	0.39	0.58 d	0.06
Xinjiang	1.33	0.74	1.03 b	0.06
Yunnan	1.41	0.76	1.16 ab	0.05

3.3. Soil Carbon and Nitrogen Ratio (C/N Ratio)

Soil C/N ratio is a sensitive indicator of soil quality. Soil C/N ratio is often considered as a sign of soil nitrogen mineralization capacity. High soil C/N ratio can slow down the decomposition rate of organic matter and organic nitrogen by limiting the soil microbial activity ability, whereas low soil C/N ratio could accelerate the process of microbial decomposition of organic matter and nitrogen, which is not conducive for carbon sequestration [13]. **Figure 2** shows that significant differences of soil C/N ratio exist among the 12 major apple production areas. Six provinces exhibited apple orchard soil C/N ratio of above 10. The highest soil C/N ratio was found in Heilongjiang (15.42), followed by Xinjiang (14.45), Ningxia (13.38), Liaoning (12.24), Yunnan (11.03), and Gansu (10.63) provinces. Soil C/N ratio was below 10 in the other six provinces, in which the highest was found in Shaanxi (9.47), Beijing (8.98), Henan (7.99), and Shanxi (7.62), and the lowest was found in Hebei (6.80) and Shandong (6.05). At present, China is facing the pressure of CO_2 emission reduction, which may be an effectively approach to make the atmosphere–CO_2 into the apple orchard soil through maintaining higher soil C/N ratio [18,19].

4. Conclusion

Average soil organic carbon, total nitrogen content, and C/N ratio was 9.52 $g{\cdot}kg^{-1}$, 0.93 $g{\cdot}kg^{-1}$, and 10.33, respectively. Therefore, more efforts should be given to increase soil C/N ratio by increasing the application of organic manure or straw to improve the level of soil organic carbon.

5. Acknowledgements

The study was conducted with the support of the Special Fund for Agro-scientific Research in the Public Interest (201103003), the Earmarked Fund for the China Agriculture Research System (CARS-28), and the Agricultural Application and Innovation Project of Shandong Province (201009).

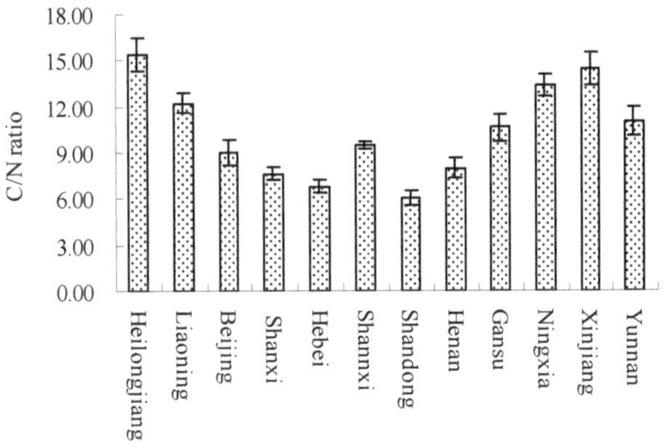

Figure 2. Soil C/N ratio of apple orchards in different areas in China.

REFERENCES

[1] Chinese Ministry of Agriculture, "China's Agricultural Yearbook, 2010," Chinese Agriculture Press, Beijing, 2010.

[2] D. Q. Lv, Y. A. Tong, B. H. Sun and O. Emteryd, "Study on Effect of Nitrogen Fertilizer Use on Environment Pollution," *Plant Nutrition and Fertilizer Science*, Vol. 4, No. 2, 1998, pp. 8-15.

[3] J. Galantini and R. Rosell, "Long-term Fertilization Effects on Soil Organic Matter Quality and Dynamics under Different Production Systems in Semiarid Pampean Soils," *Soil and Tillage Research*, Vol. 87, No. 1, 2006, pp. 72-79. doi:10.1016/j.still.2005.02.032

[4] I. Vogeler, J. Rogasik, U. Funder, K. Panten and E. Schnug, "Effect of Tillage Systems and P-Fertilization on Soil Physical and Chemical Properties, Crop Yield and Nutrient Uptake," *Soil and Tillage Research*, Vol. 103, No. 1, 2009, pp. 137-143. doi:10.1016/j.still.2008.10.004

[5] P. P. Motavalli and R. J. Miles, "Soil Phosphorus Fractions after 111 Years of Animal Manure and Fertilizer Application," *Biology and Fertility of Soils*, Vol. 36, 2002, pp. 35-42. doi:10.1007/s00374-002-0500-6

[6] W. L. Zhang, Z. X. Tian, N. Zhang and X. Q. Li, "Investigation of Nitrate Pollution in Ground Water Due to Nitrogen Fertilization in Agriculture in North China," *Plant Nutrition and Fertilizer Science*, Vol. 1, No. 2, 1995, pp. 80-871.

[7] M. Liu, Z. P. Li, T. L. Zhang, C. Y. Jiang and Y. P. Che, "Discrepancy in Response of Rice Yield and Soil Fertility to Long-Term Chemical Fertilization and Organic Amendments in Paddy Soils Cultivated from Infertile Upland in Subtropical China," *Agricultural Sciences in China*, Vol. 10, No. 2, 2011, pp. 259-266. doi:10.1016/S1671-2927(11)60003-5

[8] Y. Jiang, Q. L. Zhuang and W. J. Liang, "Soil Organic Carbon Pool and Its Affecting Factors in Farm Land Ecosystem," *Chinese Journal of Ecology*, Vol. 26, No. 2, 2007, pp. 278-285.

[9] Y. Y. Sui, X. Y. Zhang, X. G. Jiao, Q. C. Wang and J. Zhao, "Effect of Longterm Different Fertilizer Applications on Organic Matter and Nitrogen of Black Farmland," *Journal of Soil and Water Conservation*, Vol. 19, No. 6, 2005, pp. 190-192, 200.

[10] C. H. Zhang, Z. M. Wang, W. M. Ju and C. Y. Ren, "Spatial and Temporal Variability of Soil C/N Ratio in Songnen Plain Maize Belt," *Environmental Science*, Vol. 32, No. 5, 2011, pp. 1407-1414.

[11] Z. Y. Zhao, G. R. Feng, L. C. Wang, Z. L. Wang and W. G. Zhou, "The Strategic Thinkings on the Concentrating Apple Industry in The superior District of China," *Acta Agriculturae Boreali-occidentalis Sinica*, Vol. 13, No. 4, 2004, pp. 195-199.

[12] H. Eswaran, E. vander Berg and P. Reich, "Organic Car-

bon in Soils of the World," *Journal of Soil Science Society of America*, Vol. 57, No. 1, 1993, pp. 192-194. doi:10.2136/sssaj1993.03615995005700010034x

[13] H. B. Wu, Z. T. Guo and C. H. Peng, "Changes in Terrestrial Carbon Storage with Global Climate Changes since the Last Interglacial," *Quaternatry Scicences*, Vol. 21, No. 4, 2001, pp. 366-376.

[14] S. Goyal, K. Chander, M. C. Mundra and K. K. Kapoor, "Influence of Inorganic Fertilizers and Organic Amendments on Soil Organic Matter and Soil Microbial Properties under Tropical Conditions," *Biology and Fertility of Soils*, Vol. 29, No. 2, 1999, pp. 196-200. doi:10.1007/s003740050544

[15] W. J. Zhang, X. J. Wang, M. G. Xu, S. M. Huang, H. Liu and C. Peng, "Soil Organic Carbon Dynamics under Long-Term Fertilizations in Arable Land of Northern China," *Biogeosciences*, Vol. 7, No. 2, 2010, pp. 409-425. doi:10.5194/bg-7-409-2010

[16] D. K. Benbi, C. R. Biswas and J. S. Kalkat, "Nitrate Distribution and Accumulation in an Ustochrept Soil Profile in a Long Term Fertilizer Experiment," *Fertilizer Research*, Vol. 28, No. 2, 1991, pp. 173-177. doi:10.1007/BF01049747

[17] Q. R. Huang, F. Hu, S. Huang, H. X. Li, Y. H. Yuan, G. X. Pan and W. J. Zhang, "Effect of Long-Term Fertilization on Organic Carbon and Nitrogen in a Subtropical Paddy Soil," *Pedosphere*, Vol. 19, No. 6, 2009, pp. 727-734. doi:10.1016/S1002-0160(09)60168-5

[18] W. H. Schlesinger and J. Lichter, "Limited Carbon Storage in Soil and Litter of Experimental Forest Plots under Increased Atmospheric CO_2," *Nature*, Vol. 411, No. 6836, 2001, pp. 466-468. doi:10.1038/35078060

[19] S. Gunther and K. Holger, "Bulk Soil C to N Ratio as a Simple Measure of Net N Mineralization from Stabilized Soil Organic Matter in Sandy Arable Soils," *Soil Biology and Biochemistry*, Vol. 35, No. 4, 2003, pp. 629-632. doi:10.1016/S0038-0717(03)00052-X

Soil Thermal Diffusivity of a Gleyic Solonetz Soil Estimated by Different Methods in the Brazilian Pantanal

Victor H. M. Danelichen[1], Marcelo S. Biudes[1], Maísa C. Souza[1], Nadja G. Machado[1,2], Leone F. A. Curado[1], José S. Nogueira[1]

[1]Postgraduate Program in Environmental Physics, Physics Institute, Federal University of Mato Grosso (UFMT), Cuiaba, Brazil;
[2]Laboratory of Conservation Biology, Federal Institute of Mato Grosso, Cuiaba, Brazil.
Email: danelichen@fisica.ufmt.br, marcelo@pgfa.ufmt.br, maisacaldas@fisica.ufmt.br, nadja.machado@blv.ifmt.edu.br, nogueira@ufmt.br

ABSTRACT

The soil temperature is an important microclimatic factor due to the interactions between soil and plant, and the energy exchange with the atmosphere. The soil energy exchange is affected by the incident solar radiation, type of coverage and mainly by the soil thermal properties. Among the soil thermal properties, the soil thermal diffusivity is highlighted because it affects the soil temperature profile and soil heat flux transport and distribution. Thus, the aim of this study was to evaluate different estimates of soil thermal diffusivity of a Gleyic Solonetz soil in the Brazilian Pantanal. The soil thermal diffusivity was determined by the amplitude, logarithmic, arctangent and the phase methods between 0.01 and 0.03 m, 0.01 and 0.07 m and 0.01 and 0.15 m depth. The soil thermal diffusivity estimated by the four methods showed significant differences and varied over the study period as a function of volumetric soil water content. The soil thermal diffusivity estimated by logarithmic methodshowed better performance at different depths, followed by the method of phase.

Keywords: Water Content; MonodominantForest; Soil Thermal Properties; Soil Temperature Profile

1. Introduction

Brazilian Pantanal is one of the largest floodplains of the Earth with 138,183 km^2, and 35% of its area is in Mato Grosso state. Paraguay River crosses the Pantanal from North to South and becomes larger and formsa complex river system [1]. The climate is characterized by periods of floods and droughts that alternate throughout the year, allowing the formation of monodominants stands. At the other extreme, frequent drought cause local water stress in plants which regulates its phenology and mass production [2].

The soil temperature is a factor which controls microclimatic seed germination, rate and duration of plant growth and development, nutrient uptake, water and gas distribution in soil as well as the growth and development of microorganisms [3]. Considering each soil type has unique characteristics, it is necessary to identify, for each particular situation, their thermal properties, which are the result of a number of factors, including its texture and chemical composition. Regular observations of soil temperature at different depths allow the knowledge of their thermal profile variation and important soil properties such as soil thermal conductivity and soil thermal diffusivity [4]. These soil thermal properties, especially the soil thermal diffusivity, are used in studies of soil temperature simulation at different depths [5].

Soil thermal diffusivity is an important parameter that affects soil temperature distribution and soil heat flux density [6]. Furthermore, it is a property that informs the soil heating speed that relate with soil thermal conductivity and soil thermal capacity [7].

The estimation of soil thermal diffusivity can be determined by various methods, such as amplitude, phase, arctangent, logarithmic, numeric and harmonic, which utilize soil profile temperature measurements [8]. These methods are based on the soil heat transfer theory, using analytical or numerical solutions of the heat conduction equation described as uniform and homogeneous soil [5].

Therefore, the objective of this study was to evaluate different estimates of soil thermal diffusivity of a Gleyic Solonetz soil in a mono dominant forest in the Brazilian Pantanal.

2. Materials and Methods

2.1. Characterization of the Study Area

The experiment was conducted in an experimental area of Reserva Particular do Patrimônio Natural-RPPN SESC Pantanal (16°39'50"S; 56°47'50"W; 120 m of altitude) in the Barão de Melgaço, Mato Grosso, Brazil between October 2008 to February 2009, during the period when there was no flooding, which occurred in March 2009. This area is composed mainly of *Vochysiadivergens* Pohl, known locally as cambarazal, with 28 - 30 m canopy heights, forming a continuous band of approximately 25 km long and 4 km wide, parallel to the Cuiabáriver. The annual average temperature is between 24.9°C to 25.4°C, annual rainfall is on average 1400 mm with a pronounced dry season extending from May through September [9]. The topography of the floodplain is virtually flat, causing extensive flooding during the wet season. Wet-season floods are 1 to 2 m in depth, but during the dry season many of the floodplain lakes become disconnected from the river channel as the floodwater recedes [10]. The soil type of the study site is classified as a Gleyic Solonetz soil [11].

2.2. Instrumentation

Precipitation data were obtained daily by personnel of RPPN SESC-Pantanal from a manual rainfall gauge located 10 km NW of experimental area. Soil temperature was measured at 1, 3, 7, 15 and 30 cm depth, using five thermistors (108-L, Campbell Scientific, Inc., Logan, Utah, USA). These depths were chosen due the higher soil temperature amplitude from surface to 7 cm depth and practically null at 30 cm depth. Volumetric soil water content was measured with two time domain reflectometer (CS-615, Campbell Scientific, Inc., Logan, Utah, USA) at 5 and 25 cm depth. The sensors which produced signals that were stored by a datalogger (model CR10-X; Campbell Scientific, Logan, UT).

2.3. Estimation of Soil Thermal Diffusivity

The one-dimensional heat conduction in soil is given by Equation (1).

$$C\left(\frac{\partial T}{\partial t}\right)=\left(\frac{\partial}{\partial z}\right)\left(CK\left(\frac{\partial T}{\partial z}\right)\right) \tag{1}$$

where T is the soil temperature, t is the time and z is the depth. Due to the partitioning of a not uniform soil in soil layers with uniform composition and texture, the Equation (1) can be rewritten to Equation (2).

$$\frac{\partial T}{\partial t}=K_j\left(\frac{\partial^2 T}{\partial z^2}\right) \tag{2}$$

where $K_j = (k/C)_j$ is the apparent soil thermal diffusivity of j layer, delineated by z_{j-1} and z_{j+1}, with $j = 1, 2, \cdots,$ m the number of layers, not necessarily same thicknesses.

The apparent soil thermal diffusivity can be obtained by the soil temperature measurements replaced in expression of analytical solutions of Equation (2). Several methods have been used to solving this Equation [12]. In this study the soil thermal diffusivity was estimated between 0.01 and 0.03 m, 0.01 and 0.07 m and 0.01 and 0.15 m depth by the amplitude, logarithmic, phase and arctangent methods.

2.3.1. Amplitude Method

The daily soil thermal diffusivity K estimated by amplitude method [12] is shown in Equation (3).

$$K=\left(\frac{\omega}{2}\right)\left[\frac{(Z_2-Z_1)}{\ln\left(\frac{A_1}{A_2}\right)}\right]^2 \tag{3}$$

where ω is the angular velocity of the Earth and A_1 and A_2 are the temperature amplitudes in the Z_1 and Z_2 depths, respectively.

2.3.2. Phase Method

The daily estimation of soil thermal diffusivity by the phase method was obtained by the Equation (4).

$$K=\left(\frac{1}{2\omega}\right)\left[\frac{Z_2-Z_1}{\Delta t}\right]^2 \tag{4}$$

where Z_1 and Z_2 are the soil depths and $\Delta t = t_2 - t_1$, where Δt is the time interval between t_1 and t_2, which are the instant when the temperature is maximum in Z_1 and Z_2.

2.3.3. Arctangent Method

The soil temperature at the surface can be described by a sines series. The values of soil temperature measured at a specific depth can be adjusted to Fourier series [13], given by Equation (5).

$$T(t)=\overline{T}+\sum_{n=1}^{M}\left[A_n\cos(n\omega t)+B_n\sin(n\omega t)\right] \tag{5}$$

where $\overline{T}$ is the average temperature in considered time interval, M is the number of harmonics, A_n and B_n are their amplitudes, n is the number of observations, ω is the angular frequency and t is the time. If the first two harmonics are sufficient to describe one boundary condition above the level $Z = Z_1$ ($Z_1 = 0$, the surface level), K can be calculated using Equation (6) of the arctangent method.

$$K=\frac{\omega(Z_2-Z_1)^2}{2\left[\arctan\frac{(T_1-T_3)(T_2'-T_4')-(T_2-T_4)(T_1'-T_3')}{(T_1-T_3)(T_1'-T_3')+(T_2-T_4)(T_2'-T_4')}\right]^2} \tag{6}$$

where T_1, T_2, T_3 and T_4 are the soil temperatures recorded in Z_1 and T'_1, T'_2, T'_3 and T'_4 are the soil temperatures recorded in Z_2, at 0, 6, 12 and 18 hours [14].

2.3.4. Logarithmic Method

Using the same methodology of arctangent method, Seemann [15] demonstrated that K can also be calculated by Equation (7).

$$K = \left[\frac{0.121(Z_2 - Z_1)}{\ln\left\{ \frac{(T_1 - T_3)^2 + (T^2 - T_4)^2}{(T_1' - T_3')^2 + (T_2' - T_4')^2} \right\}} \right]^2 \quad (7)$$

The arctangent and logarithmic methods are analogous to the Amplitude method, but require a greater number of observations to approximate the behavior essentially non-sinusoidal [4].

2.4. Statistical Analysis

The soil thermal diffusivity values using the arctangent, logarithmic and phase methods were compared with the values obtained by the amplitude method. To evaluate these methods, we considered the mean absolute error (MAE; Equation (8)) and root mean squared error (RMSE; Equation (9)).The MAE indicates the distance (deviation) of the mean absolute values estimated from the measured values. Ideally, the MAE and RMSE values were near to zero [16].

$$\text{MAE} = \sum \frac{|P_i - O_i|}{n} \quad (8)$$

The RMSE indicates how the model fails to estimate the measurements variability around the mean and measures the change in the estimated values around the measured values [16]. The lower limit of RMSE is 0, which means that there is full compliance between the model estimates and measurements.

$$\text{RMSE} = \sqrt{\frac{\sum (P_i - O_i)^2}{n}} \quad (9)$$

where P_i is the estimated value, O_i is the observed value and O is the average of the observed values.The multivariate analysis of variance test-MANOVA was used to determine possible differences between the methods for estimating the soil thermal diffusivity.

3. Results and Discussion

3.1. Variation in Rainfall, Volumetric Soil Water Content and Soil Temperature

The monthly rainfall was 138, 121, 320, 105 and 305 mm in October, November and December 2008, and January and February 2009, and the total accumulate drain fall was 989 mm during the study period (**Figure 1**). The volumetric soil water content (VSWC) increased from October 2008 to February 2009 (**Figure 2**) as a function of increased local rainfall (**Figure 1**). The largest amount of VSWC coincides with the wettest month, and the inverse occurs with the driest month. The highest VSWC occurred during February 2009, with 0.38 and 0.47 $m^3 \cdot m^{-3}$ at 0.25 to 0.05 m depth, and the lowest VSWC occurred during October 2008, with 0.15 and 0.30 $m^3 \cdot m^{-3}$ at 0.25 to 0.05 m depth (**Figure 3**).

The average maximum soil temperature decreased from October to December, while the minimum soil temperature had less variation. The maximum soil temperatures increased from January to February 2009, but at a lower rate than October-December 2008, and also the minimum soil temperature values increased from January to February 2009 (**Figure 2**). The average maximum soil temperatures at 0.01 m depth were 32.8˚C, 31.3˚C, 29.6, 29.9˚C and 30.9˚C at 13 h, 13 h and 30 min, 15 h, 15 h and 15 h and 30 min in October, November and December 2008 and January and February 2009, respectively. The average minimum soil temperatures at 0.01 m depth were 24.0˚C, 24.1˚C, 23.9˚C, 23.7˚C and 25.1˚C at 6 h, 5 h and 30 min, 30 min and 5 h, 6 h and 6 h and 30 min in October, November and December 2008 and January and February 2009, respectively. The attenuation of the maximum soil temperature from 0.01 m to 0.30 m depth was 6.1˚C, 4.4˚C, 2.8˚C, 3.3˚C and 4.0˚C with a lag of 11 h 30 min, 10 h, 9 h 30 min, 9 h and 10 h in October, November and December 2008 and January and February 2009, respectively.

The soil temperature amplitude decreased from October to February (**Figure 4**), reflecting the increase of water content in soil, caused by increased rainfall over the period (**Figure 3**). The soil temperature amplitude reduction during the summer, during this work, is due the higher VSWC (**Figure 3**) that changes the magnitude of energy balance components, and modify the properties of soil heat conduction [17,18]. The water fills the soil pores, which contribute to increase heat flux and decrease soil temperature amplitude [19].

The lowest temperature of wet soil than dry soil is due to the higher specific heat of moist soil. The soil volumetric heat capacity of wet soil is higher than dry soil. Thus, dry soil turn heats and cools faster than wet soil which turn heats and cools slowly, requiring greater heat supply, since its specific heat is greater. Furthermore, there is a significant litter layer over the soil of experimental area that associated to the natural vegetation protects the soil. Therefore, there is a considerable coverage over the soil of experimental area, which keeps the water content in soil greater than a bare soil [20]. In covered

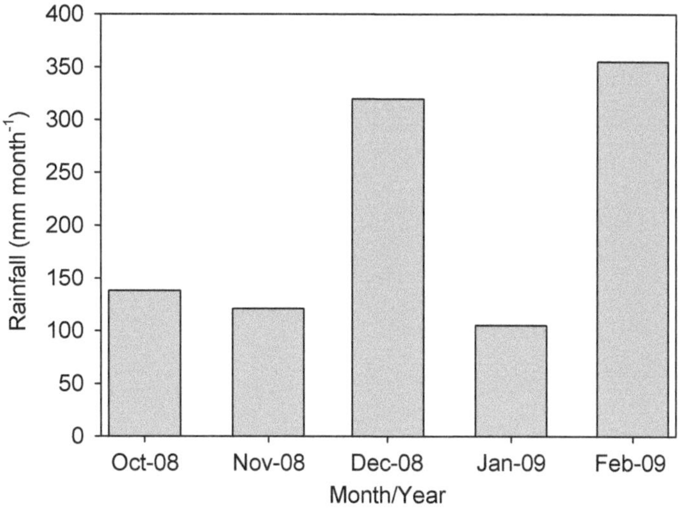

Figure 1. Monthly rainfall in the Brazilian Pantanal from October 2008 to February 2009.

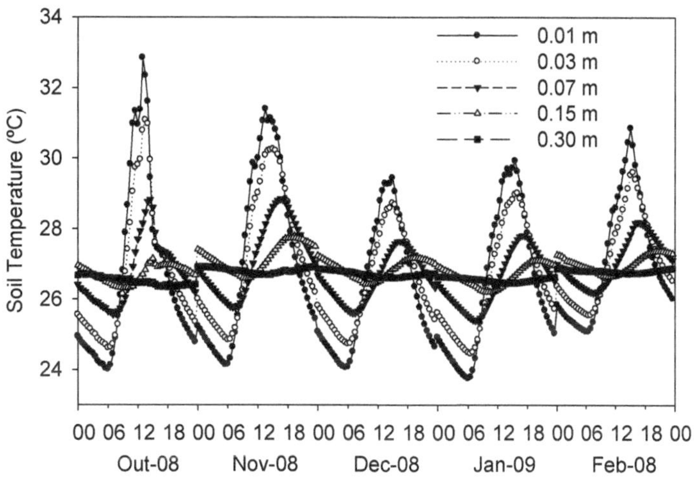

Figure 2. Mean monthly composite average of the diel soil temperature at 0.01, 0.03, 0.07, 0.15 and 0.30 m depth in the Brazilian Pantanalfrom October 2008 to February 2009.

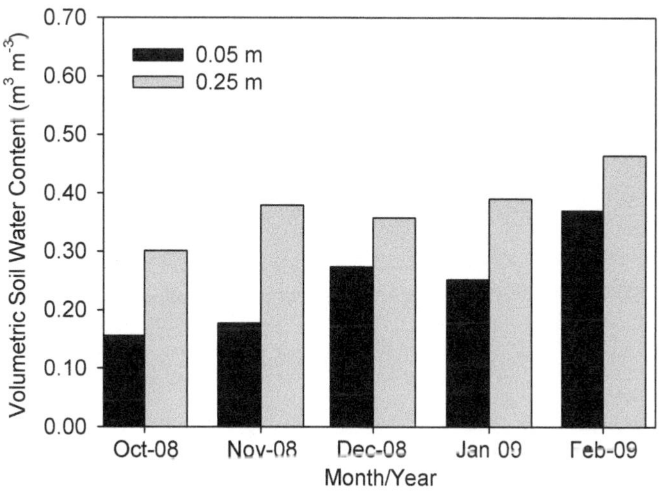

Figure 3. Monthly volumetric soil water content at 0.05 and 0.25 m depth in the Brazilian Pantanal from October 2008 to February 2009.

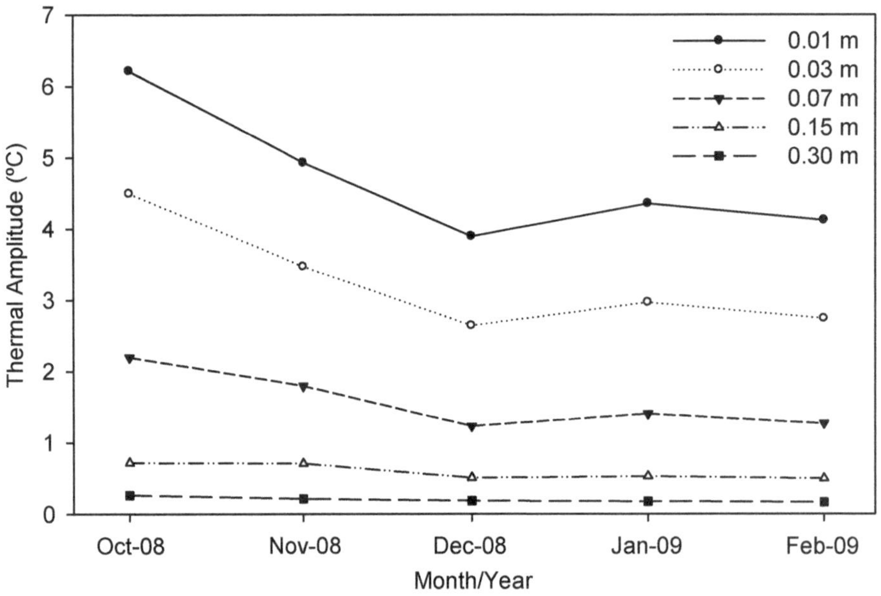

Figure 4. Mean monthly soil temperature amplitude at 0.01, 0.03, 0.07, 0.15 and 0.30 m depth in the Brazilian Pantanal from October 2008 to February 2009.

soil is assumed that there is higher VSWC and it alters the soil specific heat and soil thermal conductivity. Thus, the soil heat capacity increases with VSWC increases due to the high water specific heat (4.18 $MJm^{-3}\cdot K^{-1}$) [7].

3.2. Variation of Soil Thermal Diffusivity

The highest average of soil thermal diffusivity estimated by the amplitude method was 1.95×10^{-7} $m^2\cdot s^{-1}$ in November 2008between 0.01 and 0.15 m depth, and the lowest average was 1.02×10^{-7} $m^2\cdot s^{-1}$ in December 2008 between 0.01 and 0.03 m depth (**Figure 5(a)**).The soil thermal diffusivity estimated by amplitude method is used as reference among all used models [5,20]. The soil thermal diffusivity values calculated in this study agreeto estimates by Oliveira *et al.* [21] in Campina Grande (1.5 – 3.33×10^{-7} $m^2\cdot s^{-1}$), Souza *et al.* [22] in forest and pasture in Amazon (1.45 – 6.74×10^{-7} $m^2\cdot s^{-1}$)and Rao *et al.* [4] in Salvador (4.7 – 7.5 10 $\times 10^{-7}$ $m^2\cdot s^{-1}$).

The soil thermal diffusivity estimated by the amplitude method between 0.01 and 0.15 m depth was higher than soil thermal diffusivity between 0.01 and 0.03 and between 0.01 and 0.07 (**Figure 5(a)**). There was a tendency of increasing of the soil thermal diffusivity estimated by amplitude method as function of increasing of depth. Furthermore, the VSWC seems to be the main factor to regulate the soil thermal diffusivity because the average of VSWC during the studied period was 0.24 $m^3\cdot m^{-3}$ at 0.05 m depth and 0.37 $m^3\cdot m^{-3}$ at 0.25 m depth.

The highest average of soil thermal diffusivity estimated by the logarithmic method was 2.47×10^{-7} $m^2\cdot s^{-1}$ in December 2008 between 0.01 and 0.03 m depth, and the lowest average was 1.87×10^{-7} $m^2\cdot s^{-1}$ in January 2009 between 0.01 and 0.07 m depth (**Figure 5(d)**). The soil thermal diffusivity estimated by the logarithmic method also showed a tendency of increasing as function of increasing of depth, with higher values of soil thermal diffusivity estimated by amplitude method.

The soil thermal diffusivity estimated by phase (**Fiure 5(b)**) and arctangent(**Figure 5(c)**) methods showed similar dynamic, with decreasing from October to December 2008, and increasing from December 2008 to February 2009, and values were overestimated in relation to the amplitude and logarithmic methods (**Figures 5(a)** and **5(d)**). The soil thermal diffusivity estimated by arctangent method was higher in October 2008 (5.38×10^{-7} $m^2\cdot s^{-1}$) when the VSWC was lowest. The soil thermal diffusivity estimated by phase method was higher in January 2009 (3.55×10^{-7} $m^2\cdot s^{-1}$) when the rainfall was lowest.

The VSWC is an important factor to regulate the thermal process in the soil. The soil thermal diffusivity can be higher or lower as function of VSWC [18,23]. The increasing of VSWC becomes the soil thermal conductivity higher, because the water that replaces the air in the soil pores has higher thermal conductivity [24]. There were differences in the structure of each layer of soil studied (**Figure 6**), which influence in water retention, for example, the clay retains more water than the sand interfering in soil thermal diffusivity variation [20]. There is a maximum value to soil thermal diffusivity at a determined VSWC, while the volumetric heat capacity continues to increase with increasing VSWC [25].

Evaluation of methods for determining the apparent

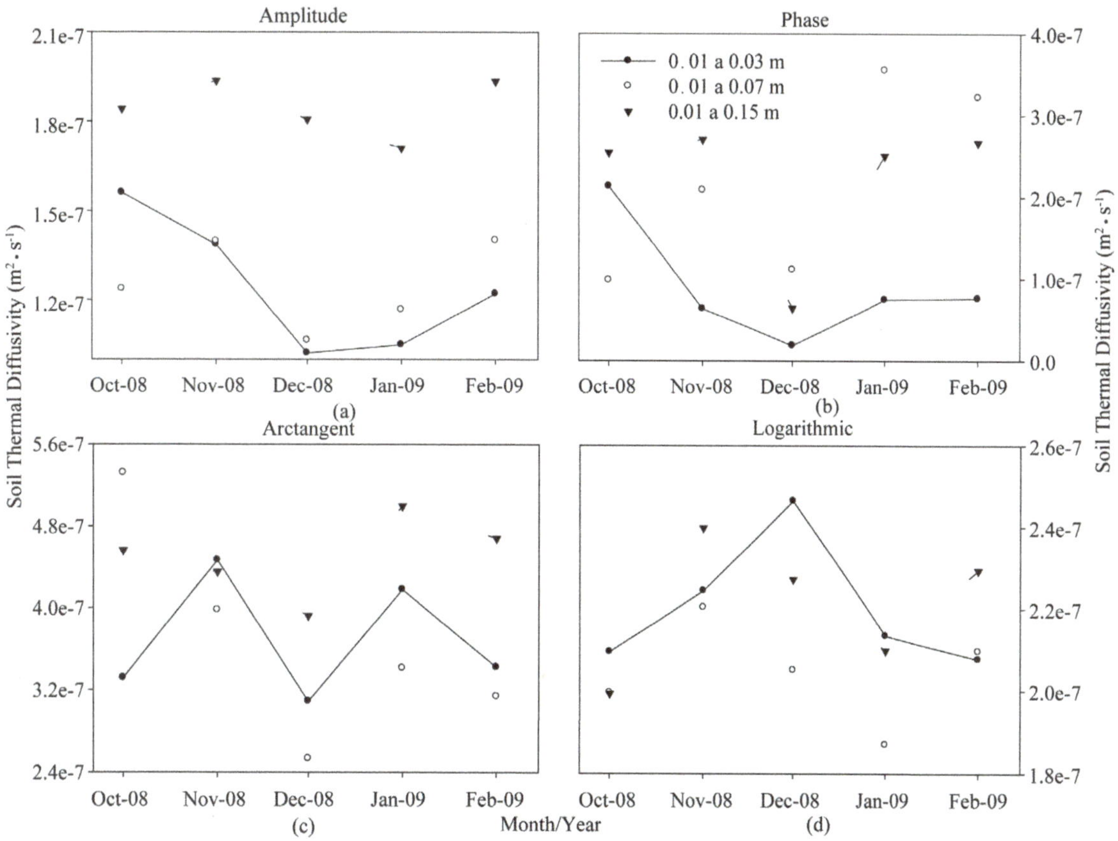

Figure 5. Mean monthly soil thermal diffusivity estimated by the amplitude, phase, logarithmic and arctangent methods between 0.01 and 0.03 m, 0.01 and 0.07 m and 0.01 and 0.15 m in the Brazilian Pantanal from October 2008 to February 2009.

Figure 6. Soil profile of the study area in the Brazilian Pantanal, 2009.

Table 1. Mean absolute error—MAE, root mean squared error—RMSE, Pearson correlation coefficient—r for soil thermal diffusivity estimated by phase, arctangent and logarithmic methods between 0.01 and 0.03 m, 0.01 and 0.07 m and 0.01 and 0.15 m depths, taking as reference values estimated by amplitude method in the Brazilian Pantanal.

Methods		Soil Layers (m)		
		(0.01 - 0.03)	(0.01 - 0.07)	(0.01 - 0.15)
Phase	r	0.81	0.35	0.31
	MAE	0.356×10^{-7}	0.938×10^{-7}	0.364×10^{-7}
	RMSE	0.614×10^{-7}	1.380×10^{-7}	0.846×10^{-7}
Arctangent	r	0.06	0.28	−0.28
	MAE	2.440×10^{-7}	2.420×10^{-7}	2.650×10^{-7}
	RMSE	2.500×10^{-7}	2.580×10^{-7}	2.650×10^{-7}
Logarithmic	r	−0.48	0.62	0.61
	MAE	0.956×10^{-7}	0.791×10^{-7}	0.366×10^{-7}
	RMSE	1.000×10^{-7}	0.797×10^{-7}	0.383×10^{-7}

thermal diffusivity of soil have been made to different locations and specific depths [4,12,22,26,27]. However, they were restricted to describe the soil thermal diffusivity values estimated by the methods studied in this work.

The normality of the soil thermal diffusivity was checked and was applied MANOVA statistical test to evaluate if there were differences between the estimates of soil thermal diffusivity at each depth at 5% probability. The soil thermal diffusivities estimated by the methods of amplitude, phase, and the logarithm of the arctangent differed ($p < 0.05$) and varied over time (**Figure 5**).

Using the soil thermal diffusivity estimated by amplitude method as reference was computed mean absolute error-MAE and the root mean squared error-RMSE between phase methods, arctangent and logarithmic. Among the three methods the phase method showed lower EMA between 0.01 and 0.03 m depth and the logarithmic method showed lower RMSE between 0.01 to 0.15 m depth. The phase method showed higher correlation ($r = 0.81$) between 0.01 and 0.03 m depth, followed by logarithmic method ($r = 0.62$) between 0.01 and 0.07 m depth and ($r = 0.61$) between 0.15 to 0.01 m depth, while the arctangent method showed no significant correlation (**Table 1**).

4. Conclusions

The soil thermal diffusivity estimated by the four methods showed significant differences ($p < 0.05$) and varied over the study period as a function of water content in the soil.

Even the methods to estimate soil thermal diffusivity are significant different the methods of phase, logarithmmic and arctangent were significantly correlated to the amplitude method.

The method of estimation of the soil thermal diffusivity that showed better performance at different depths was the logarithmic method, followed by the phase method with the highest correlation to lowest depth.

The estimation of soil thermal diffusivity by phase method showed lower mean absolute error to less depth, while the logarithmic method showed lower mean squared error to greater depth.

5. Acknowledgements

Thisresearchwassupported in partby Universidade Federal de Mato Grosso (UFMT), Conselho Nacional de Desenvolvimento Científico e Tecnológico(CNPq), andfromlogisticalsupportfromtheReserva Particular do Patrimônio Natural (RPPN) ofthe Serviço Social do Comércio (SESC)-Pantanal.The authors thank Vicente Bellaver (*in memoriam*) for help with field sampling and soil laboratory analysis.

REFERENCES

[1] M. N. Castelnou, D. Floriani, I. A. Vargas and J. B. Dias, "Sustentabilidade Socioambiental e Diálogo de Saberes: O Pantanal Mato-Grossense e Seu Espaço Vernáculo Como Referência," *Desenvolvimento e Meio Ambiente*, Vol. 7, 2003, pp. 41-67.

[2] W. J. Junk and C. N. Cunha, "Pantanal: A Large South American Wetland at a Crossroads," *Ecological Engineering*, Vol. 24, No. 4, 2005, pp. 391-401. doi:10.1016/j.ecoleng.2004.11.012

[3] E. Gasparim, P. R. Ricieri, S. S. Lima, R. Dallacort and E. Gnoatto, "Temperatura No Perfil Do Solo Utilizando Duas Densidades de Cobertura e Solo Nu," *Acta Scientiarum. Agronomy*, Vol. 27, No. 1, 2005, pp. 107-115.

[4] T. V. R. Rao, B. B. Silva andA. A. Moreira, "Caracte-

rísticas Termicas do Solo em Salvador, BA.," *Revista Brasileira de Engenharia Agrícola e Ambiental*, Vol. 9, No. 4, 2005, pp. 554-559. doi:10.1590/S1415-43662005000400018

[5] V. H. M. Danelichen and M. S. Biudes, "Avaliação da Difusividade Térmica de um Solo no Norte do Pantanal," *Ciência e Natura*, Vol. 33, No. 8, 2011, pp. 227-240.

[6] C. C. L. Beber, "Difusividade de Solos do Tipo Latossolo Vermelho em Função do Teor de Umidade," M. Sc. Thesis, Universidade Regional do Noroeste do Estado do Rio Grande do Sul, Ijuí, 2006.

[7] E. R. Schöffel and M. E. G. Mendes, "Influência da Cobertura Sobre o Perfil Vertical de Temperatura do Solo," XIV Congresso Brasileiro de Agrometeorologia, Piracicaba, 18-21 July 2005, pp. 1-2.

[8] R. Horton, E. O. Tyson and T. Ren, "A New Perspective on Soil Thermal Properties," *Soil Science Society of American Journal*, Vol. 65, No. 6, 2001, pp. 1641-1647. doi:10.2136/sssaj2001.1641

[9] M. S. Biudes, J. S. Nogueira, H. J. Dalmagro, N. G. Machado, V. H. M. Danelichen and M. C. Souza, "Mudança No Microclima Provocada Pela Conversão de Uma Floresta de Cambará em Pastagem no Norte do Pantanal," *Revista de Ciências Agro-Ambientais*, Vol. 10, 2012, pp. 61-68.

[10] C. Nunes da Cunha and W. J. Junk, "Year-to-Year Changes in Water Lavel Drive the Invasion of *Vochysiadivergens* in Pantanal grassland," *Applied Vegetation Science*, Vol. 7, No. 1, 2004, pp. 103-110. doi:10.1111/j.1654-109X.2004.tb00600.x

[11] P. Zeilhofer, "Soil Mapping in the Pantanal of Mato Grosso, Brazil, Using Multitemporal Landsat TM Data," *Wetlands Ecology and Management*, Vol. 14, No. 5, 2006, pp. 445-461. doi:10.1007/s11273-006-0007-2

[12] R. Horton, P. J. Wierenga and D. R. Nielsen, "Evaluation of Methods for Determining Apparent Thermal Diffusivity of Soils Near the Surface," *Soil Science Society of American Journal*, Vol. 47, No. 1, 1983, pp. 25-32. doi:10.2136/sssaj1983.03615995004700010005x

[13] N. R. Draper and H. Smith, "Applied Regression Analysis," John Wiley & Sons, Inc., New York, 1966.

[14] S. V. Nerpin and F. Chudnovskii, "Physics of the Soil," Keter Press, Jerusalem, 1967.

[15] J. Seemann, "Measuring Technology," In: J. Seemann, Y. I. Chirkov, J. Lomas and B. Primault, Eds., Agrometeorology, Springer-Verlag, Berlin, 1979, pp. 40-45. doi:10.1007/978-3-642-67288-0_9

[16] C. J. Willmott and K. Matsuura, "Advantages of the Mean Absolute Error (MAE) over the Root Mean Square Error (RMSE) in Assessing Average Model Performance," *Climete Research*, Vol. 30, No. 1, 2005, pp. 79-92. doi:10.3354/cr030079

[17] J. E. M. Pezzopane, G. M. Cunha, E. Arnsholz and M. Costalonga, "Temperatura do Solo em Função da Cobertura Morta por Palha de Café," *Revista Brasileira de Agrometeorologia*, Vol. 4, No. 2, 1996, pp. 7-10.

[18] L. P. Petean, C. A. Tormena and S. J. Alves, "Intervalo Hídrico Ótimo de um Latossolo Vermelho Distroférrico sob Plantio Direto em um Sistema de Integração Lavoura-Pecuária," *Revista Brasileira de Ciência do Solo*, Vol. 34, No. 5, 2010, pp. 1515-1526. doi:10.1590/S0100-06832010000500004

[19] J. Thomson, "Observations of Thermal Diffusivity and a Relation to the Porosity of Tidal Flat Sediments," *Journal of Geophysical Research*, Vol. 115, No. 6, 2010, p. C05016. doi:10.1029/2009JC005968

[20] V. Bellaver, "Diffusividade Térmica do Solo em Área Monodominante de Cambará no Norte do Pantanal Mato-Grossense," M. Sc. Thesis, Universidade Federal de Mato Grosso,Cuiabá, 2010.

[21] E. M. M. Oliveira, H. A. Ruiz, V. H. Alvarez, P. A. Ferreira, F. O. Costa and I. C. C. Almeida, "Nutrient Supply by Mass Flow and Diffusion to Maize Plants in Response to Soil Aggregate Size and Water Potential," *Revista Brasileira de Ciência do Solo*, Vol. 34, No. 2, 2010, pp. 317-327. doi:10.1590/S0100-06832010000200005

[22] J. R. S. Souza, M. Makino, R. L. C. Araújo, J. C. P. Cohen and F. M. A. Pinheiro, "Thermal Properties and Heat Fluxes in Soils under Forest and Pasture, in Marabá, PA, Brazil," *Revista Brasileira de Meteorologia*, Vol. 21, No. 3a, 2006, pp. 89-103.

[23] R. C. Santos, "Propriedades Térmicas do Solo: Um Estudo de Caso," M. Sc. Thesis, Instituto Nacional de Pesquisas Espaciais, São José dos Campos, 1987.

[24] T. O. A. Pessoa, "Avaliação da Influência da Mineralogia, Índices de Vazios e Teor de Umidade em Propriedades Térmicas," M. Sc. Thesis, Pontifícia Universidade Católica do Rio de Janeiro, Rio de Janeiro, 2006.

[25] O. T. Farouki, "Thermal Properties of Soils (Series on Rock and Soil Mechanics)," Trans Tech Publications, 1986.

[26] A. C. D. Antonino, E. V. S. B. Sampaio, A. Dall'olio and I. H. Salcedo, "Balanço híDrico em Solo com Cultivos de Subsistência no Semi-Árido do Nordeste do Brasil," *Revista Brasileira de Engenharia Agrícola e Ambiental*, Vol. 4, No. 1, 2000, pp. 29-34. doi:10.1590/S1415-43662000000100006

[27] Z. Gao, L. Wang and R. Horton, "Comparison of Six Algorithms to Determine the Soil Thermal Diffusivity at a Site in the Loess Plateau of China," *Journal Hydrology and Earth System Sciences*, Vol. 6, No. 2, 2009, pp. 2247-2274. doi:10.5194/hessd-6-2247-2009

5

Effects of Conservation Tillage on Total and Aggregated Soil Organic Carbon in the Andes

Marcela Quintero[1,2*], Nicholas B. Comerford[2]

[1]Decision and Policy Analysis Research Area, International Center for Tropical Agriculture (CIAT), Cali, Colombia; [2]University of Florida North Florida Research and Education Center, University of Florida (UF), Quincy, USA.
Email: *m.quintero@cgiar.org, nbc@ufl.edu

ABSTRACT

Many Andisols of the Andes have been disturbed by traditional potato-based rotation agriculture disrupting soil structure, water retention capacity and organic matter content. This study was undertaken to investigate the contribution of conservation farming technology or reduced tillage in potato-based rotations in the Colombian Andes in order to rehabilitate total and aggregated soil organic C in disturbed organic matter-rich Andisols. Soils were sampled from farms with 7-year of reduced tillage and farms with conventional farming practices. Ultrasound energy was applied to samples to disrupt aggregation and total soil C was determined in order to investigate the amount of carbon held inside the aggregates of different soil size classes. Results indicated that reduced tillage in potato-based crop rotations increased the soil C concentration and average C content in the whole profile (≈117 cm depth) by 50 and 33% (1636 t C ha^{-1} vs. 1224 t C ha^{-1}), respectively, as compared to conventional farming practices. Carbon content increased 177% in the subsoil (A2 horizon, 78 - 117 cm depth, from 215 to 596 t·ha^{-1}), although most of the soil C was in the A1 horizon (between 0 - 78 cm average thickness, 1097 t·ha^{-1}). These increases show that reduced tillage enhances C stores in Andisols which are already high in organic matter. In addition, C in aggregates represented more than 80% of the total organic matter and it was positively affected by conservation practices. The C increase was preferential in the smaller macroaggregates (<2 mm). The aggregate dispersion energy curves further suggested that C increase was occurring in microaggregates within the smaller macroaggregate fraction. Data suggested that smaller macroaggregates can be used in these soils to evaluate the influence of field management practices on soil C sequestration.

Keywords: Conservation Tillage; Soil Organic Carbon; Soil Aggregates; Andes; Potato

1. Introduction

Over 60% of the world's carbon (C) is in soils (over 40%) and the atmosphere (20%) [1,2]. However, soil disturbance is redistributing the carbon and augmenting the atmospheric carbon pool. Tillage disrupts soil aggregates, which hasten soil organic matter (SOM) mineralization [3]. The increased release of CO_2 from agricultural soils not only contributes to atmospheric greenhouse gasses, but also has a negative impact on agriculture and natural resource productivity and sustainability [4].

Jastrow *et al.* [5] noted that nearly 90% of SOM was found within soil aggregates. Aggregates, particularly aggregates in volcanic soils, have been classified into macroaggregates (>0.25 mm) and microaggregates (0.05 - 0.25 mm) [6,7]. Jastrow [8] and Six *et al.* [9] found that the majority of C in macro and microaggregates is mineral-associated C formed during the decomposition of particulate organic matter (POM) and this fraction was responsible for long-term C sequestration. Additionally, Kong *et al.* [10] demonstrated that the relationship between organic C input and SOC sequestration was dominated by SOC increases within macroaggregates. However, macroaggregates are susceptible to tillage disruption [11,12]. Disrupting macroaggregates exposes the microaggregate C pool to decomposers, thereby increasing SOC mineralization [13,14].

The level of physical protection provided by aggregates varies with soil management practices [15]. Generally, there is more aggregate protection in no-till soils than in cultivated ones. Thus, the fate of SOM will de-

*Corresponding author.

pend upon its decomposability and the persistence of aggregates, which relate to aggregate stability in water and resistance to other mechanical stresses (e.g. tillage). Several researchers found more macroaggregates in non-tillage or reduced tillage soils, as compared with conventional tillage soils [16,17]. Tisdall and Oades [18] found that cultivation generally resulted in reduced macroaggregate number and stability but it had little effect on microaggregate stability. As a consequence, the SOM that binds microaggregates into larger macroaggregates was suggested to be the primary source of organic matter loss from soil cultivation practices [11]. Although there are other ways to protect SOC, such as by adsorption to clay minerals and by isolation in soil micropores [19], aggregate-associated SOC may increase simply through changes in cultivation practices.

Management systems involving high C inputs and reduced tillage should favor C storage directly by reducing aggregate breakdown and by enhancing SOM-mediated aggregation. For example, establishment of perennial grasses or legumes in poorly-structured soils contributed to macroaggregation, which favor the protection of labile C and, as a consequence, long-term C storage [20]. Similar results have been obtained by Carter [16] and Beare *et al.* [15] when no-tillage was practiced. Therefore, similar soil conservation practices are recommended in order to increase SOC sequestration. In addition, practices such as non-tillage and reduced tillage are increasingly adopted by farmers because it has the potential to reduce production costs [21,22] and use of fossil fuels [23].

Beginning in 1999, Colombia's Technical German Cooperation (GTZ) and the regional environmental authority (CAR) adopted reduced tillage and cover crop conservation agricultural practices. This system is referred to as conservation tillage (RT) in this document. In this form of tillage, soil loss is reduced compared to traditional tillage and crop residues are left unincorporated [24]. Experience has shown that reduced tillage with cover crops is protecting watersheds from run-off and increasing farmer income [25]. However, there are no studies with reference to the impact of RT on soil C sequestration in the Andes, with particular reference to the páramos soils of Colombia. Thus, the objectives of this research were to 1) determine if SOC and SOM can be increased in already organic matter-rich soils through conservation tillage in order to bring these soil back to their characteristically high SOM levels; and 2) estimate the amount of aggregated organic matter (AOM) in stable, soil macroaggregates under potato-based rotations using conventional tillage vs. reduced tillage with cover crops in Colombia's upper Lake Fuquene watershed.

The first objective relied on the hypothesis that, in spite of the high OM content of these high elevation páramos soils, conservation tillage would still increase SOM with respect to conventional tillage. The rationale was that, in the past, these soils were disrupted by intensive tillage and were not enriched with additional sources of OM, and that soil C had been released to the atmosphere and would not be recuperated until additional C inputs were incorporated (*i.e.* oats cover crop residues in this study). The unique aspect of this objective is that the soils selected for this study already have very high SOM to at least 1 meter depth; so the potential to increase SOM was unknown.

For the second objective, the hypotheses were: 1) Conservation agriculture increases OM in the aggregate organic matter pool for these high OM soils, and 2) The OM contained in aggregates is different across size classes, being greater in smaller macroaggregates. The first hypothesis is based on results from other studies that reported improved soil aggregation and increased SOC levels with no-till compared with conventional tillage, e.g., [9,16,26,27], yet no information exists for these unique soils. The second hypothesis is expected since the distribution of SOM among aggregate size classes can be heterogeneous [14] while Kong *et al.* [10] and Denef *et al.* [28] found—although for other soil types—that with non-tillage the AOM is higher in smaller macroaggregates than in larger ones.

2. Materials and Methods

2.1. Study Sites

This study was conducted in the upper part of the Lake Fuquene watershed (2985 - 3070 m elevation) located in the valleys of Ubaté and Chiquinquirá, north of Bogotá, the capital of Colombia (South America) (N 05°20' W 73°51'). The monthly temperatures do not have large variations throughout the year and the mean annual values are between 12.0°C and 13.2°C. The mean monthly relative humidity varies between 70% and 80%. The annual mean precipitation is 610 mm [29,30]. The study sites, located in the Carmen de Carupa Municipality, corresponded to cropping areas that formerly were páramo —alpine Andean grasslands, a typical Andean natural ecosystem. Andisols are the primary soil order in this region, and the field soils were classified as Lithic Hapludands [31]. Basic physical properties of páramo soils include a very high porosity due to the high organic matter content (up to 40% of OM) with an open and porous structure, a rapid hydraulic conductivity and high water content (2 $g \cdot g^{-1}$) under saturated and 1500 KPa suction conditions, respectively). The high water retention is attributed to the presence of amorphous clay minerals such as allophane and imogolite and the high OM content [32,33]. Bulk density ranges from as low as 0.15 $g \cdot cm^{-3}$ in wet, weathered soils to about 0.9 $g \cdot cm^{-3}$ in dryer younger soils [33]. However, much of the native soil characteristics have been greatly modified by conventional tillage.

Two types of soil management were selected for this study: 1) parcels where conventional tillage (CT) had been practiced for over 7 years and 2) parcels where reduced tillage and cover crops (denoted in this manuscript as RT—for reduced tillage) had been practiced within the last 7 years following prior conventional tillage similar to that in parcel type CT. **Table 1** describes the sequence of the rotations with CT and RT and respective planting dates. The CT rotation (potato-potato-pasture-pasture) was two potato seasons rotated with 2-year of pasture (ryegrass). The pasture was not fertilized and relied on the residual fertility left after potato cropping. The RT rotation (potato-oats-potato-oats-potato-pasture-pasture) was potatoes with an oat cover crop previous to potato sowing. The RT treatment started with potatoes sowed into a degraded pasture that pertained to a prior CT rotation. In these parcels, oat and potatoes were also rotated with pastures as in CT. In CT and RT treatments potato was generally grown in March and September. RT is distinguished from CT because it left unincorporated oats residues and practiced minimum tillage. Minimum tillage here involves non-inverted deep tillage and differed from the CT treatment where conventional plowing inverted the soil, more machinery passes were applied and weed control was mechanical. At 70 days, soil was put into beds. In the CT treatment, a disc plow was passed twice to 30 cm depth to invert the soil and rotovated with two to three passes to 15 - 20 cm depth. In addition to these practices there is mechanical weed control one month after planting and soil is put into beds at day 70. In the RT system, there was an initial chemical control of weeds followed by 1 to 3 passes with a chisel plow along with disking to 30 cm depth. There was no need to control weeds up to 35 days after potato planting. At day 75, soil was mounded around the plant base. In addition and prior to potato, oats were grown as a cover crop. For this, chemical weed control was applied and a chisel plow pass was used for land preparation. Each system was represented by 3 sites. The 6 sites were selected based on similar: 1) landscape position; 2) land use; 3) slope; and 4) rainfall. Thus all sites were located on backslope positions, with linear, moderate slopes, under potato-based rotations and the same rainfall regimen.

In each site, two soil pits were dug in May 2007. Soil horizons were identified in each of the 12 soil profiles (12 pits), and one soil sample (500 g) was removed from the middle of each of the identified horizons (**Table 2**) for aggregation, carbon and texture analyses. Thirty-five soil samples were collected in total. In addition, soil samples per horizon were removed using 3 cylinders (5 × 5 cm.) per horizon (105 soil core samples in total) to determine the bulk density. In general, three horizons were found in most of the profiles with an average thickness of 78 cm (horizon A1, top), 39 cm (horizon A2) and 49 cm (horizon C, bottom). The main differences between horizon descriptions were that the percent of clay increased with depth and the color was very dark in the first two horizons while the third one was mostly yellowish. In **Table 2**, the main characteristics of the 12 soil profiles are shown.

2.2. Soil Analyses

The soil organic C concentration was determined using the method of Walkley and Black as described by Nelson and Sommers [34]. Soil organic matter (SOM) was determined by loss on ignition [35]. The soil carbon content was estimated using the equivalent soil mass approach (mass-depth) to correct for differences in soil bulk density, thereby allowing for more precise and accurate quantitative comparisons of soil constituents [36]. Average soil mass was calculated per site according to bulk density and soil depths. With these averaged soil masses the soil depth was adjusted in order to ensure that final C contents were representing the same amount of soil mass. Undisturbed soil cores were used to determine soil bulk density by the cylinder method [37,38].

For aggregate soil organic matter determination, 35 fresh samples were segregated and classified by size using "dry" sieving with a nest of sieves representing five different size classes: 5, 2 (large macroaggregates), 1, 0.5 and <0.5 mm (small macroaggregates) screen sizes.

Macroaggregates from each size class were labeled as: SF1 (>5 mm size class), SF2 (2 - 5 mm size class), SF3 (1 - 2 mm size class), SF4 (0.5 - 1 mm size class) and SF5 (<0.5 mm size class). Field-moist samples were used to avoid changes associated with drying [7]. The samples

Table 1. Conventional and conservation tillage rotations.

	Yr 1		Yr 2		Yr 3		Yr 4		Yr 5		Yr 6		Yr 7	
CT	Potato	Potato	Potato	Rye grass	Rye grass	Rye grass	Rye grass	Potato	Potato	Potato	Rye grass	Rye grass	Rye grass	Rye grass
RT	*Oats*	*Potato*	*Oats*	*Potato*	*Oats*	*Potato*	*Rye grass*	*Rye grass*	*Rye grass*	*Oats*	*Potato*	*Oats*	*Potato*	*Oats*

†CT: Conventional tillage rotation. RT: Conservation tillage rotation, ‡Planting dates for potato and oats were generally March 15 and September 15, *Ryegrass was neither planted nor fertilized.

Table 2. Characteristics of soil profiles described for 6 sites and two treatments (Conservation and conventional tillage).

Pedon depth (cm)	Horizon		Texture		Color			Bulk density (g·cm^{-3})	Organic matter (g·g^{-1})
	No.	Lower depth (cm)	USDA Class	Clay %	Hue	Value	Chroma		
				Conservation tillage—profile 1					
140	1	55	sil	19	10 YR	2.5	1	0.98	0.20
	2	105	sl	9	7.5 YR	3	2	0.69	0.17
	3		c	51	10 Y	5	4	0.98	0.07
				Conservation tillage—profile 2					
140	1	34	l	26	10 YR	3	2	0.89	0.16
	2	89	sil	22	10 YR	2.5	1	0.89	0.20
	3	120	c	46	10 YR	3	2	0.89	0.10
	4		cl	39	10 YR	5	4	1.12	0.06
				Conservation tillage—profile 3†					
140	1	140	sil	12	10 YR	2.5	1	0.79	0.25
				Conservation tillage—profile 4†					
130	1	130	sil	13	10 YR	2.5	1	0.73	0.31
				Conservation tillage—profile 5					
150	1	72	sil	23	10YR	2.5	1	0.81	0.22
	2	123	sic	42	10 YR	4	3	1.09	0.08
	3		sic	46	10 YR	6	1	1.40	0.04
				Conservation tillage—profile 6					
150	1	110	sil	22	10YR	2.5	1	0.74	0.27
	2	135	l	17	10 YR	2.5	1	0.64	0.19
	3		c	48	10 YR	4	3	0.85	0.11
				Conventional tillage—profile 7					
144	1	70	sil	25	10YR	2.5	1	0.82	0.24
	2	90	c	54	10 YR	3	2	0.99	0.09
	3		sic	44	10 YR	6	5	1.43	0.05
				Conventional tillage—profile 8					
150	1	93	sil	22	10YR	2.5	1	0.72	0.26
	2	123	cl	30	7.5 YR	4	2	0.60	0.17
	3		sic	42	7.5 YR	5	2	1.32	0.08
				Conventional tillage—profile 9					
150	1	55	l	18	10YR	2.5	1	0.79	0.22
	2	75	cl	39	10 YR	3	2	1.28	0.09
	3		l	21	5 YR	7	1	1.64	0.04
				Conventional tillage—profile 10					
150	1	60	sil	17	10YR	3	1	0.78	0.24

Continued

	2	123	l	24	10 YR	2.5	1	0.84	0.24
	3		cl	35	10 YR	3	3	1.07	0.09
				Conventional tillage—profile 11					
150	1	25	l	25	10YR	3	2	0.98	0.12
	2	35	cl	36	10 YR	2.5	1	1.33	0.10
	3	150	c	65	10 YR	5	6	1.38	0.10
				Conventional tillage—profile 12					
130	1	25	l	26	10YR	3	2	1.06	0.12
	2	35	c	45	10 YR	2.5	1	1.28	0.12
	3	130	c	65	10 YR	5	6	1.39	0.10

[†]For this profile, only one deep horizon was identified. However two set of samples were taken each at a different depth (10 cm and 60 cm). Data shown for this horizon is an average of the subsamples.

were sieved in a mechanical shaker for 5 minutes. To examine the AOM content, a sonication procedure was used for each soil size class [39-42]. It consisted in applying different ultrasound energy inputs to aggregates in order to cause their dispersion and release of SOM. Thus, this procedure permitted measurement of the amount of carbon associated with aggregation at different dispersive energy levels [42]. For this, a Sonic Dismembrator (Fisher Scientific, model 550) was used and 10 energy levels were obtained by combining different amplitudes (20% to 60%) and time periods (1 to 5 minutes). The pulse method (60 sec ON and 30 sec OFF) was used to avoid an excessive rise in temperature.

Up to 10 sub-samples of approximately 5 g each were tested per each of the 5 aggregate size classes derived from each of the 35 fresh soil samples (1750 samples in total). To each sub-sample, one of the 10 energy levels was applied. Energy was applied starting from the lowest level (5.1 kJ) and incrementing it until reaching the maximum level (11 kJ) or before in those cases where a sub-sample's aggregation was completely destroyed. The actual energy inputs ($J \cdot mL^{-1}$) were calculated based on the particle size density ($g \cdot cm^{-3}$) and the calculated initial soil mass (g) of each sample (adjusted based on initial moisture content), the water volume used for sonication (mL), and the energy output (Joules) from the sonicator for each run. A correction factor (0.7) was included that corresponds to the ratio of energy output represented by the sonicator energy output and calculated by the rise in temperature of a given mass of water [42].

Before sonicating, each aggregate suspension was prepared in a 250 ml beaker containing 100 ml of water. The sonicator probe was immersed into the beaker at a depth of 8 cm. After sonicating, the suspension was passed through the sieve corresponding to the original size class of the sub-sample. In all cases, one of the ten subsamples was suspended in water and not subjected to sonication in order to estimate the OM that was water-dispersible.

In this study, the SOM extracted from the aggregate by sonication (passing through the sieve) was called AOM, as it contained organic matter from inside aggregates. Organic matter remaining after the highest sonication, when no more OM was being lost, was termed particulate organic matter (POM).

The OM passing through the sieve (AOM) and OM remaining in the sieve (POM + remaining AOM) were measured by the loss-on-ignition procedure. The AOM of each size class was calculated as a percentage of total SOM (AOM + POM). Using the overall results it was possible to relate both, percentage of AOM and the actual AOM concentration, with different energy inputs for both conventional and conservation agriculture systems. At the end, it was possible with this procedure to measure the proportion of AOM in total soil SOM across different aggregate size classes.

2.3. Statistical Analyses

2.3.1. Soil Organic Carbon

Since, SOM and SOC data were shown to have near normal distributions, an analysis of variance (ANOVA) was applied using STATISTICA (Version 7; 2004) with a factorial design. The main effects were type of tillage (reduced tillage vs. conventional) and horizons (A1, A2, and C). Preliminary analyses indicated that Horizon C was not different among treatments (data not shown), therefore the ANOVA for SOM and SOC involved only the two upper horizons. Significant ($P \le 0.05$) main and interaction effects were compared with the Duncan post hoc mean separations test.

2.3.2. Aggregated Organic Carbon

Since the measured energy inputs showed a minimum

variance across soil samples (data not shown), the energy (J) effect on %AOM was analyzed using energy class category comparisons, ranging from 1 (3.5 $J \cdot mL^{-1}$) to 10 (75.5 $J \cdot mL^{-1}$). The effect of different energy levels, treatment, and size classes were analyzed for the percentage of AOM released after sonication by applying an ANOVA analysis. Each of the 2 surface horizons was analyzed separately. In addition, the 5 mm soil class was excluded from all analyses because of the low number of observations in this class.

Further statistical analysis was performed separately for %AOM of each size class in order to analyze the effects of energy level with treatment. A post hoc comparison procedure with the Duncan adjustment was used to compare %AOM of size class where treatment had a significant effect. Since energy levels showed a significant effect on %AOM in all size classes, a post hoc Duncan analysis was used to compare %AOM means across the different energy levels for each size class.

The %AOM values were transformed to actual AOM ($g \cdot g^{-1}$) released after sample sonication. An analysis of variance (ANOVA) was performed to test the effect of treatment, energy, soil horizon and size class on AOM ($g \cdot g^{-1}$). Since there were significant interactions Treatment * Size class; Treatment * Energy level and Size class * Energy level for the two horizons, a further analysis of variance was conducted per size class. A post hoc Duncan means separation was used to test main effects and significant interactions for each size class. The AOM ($g \cdot g^{-1}$) and dispersive energy inputs ($J \cdot mL^{-1}$) were plotted per size class and treatment.

3. Results

3.1. Effect of Horizons and Tillage Systems on Soil Organic Carbon and Organic Matter

The soil bulk density, SOM and SOC concentrations showed no interactions between main effects (data not shown). Horizons were different for SOC and SOM (**Table 3**), and the tillage system had a significant effect on bulk density, SOM and SOC. The bulk density was lower in conservation tillage while the SOM and SOC were higher under conservation tillage (**Table 4**).

The soil C content was estimated using the mass equivalent method. The calculated average soil mass across sites was 6635 $t \cdot ha^{-1}$ and 3943 $t \cdot ha^{-1}$ for horizons A1 and A2, respectively. The carbon content calculated for these soil masses was different by treatment at $P < 0.1$ and by horizon at $P < 0.05$. The A1 horizon had the highest C content with 1097 t C ha^{-1}; while the A2 had an average content of 406 t C ha^{-1} (**Table 3**). With respect to treatment, RT sites had an average carbon content in the soil profile of 1636 t C ha^{-1} vs. 1224 t C ha^{-1} in conventional tillage; a 33% increase due to conservation tillage (**Table 4**).

3.2. Effect of Horizons, Tillage Systems and Size Classes on Aggregated Organic Carbon

Size Class × Energy Level and Size Class × Treatment interactions for Horizon A1 existed and horizon A2 was characterized by a three way interaction among all main effects. Analysis of variance for each size class showed that, for both horizons, as the ultrasonic energy applied to the soil increased, more aggregates were destroyed, increasing the amount of AOM removed (**Figures 1** and **2**). SF2, in both horizons, was directly influenced by tillage treatment. The %AOM, across all energy levels, was uniformly higher with RT (**Table 4**). In contrast, SF3 for Horizon A1 and SF3/SF4 for Horizon A2 showed significant interaction of treatment with the Energy level (**Table 5**). At lower energy levels, treatment differences were small. In comparison, for SF3 of the two horizons, conservation agriculture had higher %AOM when applying energy levels 7 and 8 (17 and 32 $J \cdot mL^{-1}$, respectively). In

Table 3. Comparison of bulk density, soil organic matter (SOM) and soil organic carbon (SOC) across soil horizons (n = 6).

Horizon	Bulk density ($g \cdot cm^{-3}$)	SOM ($g \cdot kg^{-1}$)*	SOC ($g \cdot kg^{-1}$)*	C ($Mg \cdot ha^{-1}$)*
Horizon A1 (0 - 78 cm)‡	0.85 (a)†	230 (a)	160 (a)	1097(a)
Horizon A2 (78 - 117 cm)‡	0.92 (a)	150 (b)	80 (b)	406 (b)

†Within a soil characteristic, the means followed by different letters are statistically different and show the effect of horizon or treatment. ‡Averaged depth across all sites. *Significantly different at $P < 0.05$

Table 4. Comparison of bulk density, soil organic matter (SOM), soil organic carbon (SOC) and aggregated organic matter (AOM) across treatments (n = 6).

Treatment	Bulk densit ($g \cdot cm^{-3}$)*	SOM ($g \cdot g^{-1}$)**	SOC ($g \cdot g^{-1}$)*	C ($t \cdot ha^{-1}$)*‡	Horizon A1 %AOM§ Mean**	Horizon A2 %AOM Mean**
Conservation tillage	0.81 (a)†	0.22 (a)	0.15 (a)	1636(a)	33.0 (a)	41.6 (a)
Conventional tillage	0.96 (b)	0.17 (b)	0.10 (b)	1224(b)	27.8 (b)	33.9 (b)

†Within a soil characteristic, the means followed by different letters are statistically different and show the effect of horizon or treatment. ‡Calculated for the equivalent soil mass at 118 cm, which was the averaged depth across sites. §AOM (aggregated organic matter) for size class 2 (2 - 5 mm). **Significantly different at $P < 0.05$, *significantly different at $P < 0.1$.

Table 5. Analysis of variance of %AOM and AOM ($g{\cdot}g^{-1}$) (aggregated organic matter)‡ and energy levels per size fraction classes in Horizon A1 and A2.

	p-values							
	AOM (%)				AOM ($g{\cdot}g^{-1}$)			
	SF1 (>5 mm)	SF2 (2 - 5 mm)	SF3 (1 - 2 mm)	SF4 (0.5 - 1 mm)	SF1 (>5 mm)	SF2 (2 - 5 mm)	SF3 (1 - 2 mm)	SF4 (0.5 - 1 mm)
Horizon A1								
Treatment†	0.06	0.019*	0.013*	0.14	0.774	0.001*	0.000*	0.000*
E level	0.000*	0.000*	0.000*	0.000*	0.000*	0.000*	0.000*	0.000*
Treatment* E level	0.70	0.41	0.024*	0.28	0.984	0.06	0.098	0.614
Horizon A2								
Treatment†	0.52	0.026*	0.000*	0.000*	0.102	0.000*	0.000*	0.000*
E level	0.000*	0.000*	0.000*	0.000*	0.000*	0.000*	0.000*	0.000*
Treatment* E level	0.62	0.63	0.036*	0.017*	0.953	0.164	0.010*	0.035*

†Treatments: Conservation and conventional tillage; ‡%AOM is the percentage of total SOM and AOM (g/g) is the actual AOM concentration; *Significant at 5% ($P < 0.05$).

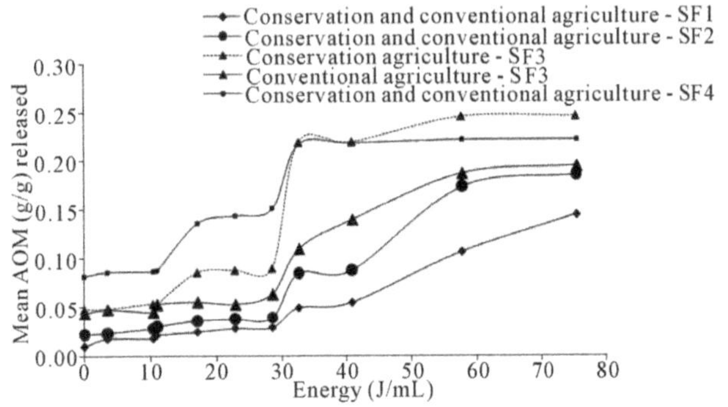

Figure 1. Aggregated organic matter of all aggregates size classes from horizon A1 (top horizon), released with different energy inputs in two potato management systems (Tillage system had a significant effect only on SOM from size class 3 (SF3)).

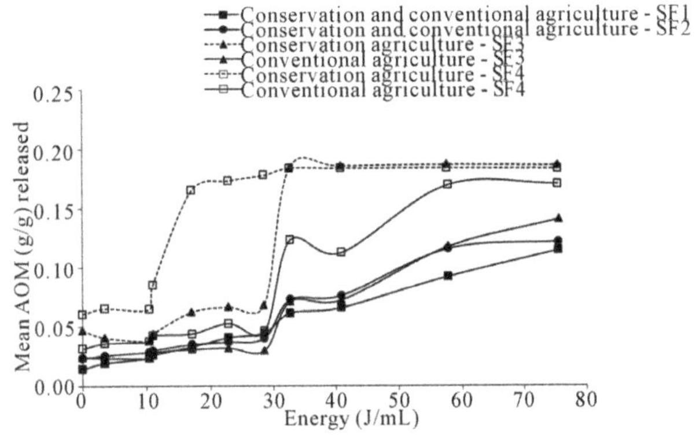

Figure 2. Aggregated organic matter of all size class aggregates for horizon A2 released with different energy inputs from the soil of two potato management systems (Tillage system had a significant effect only on SOM from size classes 3 and 4 (SF3 and SF4)).

SF4 of horizon A2, the %AOM was also higher in conservation agriculture at energies of 17, 22 and 28 $J{\cdot}mL^{-1}$ (**Figures 1** and **2**). SF1 in both horizons and SF4 from horizon A1 were unaffected by the tillage system (**Table 5**) The %AOM * Energy ($J{\cdot}mL^{-1}$) curves indicated that, for all size classes, the curve eventually flattened; indicating that all AOM was released with the exception of SF1 (>5 mm) which did not reach a plateau within the energy range used in this study (**Figures 1** and **2**). After converting the %AOM to AOM ($g{\cdot}g^{-1}$), the analysis of variance showed that the size classes 2, 3, and 4 had a significantly higher concentration of AOM in conservation tillage samples from both horizons (**Table 5**, **Figures 1** and **2**). For SF1 there was no treatment effect. Also, size classes 3 and 4 of horizon A2 had exhibited a significant Treatment * Energy ($J{\cdot}mL^{-1}$) interaction (**Table 5**), showing that the main difference was that AOM ($g{\cdot}g^{-1}$) was released differentially with Energy Level. In SF 3, the differences were found in levels 7, 8 and 9. In SF 4 the differences were in energy levels 5, 6, 7 and 8.

Total aggregated organic matter, which corresponds to the maximum %AOM released after applying the highest Energy Levels (8.4 - 11 kJ), was high on most soil samples, being in the range of 80% to 100% of total organic matter), and only 17% of soil samples released <80% of the total organic matter (between 12% and 79%) (data not shown). This means that about 80% of the total organic carbon was in the aggregate pool.

3.3. Aggregation Hierarchy

All the aggregate energy dispersion curves exhibited a step-wise pattern. All curves have the same step at the 32 $J{\cdot}mL^{-1}$ Energy Level. In some curves, especially for SF3 and SF4 a previous step is recognized at about 17 J/mL^{-1}. A third step at 57 J·mL^{-1} was seen for SF2 (**Figures 1** and **2**) Conservation agriculture, as mentioned above, produced different curves for SF3 in both horizons and SF4 of horizon A2. In SF3, the effect of conservation agriculture was to accentuate the second step of the curve (32 J·mL^{-1}). It is worth noting that this second step was not pronounced in the conventional agriculture sites (**Figures 1** and **2**). For SF4 in the horizon A2, the effect of conservation agriculture was mainly on the first step (17 J·mL^{-1}) making it more pronounced (**Figure 2**).

4. Discussion

The greater amount of %AOM, SOC (g·g^{-1}) and SOM (g·g^{-1}) from soils under conservation agriculture was aligned with results from other studies that also found greater SOC in no-till compared to conventional tillage for a variety of soils types (e.g. Alfisol, Oxisol, Mollisol, Ultisol; [19,27]. This is probably related to a positive linear relationship between SOC and the proportion of crop residues returned to soil described in other studies, *i.e.*, [43-45]. However, most of the SOM and SOC similar studies were concentrated in superficial soil horizons (e.g. 20 cm depth; [19]) and the SOC at greater depths is frequently not studied. Our study explored differences to an average depth of 117 cm (horizons A1 and A2) and the increased higher values of SOM (g·g^{-1}) with conservation tillage suggested that SOM improvements were promoted throughout the depth of the first two soil horizons.

4.1. Rehabilitation of Soil Organic Carbon and Soil Organic Matter by Conservation Tillage in Andisols

Soil C concentrations in conservation and conventional tillage sites are in the range reported for undisturbed or recolonized humid páramos. In Northern Ecuador and Colombia the A horizon has been reported to have from 0.10 to 0.21 g C g^{-1} [46,47; Amezquita *et al.*, unpublished data, 2005] while this study recorded 0.15 and 0.18 g·g^{-1} in conventional and conservation tillage sites, respectively. Although our conventional tillage values are in that normal range, conservation tillage showed a marked improvement in the C concentration and in the average C content for the soil profile to 117 cm depth with respect to conventional tillage. The average C content had changed from 1224 to 1636 t·ha^{-1}. Particularly interesting was a 177% increase in the deeper A2 horizon (from 215 to 596 t·ha^{-1}) although most of the C was stored in the top A1 horizon (1097 t·ha^{-1}). This increase in C content and AOM, with depth caused by conservation tillage illustrates redistribution of C occurring throughout the soil profile; and is consistent with other reports, *i.e.*, [15,48,49]. This clearly visible effect can be attributed to the effect of the oat cover crop roots that grow to depth in these soils. Considering aggregation formation processes, the penetration of these roots deep into the soil (personal observations) is expected to cause a flush of microbial activity at a lower depth, causing the formation of aggregate binding agents at the scale of macroaggregates [18,19,50]; and therefore the physical protection of OM by these aggregates.

The general improvement in SOC indicates that cover crops and reduced tillage (conservation tillage) has decreased the gap between SOC in conventional tillage and in undisturbed páramos—a 33% C stock increase over conventional tillage. Edwards *et al.* [51] found that conversion from conventional tillage to conservation tillage in soybeans and corn systems rotated with wheat during winter in a Hapludult of Southern USA also increased soil organic carbon on average by 31% over a 10 yr period. Although the initial organic matter conditions are different between the soils studied by Edwards *et al.* [51] and our Andisols (0.001 g·g^{-1} vs. 0.17 g·g^{-1} under conventional tillage), conservation tillage improved those initial levels of organic matter, even in OM-rich soils. This confirms this study's first objective showing that conservation tillage is effective in increasing SOM even in soils that have a high OM content (15% of OM in soils under conventional agriculture); at least in these unique páramos ecosystems. This counters the results of others who reported that for high OM soils varying C inputs did not have any effect on SOC levels; indicating a state of soil C saturation [52,53]). It highlights the question: What is C saturation?

With respect to organic matter found in other disturbed páramos, the average organic matter concentration of conventional tillage sites (0.17 g·g^{-1}, **Table 4**) is greater than the high OM concentrations found in other potato systems in Southern Colombian páramos (0.10 - 0.13 g·g^{-1}, [54]) and other parcels located in the Fuquene watershed (0.15 g·g^{-1}, Amezquita *et al.*, unpublished data, 2005). The average organic matter content of conservation tillage sites (0.22 g·g^{-1}, **Table 4**) was instead similar to the content reported also by Diaz & Paz [54] of 0.17 - 0.24 g·g^{-1} in sites that were previously pastures and that were recently cultivated with potato. These authors attributed this to the remaining effect of pasture roots on the organic matter content that may be similar to the effect of the oat roots in our conservation tillage system. The organic matter content was also similar to the average organic matter reported by Amezquita *et al.* (unpublished data, 2005) for undisturbed páramos of the Fuquene watershed (0.24 g·g^{-1}).

Thus, conservation tillage presented higher C concentrations (and organic matter) similar to undisturbed páramos; indicating that conservation tillage can be used to rehabilitate soil carbon under potato production in the region. This positive effect of conservation tillage on the organic matter and carbon contents has been reported by many studies, *i.e.*, [55,56]. It has been suggested that the increase in soil organic carbon associated with the adoption of conservation tillage will continue for a period of 25 to 50 yr depending on climatic conditions, soil characteristics, and production management practices [26, 57-60]. It is unclear how long the increase in SOM will continue in these soils under conservation tillage.

The relevance of these results lies in the fact that while most soil páramos studies have reported how land use changes modify the unique properties of páramos soils, none have explored how better management practices in agriculture can rehabilitate them. The question left unanswered by this study relates to the time frame for which improvements on SOC and organic matter will be achieved with conservation tillage, and also under which baseline conditions conservation tillage could improve disturbed soil properties in páramos. It only suggests that these changes can be brought about in as little as 7 years.

4.2. Higher AOM and SOM in Smaller Macroaggregates

Bossuyt *et al.* [19] found in Ultisols, that microaggregate-protected and micro within macroaggregate-protected C was higher in no-till (NT) systems than in conventional tillage systems. This was a result of disruption avoidance of macroaggregates with no-till systems. The reason is that when disruption of macroaggregates is avoided, residue that forms the center of a macroaggregate decomposes into finer organic matter that gradually becomes encrusted with clay particles and microbial products, forming microaggregates within macroaggregates [61]. When macroaggregates are disrupted by conventional tillage, OM is released and does not have the time to form microaggregates, resulting in a much smaller amount of microaggregates within macroaggregates [3,9, 17]. Thus although these results suggest that microaggregates are important for ensuring SOM protection, macroaggregates stabilization is important for this protection to occur [19]. In the same sense, Six *et al.* [9] suggested that a reduced rate of macroaggregate turnover under no-till increases the formation of microaggregates in which C is stabilized and sequestered in the long-term. The same author found that the amount of microaggregates protected in macroaggregates was two times greater with no-till compared to conservation tillage [17]. Similarly, Denef *et al.* [27] reported for a Mollisol an increase of microaggregates within smaller macroaggregates (0.25 - 2 mm size) between 20% - 39% with no-till compared to conventional treatment. We postulate that higher amounts of SOM in SF2, SF3 and SF4 found in this study, and specially a relative greater increase of AOM in SF3 and SF4 (0.5 - 2 mm size) with respect to conventional agriculture, may be related to an increase of microaggregates within these smaller macroaggregates.

The rationale behind the above thought is that the AOM in SF3 and SF4 was released at specific energy levels noted clearly by a step-wise curve denoting a hierarchical order of aggregation. According to Duiker [62], when increasing levels of energy are applied, aggregates of most soils fall apart into smaller aggregates in a step-wise manner. Conservation agriculture produced different curves for SF3 of both horizons and SF4 of horizon A2. In both cases, some steps of the curve were accentuated as more AOM was released after applying the energy levels corresponding to these steps. We inferred that those specific energy levels (17 $J \cdot mL^{-1}$ and 32 $J \cdot mL^{-1}$) correspond to the energy levels where the macroaggregates broke down into microaggregates. This is supported by the pattern of aggregation explained by Oades and Waters [63] for Mollisols and Alfisols when they described a hierarchical structure where larger, weaker aggregates break down to release smaller, stronger aggregates before breaking down into primary particles. In consequence, these distinct units can be separated as defined by our aggregate dispersion energy curves.

Denef *et al.* [27] found in different soil types that 91% of the difference in total SOC between no-till and conventional tillage was explained by the C associated with microaggregates that were isolated from smaller macroaggregates (0.25 - 2 mm). Similarly, considering that our study found more than 80% of the total carbon in aggregates, we suggest that the 29% difference in Total SOM ($g \cdot g^{-1}$) between conservation tillage and conventional tillage is explained by increments of AOM in SF2, SF3 and SF4—classes where AOM increases were detected. Kong *et al.* [10] also found that the majority of the accumulation of SOC due to additional C inputs in agricultural lands was preferentially sequestered in the microaggregates within-small-macroaggregates (mM). For this reason they proposed the use of the mM class (microaggregates within small macroaggregates) as an indicator for C sequestration potential in agroecosystems. This corresponds to the same macroaggregate class for which we found improvements of AOM. In horizon A1 (0 - 78 cm) the differences between conservation agriculture and conventional agriculture were 37%, 33% and 30% for SF2, SF3 and SF4 respectively, and 58%, 99% and 98% in SF2, SF3 and SF4 of horizon A2 (78 - 117 cm).

Thus, this study confirms the findings of Denef *et al.* [27] and Kong *et al.* [10] in that most changes in total SOC were explained by differences in AOM caused by no-till (in our case reduced tillage and cover crops) in smaller macroaggregates (0.25 - 2 mm). Therefore our results tentatively support Kong *et al.* [10] in the use of the microaggregate-within small macroaggregate class as a potential indicator of long-term C sequestration in agricultural lands. The protection mechanisms—including the effect of mineralogy and climate, invoked by this relationship as well as the separate effect of reduced tillage and cover crops at different soil depth should be the topic of future studies.

4.3. Other Considerations

The higher values of %AOM derived from smaller macroaggregates (SF3 and SF4) suggest that in these classes the C has a slower turnover than the C in bigger macroaggregates (>2 mm). Based on Kong *et al.* [10] findings, where increases on C stabilization in the smaller macroaggregates were associated to greater aggregate stability and long-term sequestration, we suggest in the same direction that the higher AOM and SOM in smaller macroaggregates in our soils is linked to greater C and aggregate stability and in consequence is contributing to long-term C sequestration in the Andes. In addition, increases of AOM may be related to improvement of soil structure. The conservation agriculture curves for SF3 and SF4 had better defined hierarchal steps than did the conventional agriculture curves. Since well-defined steps indicate well developed structure, we suggest that conservation agriculture in these Andean soils also improves structure.

5. Conclusions

Reduced tillage and cover crops in potato-based systems improved in a 7-year period the soil organic matter and carbon content in disturbed soils of the páramos of Colombia. The soil carbon concentration in the whole profile was 29% higher under conservation tillage than under conventional tillage sites and the carbon content was higher by 33%. C content improvement was the greatest in the subsoil (A2 horizon) increasing by 177% although most of the C is stored in the top A1 horizon. This may be mostly an effect of cover crops. These improvements reflect that conservation tillage (reduced tillage and cover crops) is allowing the rehabilitation of soil carbon compared to conventional tillage systems.

In the Andisols analyzed in this study, the soils had more than 80% of total OM as AOM, and conservation agriculture involving reduced tillage and cover crops in these Andean soils increased AOM. This study was able to evaluate the effects of conservation and conventional agriculture by studying differences in AOM. The major differences in AOM were seen to occur in smaller macroaggregates (0.5 - 2 mm size classes). The aggregate dispersion energy curves further suggest that this is happening in microaggregates within the smaller macroaggregates class. Similar results have been obtained for other soils suggesting that smaller macroaggregates can be used to evaluate potential of long-term C sequestration in some Alfisols, Mollisols, Ultisols and now Andisols.

6. Acknowledgements

The authors acknowledge the CGIAR Challenge Program on Water & Food for support in implementing this study. Also thank the many contributors to this analysis, Arnulfo Rodríguez (CIAT) for helping during soil field sampling and lab analysis, Elena Azuaje (UF soils lab) for training on sonication methods, Wilson Otero (FUNDESOT) and Diego Perez (CAR—Regional Autonomous Corporation) and its conservation tillage extension agents for describing the systems and helping to identify the selected farms for this study, farmers of Carmen de Carupa municipality for allowing the field sampling, Drs. Cheryl Mackowiack and David Wright (NFREC-UF) for comments on earlier drafts.

REFERENCES

[1] E. T. Sundquist, "The Global Carbon Dioxide Budget," *Science*, Vol. 259, No. 5097, 1993, pp. 936-961. http://dx.doi.org/10.1126/science.259.5097.934

[2] F. J. Stevenson, "Humus Chemistry: Genesis, Composition, Reactions," 2nd Edition, John Wiley and Sons Ltd., New York, 1994.

[3] J. Six, E. T. Elliott, K. Paustian and J. W. Doran, "Aggregation and Soil Organic Matter Accumulation in Cultivated and Native Grassland Soils," *Soil Science Society of America Journal*, Vol. 62, No. 5, 1998, pp. 1367-1377. http://dx.doi.org/10.2136/sssaj1998.03615995006200050032x

[4] M. Robbins, "Carbon Trading, Agriculture and Poverty," World Association of Soil and Water Conservation (WASWC), Special Publication No. 2, 2004, Beijing.

[5] J. D. Jastrow, T. W. Boutton and R. M. Millar, "Carbon Dynamics of Aggregate-Associated Organic Matter Estimated by Carbon-13 Natural Abundance," *Soil Science Society of America Journal*, Vol. 60, No. 3, 1996, pp. 801-807. http://dx.doi.org/10.2136/sssaj1996.03615995006000030017x

[6] A. P. Edwards and J. M. Bremmer, "Microaggregates in Soils," *Journal of Soil Science*, Vol. 18, No. 1, 1967, pp. 64-73. http://dx.doi.org/10.1111/j.1365-2389.1967.tb01488.x

[7] T. Maeda, H. Takenaka and B. P. Warkentin, "Physical Properties of Allophane Soils," *Advances in Agronomy*, Vol. 29, 1977, pp. 229-264. http://dx.doi.org/10.1016/S0065-2113(08)60220-5

[8] J. D. Jastrow, T. W. Boutton and R. M. Millar, "Carbon Dynamics of Aggregate-Associated Organic Matter Estimated by Carbon-13 Natural Abundance," *Soil Science Society of America Journal*, Vol. 60, No. 3, 1996, pp. 801-807. http://dx.doi.org/10.2136/sssaj1996.03615995006000030

017x

[9] J. Six, E. T. Elliott and K. Paustian, "Aggregate and Soil Organic Matter Dynamics under Conventional and No-Tillage Systems," *Soil Science Society of America Journal*, Vol. 63, No. 5, 1999, 1350-1358. http://dx.doi.org/10.2136/sssaj1999.6351350x

[10] A. Y. Kong, J. Six, D. C. Bryant, R. F. Denison and C. van Kessel, "The Relationship between Carbon Input, Aggregation, and Soil Organic Carbon Stabilization in Sustainable Cropping Systems," *Soil Science Society of America Journal*, Vol. 69, No. 4, 2005, pp. 1078-1085. http://dx.doi.org/10.2136/sssaj2004.0215

[11] E. T. Elliott, "Aggregate Structure and Carbon, Nitrogen and Phosphorus in Native and Cultivated Soils," *Soil Science Society of America Journal*, Vol. 50, No. 3, 1986, pp. 627-633. http://dx.doi.org/10.2136/sssaj1986.03615995005000030017x

[12] C. A. Cambardella and E. T. Elliott, "Methods for Physical Separation and Characterization of Soil Organic Matter Fractions," *Geoderma*, Vol. 56, No. 1-4, 1993, pp. 449-457. http://dx.doi.org/10.1016/0016-7061(93)90126-6

[13] R. M. Bajracharya, R. Lal and J. M. Kimble, "Soil Organic Carbon Distribution in Aggregates and Primary Particle Fractions as Influenced by Erosion Phases and Landscape Positions," In: R. Lal, J. M. Kimble, R. F. Follet and B. A. Stewart, Eds., *Soil Processes and the Carbon Cycle*, CRC Press, Boca Raton, 1997, pp. 353-368.

[14] D. A. Angers and C. Chenu, "Dynamics of Soil Aggregation and C Sequestration," In: R. Lal, J. M. Kimble, R. F. Follet and B. A. Stewart, Eds., *Soil Processes and the Carbon Cycle*, CRC Press, Boca Raton, 1997, pp. 199-206.

[15] M. H. Beare, P. F. Hendrix and D. C. Coleman, "Water-Stable Aggregates and Organic Matter Fractions in Conventional and No-Tillage Soils," *Soil Science Society of America Journal*, Vol. 58, No. 3, 1994, pp. 777-786. http://dx.doi.org/10.2136/sssaj1994.03615995005800030020x

[16] M. R. Carter, "Influence of Reduced Tillage Systems on Organic Matter, Microbial Biomass, Macro-Aggregate Distribution and Structural Stability of the Surface Soil in a Humid Climate," *Soil & Tillage Research*, Vol. 23, No. 4, 1992, pp. 361-372. http://dx.doi.org/10.1016/0167-1987(92)90081-L

[17] J. Six, E. T. Elliott and K. Paustian, "Soil Macroaggregate Turnover and Microaggregate Formation: A Mechanism for C Sequestration under No-Tillage Agriculture," *Soil Biology & Biochemistry*, Vol. 32, No. 14, 2000, pp. 2099-2103. http://dx.doi.org/10.1016/S0038-0717(00)00179-6

[18] J. M. Tisdall and J. M. Oades, "Organic Matter and Water-Stable Aggregate and Soil Organic Aggregates in Soils," *Journal of Soil Science*, Vol. 33, No. 2, 1982, pp. 141-163. http://dx.doi.org/10.1111/j.1365-2389.1982.tb01755.x

[19] H. Bossuyt, J. Six and P. F. Hendrix, "Aggregate-Protected Carbon in No-tillage and Conventional Tillage Agroecosystems Using Carbon-14 Labeled Plant Residue," *Soil Science Society of America Journal*, Vol. 66, No. 6, 2002, pp. 1965-1973. http://dx.doi.org/10.2136/sssaj2002.1965

[20] D. A. Angers, "Changes in Soil Aggregation and Organic carbon under Corn and Alfalfa," *Soil Science Society of America Journal*, Vol. 56, No. 4, 1992, pp. 1244-1249. http://dx.doi.org/10.2136/sssaj1992.03615995005600040039x

[21] J. S. Kern and M. G. Johnson, "Conservation Tillage Impacts on National Soil and Atmospheric Carbon Levels," *Soil Science Society of America Journal*, Vol. 57, No. 1, 1993, pp. 200-210. http://dx.doi.org/10.2136/sssaj1993.03615995005700010036x

[22] I. C. Burke, E. T. Elliott and C. V. Cole, "Influence of Macroclimate, Landscape Position and Management on Soil Organic Matter in Agroecosystems," *Ecological Applications*, Vol. 5, No. 1, 1995, pp. 124-131. http://dx.doi.org/10.2307/1942057

[23] R. Nelson, C. M. Hellwinckel, C. C. Brandt, T. O. West, D. De la Torre Ugarte and G. Marland, "Energy Use and Carbon Dioxide Emissions from Cropland Production in the United States, 1990-2004," *Journal of Environmental Quality*, Vol. 38, No. 2, 2009, pp. 418-425. http://dx.doi.org/10.2134/jeq2008.0262

[24] D. H. Muller, T. C. Daniel and R. C. Wendt, "Conservation Tillage: Best Management Practice for Nonpoint Runoff," *Environmental Management*, Vol. 5, No. 1, 1981, pp. 33-53. http://dx.doi.org/10.1007/BF01866608

[25] J. Rubiano, M. Quintero, R. D. Estrada and A. Moreno, "Multiscale Analysis for Promoting Integrated Watershed Management," *Water International*, Vol. 31, No. 3, 2006, pp. 398-411. http://dx.doi.org/10.1080/02508060608691941

[26] A. J. Franzluebbers, F. M. Hons and D. A. Zuberer, "Tillage and Crop Effects on Seasonal Dynamics of Soil CO_2 Evolution, Water Content, Temperature, and Bulk Density," *Applied Soil Ecology*, Vol. 2, No. 2, 1995, pp. 95-109. http://dx.doi.org/10.1016/0929-1393(94)00044-8

[27] K. Paustian, J. Six, E. T. Elliott and H. W. Hunt, "Management Options for Reducing CO_2 Emissions from Agricultural Soils," *Biogeochemistry*, Vol. 48, No. 1, 2000, pp. 147-163. http://dx.doi.org/10.1023/A:1006271331703

[28] K. Denef, J. Six, R. Merckx and K. Paustian, "Carbon Sequestration in Microaggregates of No-Tillage Soils with Different Clay Mineralogy," *Soil Science Society of America Journal*, Vol. 68, No. 6, 2004, pp. 1935-1944. http://dx.doi.org/10.2136/sssaj2004.1935

[29] Japanese International Cooperation Agency—JICA, "El Estudio Sobre Plan de Mejoramiento Ambiental Regional Para la Cuenca de la Laguna de Fuquene," CTI Engineering International co., Ltd., 2000.

[30] Instituto de Hidrología, "Meteorología y Estudios Ambientales (IDEAM), Climatic Data," Tausa municipality Station, Colombia, 2004.

[31] Instituto Geográfico Agustín Codazzi (IGAC), "Estudio General de Suelos y Zonificación de Tierras del Departamento de Cundinamarca," Bogota, Colombia, 2000.

[32] W. Buytaert, R. Celleri, B. De Bièvre, R. Hofstede, F. Cisneros, G. Wyseure and J. Deckers, "Human Impact on

the Hydrology of the Andean Páramos," *Earth Science Reviews*, Vol. 79, No. 1-2, 2006, pp. 53-72. http://dx.doi.org/10.1016/j.earscirev.2006.06.002

[33] W. Buytaert, J. Sevink, B. D. Leeuw and J. Deckers, "Clay Mineralogy of the Soils in the South Ecuadorian Páramo Region," *Geoderma*, Vol. 127, No. 1-2, 2005, pp. 114-129. http://dx.doi.org/10.1016/j.geoderma.2004.11.021

[34] D. W. Nelson and L. E. Sommers, "Total Carbon, Organic Carbon, and Organic Matter," In: D. L. Sparks, Ed., *Methods of Soil Analysis. Part 3. Chemical Methods*, SSSA Book Ser. 5. SSSA, Madison, 1996, pp. 961-1010.

[35] E. E. Schulte and B. G. Hopkins, "Estimation of Soil Organic Matter by Weight Loss-on-Ignition," In: F. R. Magdoff, *et al.*, Eds., *Soil Organic Matter: Analysis and Interpretation*, SSSA Spec. Publ. 46, Madison, 1996, pp. 21-31.

[36] B. H. Ellert, H. H. Janzen and T. Entz, "Assessment of a Method to Measure Temporal Change in Soil Carbon Storage," *Soil Science Society of America Journal*, Vol. 66, No. 5, 2002, pp. 1687-1695. http://dx.doi.org/10.2136/sssaj2002.1687

[37] E. Elliot, J. Heil, E. Kelly and H. C. Monger, "Soil Structural and Other Physical Properties," In: G. P. Robertson, D. Coleman, C. Bledsoe and P. Sollins, Eds., *Standard Soil Methods for Long-term Ecological Research*, LTER, Oxford University Press, New York, 1999.

[38] A. Klute, "Methods of Soil Analysis. Part 1, Physical and Mineralogical Methods," 2nd Edition, Agronomy 9, American Society of Agronomy, Madison, 1986.

[39] P. F. North, "Towards an Absolute Measurement of Soil Structural Stability Using Ultrasound," *Journal of Soil Science*, Vol. 27, No. 4, 1976, pp. 451-459. http://dx.doi.org/10.1111/j.1365-2389.1976.tb02014.x

[40] J. Six, G. Guggenberger, K. Paustian, L. Haumaier, E. T. Elliott and W. Zech, "Sources and Composition of Soil Organic Matter Fractions between and within Soil Aggregates," *European Journal of Soil Science*, Vol. 52, No. 4, 2001, pp. 607-618. http://dx.doi.org/10.1046/j.1365-2389.2001.00406.x

[41] C. W. Swanston, M. S. Torn, P. J. Hanson, J. R. Southon, C. T. Garten, E. M. Hanlon and L. Ganio, "Initial Characterization of Processes of Soil Carbon Stabilization Using Forest Standlevel Radiocarbon Enrichment," *Geoderma*, Vol. 128, No. 1-2, 2005, pp. 52-62. http://dx.doi.org/10.1016/j.geoderma.2004.12.015

[42] D. V. Sarkhot, N. B. Comerford, E. J. Jokela, J. B. Reeves III and W. G. Harris, "Aggregation and Aggregate Carbon in a Forested Southeastern Coastal Plain Spodosol," *Forest, Range and Wildland Soils*, Vol. 71, No. 6, 2007, pp. 1779-1787.

[43] S. Kuo, U. M. Sainju and E. J. Jellum, "Winter Cover Crops Effects on Soil Organic Carbon and Carbohydrate in Soil," *Soil Science Society of America Journal*, Vol. 61, No. 1, 1997, pp. 145-152. http://dx.doi.org/10.2136/sssaj1997.03615995006100010022x

[44] P. E. Rasmussen, R. R. Allmaras, C. R. Rohde and N. C. Roager Jr., "Crop Residue Influences on Soil Carbon and Nitrogen in Wheat-Fallow System," *Soil Science Society of America Journal*, Vol. 44, No. 3, 1980, pp. 596-600. http://dx.doi.org/10.2136/sssaj1980.03615995004400030033x

[45] C. V. Cole, K. Flach, J. Lee, D. Sauerbeck and B. Stewart, "Agricultural Sources and Sinks of Carbon," *Water, Air, and Soil Pollution*, Vol. 70, No. 1-4, 1993, pp. 111-122. http://dx.doi.org/10.1007/BF01104991

[46] P. Podwojewski, L. Poulenard, T. Zambrana, and R. Hofstede, "Overgrazing Effects on Vegetation Cover and Properties of Volcanic Ash Soil in the Páramo of Llangahua and La Esperanza (Tungurahua, Ecuador)," *Soil Use and Management*, Vol. 18, No. 1, 2002, pp. 45-55. http://dx.doi.org/10.1079/SUM2002100

[47] J. Poulenard, P. Podwojewski and A. J. Herbillon, "Characteristics of Non-Allophanic Andisols with Hydric Properties from the Ecuadorian Paramos," *Geoderma*, Vol. 117, No. 3-4, 2003, pp. 267-281. http://dx.doi.org/10.1016/S0016-7061(03)00128-9

[48] M. R. Carter and D. A. Rennie, "Changes in Soil Quality under Zero Tillage Farming Systems: Distribution of Microbial Biomass and Mineralizable C and N Potentials," *Canadian Journal of Soil Science*, Vol. 62, No. 4, 1982, pp. 587-597. http://dx.doi.org/10.4141/cjss82-066

[49] A. J. VandenBygaart, X. M. Tang, B. D. Kay and J. D. Aspinall, "Variability in Carbon Sequestration Potential in No-Till Soil Landscapes of Southern Ontario," *Soil and Tillage Research*, Vol. 65, No. 2, 2002, pp. 231-241. http://dx.doi.org/10.1016/S0167-1987(02)00003-X

[50] C. A. Cambardella, "Aggregation and Organic Matter," In: R. Lal, Ed., *Encyclopedia of Soil Sciences*, School of Natural Resources, The Ohio State University, Marcel Dekker, Inc. New York, 2002, pp. 52-55.

[51] J. H. Edwards, C. W. Wood, D. L. Thurlow and M. E. Ruf, "Tillage and Crop Rotation Effects on Fertility Status of a Hapludult Soil," *Soil Science Society of America Journal*, Vol. 56, No. 5, 1992, pp. 1577-1582. http://dx.doi.org/10.2136/sssaj1992.03615995005600050040x

[52] C. A. Campbell, K. E. Bowren, M. Schnitzer, R. P. Zentner and L. Townley-Smith, "Effect of Crop Rotations and Fertilization on Soil Biochemical Properties in a Thick Black Chernozem," *Canadian Journal of Soil Science*, Vol. 71, No. 3, 1991, pp. 377-387.

[53] J. Six, R. T. Conant, E. A. Paul and K. Paustian, "Stabilization Mechanisms of Soil Organic Matter: Implications for C-Saturation of Soils," *Plant and Soil*, Vol. 241, No. 2, 2002, pp. 155-176. http://dx.doi.org/10.1023/A:1016125726789

[54] E. B. Díaz and L. P Paz, "Evaluación del Régimen de Humedad del Suelo Bajo Diferentes Usos, en los Páramos Las Animas (Municipio de Silvia) y Piedra de León (Municipio de Sotará). Departamento del Cauca," MSc Thesis, Fundacion Universitaria de Popayan, Popayán, 2002.

[55] F. Grant, "Changes in Soil Organic Matter under Different Tillage and Rotations: Mathematical Modeling in Ecosystems," *Soil Science Society of America Journal*, Vol. 61, No. 4, 1997, pp. 1159-1175. http://dx.doi.org/10.2136/sssaj1997.03615995006100040

023x

[56] A. Black and D. Tanaka, "A Conservation Tillage Cropping Systems Study in the Northern Great Plains of the United States," In: E. Paul, K. Paustien, E. Elliott and C. Cole, Eds., *Soil Organic Matter in Temperate Agroecosystems*, CRC Press, Boca Raton, 1997, pp. 335-342.

[57] A. Franzluebbers, "Soil Microbial Biomass and Mineralizable Carbon of Water-Stable Aggregates," *Soil Science Society of America Journal*, Vol. 61, No. 4, 1997, pp. 1090-1097. http://dx.doi.org/10.2136/sssaj1997.03615995006100040015x

[58] P. G. Hunt, "Changes in Carbon Content of a Norfolk Loamy Sand after 14 Years of Conservation or Conventional Tillage," *Journal of Soil and Water Conservation*, Vol. 51, No. 3, 1996, pp. 255-258.

[59] C. Wood, J. Edwards and C. Cummins, "Tillage and Crop Rotation Effects on Soil Organic Matter in a Typic Hapludalt of Northern Alabama," *Journal of Sustainable Agriculture*, Vol. 2, No. 2, 1992, pp. 31-41. http://dx.doi.org/10.1300/J064v02n02_04

[60] T. Zobeck, N. Rolong, D. Fryear, J. Bilbro and B. Allen, "Properties of Recently Tilled Sod, 70-Year Cultivated Soil," *Journal of Soil and Water Conservation*, Vol. 50, No. 2, 1995, pp. 210-215.

[61] J. M. Oades, "Soil Organic Matter and Structural Stability: Mechanisms and Implications for Management," *Plant and Soil*, Vol. 76, No. 1-3, 1984, pp. 319-337. http://dx.doi.org/10.1007/BF02205590

[62] S. W. Duiker, "Aggregation," In: R. Lal, Ed., *Encyclopedia of Soil Sciences*, School of Natural Resources, The Ohio State University, Marcel Dekker, Inc., New York, 2002, pp. 49-51.

[63] J. M. Oades and A. G. Waters, "Aggregate Hierarchy in Soils," *Australian Journal of Soil Research*, Vol. 29, No. 6, 1991, pp. 815-828. http://dx.doi.org/10.1080/028418501127346846

6

Distribution and Mobility of Zn, Pb and Cd in a Sewage Sludge-Amended Soil

Dominique Proust, Vivien Mathé, François Lévêque

LIENSs UMR 7266 CNRS-Université de La Rochelle, La Rochelle, France.
Email: dominique.proust@univ-lr.fr

ABSTRACT

The distribution of the metallic trace elements (MTE) Zn, Pb and Cd in a sludge-amended soil and their partitioning in specific soil microystems are studied by comparing their contents in amended and control soils. This comparison is achieved at the metric scale of the bulk soil horizons and at the micrometric scale of the weathering microsites (weathering rock-forming minerals and their specific weathered products). The chemical analyses of the MTE in the bulk samples do not show any anthropic contamination of the amended soil with repeated sewage sludge spreading. The chemical analyses of the bulk < 2 μm clay fractions indicate the occurrence of local MTE concentrations and vertical migration in deeper soil horizons. Precise chemical analyses in the weathering microsites indicate that, as a general rule and whatever microsite is considered, Zn, Pb, and Cd accumulate in clay minerals from surface horizons where the sludge was spread. On the contrary, the vertical MTE migration is restricted to the connected macroporosity of the fissural system filled with clay minerals and does not affect the weathering clays of rock-forming minerals. Such MTE mobility through the fissural system gives rise to two main environmental problems: 1) Zn, Pb, and Cd have the potential to move several meters deep along fissures in the soil profiles and may represent potential contaminants for unconfined aquifer and 2) because plant root system grows preferentially along soil fissural pattern, it may adsorb migrating MTE.

Keywords: Soil; Metallic Trace Elements; Heavy Metals; Clay Minerals

1. Introduction

As a result of the substantial increase in waste water treatment plants in Europe over the past thirty years, problems have emerged in disposing of sewage sludges in an environmentally sound manner. Sludges landfilling, incinerating or landspreading are the most currently used practices. Properly constructed and monitored landfillings are very expensive and will be forbidden in Europe in 2015 because they may generate point-source contaminations due to waste concentrations. Sludge incinerating is also expensive due to the need for sludge dehydration prior to burning and the cost of incinerators preventing air pollution. These practices have thus decreased in the last decade in favor of sludge spreading on cropland, especially in France [1]. This spreading method is considered both a low cost disposal method as well as a source of nutrient-rich fertilizer. However, the use of sewage sludge as a fertilizer in soil remains a debated subject. As a source of plant nutrients (especially P and N) and organic matter, sewage sludge is a beneficial soil amendment. Nevertheless, these sludges originate from industrial and urban by-products treatments and may contain MTE that can be toxic at very low concentration: for instance, Cd concentration cannot exceed 2 $mg{\cdot}kg^{-1}$ in non polluted soils [2-4].

The MTE in soils result in 1) the natural geochemical stock produced by the weathering of the parent rock [5-7], and 2) the anthropogenic contamination by repeated sludge spreading [8,9] and/or atmospheric deposition [10,11]. In such a pedological context, the chemical reactions of weathering will control the solubility, mobility and, as a consequence, the concentration of MTE in soils. The weathering reactions result in 1) the release of the MTE from the dissolving primary minerals and 2) their sorption onto clay minerals, hydrous oxides, and organic matter. The pH, together with the redox potential is also active parameter on the mobility of MTE in soils: a decrease in soil pH and in redox potential increases MTE release from soil [12-14].

Several mineralogical studies showed that the chemical reactions of weathering and their associated metal release were operating in specific soil microsystems with their own solid and solution chemical properties [15-17]. Such microsystems are typically composite aggregates of weathering rock-forming minerals, newly-formed clay minerals, oxi-hydroxides, and organic matter moistened with soil solutions. Even though MTE bulk concentrations could be indicators of global soil contamination, they result from the partial MTE concentrations in each of these microsystems. In that meaning, the detailed analysis of contaminated soils with MTE requires to know the bulk metals content in the soil but also their partial contents in the soil microsystems in which they are retained.

The present study was undertaken to estimate the impact of long-term sludge application upon Zn, Pb, and Cd distribution and migration in soil. In the bulk soil study, MTE concentrations are compared in control and amended soils to characterize the extent to which spreaded metals may have moved throughout the soil profile. In the micrometric study, MTE are analysed in the weathering microsystems to detect possible zones of metals accumulation within soil and weathered parent rock.

2. Materials and Methods

2.1. Soils and Sampling Process

The soils were sampled in a dioritic massif 25 km in the South of Limoges, Haute-Vienne (**Figure 1(a)**). They are inceptisols covering a hill 305 meters high at Labrousse locality. The two soils were sampled on the same dioritic parent-rock (**Figures 1(b)** and **(c)**): 1) the Amended Soil profile (AS) that received for ten years wet sewage sludges heavily loaded with Zn, Pb, and Cd, and 2) the Control Soil (CS) that is free of sludge spreading.

The two soils are moderately drained, deep (1.00 to 1.50 m to C horizon) inceptisols with the typical A, Bw, and C horizons sequence. The R horizon, *i.e.* unweathered parent rock (**Figure 1(d)**), is a diorite which appears as a "salt and pepper" coarse-grained rock made up of major amphiboles, felspars, and minor quartz.

Each soil sample was carefully collected as undisturbed block using plastic core samplers (20 cm length) with cutting edges, smoothly and continuously pushed into the soil to preserve the original fabric of the soil and to prevent the sample metal contamination. Each sample was further divided into two subsamples. The first was devoted to the bulk chemical analysis and the X-ray diffraction (XRD) study whereas the second part was used for thin sections preparation.

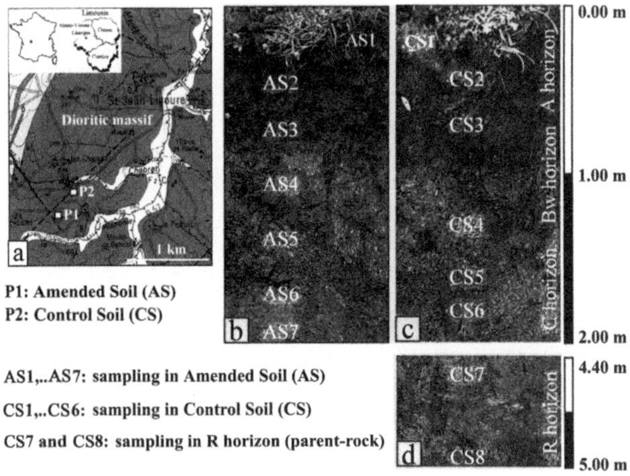

Figure 1. Location and sketch of the soil profiles. (a) Geological setting; (b) Amended soil (AS); (c) Control soil (CS); (d) R horizon (parent-rock).

2.2. Bulk Chemical and X-Ray Diffraction Analyses

An aliquot quantity of the first subsample (5 g) was crushed into agate mortar down to less than 50 μm size and 300 mg of total powder (fused with $LiBO_2$ and dissolved in 1N HNO_3) was used for the bulk chemical analysis of major and trace elements with ICP-AES and ICP-MS (PERKIN ELMER 5000 spectrometer) by the Service d'Analyse des Roches et des Minéraux, (SARM), CRPG-CNRS (Vendoeuvre-lès-Nancy, France). The major and trace element concentrations were expressed in wt.% and $mg \cdot kg^{-1}$, respectively. In order to calculate geochemical mass balances from the bulk chemical analyses, the apparent density of each sample was obtained using the hydrostatic balance.

The rock-forming minerals and their weathering products were identified using XRD on random powders with a PHILIPS PW 1730 diffractometer (40 kV, 40 mA), Fe-filtered Co$k\alpha$ radiation and a stepping motor-driven with a DACO-MP recorder and the Diffrac-AT software (Socabim). The clay minerals were exracted from the soil samples by ultrasonic treatment and centrifugation of the supernatant in order to separate the less than 2 μm fraction. These clay minerals were then identified by XRD on random powders, Ca-oriented and Ca-glycolated oriented preparations.

2.3. Electron Probe Micro Analyses

The second subsample was devoted to the Electron Probe Micro Analyses (EPMA). Thin sections were prepared following the method of Camuti and McGuire [18]: after hardening using vacuum impregnation unit (BROT 1.04.05, 100 mbars) with ARALDITE 2020 epoxy resin and 20% acetone thinner, samples were cut and thin sections obtained through polishing with silicon carbide

(17 and 9 µm particle sizes) and diamond calibrated powders (6, 3, 1 and 0.25 µm particle sizes).

Representative pedofeatures as well as weathering microsites were first located on the thin sections, marked with black circle under an optical microscope and then "*in situ*" analysed for major and trace elements. EPMA were obtained using a CAMECA SX 50 electron microprobe (Service CAMPARIS, Université Paris VI) equipped with wavelength-dispersive spectrometers (WDS). The microprobe was calibrated using synthetic and natural oxides. Corrections were made with a ZAF program. A specific trace program with the electron microprobe was developed [19] to analyse MTE in small volumes with detection limits closed to 6 - 8 mg·kg^{-1}. Major elements analyzes were performed at 15 kV and 4 nA prior to trace element analyzis performed at higher voltage (30 kV) and higher beam current (100 nA and 500 nA for small and large minerals, respectively) in order to improve detection limits. The spot sizes varied from 1 to 5 - 10 µm to limit beam damage. The counting time was 10 s per major element and 1000 s per trace element. Major and trace elements are expressed respectively in wt% and mg·kg^{-1}.

3. Results

3.1. Parent Rock Mineralogy

The parent rock mineralogy was studied from the CS8 sample of the "R" horizon (**Figure 1(d)**). This sample is the unaltered fresh rock with the highest apparent density and low loss on ignition (L.O.I., **Table 1**). The modal composition obtained from 3000 points counting gives a typical dioritic mineralogical composition with amphiboles (38%), plagioclases (32%), orthoclase (12%) with minor quartz (8%), albite (7%), and titanomagnetite (3%). The EPMA of the major elements in the rock-forming minerals (**Table 2**) shows that the amphiboles are calcic hornblende whereas plagioclases are andesine with 48 % anorthite content. The EPMA of the MTE (**Table 2**) shows that amphiboles are the major MTE-bearing minerals with high Zn and Pb concentrations (180 mg·kg^{-1} for Zn and 73 mg·kg^{-1} for Pb) and very low Cd concentrations (6 mg·kg^{-1}). The feldspars show significantly lower MTE contents with almost similar concentrations in orthoclase, albite and plagioclase (14 - 20 mg·kg^{-1} for Zn, 19 - 27 mg·kg^{-1} for Pb, and 2 - 4 mg·kg^{-1} for Cd).

3.2. Weathering Mass Balances

Many mass balance models have been published in the literature to investigate and quantify the chemical change in rock during its interaction with altering fluids [20-23].

Most of these models are based upon the general equation of Gresens [20] where bulk densities and chemical data of altered rocks are compared to a reference rock in order to quantify the losses or gains of elements resulting

Table 1 Bulk chemical analyses of major elements (wt.%) and heavy metals (Zn, Pb, Cd, mg·kg^{-1}), and apparent density (d, g·cm^{-3}) of control soil (CS) and amended soil (AS) samples.

Depth	CS8	CS7	CS6	CS5	CS4	CS3	CS2	CS1	AS7	AS6	AS5	AS4	AS3	AS2	AS1
(m)	4.85	4.50	1.80	1.60	1.30	0.65	0.40	0.20	1.90	1.70	1.40	1.10	0.70	0.40	0.20
SiO_2	48.79	47.24	47.99	46.69	46.03	46.76	50.12	46.07	50.90	50.21	51.27	52.34	50.83	51.23	50.95
TiO_2	1.08	1.27	1.14	1.29	1.27	1.23	1.29	1.37	1.28	1.24	1.27	1.31	1.42	1.48	1.51
Al_2O_3	18.78	18.25	19.10	18.95	18.32	18.78	21.60	18.98	18.56	18.47	18.41	17.93	18.67	17.53	18.74
Fe_2O_3[a]	9.58	11.04	10.03	10.88	11.68	10.97	7.42	9.75	8.07	8.53	7.99	7.97	8.06	8.55	8.00
MnO	0.17	0.20	0.18	0.19	0.20	0.19	0.11	0.16	0.14	0.15	0.15	0.12	0.11	0.12	0.11
MgO	4.47	4.86	4.20	4.60	4.96	4.76	2.47	4.09	3.25	3.44	3.19	3.29	3.36	3.36	3.08
CaO	7.52	8.06	6.97	7.50	7.91	8.01	4.38	6.26	5.70	5.83	5.24	2.11	3.55	2.41	4.26
Na_2O	3.54	3.17	3.36	3.06	2.80	3.08	4.10	2.82	3.19	3.08	3.19	3.50	2.96	3.18	3.25
K_2O	1.44	1.35	1.50	1.16	1.31	1.20	1.59	1.67	2.54	2.35	2.69	3.70	2.86	3.14	2.63
P_2O_5	0.59	0.67	0.60	0.65	0.67	0.63	0.50	0.57	0.00	0.00	0.00	0.00	0.00	0.00	0.00
LOI[b]	3.77	3.46	4.68	5.10	4.80	4.37	6.86	8.53	5.91	6.48	6.19	7.77	7.74	8.49	6.72
Total	99.73	99.57	99.75	100.06	99.95	99.96	100.43	100.26	99.53	99.79	99.59	100.04	99.56	99.49	99.24
Zn	111.18	124.46	113.06	119.32	126.08	120.42	88.24	115.56	90.11	93.04	90.64	82.00	94.64	95.27	99.29
Pb	16.46	22.46	15.56	18.68	14.84	20.18	18.28	38.24	20.08	17.00	19.09	16.24	21.00	29.00	39.08
Cd	<d. l.	<d. l.	<d. l.	<d. l.	<d. l.	<d. l.	<d. l.	<d. l.	0.13	0.20	0.26	0.25	0.32	0.30	0.33
d	2.92	2.55	2.65	2.49	2.39	2.43	2.13	2.07	2.56	2.37	2.34	2.25	2.14	2.09	2.05

[a]Total iron expressed as Fe_2O_3; [b]Loss on ignition, < d. l.: lower than detection limit.

Table 2. EPMA of major elements (wt.%), MTE (mg·kg^{-1}) and structural formulae of rock-forming minerals from control soil CS8 sample.

	Amphibole	Orthoclase	Albite	Plagioclase
	Mean	Mean	Mean	Mean
SiO_2[a]	42.54 (0.88)	61.91 (2.19)	67.87 (0.92)	58.24 (1.25)
TiO_2	1.31 (0.36)	0.04 (0.05)	0.05 (0.06)	0.06 (0.19)
Al_2O_3	11.00 (0.61)	18.71 (0.45)	19.59 (0.75)	26.19 (2.68)
FeO[b]	17.98 (0.73)	0.25 (0.40)	0.10 (0.08)	0.09 (0.09)
MnO	0.40 (0.12)	0.04 (0.07)	0.04 (0.07)	0.05 (0.08)
MgO	9.79 (0.55)	0.08 (0.20)	0.02 (0.02)	0.27 (1.55)
CaO	11.34 (0.32)	0.04 (0.07)	0.37 (0.25)	7.21 (0.61)
Na_2O	1.31 (0.14)	0.39 (0.45)	11.22 (1.02)	7.76 (0.71)
K_2O	1.05 (0.13)	15.14 (1.20)	0.26 (0.46)	0.16 (0.09)
Total	96.72	96.59	99.51	100.03
	Mean	Mean	Mean	Mean
Zn	180 (16)	20 (5)	14 (7)	17 (6)
Pb	73 (26)	23 (4)	19 (4)	27 (17)
Cd	6 (1)	4 (3)	3 (2)	2 (2)
		Structural formulae (atoms)[c]		
Si	6.75 (0.07)	3.09 (0.03)	3.17 (0.03)	2.78 (0.04)
^{IV}Al	1.25 (0.07)	0.91 (0.03)	0.91 (0.03)	1.22 (0.04)
^{VI}Al	0.50 (0.10)	0.03 (0.03)	0.08 (0.04)	0.03 (0.04)
Ti	0.21 (0.06)	0.00 (0.00)	0.00 (0.00)	0.00 (0.01)
Fe^{2+}	2.86 (0.13)	0.01 (0.02)	0.00 (0.00)	0.00 (0.00)
Mn	0.06 (0.02)	0.00 (0.00)	0.00 (0.00)	0.00 (0.00)
Mg	1.55 (0.08)	0.00 (0.01)	0.00 (0.00)	0.01 (0.08)
Ca	1.80 (0.05)	0.00 (0.01)	0.02 (0.01)	0.40 (0.03)
Na	0.21 (0.02)	0.02 (0.02)	0.88 (0.04)	0.37 (0.04)
K	0.17 (0.02)	0.94 (0.02)	0.01 (0.02)	0.15 (0.02)

[a]Values in brackets are the standard errors; [b]Total iron expressed as FeO; [c]Formulae calculated on 8 oxygens basis, except amphiboles calculated on 23 oxygens.

from rock-fluid interactions:

$$X_n = \left[F_v - \left(da_x/da_y\right)C_x\right] - C_y \quad (1)$$

where X_n is the amount (g) of the oxide n gained ($X_n > 0$) or lost ($X_n < 0$) per 100 grams of rock, F_v is the volume factor, da_x and da_y, C_x, and C_y are the apparent densities and the concentrations of the oxide n (wt.%) in the reference and altered rocks respectively. Geochemical mass balances are calculated using either rock volume or concentration of a given element as invariant. Rock alteration usually operates with increasing volume and mass balances are thus currently calculated using Al_2O_3 or TiO_2 as invariant. In this case, a volume factor (F_{vc}) required for a given oxide n to be constant during alteration can be calculated setting $X_n = 0$ in Equation (1). The F_{vc} values calculated with TiO_2 as invariant remain very close to unity throughout the two soils (**Figure 2(a)**) and indicate that TiO_2 could be used as the invariant reference for the mass transfer calculation. The CS8 sample, with low loss on ignition and high apparent density (**Table 1**) has suffered very low weathering and was selected as the reference rock. The first chemical change observed in the weathering chemical balances (**Figures 2(b)** and **(c)**) is the general increase in LOI from bottom to top of the two soil profiles (3.77 to 8.53 wt.%). After organic matter destruction and in the absence of carbon dioxide (no carbonate was detected), increasing LOI reflects a general increase in hydrate content with increasing weathering and, more precisely, with increasing hydrated clay minerals and oxyhydroxides formation. Alternatively,

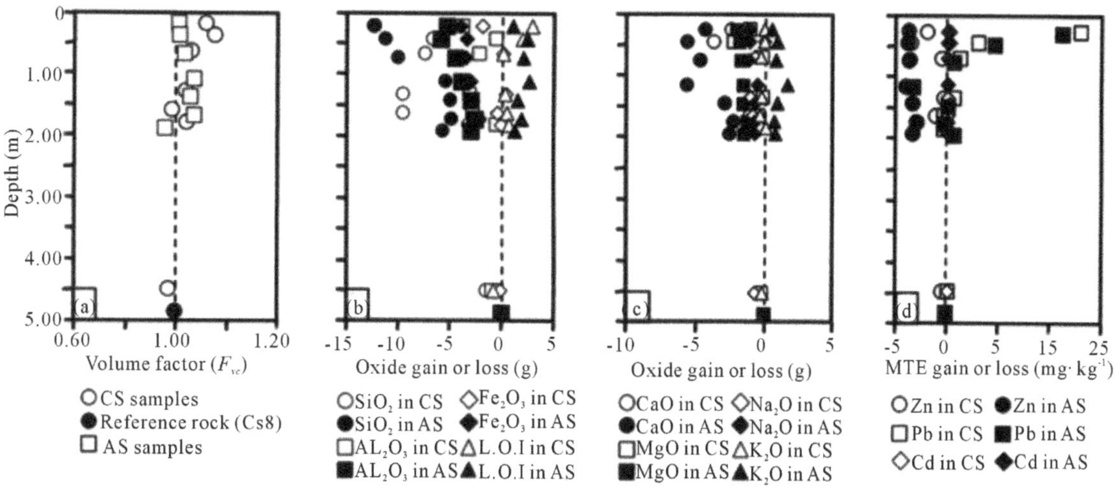

Figure 2. Bulk weathering mass balances calculated in amended and control soils using TiO_2 as invariant: (a) Variations of volume factor; (b) and (c) Mass balances for major elements (LOI: Loss on Ignition); (d) Mass balance for Zn, Pb, and Cd.

all major primary mineral components, except potassium, are continuously leached from rock to soil in variable amounts. The highest leaching rate is given by the SiO_2 compound, up to 12%, Al_2O_3, Fe_2O_3, and CaO have medium leaching rates about 5% whereas MgO and Na_2O display the lowest rates, about 1%. The leaching of SiO_2, Al_2O_3, CaO and Na_2O results for a large part from the dissolution of plagioclases whereas moderate MgO leaching points to low amphibole alteration. The constant to low increasing K_2O concentration in saprolite and soil can be attributed to unweathered orthoclase remnants.

Although spread sludges were heavily loaded with Zn, Pb and Cd (498, 330 and 20 mg·kg^{-1} as mean concentrations respectively), the MTE vertical profiles obtained from the mass balances (**Figure 2(d)**) do not show Zn and Cd anthropic contamination of the amended soil when compared to the control soil. The bulk Zn concentrations show large variations in the two soils (**Table 1**), from 82 to 126 mg·kg^{-1}. The calculated mass balance indicates low Zn depletion in the upper surface horizons of the two soils that may suggest migration of this metal in depth, as already observed [24]. The bulk Cd concentrations given in **Table 2** are under the detection limit in the control soil and range from 0.13 to 0.33 mg·kg^{-1} in the amended soil, well below the contamination threshold of 2 mg·kg^{-1} [2]. The mass balance calculated for Cd in the amended soil does not show any apparent depletion or accumulation which could induce local contaminations at the scale of the soil or saprolite horizons. The behavior of Pb strongly differs from Zinc and Cd: the bulk Pb concentrations are similar in the two soils (**Table 1**), more than ten times lower than in spread sludges, but increase in both upper horizons CS1 and AS1 samples at 38 and 39 mg·kg^{-1}. This behavior is better evidenced in the vertical mass balance profiles where Pb accumulates in the first surface horizons at similar concentrations in the two soils (**Figure 2(d)**). These similar surface accumulation rates in the two soils suggest that Pb concentrations are not modified by sludge spreading and that surface accumulation could result from atmospheric fallout, as already described [25].

3.3. Soil Clay Mineralogy and Chemistry

The <2 µm clay fractions extracted from the bulk soil samples have similar mineralogy in the control and amended soils. The smectite is the dominant clay mineral (about 80% of the clayey fraction) with XRD (001) reflections at 15.10 Å in air-dried state shifting to 17.05 Å in ethylene-glycol solvated state. Kaolinite is observed in smaller quantity (about 20% of the clayey fraction) with (001) and (002) reflections at 7.15 Å and 3.55 Å in airdried state that disappear after 450˚C heating.

The EPMA of the MTE in the clay fractions from the two soils (**Table 3**) are plotted versus depth in **Figure 3**. The first observation is that the clay minerals from the amended soil are contaminated with Zn and Cd when compared to the control soil. The Zn contents in the amended soil range from 144 to 203 mg·kg^{-1} and are always higher than in control soil (10 to 24 mg·kg^{-1}). The Zn distribution with depth is typically that of a downward migrating element. The highest Zn contents are measured in the upper horizons of the amended soil (AS2 and AS1 samples with 203 and 195 mg·kg^{-1} at 0.40 m and 0.20 m depths, respectively). These concentrations decrease progressively to 150 mg·kg^{-1} at 1.90 m depth (AS7 sample) whereas Zn contents at this depth in the control soil is almost ten times lower with 14 mg·kg^{-1} at 1.80 m depth. The Cd distribution is similar to that of Zn but with much lower values, 17 to 25 mg·kg^{-1} in amended soil samples versus 5 to 9 mg·kg^{-1} in control soil samples. Alternatively, Pb concentrations measured in the amended and control soils do not show any sig-

Table 3. EPMA of MTE (Zn, Pb, Cd, mg.kg^{-1}) in the bulk clay fractions of control soil (CS) and amended soil (AS) samples.

Depth	CS8	CS7	CS6	CS5	CS4	CS3	CS2	CS1	AS7	AS6	AS5	AS4	AS3	AS2	AS1
(m)	4.85	4.50	1.80	1.60	1.30	0.65	0.40	0.20	1.90	1.70	1.40	1.10	0.70	0.40	0.20
Zn[a]	15 (1.6)	14 (1.3)	14 (1.8)	17 (2.4)	18 (2.9)	19 (3.2)	24 (4.5)	10 (3.2)	150 (12.3)	158 (11.9)	144 (10.5)	153 (9.8)	177 (8.6)	203 (12.6)	195 (11.8)
Pb	6 (2.5)	7 (2.5)	5 (2.4)	8 (2.3)	9 (4.4)	5 (3.4)	6 (2.8)	4 (1.8)	5 (2.4)	9 (2.1)	11 (1.8)	5 (1.5)	7 (1.6)	10 (1.8)	8 (2.1)
Cd	7 (0.3)	4 (0.2)	5 (0.9)	8 (0.6)	9 (0.5)	7 (0.5)	6 (0.4)	5 (0.8)	17 (1.9)	19 (1.8)	20 (2.1)	24 (1.8)	16 (2.2)	22 (2.1)	25 (1.8)

[a]Each analysis is the result of 10 measurements on the same crystal, values in brackets are the standard errors.

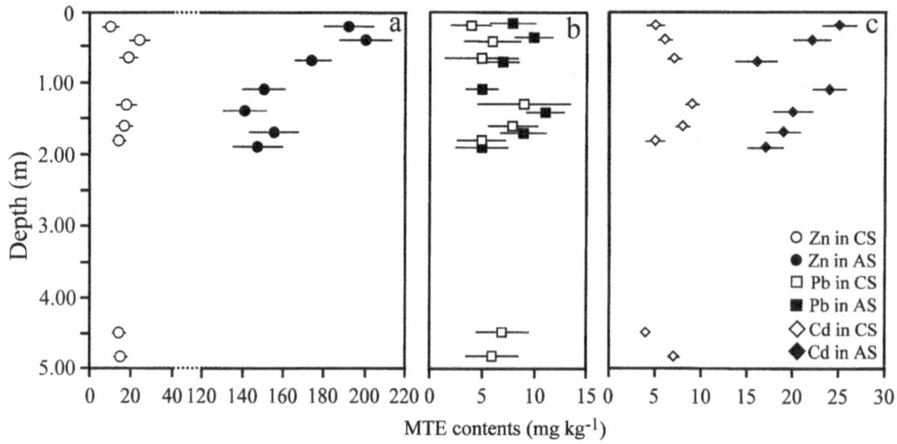

Figure 3. Vertical distribution profiles of (a) Zn, (b) Pb and (c)Cd concentrations in the clay fractions from the amended soil (AS) and the control soil (CS); error bars are showing the standard errors.

nificant evolution in the profiles when standard errors are considered.

3.4. MTE in Amphiboles

Representative MTE contents in the amphiboles and their weathering clay minerals are given in **Tables 2** and **4**. All data are distributed according to the depth in **Figures 4(a)-(c)**. Early amphibole weathering produces magnesian smectite of saponite-type (sample CS7 in **Table 4**) and leads to the leaching of Zn from 180 mg·kg^{-1} in unaltered amphibole to 85 mg·kg^{-1} in the magnesian smectite. Such a leaching does not operate for Pb whose concentrations are very closed in amphiboles and clay (73 and 71 mg·kg^{-1}), nor for Cd whose contents are close to the detection limits (6 mg·kg^{-1} and 3 mg·kg^{-1}).

More intense amphibole weathering stages in C, Bw and A soil horizons produce aluminium smectite of montmorillonite-type and result in very different MTE distribution as the control soil or the amended soil is considered (**Figures 4(a)-(c)**). Zn, Pb and Cd contents in the control soil do not exhibit important variations from the bottom to the top of the profile.

More complicated is the MTE distribution in the amended soil. First of all, the concentrations do not change in the levels where amphibole weathers into saponite, from 4.50 m to 1.70 m depths and, in the second place, concentrations increase in the horizons where amphibole weathers into montmorillonite, from 1.40 m depth to the top of the amended soil. Zn increases from 81 mg·kg^{-1} to 220 mg·kg^{-1}, Pb from 70 mg·kg^{-1} to 125 mg·kg^{-1} and Cd from 9 mg·kg^{-1} to 42 mg·kg^{-1}.

3.5. MTE in Plagioclases

Some typical MTE contents in the plagioclases and in their weathering products are listed in **Tables 2** and **4** and the whole set of data is correlated with depth in **Figures 4(d)-(f)**. The first weathering stage of plagioclases at the bottom the control soil produces aluminium smectite of montmorillonite-type (sample CS7 in **Table 4**). This weathering process operates through an increase in Zn and Pb contents from plagioclases to montmorillonites (**Figures 4(d)-(e)**). Zn concentrations increase from 17 mg·kg^{-1} in the plagioclase crystal to 66 - 78 mg·kg^{-1} in its specific montmorillonite clay mineral, whereas Pb increases from 27 mg·kg^{-1} in the plagioclase to 48 - 43 mg·kg^{-1} in the montmorillonite. The Cd contents in plagio clases and montmorillonites remain very close without any significant variations (from 2 mg·kg^{-1} to 3 - 4 mg·kg^{-1}). The weathering of plagioclases in the soil horizons leads to the crystallization of three clay mineral

Table 4. Representative EPMA of heavy metals (Zn, Pb, Cd, mg·kg^{-1}) in clays as weathering products of rock-forming mineral from control soil (CS) and amended soil (AS) samples.

	Clay minerals in amphiboles						Clay minerals in plagioclases					
Depth	CS7	CS4	CS1	AS7	AS4	AS1	CS7	CS4	CS1	AS7	AS4	AS1
(m)	4.50	1.30	0.20	1.90	1.10	0.20	4.50	1.30	0.20	1.90	1.10	0.20
Zn[a]	85 (7.25)	77 (4.20)	75 (4.40)	72 (7.60)	105 (4.05)	220 (6.20)	66 (10.20)	37 (4.20)	27 (2.40)	67 (7.60)	36 (4.05)	30 (2.10)
Pb	71 (4.75)	75 (6.50)	76 (5.15)	62 (4.60)	94 (3.50)	125 (8.70)	43 (4.75)	41 (3.85)	40 (2.10)	45 (4.60)	39 (3.50)	38 (2.00)
Cd	3 (0.55)	3 (0.60)	3 (0.50)	3 (0.40)	14 (1.20)	42 (2.05)	4 (0.60)	3 (0.90)	2 (0.50)	24 (1.35)	15 (1.20)	12 (2.00)

[a]each analysis is the result of 10 measurements on the same crystal, values in brackets are the standard errors.

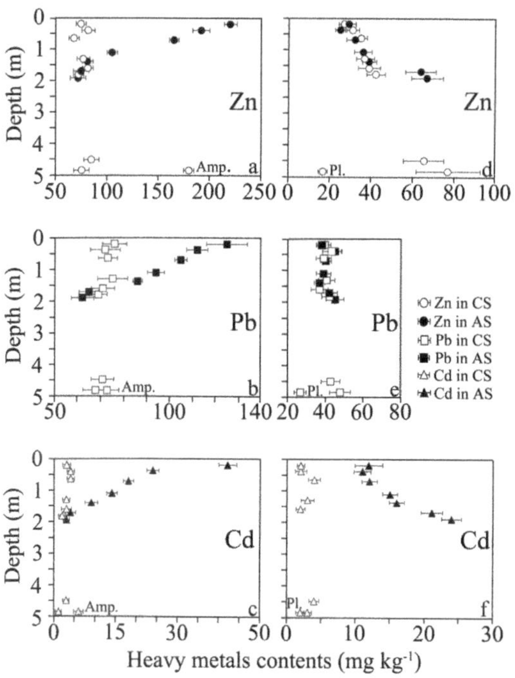

Figure 4. Distribution of the MTE Zn, Pb, and Cd in the control (open symbols) and amended soil (black symbols) as a function of the amphibole and plagioclase weathering sequences; (a), (b), (c): Zn, Pb, and Cd distributions during the weathering of amphibole (Amp.); (d), (e), (f): Zn, Pb, and Cd distributions during the weathering of plagioclase (Pl.); error bars are showing the standard errors.

species with well-defined formation horizons. Montmorillonite, already observed at the bottom of the control soil, is present in the C horizon, (1.90 m to 1.40 m depth). The Bw horizon is typified by two random kaolinite/smectite mixed layers (KS) with different kaolinite and smectite components contents. The first K/S with 65% smectite component is found at the bottom of the Bw horizon (1.40 m depth) and the second with only 5% smectite component is found at the top of the Bw horizon (0.70 m depth). The A horizon (from 0.40 m depth to surface) is characterized by the crystallization of kaolinite.

The behaviour of the MTE analysed in the plagioclase weathering products from C, Bw and A horizons depends on 1) the chemical species considered, 2) the type of soil studied (control or amended soil) and 3) the clay mineral species formed in the studied horizon. The Zn contents in the control soil horizons decrease regularly from 42 mg·kg^{-1} in the C horizon (1.80 m depth) to 27 mg·kg^{-1} in the A horizon at the soil surface **(Figure 4(d))**. Same Zn evolution appears in the amended soil, but with higher contents in the C horizon (67 mg·kg^{-1}) when compared to the control soil. This general Zn decrease from bottom to top follows the decrease in smectite contents in the plagioclase weathering products from C, Bw and A horizons (smectite-K/S—kaolinite weathering sequence). The Pb concentrations in the plagioclase weathering products from control soil and amended soil (**Figure 4(e)**) remain very close (42 to 37 mg·kg^{-1} in the control soil and 38 to 45 mg·kg^{-1} in the amended soil). The small differences in concentrations fall in the range of the standard error (2.00 - 5.30 mg·kg^{-1}) and no significant Pb evolution from bottom to top of the soil is observed. The Cd concentrations in the plagioclase weathering products are very different in the two soils (**Figure 4(f)**). The Cd concentrations in the control soil are very low (2.00 to 5.00 mg·kg^{-1}) and close to the limits of quantification. No significant variation is observed from bottom to top.

The amended soil shows higher Cd concentrations in the clay minerals derived from the plagioclases (11 to 24 mg·kg^{-1}) with a regular decrease from bottom (24 mg·kg^{-1}) to top (12 mg·kg^{-1}). This Cd decrease is similar to Zn behaviour and follows the same mineralogical trend, *i.e.* the decrease in smectite contents from the C horizon to the A horizon.

4. Discussion

The MTE concentrations in the bulk soil samples (**Table 1**) are always below the baseline values taken as the higher limits for non-contaminated soils in the new French regulations (NF French Standard U 44-041 modified by Ordinance 97-1133): Zinc concentrations are from 82.00 to 126.08 mg·kg^{-1} with 300 mg·kg^{-1} baseline, Pb values are from 14.84 to 39.08 mg·kg^{-1} with 100 mg·kg^{-1} baseline, and Cd ranges from values below analytical detection limit to 0.33 mg·kg^{-1} with 2.00 mg·kg^{-1} baseline. These results and the calculated mass balances demonstrate that sludge spreadings on the area did not induce Zn, Pb or Cd contamination. These data, however, were collected at the scale of the soil sample and a thorough study of soil components indicates that local Zn, Pb, and Cd concentrations may occur in specific soil particle-size fractions, such as <2 µm clay fraction and weathering microsites at rock-forming mineral scale.

The comparison of the MTE contents in the <2 µm clay fractions in the two soils (**Table 3** and **Figure 3**) reveals clearly anthropic Zn and Cd contaminations of the clay minerals in the amended soil surface horizons (0.20 to 0.40 m depth). This surface contamination is associated with downward migration of Zn and Cd down to 1.90 m depth within the amended soil. Lead concentrations are slighty higher in amended soil without clear evidence for downward migration. Since bulk clay fractions are the collection of the clay minerals formed during the weathering of primary rock-forming minerals, the study of these MTE concentrations is greatly improved by considering clay minerals in their formation sites, *i.e.* the weathering microsites.

The MTE concentration profiles collected from the amphiboles weathering microsites (**Figure 4**) point out a general anthropic contamination of the clay minerals with Zn, Pb, and Cd within the amended soil surface horizons. This surface contamination can be attributed to the clay mineral montmorillonite which appears in large amount from the amphiboles weathering and has a high sorption capacity of 90 to 127 cmol·kg^{-1} [26,27]. This surface accumulation is followed in the deeper horizons by a regular decrease in MTE contents, in relation with the decrease in amphibole weathering intensity which produces a decrease in the smectite content, and, as a consequence, a decrease in sorption capacity with depth. Conversely, the MTE concentration profiles obtained from the plagioclases weathering microsites (**Figure 4**) show a Zn and Cd depletion of the clay minerals within the amended soil surface horizons, whereas Pb concentrations do not show any significant evolution from the top to the bottom of the soils. This Zn and Cd surface depletion appears to be controlled by the clay mineral weathering sequence of plagioclases: montmorillonite with high sorption capacity (see above) at the bottom and kaolinite with low sorption capacity of 16 - 34 cmol·kg^{-1} [28,29] at the top of the soil profiles. As a consequence, Zn and Cd are not sorbed in the surface horizon and can migrate slowly downwards to more smectitic assemblages (K/S) and finally to montmorillonite where they are retained (1.90 m depth).

5. Conclusions

The results presented in this paper demonstrate that clay minerals are, together with oxides and organic matter, important soil components which play a major role in the mobility of MTE in contaminated soils.

Clays study is essential to understand the sorption reactions occurring at the solid/solution interface in soils. Soils are complex assemblages of clay minerals which formed in weathering microsystems and have a wide range of specific sorption properties. Therefore, there is a need for accurate studies at the scale of these weathering microsystems to identify each clay mineral species and thus to estimate the sorption capacity of the weathering microsite.

This work demonstrates that the micrometric mineralogical approach is relevant to describe the importance of the clay minerals species in the MTE fixation in soil. This mineralogical approach could be of a great interest to improve the modeling of the MTE surface complexation in soils [30].

REFERENCES

[1] G. Amon, O. Aznar and D. Vollet, "Why Are Some French Farmers Sludge-Takers? Some Agronomic and Socioeconomic Explanations," *International Journal of Agricultural Resources, Governance and Ecology*, Vol. 5, No. 2-3, 2006, pp. 289-308.

[2] D. Baize, "Teneurs Totales en Eléments Traces Métalliques dans les sols (France)," INRA Editions, Paris, 1997.

[3] F. M. G. Tack, M. G. Verloo, L. Vanmechelen and E. Van Ranst, "Baseline Concentration Levels of Trace Elements as a Function of Clay and Organic Carbon Contents in Soils in Flanders (Belgium)," *The Science of the Total Environment*, Vol. 201, No. 2, 1997, pp. 113-123. http://dx.doi.org/10.1016/S0048-9697(97)00096-X

[4] P. Planquart, G. Bonin, A. Prone and C. Massiani, "Distribution, Movement and Plant Availability of Trace Metals in Soils Amended with Sewage Sludge Compost: Application to Low Metal Loadings," *The Science of the Total Environment*, Vol. 241, No. 1-3, 1999, pp. 161-179. http://dx.doi.org/10.1016/S0048-9697(99)00338-1

[5] R. A. Klassen, "Geological Factors Affecting the Distribution of Trace Metals in Glacial Sediments of Central Newfoundland," *Environmental Geology*, Vol. 33, No. 2-3, 1998, pp. 154-169. http://dx.doi.org/10.1007/s002540050235

[6] D. Baize and T. Sterckeman, "Of the Necessity of Knowledge of the Natural Pedo-Geochemical Background Content in the Evaluation of the Contamination of Soils by Trace Elements," *The Science of the Total Environment*, Vol. 264, No. 1-2, 2001, pp. 127-139. http://dx.doi.org/10.1016/S0048-9697(00)00615-X

[7] T. Sterckeman, F. Douay, D. Baize, H. Fourier, N. Proix

and C. Schvartz, "Trace Elements in Soils Developed in Sedimentary Materials from Northern France," *Geoderma*, Vol. 136, No. 3-4, 2006, pp. 912-929. http://dx.doi.org/10.1016/j.geoderma.2006.06.010

[8] I. Walter and G. Cuevas, "Chemical Fractionation of Heavy Metals in a Soil Amended with Repeated Sewage Sludge Application," *The Science of the Total Environment*, Vol. 226, No. 2-3, 1999, pp. 113-119. http://dx.doi.org/10.1016/S0048-9697(98)00374-X

[9] M. O. Mbila, M. L. Thompson, J. S. C. Mbagwu and D. A. Laird, "Distribution and Movement of Sludge-Derived Trace Metals in Selected Nigerian Soils," *Journal of Environmental Quality*, Vol. 30, No. 5, 2001, pp. 1667-1674. http://dx.doi.org/10.2134/jeq2001.3051667x

[10] E. Steinnes, R. O. Allen, H. M. Petersen, J. P. Rambaek and P. Varskog, "Evidence of Large Scale Heavy-Metal Contamination of Natural Surface Soils in Norway from Long-Range Atmospheric Transport," *The Science of the Total Environment*, Vol. 205, No. 2-3, 1997, pp. 255-266. http://dx.doi.org/10.1016/S0048-9697(97)00209-X

[11] L. Hernandez, A. Probst, J. L. Probst and E. Ulrich, "Heavy Metal Distribution in Some French Forest Soils: Evidence for Atmospheric Contamination," *The Science of the Total Environment*, Vol. 312, No. 1-3, 2003, pp. 195-219. http://dx.doi.org/10.1016/S0048-9697(03)00223-7

[12] O. Sukreeyapongse, P. E. Holm, B. W. Strobel, S. Panichsakpatana, J. Magid and H. C. B. Hansen, "PH-Dependent Release of Cadmium, Copper, and Lead from Natural and Sludge-Amended Soils," *Journal of Environmental Quality*, Vol. 31, No. 6, 2002, pp. 1901-1909. http://dx.doi.org/10.2134/jeq2002.1901

[13] J. Bang and D. Hesterberg, "Dissolution of Trace Element Contaminants from Two Coastal Plain Soils as Affected by pH," *Journal of Environmental Quality*, Vol. 33, No. 3, 2004, pp. 891-901. http://dx.doi.org/10.2134/jeq2004.0891

[14] M. C. Chuan, G. Y. Shu and J. C. Liu, "Solubility of Heavy Metals in a Comtaminated Soil: Effects of Redox Potential and pH," *Water, Air, and Soil Pollution*, Vol. 90, No. 3-4, 1996, pp. 543-556. http://dx.doi.org/10.1007/BF00282668

[15] M. J. Wilson, "Weathering of the Primary Rock-Forming Minerals: Processes, Products and Rates," *Clay Minerals*, Vol. 39, No. 3, 2004, pp. 233-266. http://dx.doi.org/10.1180/0009855043930133

[16] J. Caillaud, D. Proust and D. Righi, "Weathering Sequences of Rock-Forming Minerals in a Serpentinite: Influence of Microsystems on Clay Mineralogy," *Clays and Clay Minerals*, Vol. 54, No. 1, 2006, pp. 87-100. http://dx.doi.org/10.1346/CCMN.2006.0540111

[17] D. Proust, J. Caillaud and C. Fontaine, "Clay Minerals in Early Amphibole Weathering: Tri- to Dioctahedral Sequence as a Function of Crystallization Sites in the Amphibole," *Clays and Clay Minerals*, Vol. 54, No. 3, 2006, pp. 351-362. http://dx.doi.org/10.1346/CCMN.2006.0540306

[18] K. S. Camuti and P. T. McGuire, "Preparation of Polished thin Sections from Poorly Consolidated Regolith and Sediment Material," *Sedimentary Geology*, Vol. 128, No. 1-2, 1999, pp. 171-178. http://dx.doi.org/10.1016/S0037-0738(99)00073-1

[19] M. Fialin, H. Rémy, C. Richard and C. Wagner, "Trace Element Anlaysis with the Electron Microprobe: New Data and Perspectives," *American Mineralogist*, Vol. 84, No. 1-2, 1999, pp. 70-77.

[20] R. L. Gresens, "Composition-Volume Relationships of Metasomatism," *Chemical Geology*, Vol. 2, No. 1, 1967, pp. 47-65. http://dx.doi.org/10.1016/0009-2541(67)90004-6

[21] D. L. Biddle, D. J. Chittleborough and R. W. Fitzpatrick, "An Algorithm to Model Mass Balances Quantitatively," *Computers and Geosciences*, Vol. 24, No. 1, 1998, pp. 77-82. http://dx.doi.org/10.1016/S0098-3004(97)00125-8

[22] P. Baveye, M. B. McBride, D. Bouldin, T. D. Hinesly, M. S. A. Dahdoh and M. F. Abdel-sabour, "Mass Balance and Distribution of Sludge-Borne Trace Elements in a Silt Loam Soil Following Long-Term Applications of Sewage Sludge," *The Science of the Total Environment*, Vol. 227, No. 1, 1999, pp. 13-28. http://dx.doi.org/10.1016/S0048-9697(98)00396-9

[23] B. Minasny, A. B. McBratney and S. Salvador-Blanes, "Quantitative Models for Pedogenesis—A Review," *Geoderma*, Vol. 144, No. 1-2, 2008, pp. 140-157. http://dx.doi.org/10.1016/j.geoderma.2007.12.013

[24] P. O. Scokart, K. Meeus-Verdinne and R. De Borger, "Mobility of Heavy Metals in Polluted Soils near Zinc Smelters," *Water, Air, and Soil Pollution*, Vol. 20, No. 4, 1983, pp. 451-463. http://dx.doi.org/10.1007/BF00208519

[25] G. Merrington and B. J. Alloway, "The Flux of Cd, Cu, Pb and Zn in Mining Polluted Soils," *Water, Air, and Soil Pollution*, Vol. 73, No. 1, 1994, pp. 333-344. http://dx.doi.org/10.1007/BF00477997

[26] F. Ayari, E. Srasra and M. Trabelsi-Ayadi, "Characterization of Bentonitic Clays and Their Use as Adsorbent," *Desalination*, Vol. 185, No. 1-3, 2005, pp. 391-397. http://dx.doi.org/10.1016/j.desal.2005.04.046

[27] J. Srodon and D. K. McCarty, "Surface Area and Layer Charge of Smectite from CEC and EGME/H_2O-Retention Measurements," *Clays and Clay Minerals*, Vol. 56, No. 2, 2008, pp. 155-174. http://dx.doi.org/10.1346/CCMN.2008.0560203

[28] A. P. Ferris and W. B. Jepson, "The Exchange Capacities of Kaolinite and the Preparation of Homoionic Clays," *Journal of Colloid and Interface Science*, Vol. 51, No. 2, 1975, pp. 245-259. http://dx.doi.org/10.1016/0021-9797(75)90110-1

[29] C. Ma and R. A. Eggleton, "Cation Exchange Capacity of Kaolinite," *Clays and Clay Minerals*, Vol. 47, No. 2, 1999, pp. 174-180. http://dx.doi.org/10.1346/CCMN.1999.0470207

[30] J. Hizal and R. Apak, "Modeling of Copper (II) and Lead (II) Adsorption on Kaolinite-Based Clay Minerals Individually and in the Presence of Humic Acid," *Journal of Colloid and Interface Science*, Vol. 295, No. 1, 2006, pp. 1-13. http://dx.doi.org/10.1016/j.jcis.2005.08.005

7

Land Use and Soil Texture Effects on Organic Carbon Change in Dryland Soils, Senegal

Assize Touré[1], Emile Temgoua[2*], Claire Guenat[3], Bo Elberling[4]

[1]Centre de Suivi Ecologique (CSE), Dakar-Fann, Sénégal; [2]Faculty of Agronomy and Agricultural Sciences, University of Dschang, Dschang, Cameroon; [3]Swiss Federal Institute of Technology EPFL, ENAC IIE ECOS, Lausanne, Switzerland; [4]Institute of Geography, Copenhagen, Denmark.
Email: *etemgoua1@yahoo.com

ABSTRACT

Soil organic carbon (SOC) losses due to poor soil management in dryland are now well documented. However, the influence of soil properties on organic carbon change is not well known. The groundnut plant (*Arachis hypogaea* L.), and the dominant crop system in the Senegal's Soudanian zone, have been compared with semi-natural savanna. Leaves, stems and roots biomass were measured, and soil characteristics were analysed. The total leaves and stems biomass was 1.7 and 2.7 Mg ha^{-1} dry matter in groundnut fields and savanna respectively. Total SOC stocks were low (8 to 20 Mg C·ha^{-1} within upper 0.2 m depth, 20 to 64 Mg C·ha^{-1} within upper 1 m depth) and were significantly lower ($P < 0.05$) in sandy soils than in sandy clayey soils, and lower (approximately 27% - 37%) in groundnut fields than in savanna soils. δ^{13}C values show that SOC quality is transformed from the savanna plants (C4/C3 mixed-pools) to C3-pools in groundnut cultivated zone, with the organic matter signature more preserved in the clayey soils. This study confirms that converting woodland to groundnut fields provokes texture transformation and SOC loss. The results call for the extreme necessity to regenerate the wooded zone or encourage practices that favour SOC restitution.

Keywords: δ^{13}C; Dryland; Land Use; Soil Organic Carbon; Soil Texture

1. Introduction

The establishment of the Kyoto Protocol has resulted in a focus on biomass and soils for C sequestration. In order to make C credited meaningful, sustained C sequestration for decades or longer period is required.

Soils contain 1500 - 2000 Pg C, representing the third reservoir of C after ocean and fossil fuels [1]. Soil organic carbon (SOC) storage is chiefly controlled by primary production input and organic matter evolution. Changes in land use affect the amount of carbon stored in vegetation and soils and, hence, the flux of carbon between land and the atmosphere [1]. Increase of primary production will result in an increase of C storage, whereas increasing decomposition (*i.e.*, reduce C turnover time) will have an opposite effect [2]. The organic matter input to arable soils is lower than to forest soils because a greater proportion of the produced biomass is harvested in the former.

Many factors (climate, land use, soil moisture, soil texture, clay content, mineral, erosion, leaching, fire, etc.) play a role in regulating the accumulation and loss of SOC storage [3,4]. The dynamic of soil carbon storage and release is complex and still not well understood [2]. Delta 13C (δ^{13}C) values have been utilized to document vegetation change and to quantify soil organic matter turnover [5]. The majority of tropical fodder and grass of savanna or steppe are C4-plants with δ^{13}C values in the range of −8% to −18%. Most notable are corn, sorghum, millet and sugar cane [6,7]. C3-plants are more characteristic of the tropical forests where the δ^{13}C varies between −22% and −35% [7], including groundnut plants. Delta 13C values are not altered significantly during decomposition and soil organic matter formation, essentially into the fine-silt-sized fraction [6,8,9]. Consequently, δ^{13}C values of SOC reflect the relative contribution of plant species to net primary production.

In the West Africa region, it has been established that intensive cropping regimes lead to a reduction in the vegetation productivity and an impoverishment of the soil (acidification, reduction in structural stability, etc.)

*Corresponding author.

[10]. This degradation is often related to a reduction in organic carbon stocks [11]. Many studies recently outlined information regarding changes in SOC distribution and the total mass stored for a dryland cropping system in semiarid regions [10-14]. The specific importance for the Sahel is the high vulnerability to climate change in already impoverished rural societies. A carbon market is now operational but without substantial inclusion or participation of Sahelian projects or countries. Many activities are now promoted to determine developing countries's biophysical and socio-economic potential for carbon sequestration in soils with the ultimate objective to promote sustainable agriculture, restore degraded environments, and enhance food security and livelihoods among people [12,13]. SOC quality studies are very scarce [13,14]. Up-to-date regional, integrated soil and woodland inventories are required for planning improved land-use management and restoration [13]. This study aims to determine SOC stocks in the semiarid central Senegal and to link SOC quality variations to the effects of soil texture (sandy versus sandy clayey) and land use (savanna versus groundnut cultivation).

2. Materials and Methods

2.1. General Aspects of the Study Area in Kaffrine

The Division of Kaffrine (total area around 1.1×10^6 ha) is located in central Senegal (13°49'N - 15°19'W). The climate is dry tropical, and rainfall occurs between June and October, averaging 600 - 700 mm per annum (yr 1960-2000). The average annual temperature is 26°C. The relief is flat and situated at 20 - 50 m above sea level. The geological substratum consists of sandy-clayey to clayey-sandy sandstone, with ferruginous cement, sometimes capped of duricrust [15]. The plateau, on which slash-and-burn agriculture is practised, is still covered with shrubby undergrowth, savanna and old fallows. This vegetation is semi-natural due to human exploitation for fire wood. Resprouting Combretaceae (mostly *Combretum geitonophyllum* Diels, *C. glutinosum* Perr. and *Terminalia macroptera* G. et Perr.) are the major component of the woody vegetation. Bush fields at the edge of the plateau are devoted to the cropping field. Groundnut (*Arachis hypogaea* L.) is the main crop. Tillage is advised for groundnut cultivation: soil quite ventilated to support the fixing of nitrogen, well-drained to avoid the rot of the seeds before the lifting and well structured to facilitate the development of the gynophores [16]. The farmers always practise at least one ploughing with draught animals to 10 - 15 cm depth. The tree cover on the groundnut fields of Kaffrine is less than 3%. After harvest, the entire groundnut plant, including roots, is removed from soil and stored for animals forage. Land use management consists of clearing savanna vegetation to establish groundnut fields, occasionally alternating with other crops (millet, rice, sorghum and cotton) and fallow. Most land is cropped, so the (semi-) natural savanna soils were very scarce.

The dominant soil type belongs to the order of ferric Luvisols with luvic Arenosols and gleyic Luvisols [17] and is sandy with a clay content of less than 20%. Iron accumulations are often observed in the form of well delimited spots or hardened concretions below the horizon of argillaceous accumulation, under 20 cm depth [18]. Chemical analysis of similar soils reveals clay mineral dominated by kaolinite [15], lesser CEC clay mineral.

2.2. Study Sites

Eleven villages of the division were selected based on soil characteristics (occurrence, area, type of soil) and distribution from previous studies within the division. In the selected villages, the soil surveyed was targeted at two types of land use: groundnut field and savanna. Study sites were selected according to age (years after land clearing) and soil texture. The soils were classified as ferric Luvisols [17]. The soil profile was homogeneous, with massive structure and without coarse elements. They are formed on sandy material of dunes or sandstone deposits. Two textural classes were selected, sandy soils (SAS) and sandy clayey soils (SCS). Furthermore, these two soil texture classes represented the range of soils in terms of land use and texture found within the study area (**Table 1**). 26 sites were surveyed, with 5 per cultivation duration and soil texture type and 6 sites in wooded savanna.

2.3. Leaves, Stems and Roots Biomass Estimation

The leaf biomass of ligneous plants was evaluated in each of the selected villages, in four circular plots along a 200 m transects, located far from human constructions. The size of each plot was selected depending on tree density and it varied between 1/16 and 1 ha. Ten small branches of each woody species among the most dominant in each plot were defoliated. The total fresh leaf biomass was weighted, and approximately 200 g of this leaf mass from each species were dried at 110°C over 48 h. Stems samples were treated similarly. The analyses were carried out on 70 plant samples (**Table 1**).

The leaf biomass, in grams per individual tree, was estimated based on the formula:

$$\text{Leaf biomass} = a * C^{b} \quad (1)$$

where a and b are constants depending on species (proposed by [19] and used by [20]; a is between 0.34 and 14, and b is between 1.18 and 2.71; C = girth of the considered trunk in centimetres at 1.3 metres height.

Leaves and stems biomass were measured in groundnut fields after harvest and plant uprooting.

Below-ground biomass was not quantified, but it is assumed to be 70% of total biomass, representing the average value of those reported by [20] for a similar Senegalese area. On this basis, the proportion of root biomass and the equivalent mass of carbon were calculated for each soil and land use type.

The rate of annual C refund was estimated according to [21], 1% and 20% of the total biomass for cultivated dryland and savanna fields respectively.

2.4. Soil Sampling and Analysis

Twenty-six soil profiles were dug on groundnut fields and areas representing semi-natural savanna (20 and 6 respectively). Eleven profiles were on SAS soil and fifteen on SCS soil (**Table 1**), with at least two soil profiles per selected village. Soil samples were collected at 0 - 10, 10 - 20, 20 - 50, 50 - 80 and 80 - 100 cm soil depths. Samples were air-dried and sieved through a 2-mm stainless steel sieve before they were analysed for organic C and N content, pH, phosphorous as well as particle-size distribution.

Soil organic carbon was analysed by conductimetric measurement of CO_2 after combustion (apparatus Casumat 8-Adge Wöstoff) and nitrogen according to the Kjeldahl method (Technicon apparatus). Soil pH was measured in a 1:2.5 soil/water solution. Phosphorous was analysed by Bay II method. Non-perturbed soil samples per 10 cm layer, down to 1 m depth, were taken using a cylindrical metal sampler (100 cm^3 in volume). Six replicates were done. The bulk densities were determined based on the mass of dry soil/specific volume of samples. C concentrations were converted to total content for each of the 5 selected depths as the product of concentration, depth of sample and bulk density, and expressed in units of kg $C \cdot m^{-2}$. Total SOC stock in the upper one metre was obtained as the sum of the SOC stock of the 5 depth intervals, as used by Chen *et al.* [4].

$$\text{Stock}\left(\text{Mg C}\cdot\text{ha}^{-1}\right)=\text{SOC concentration}\left(\text{g C}\cdot\text{kg}^{-1}\right)\times\text{bulk density}\left(\text{g}\cdot\text{cm}^{-3}\right)\times\text{depth}\left(\text{cm}\right)\times\left(1-\text{stone content}\left[\text{kg kg}^{-1}\right]\right)\times 10 \quad (2)$$

Soil organic-C stocks in the top 20 cm were calculated by summing SOC stocks in the 0 - 10 and 10 - 20 and those in the 100 cm by summing SOC stocks in the top 20 cm, 20 - 50, 50 - 80, and 80 - 100 cm depth intervals.

Soil samples for isotopic analysis were priory freeze-dried, and were crushed (<200 µm). Organic carbon content was measured by autoanalysor Carlo Erba NA 1500. The isotopic ratios were measured on CO_2 obtained by oxidation at 850°C in a quartz tube sealed in the presence of CuO. CO_2 was analyzed on a mass spectrometer VG SIRA by collecting triple and double introduction. The natural abundances are expressed in units of $\delta^{13}C$.

$$\delta^{13}C(\%)=\left[{}^{13}C/{}^{12}C_{\text{sample}}-{}^{13}C/{}^{12}C_{\text{reference}}\right]\Big/\left[{}^{13}C/{}^{12}C_{\text{reference}}\right]\times 1000 \quad (3)$$

The reference used is the International Pee Dee Belemnite (PDB) standard.

2.5. Statistics Analysis and Error Estimation

Statistical analyses were conducted using the SPSS software (version 10.0) for Windows. Non-parametric Pearson test and t-test at 5% level were performed to test differences between mean C content of each soil depth.

3. Results

3.1. Evaluation of the Leaves, Stems and Roots Biomass

The leaves, stems and roots biomass as well as the proportions of organic carbon per land use are summarized in **Table 2**. The total leaves, stems and roots biomass in Kaffrine is 4.7 Mg dry matter (dm) ha^{-1} for the semi-natural savanna. Biomass was smaller under groundnut field (3.0 Mg $dm \cdot ha^{-1}$).

The calculated rate of annual C refund was 0.42 Mg $ha^{-1} \cdot yr^{-1}$ in semi-natural savanna (**Table 2**) while the groundnut soil gained 0.01 Mg $C \cdot ha^{-1} \cdot yr^{-1}$.

3.2. Soil Characterization

The ferric Luvisols appear similar in terms of colour, pH and bulk density. The moist colour at near-surface layers varied from dull yellowish brown (10YR 4/3) to dark

Table 1. Different sites surveyed in Kaffrine and surface area.

Soil texture	Area (ha)	Soil proportion (% total area)	Number of sites surveyed according to groundnut cultivation duration				Number of sites surveyed in the wooded savanna
			0 - 2 year	3 - 5	6 - 10	11 - 25	
Sandy soils (SAS)	189,396	16.8	1	2	2	2	4
Sandy clayey soils (SCS)	130,854	11.6	4	3	3	3	2

Table 2. Tree-leaves, stems and roots biomass in Kaffrine (mean ± SD).

Land use	Leaf and herbaceous		Root		Total dry biomass		Annual C refund into soil
	Dry Biomass	Carbon	Biomass	Carbon	Biomass	Carbon	
	($Mg\ dm \cdot ha^{-1}$)	($Mg\ C \cdot ha^{-1}$)	($Mg\ dm \cdot ha^{-1}$)	($Mg\ C \cdot ha^{-1}$)	($Mg\ dm \cdot ha^{-1}$)	($Mg\ C \cdot ha^{-1}$)	($Mg\ ha^{-1} \cdot yr^{-1}$)
Groundnut field (n = 53)	1.7 ± 0.2	0.7 ± 0.07	1.3 ± 0.1	0.6 ± 0.01	3.0	1.3	0.01
Wooded savanna (n = 70)	2.7 ± 0.3	1.2 ± 0.03	2.0 ± 0.1	0.9 ± 0.01	4.7	2.1	0.42

n = number of samples analysed, SD = standard deviation, dm: vegetation dry matter.

brown (7.5YR 3/4), yellowish brown (10YR 5/8) to bright brown (7.5YR 5/8) at greater depths. The pH (water) values were 4.3 to 5.8 and increased with depth. Values of pH (KCl) were 0.1 to 0.2 pH units lower than pH (water). Bulk density ranged from 1.40 to 1.53 $g \cdot cm^{-3}$.

Table 3 shows the particle-size distribution of the soils studied. The sandy soils (SAS) were slightly structured; near-surface concentrations of nitrogen and phosphorus were between 0.113 - 0.325 $g \cdot kg^{-1}$ and 0.029 - 0.087 $g \cdot kg^{-1}$, respectively. Those of sandy clayey soils (SCS) were between 0.162 - 0.429 $g \cdot kg^{-1}$ and between 0.044 - 0.072 $g \cdot kg^{-1}$ near the surface, respectively. At 20 - 100 cm depth, SCS soils had higher clay contents (18% - 35%) than SAS soils (11% - 15%). Statistical analyses of these results show that only the content of silt and silt + clay distinctly defined the two soil texture classes (<12 wt%-silt and <19-wt%-silt + clay for SAS). For the same ferric Luvisols, the sand proportion in 0 - 20 cm depth was evidently higher in groundnut fields than in semi-natural savanna.

3.3. Soil Organic Carbon

Figure 1 shows the SOC distribution to 1 m depth, at the Kaffrine study site. The detailed distribution is shown in **Tables 4** and **5**. The semi-natural savanna soil contains 64 (±7 SD) $Mg\ C \cdot ha^{-1}$ in sandy clayey soil, of which 20 $Mg\ C \cdot ha^{-1}$ (31%) was located near the surface (0 - 20 cm), 31 (±3) and 11 $Mg\ C \cdot ha^{-1}$ (35%) in 0 - 100 and 0 - 20 cm depths, respectively. The total C stock in 0 - 1 m depth in groundnut soil was about 40 (±5) and 20 (±2) $Mg\ C \cdot ha^{-1}$ in the SCS and SAS, respectively. The SOC concentration and SOC stocks measured were higher under semi-natural savanna compared to the groundnut fields. T-test revealed that the differences between SOC means in Kaffrine are significant with land use parameter. This parameter shows significant ($P < 0.05$) mean-difference only in the <10 years cultivated soils (**Figure 2(a)**). For texture, the mean C content difference is statistically significant at the bottom of soil profiles (**Figure 2(b)**). The variation in C content is always smaller in SAS soils compared to SCS soils. For the same soil, the correlation coefficient (r) between organic carbon and particle-size were established as follows: carbon-clay (r = 0.62), carbon-silt and clay (r = 0.77).

SOC quality and origin were evaluated using the C/N ratio and $\delta^{13}C$ values. The C/N ratio was smaller (1 - 3 units) in 0 - 20 cm depth for SAS than for SCS (**Table 4**). According to land use, the C/N ratio was not highly different, 0.1 - 0.4 units in 0 - 10 cm depth and 0.8 - 1.3 units in 10 - 20 cm depth. The tendency is reversed at greater depth and is more marked in cultivated soils (20 cm under culture and 50 cm under savanna). **Figure 3** shows the $\delta^{13}C$ of the soils studied. Within the top 20 cm depth, values of $\delta^{13}C$ were between −20% and −25% in the cultivated sandy soil, and between −15% and −20% for all the other soils. Below 20 cm, values of $\delta^{13}C$ approached between −17% and −18%. The isotopic ratio of the near-surface organic C was 6% smaller under groundnut fields than those of other soils, and close to the C3-plant $\delta^{13}C$ values. Under semi-natural savanna, SOC is slightly impoverished in ^{13}C at the bottom of the profile. At the upper levels, this organic matter shows a predominance of graminaceous C4-species, but some ligneous species were described on the site such as *Guiera* and *Combretum*.

4. Discussion

4.1. Carbon Stocks in Senegal's Dryland Zone

Measured biomass represents the short-term source of soil carbon. Analysed total biomass ranged between 3 and 5 $Mg\ dm \cdot ha^{-1}$. The mean C values were between 1.3 and 2.1 $Mg\ C \cdot ha^{-1}$. Despite the fact that these contents do not include the carbon contained in wood that constitutes 8% to 9% of the total above-ground carbon in Delbi and Paniates forests near Kaffrine [20], 34% of total biomass in tropical Africa open forest and more for closed tropical forest [23], these contents are similar to those usually noted (2 and 4 $Mg\ dm \cdot ha^{-1}$) in the degraded savanna of the West African semiarid region [11,24]. This vegetation contributes to 0.42 $Mg\ C/ha/yr^{-1}$ of organic carbon refund into soil. These values are slightly high, considering the fact that in Eastern North American soils, after five years of afforestation, accumulation rates are between −0.85 and +0.58 $Mg\ C \cdot ha^{-1} \cdot yr^{-1}$ [25]. They can be

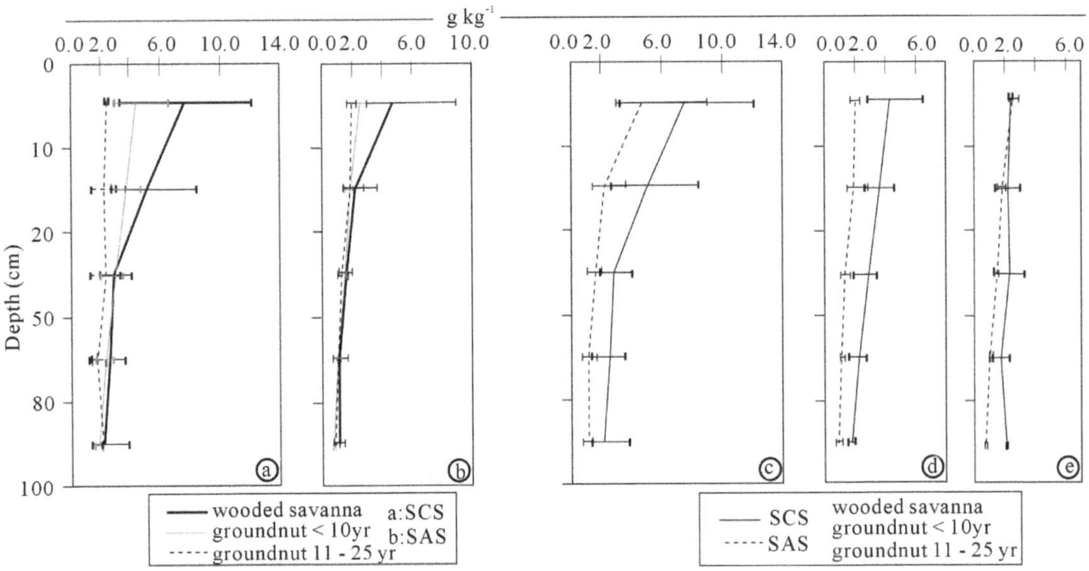

Figure 1. Soil organic C contents and vertical distribution in Sandy soils (SAS) and Sandy clayey soils (SCS), for the wooded savanna and groundnut field. Error bars represent ± one standard deviation.

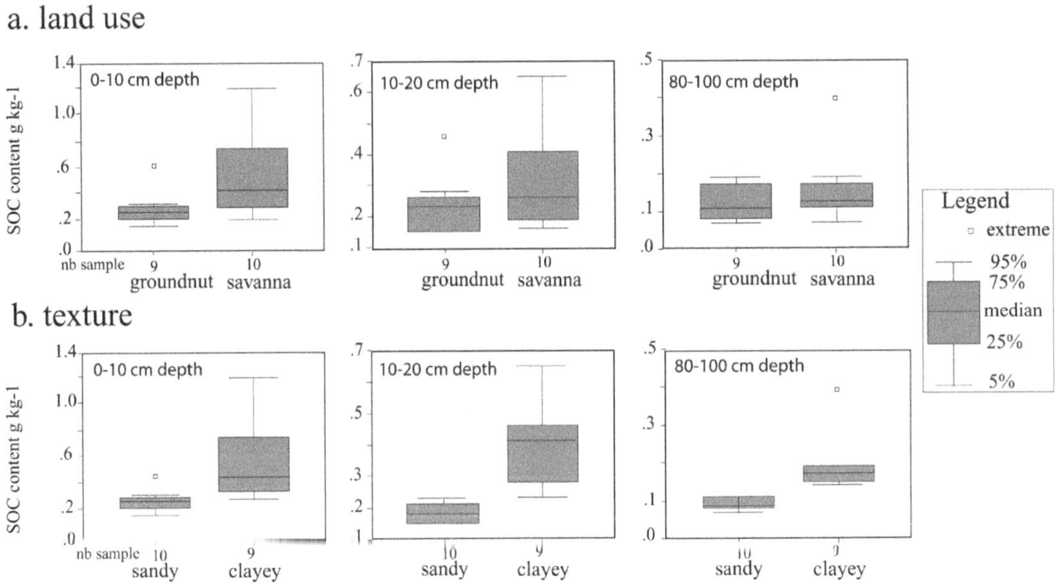

Figure 2. SOC content according to land use (a) and soil texture (b). From near-surface horizon (0 - 10 cm) to greater depth (80 - 100 cm), the SOC content is continuously more differentiated with texture so that intermediate horizons (not shown) presented intermediate situations. There is no significance with land use (savanna and up to 10 yr cultivation).

explained by abundant root-hairs in the West African region. The root biomass in Kaffrine was 2.0 Mg dm·ha^{-1} which was lower than 5 - 10 Mg dm·ha^{-1} and 2.5 - 5 Mg ha^{-1} reported by [11,24].

For a fire-protected and animal-free savanna in Senegal (500 mm mean annual precipitation), the herbaceous biomass was estimated to be 1.6 Mg dm·ha^{-1} [26]. The low total biomass in Kaffrine is probably due to the degraded savanna after decades of human impact. With equal organic carbon input, these soils do not preserve the same organic carbon stocks with cultivation. Thus, the organic carbon input to semi-natural savanna soils is probably also affected by human activities. The average density of SOC published by Batjes [27] in the first 1 m of soil for West Africa was significant, about 42 - 45 Mg C·ha^{-1}. The level of SOC density obtained in Kaffrine is near the previous range (**Table 5**). Elberling *et al.* [14] found 3.2 and 2.7 kg C·m^{-2} in the woodland savanna in Tiel and Dahra, far north of Kaffrine. In southern Senegal, Manlay [22] found 2.78 kg C·m^{-2} in groundnut crops and 5.47 kg C·m^{-2} in >10-years fallows. Our results are then correlated to those of sites situated far north or far

Table 3. Average particle-size analysis of soil in wooded savanna and groundnut field.

	vegetation	Depth (cm)	Clay (Wt-%)	Silt (Wt-%)	Sand (Wt-%)	Bulk density (g kg^{-1})	pH (water)
Sandy soils (SAS)	Wooded savanna	0 - 10	6	12	82	1.47 ± 0.07	4.3
		10 - 20	7	12	81	1.53 ± 0.03	5.1
		20 - 50	11	10	79	1.40 ± 0.05	5.3
		50 - 80	12	10	77	1.44 ± 0.09	5.5
		80 - 100	13	13	74	1.45 ± 0.5	5.8
	Groundnut field	0 - 10	3	6	91		
		10 - 20	5	9	86		
		20 - 50	9	8	83		
		50 - 80	12	8	80		
		80 - 100	15	8	77		
		100 - 105	14	8	78		
Sandy clayey soils (SCS)	Wooded savanna	0 - 10	8	19	73	1.42 ± 0.17	4.4
		10 - 20	15	18	67	1.53 ± 0.07	5.2
		20 - 50	18	16	67	1.46 ± 0.03	5.5
		50 - 80	32	13	55	1.48 ± 0.04	5.7
		80 - 100	35	15	50	1.54 ± 0.05	5.8
	Groundnut field	0 - 10	6	19	76		
		10 - 20	11	17	72		
		20 - 50	13	17	70		
		50 - 80	18	14	68		
		80 - 100	26	12	62		

± Standard deviation.

south of Kaffrine, related to a north-south climatic gradient across the region; indeed, C content decreases with decreasing precipitation towards the north [3,14]. Also, the average SOC density in West Africa is lower than the average found for the whole of Africa (64 - 67 Mg C·ha^{-1}) [27], mainly as a result of the less favourable agro-ecological conditions.

4.2. Land Use Effect on SOC Storage and Distribution

Crop cultivation led to a decrease in soil organic C content. This is noted for C input due to wood uprooting and for SOC. Compared to the semi-natural savanna, SOC storage in groundnut fields is approximately 27% - 35% less for the upper 20 cm depth. This reduction cannot be attributed to soil texture variation because the effect is noted on both SCS and SAS texture. Also, in West Africa, 50% to 70% of the soils are exploited with minimal carbon restitution [11]. Groundnut cropping is considered an extreme case in relation to soil C losses [28], particularly during the first years of cultivation [14]. This is related to the fact that groundnuts are harvested by pulling up the entire plants with roots to save as feed for livestock. The accompanying impoverishment of C would therefore be related to cultivation through increased decomposition of plant residue by soil mixing on the surface [6], decreasing of organic matter input, indicated in this study by biomass estimated.

4.3 Soil Texture Effect on SOC Storage and Distribution

For the Ferric Luvisols in the Kaffrine area, the two types of soil texture are basically not differentiated with the content of clay but with those of silt and silt + clay. Feller and Beare [3] obtained a correlation of carbon-clay to 0.86 in tropical ferruginous soil. These authors have shown from different sites in the tropics (West and Central Africa, Antilles, Brazil, south India) that clay (or clay

Table 4. Soil organic C content, C/N and vertical distribution in each soil and land use (mean ± SD).

Depth (cm)	Land use	C (g kg^{-1})		C/N	
		Sandy soils (SAS) Average	Sandy clayey soils (SCS) Average	SAS	SCS
	Wooded savanna	4.7 ± 1.6	7.6 ± 4.6	12.6	14.1
0 - 10	Groundnut < 10 yr	2.0 ± 0.3	4.2 ± 2.2	12.2	14.2
	Groundnut > 10 yr	2.5 ± 0.4	2.4 ± 1.0	nd	nd
	Wooded savanna	2.2 ± 1.4	5.0 ± 1.5	11.8	14.1
10 - 20	Groundnut <10 yr	1.9 ± 0.9	3.6 ± 1.0	10.5	13.3
	Groundnut >10 yr	1.8 ± 0.3	2.2 ± 0.8	nd	nd
	Wooded savanna	1.6 ± 0.3	2.8 ± 1.3	11.8	12.7
20 - 50	Groundnut < 10 yr	1.4 ± 0.3	2.8 ± 0.6	12.2	10.7
	Groundnut > 10 yr	1.5 ± 0.1	2.3 ± 1.5	nd	nd
	Wooded savanna	1.2 ± 0.3	2.6 ± 1.0	11.2	9.7
50 - 80	Groundnut < 10 yr	1.1 ± 0.2	2.2 ± 0.5	11.4	8.7
	Groundnut > 10 yr	1.1 ± 0.1	1.8 ± 0.6	nd	nd
	Wooded savanna	1.1 ± 0.2	2.2 ± 1.7	9.2	6.9
80 - 100	Groundnut < 10 yr	0.9 ± 0.3	1.8 ± 0.1	9.9	8.5
	Groundnut > 10 yr	0.7 ± 0.1	1.1 ± 0.3	nd	nd

nd: not determined.

Table 5. Soil organic C stocks in 0 - 20 and 0 - 100 cm depth (mean ± SD).

		Groundnut field		Wooded savanna		Loss (%)*
Depth (cm)		Stock (Mg C·ha^{-1})	Standard deviation (%)	Stock (Mg C·ha^{-1})	Standard deviation (%)	
0 - 20	SAS	08 ± 3	41	11 ± 5	45	35
	SCS	13 ± 3	26	20 ± 6	29	27
0 - 100	SAS	20 ± 2	12	31 ± 3	08	37.5
	SCS	40 ± 5	13	64 ± 7	10	35

SAS = Sandy soils, SCS = Sandy clayey soils, *with setting in culture.

+ silt) content is an important determinant of SOM levels in soils. The positive relationship (correlation coefficient) obtained in Kaffrine was 0.62 (and significant, $P < 0.05$). Significantly lower SOC stocks within the upper 20 cm were noted for soils with a low clay content compared to soils with higher clay content. Differentiation is more significant ($P < 0.05$) between the SOC content of SCS and that of SAS in semi-natural savanna and during the first 10 years of cultivation. The cultivation system used in the division of Kaffrine involves surface tillage. Feller and Beare [3] show that tillage provokes eluviation of fine particles, leading to a relative surface enrichment of sand.

On the other hand, under groundnut fields, SAS soils show more or less positive isotopic gradient from top to bottom. Groundnuts are cultivated in alternation with millet (C4-plant) but the cultivation of the latter is very occasional and seems not to reduce the progression of the SOC towards C3 of groundnut. The signature of the organic matter resulting from the savanna plants (C4/C3 mixed-pools) is preserved in the cultivated clayey soils, on the other hand kept developing to the signature of the C3-pools in the first 20 centimetres of the cultivated sandy soils. $\delta^{13}C$ values of –19.8% and –18.0% are attributed to a strong contribution of C3-plants [29]. Significant depletion in $\delta^{13}C$ occurring in near-surface SOC

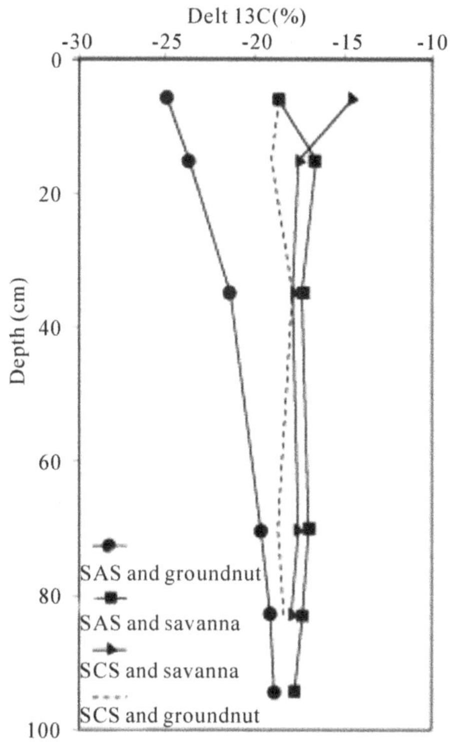

Figure 3. $\delta^{13}C$ values of Sandy soils (SAS) and Sandy clayey soils (SCS) analysed in cultivated (up to 10 yr) and semi-natural savanna soils.

with a shift towards greater C3 inputs has been reported in grazed mixed-grass prairies [29]. The question is whether the change is related to the organic quality or to the transformation by humification. The decomposition and humification of OM is done with only relatively weak isotopic fractionation (0% to +2% approximately, [5]). Six *et al.* [9] showed differences among soil fractions for a soil of up to 5.3% but the authors have also agreed a selective stabilization of residual plant-derived compounds in the fine-silt fraction. We can thus tie the disappearance of the delta ^{13}C values of mixed-C4/C3 savanna plants to the fast incorporation of the organic matter in sandy soils (this remains recognizable below 20 cm of depth) and to its replacement by C3-groundnuts plants put in cultivation. Since the installation of the groundnut area "bassin arachidier" 40 years ago (year 1965), the near-surface $\delta^{13}C$ value has increased by 6% towards the negative values of C3. That was not the case of the sandy clayey soils in which clay protects the organic matter. The fact that $\delta^{13}C$ value converges with increasing depth indicates the actual soil volume responding to almost 40 years of cultivation. The role of clay fraction in SOC protection against decomposition has been outlined by many authors (e.g., [3,30]). For the groundnut field in Kaffrine, delta 13C ($\delta^{13}C$) values show that the signature of organic matter is more preserved in the clayey soils.

5. Conclusions

The results reveal that, in addition to management associated factors such as overgrazing and intensive farming, the impoverishment of carbon in West African soils may be attributed to soil cultivation and loss of fine texture by tillage. Average annual C refund is 0.1 - 0.2 Mg C·ha^{-1}, which is the usually proposed value for West Africa soils. The effect of land use is limited to the upper 20 cm depth while the texture effect is found all along the soil profile.

Texture influences the SOC stocks. In turn, cultivation influences the texture by deepening the particulate component; near-surface level becomes more particulate and advisable for organic matter mineralization. The SOC content was similar for all soils at 20 - 100 cm depth. Compared to the semi-natural savanna, SOC storage in groundnut fields is approximately 27% - 35% less for the upper 20 cm depth. However, it is very less in SAS compared to SCS. Also, SOC evolves from C4 toward groundnut C3 pool in the 0 - 20 cm depth of SAS.

For better C sequestration, practices should focus on the improvement of the fine textures stability and on the regeneration of the woody zones. However, in Senegal, extension of the woody zones or conversion into rangeland conflicts with the requirements for land for cultivation for ever-growing population. The most important factors to protect organic carbon are either natural conservative vegetation or the more clayey soils or both. Awareness of these variable soil parameters and farm residue restitution into the soil is important in order to enhance soil organic carbon sequestration, agricultural production without soil degradation and reduce emission of greenhouse gases.

6. Acknowledgements

The authors owe thanks to CSE staff for the assistance in the field: Dr. Magatte Bâ and M. Samba Ndao; to Isotope Laboratory of Biology Department, Augustana College, Soux Falls, South Dakota for $^{13}C/^{12}C$ analysis, and to Dr. Rokhaya Samba of "Direction des Mines et de la Géologie"-Senegal. Thanks to Dr TABI Fritz OBEN of FASA-Dschang for his comments.

REFERENCES

[1] R. A. Houghton, "The Contemporary Carbon Cycle," In: J. I. Drever, H. D. Holland and K. K. Turekian, Eds., *Treatise on Geochemistry*, Vol. 8. *Biogeochemistry*, Elsevier, Oxford, 2003, pp. 473-512.

[2] Y. Wang and Y. P. Hiesh, "Uncertainties and Novel Prospects in the Study of the Soil Carbon Dynamics," *Chemosphere*, Vol. 49, No. 8, 2002, pp. 791-804. http://dx.doi.org/10.1016/S0045-6535(02)00381-8

[3] C. Feller and M. H. Beare, "Physical Control of Soil Organic Matter Dynamics in the Tropics," *Geoderma*, Vol.

79, No. 1-4, 1997, pp. 69-116. http://dx.doi.org/10.1016/S0016-7061(97)00039-6

[4] H. Chen, S. Marhan, N. Billen and K. Stahr, "Soil Organic-Carbon and Total Nitrogen Stocks as Affected by Different Land Uses in Baden-Württemberg (Southwest Germany)," *Journal of Plant Nutrition and Soil Science*, Vol. 172, No. 1, 2009, pp. 32-42. http://dx.doi.org/10.1002/jpln.200700116

[5] T. W. Boutton, S. R. Archer, A. J. Midwood, S. F. Zitzer and R. Bol, "13C Values of Soil Organic Carbon and Their Use in Documenting Vegetation Change in a Subtropical Savannah Ecosystem," *Geoderma*, Vol. 82, No. 1-3, 1998, pp. 5-41. http://dx.doi.org/10.1016/S0016-7061(97)00095-5

[6] J. Balesdent, E. Besnard, D. Arrouays and C. Chenu, "The Dynamics of Carbon in Particle-size Fractions of Soil in a Forest-cultivation Sequence," *Plant Soil*, Vol. 201, No. 1, 1998, pp. 49-57. http://dx.doi.org/10.1023/A:1004337314970

[7] S. C. Clifford, I. M. Stronach, A. D. Mohamed, S. N. Azam-Ali and N. M. J. Crout, "The Effects of Elevated Atmospheric Carbon Dioxide and Water Stress on Light Interception, Dry Matter Production and Yield in Stands of Groundnut (*Arachis hypogaea* L.)," *Journal of Experimental Botany*, Vol. 44, No. 12, 1993, pp. 1763-1770. http://dx.doi.org/10.1093/jxb/44.12.1763

[8] J. D. Tenhunen, L. C. Tenhunen, H. Ziegler, W. Stichler and O. L. Lange, "Variation in Carbon Isotope Ratios of Euphorbia Species from Different Habitats of Teneriffe in Spring," *Flora*, Vol. 173, No. 5-6, 1983, pp. 363-370.

[9] J. Six, R. Merckx, K. Kimpe, K. Paustian and E. T. Elliott, "A Re-Evaluation of the Enriched Labile Soil Organic Matter Fraction," *European Journal of Soil Science*, Vol. 51, No. 2, 2000, pp. 283-293. http://dx.doi.org/10.1046/j.1365-2389.2000.00304.x

[10] P. A. Sanchez and T. J. Logan, "Myths and Science about the Chemistry and Fertility of Soils in the Tropics," In: R. Lal and P. A. Sanchez, Eds., *Myths and Science of Soils of the Tropics*, Soil Science Society of America, Madison, 1992, pp. 35-46.

[11] H. Tiessen, C. Feller, E. V. S. B. Sampaio and P. Garin, "Carbon Sequestration and Turnover in Semiarid Savannahs and Dry Forest," *Climatic Change*, Vol. 40, No. 1, 1998, pp. 105-117. http://dx.doi.org/10.1023/A:1005342932178

[12] L. L. Tieszen, G. G. Tappan and A. Touré, "Sequestration of Carbon in Soil Organic Matter in Senegal: An Overview," *Journal of Arid Environments*, Vol. 59, No. 3, 2004, pp. 409-425. http://dx.doi.org/10.1016/j.jaridenv.2004.04.002

[13] S. A. Alam, M. Starr and B. J. F. Clark, "Tree Biomass and Soil Organic Carbon Densities across the Sudanese Woodland savannah: A Regional Carbon Sequestration Study," *Journal of Arid Environments*, Vol. 89, 2013, pp. 67-76. http://dx.doi.org/10.1016/j.jaridenv.2012.10.002

[14] B. Elberling, A. Touré and K. Rasmussen, "Changes in Soil Organic Matter Quantity and Quality Following Groundnut-Millet Cropping at three Locations in Semi-Arid Senegal, West Africa," *Agriculture, Ecosystems & Environment*, Vol. 96, No. 1-3, 2003, pp. 37-47. http://dx.doi.org/10.1016/S0167-8809(03)00010-0

[15] R. Fauck, "Contribution à l'Etude des Sols des Régions Tropicales-Les Sols Rouges sur Sables et sur Grès d'Afrique Occidentale," Ph.D. Thesis, Sciences Naturelles, Université de Strasbourg. 1971, 377p.

[16] P. Cattan, P. Letourmy, B. Zagre, A. Minougou and E. Compaoré, "Rendement de l'Arachide et du Sorgho en Rotation sous Différents Itinéraires Techniques au Burkina Faso," *Cah. études recher. Franco./Agricultures*, Vol. 10, No. 3, 2001, pp. 159-172.

[17] FAO, "World Reference Base for Soil Resources," *World Soil Resources Reports*, Vol. 84, Food and Agriculture Organization, Rome, 1998, 98p.

[18] A. Toure, "Contribution à l'Etude de l'Evolution des Réservoirs de Carbone en Zone Nord-Soudanienne au Sénégal," Ph.D. Thesis, no. 2585, EPF Lausanne, Switzerland, 2002

[19] M. I. Cissé, "The Browse Production of Some Trees of the Sahel: Relationships between Maximum Foliage Biomass and Various Physical Parameters," In: Le Houérou, Ed., *Browse in Africa, The Current State of Knowledge*, International Symposium on Browse in Africa, Addis Ababa, 8-12 April 1980, pp. 205-210.

[20] A. Touré, K. Rasmussen, O. Diallo and A. Diouf, "Actual and Potential Carbon Stocks in the North-Sudanian Zone. Case Study: The Forest of Delbi and Paniates in Senegal," *Danish Journal of Geography*, Vol. 103, No. 1, 2003, pp. 63-70.

[21] E. Roose, "L'Effet du Couvert Végétal," In: Introduction à la Gestion Conservatoire de l'Eau, de la Biomasse et de la Fertilité des Sols (GCES), Service des Sols-Ressources, Aménagement et Conservation, Division de la Mise en Valeur des Terres et des Eaux," *Bulletin Pédologique FAO*, Vol. 70, 1994, 420p.

[22] R. Manlay, "Organic Matter Dynamics in Mixed-Farming Systems of the West African Savannah: A Village Case Study from South Senegal," Ph.D. Thesis, IRD Centre, Montpellier, 2000, 192p.

[23] S. Brown and A. E. Lugo, "Biomass of Tropical Forests: A New Estimate based on Forest Volumes," *Science*, Vol. 223, No. 4642, 1984, pp. 1290-1293. http://dx.doi.org/10.1126/science.223.4642.1290

[24] J. César and Z. Coulibaly, "Conséquence de l'Accroissement Démographique sur la Qualité de la Jachère dans le Nord de la Côte d'Ivoire," In: C. Floret and G. Serpantié, Eds., *La Jachère en Afrique de l'Ouest*, Atelier International, Montpellier, 2-5 December 1991.

[25] E. A. Paul, S. J. Morris, J. Six, K. Paustian and E. G. Gregorich, "Interpretation of Soil Carbon and Nitrogen Dynamics in Agricultural and Afforested Soils," *Soil Science Society of America Journal*, Vol. 67, No. 5, 2003, pp. 1620-1628. http://dx.doi.org/10.2136/sssaj2003.1620

[26] F. Bernhart-Reverst, "Le Recyclage des Eléments Minéraux par la Strate Herbacée dans un Peuplement Naturel à Acacia et dans une Plantation d'Eucalyptus au Sénégal," *Acta Oecologica Oecol. Gener*, Vol. 7, 1986, pp. 353-364.

[27] N. H. Batjes, "Options for Increasing Carbon Sequestra-

tion in West Africa Soils: an Exploration Study with Special Focus on Senegal," *Land Degradation & Development*, Vol. 12, No. 2, 2001, pp. 131-142. http://dx.doi.org/10.1002/ldr.444

[28] R. Lal and J. M. Kimble, "Tropical Ecosystems and the Global C Cycle," In: R. Lal, J. M. Kimble and B. A. Stewart, Eds., *Global Climate Change and Tropical Ecosystems, Advances in Soil Sciences*, CRC Press, New York, 2000, pp. 3-32.

[29] D. C. Henderson, B. H. Ellert and M. A. Naeth, "Utility of ^{13}C for Ecosystem Carbon Turnover Estimation in Grazed Mixed Grass Prairie," *Geoderma*, Vol. 119, No. 3-4, 2004, pp. 219-231. http://dx.doi.org/10.1016/j.geoderma.2003.08.001

[30] J. Hassink, L. A. Bouwman, K. B. Zwart, J. Bloem and L. Brussaard, "Relationships between Soil Texture, Physical Protection of Organic Matter, Soil Biota, and C and N Mineralization in Grassland Soil," *Geoderma*, Vol. 57, No. 1-2, 1993, pp. 105-128. http://dx.doi.org/10.1016/0016-7061(93)90150-J

8

Physicochemical and Biochemical Reclamation of Soil through Secondary Succession

Kamala Haripal, Sunanda Sahoo

Ecology Section, School of Life Sciences, Sambalpur University, Burla, India.
Email: drsunsnda_sahoo@yahoo.com

ABSTRACT

Conversion of forest to agricultural fields has become a common practice in India. Very often these fields have been abandoned due to lack of sustainable production. In course of time these fallow lands undergo natural secondary succession. The present study was carried out to find out the restoration of soil physicochemical and biochemical properties in a chronosequence of 2 yr, 4 yr, 6 yr, 11 yr, and 15 yr fallow lands. Soil enzyme activities play key roles in the biochemical functioning of soils, including soil organic matter formation and degradation, nutrient cycling, and reflect the change in soil management and land use. There was gradual improvement in the physical condition and nutrient status along with increase in soil amylase, cellulase, dehydrogenase, phophatase, and urease activity in the present study with the progress of fallow age which indicates the importance of natural secondary succession in soil restoration. However the PCA analysis indicated that natural vegetational succession could reclaim the soil quality and promote ecosystem restoration but it required a long time under the present local climatic condition.

Keywords: Sustainable Production; Secondary Succession; Soil Restoration; Land Use

1. Introduction

India has 329 million ha of total land area of which 43% is under cropping and 23% is under forest [1]. From the beginning human has exploited the natural resources for agricultural activities. As a result most of the forests have been converted into agricultural lands. Over recent years there have been intensive agricultural practice basically aimed to enhance the productivity to meet the food demands of huge population throughout the world. In spite of increased population, the productive potential has been decreased in many areas of tropical countries. Intensive cultivation leads to reduced soil fertility and increased soil erosion in many areas of tropics [2]. Specific detrimental effects on biological and biochemical characters of soil quality have also been noted due to changes in microclimate at the soil surface by tillage and on the rate and quality of organic matter input to the soil. Therefore, most of the agro-ecosystems are often abandoned due to unsustainable agricultural production. Maintenance of sustainable agricultural production has become a matter of great concern for the scientists all over the world. In many instances, the disturbed areas undergo natural recovery of vegetation within 5 to 20 years depending on population pressure and land availability [3].

Soil enzymes (microbial exoenzyme) are recognized as sensitive indicators of soil health and quality [4] due to the rapid response to changes in soil management. In particular, enzyme activities are especially significant in soil quality assessments because of their major contribution to degrading organic matter [5]. In fact they have been related to soil physico-chemical characters, microbial community structure and disturbance [6]. It has been reported that any change in soil management and land use is reflected in the soil enzyme activities, and that they can anticipate changes in soil quality before they are detected by other soil analyses [7]. The studies of soils from different regions indicated that enzyme activities are sensitive to soil changes due to tillage [8], cropping systems [9], and land use [10]. Most of the studies on soil enzyme activities in different land use systems have focused on temperate regions. Few literatures are available regarding the enzyme activities due to land use changes in tropics [11,12].

The present study of enzyme activity in soil along a chronosequence of vegetation regrowth in the fallow lands can give insight into the role of soil enzymes in restoring soil fertility during secondary succession. Currently, no information is available on microbial biomass and enzyme activity that affected by conversion of forest into agricultural land, and then into fallow land in India except the work done by Maithani *et al.* [3] on microbial biomass in 7, 13, and 16 year regrowth of a disturbed subtropical humid forest in Meghalaya, India, and by Ralte *et al.* [13] on microbial biomass and activity in relation to shifting cultivation and horticultural practices in subtropical evergreen forest ecosystem of North-East India. Thus the present investigation was aimed to assess some key enzyme activities along with some of the selected physico-chemical characters involved in soil restoration in a chronosequence of abandoned rice fields in Western Odisha, India.

2. Materials and Methods

2.1. Study Sites and Climate

The present study was confined to the revenue district Sambalpur, located in the Western part of Odisha in India. Prior to 1950s the study sites were dense forest forming a part of Barapahar forest range. After the establishment of multipurpose hydroelectric Hirakud Dam project and subsequent industrialization, urbanization and increase in population pressure, the district forest coverage has been reduced to 30% [14]. Most of the forest lands were cleared and used for production of agricultural crops (rice, maize, beans and sugarcane). The climate is tropical monsoonal. The annual average rainfall during the study period (August, 2010-July 2011) was 1597 mm out of which about 80% fell during rainy season (July-October). The mean air temperature varied from 5˚C (during December) to 44˚C (during May). The soils of this district belong to mixed red and black soil, Red sandy soil, mixed red and yellow lateritic soil [15]. The climatological data during the study period are shown in **Figure 1**. In many places of the study area the rice fields were derived by clearing natural forest which had been subjected to abandonment by farmers due to lack satisfactory agricultural production. All the sampling sites were located on the same topographic situation and with the area of nearly 900 sq meters except 11 yr fallow land which covered 12000 sq meters. The age of the fallow period was ascertained by asking the land owners. By interviewing the elderly persons of the locality it was known that the peoples of these localities were practicing rain fed paddy cultivation since 1975. The selected study sites were paddy fields abandoned since 1995 (15 yr·F), 1999 (11 yr·F), 2004 (6 yr·F), 2006 (4 yr·F), and since 2008 (2 yr·F).

The 4-year-old field was adjacent to the 6-year-old field, while the 11-year-old field was adjacent to the 15-year-old field. All these sites were located at the bottom of hill, Chandili Dunguri, 1/2 Km towards south west from Sambalpur University campus at Burla (20˚43'N-20˚11'N and 82˚39'E-85˚13'E longitude, latitude at 263 m above mean sea level) in Sambalpur district of Odisha, India.

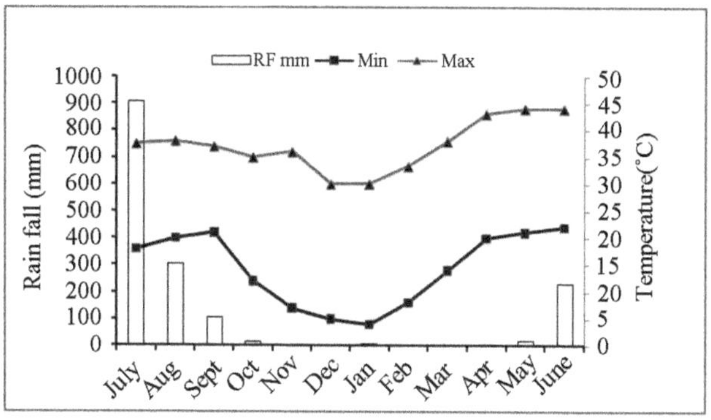

Figure 1. Climate pattern of study sites.

2.2. Soil Physicochemical Analysis

The soils were sampled from the experimental plots bimonthly during August 2010- June 2011. The soil samples were taken by using a cylindrical soil sampler having a diameter of 20 cm. and five random samples were taken from 0 - 10 cm, 10 - 20 cm, and 20 - 30 cm depths. The soil samples were packed in plastic zipper bag and brought into the laboratory and stored at 4˚C before analysis. The enzyme analyses were made within four weeks after sampling because storage beyond two weeks can cause decline in enzyme activities [16]. The soil samples were air dried, gently crushed and sieved through 2 mm sieve and then used for different physicochemical analysis.

The laboratory analyses were conducted for bulk density and water holding capacity following the method prescribed in TSBF hand book [17], soil pH by pH meter using 1:5 soil water suspension, the organic carbon content by Walkely and Black's titration method [18], total nitrogen (TN) by Kjeldhal method [19], the total phosphorus by Bray and Kurtz [20] .Soil microbial biomass carbon was calculated by chloroform fumigation-extraction method [21].

2.3. Soil Enzyme Activity

Soil amylase & cellulase activity was determined following Mishra *et al.* [22]. 3 g of soil sample was incubated with 0.2 ml tolune, 6 ml substrate solution (starch for amylase and carboxymethyl cellulose for cellulase) and 6 ml Sorenson's buffer (pH = 5.9) for 24 hr at 34˚C. After incubation the reducing sugar test was performed by adding 2 ml of 3, 5 dinitrosalicylic acid to 1 ml of supernatant of incubated mixture in a test tube and allowed to stand for 5 minutes for colour development.

The enzyme activity was quantified by spectrophotometer at 540 nm and expressed in µg glucose g^{-1} dry soil hr^{-1}.

Soil dehydrogenase was measured according to Casida *et al.* [23]. 5 g of soil was incubated with 2 ml of 1% soluble 2, 3, 5-triphenyl tetrazolium chloride (TTC) in a sealed screw cap test tube at 32˚C for 24 hr. The triphenyl formazon (TPF) formed in the incubated sample was extracted by methanol. The extrantant triphenyl formazon (TPF) was determined by spectrophotometer at 485 nm, using methanol as control.

Acid phosphatase was estimated following the method of Tabatabai and Bremnar [24] using paranitrophenyl phosphate (p-NPP) as a substrate. A mixture of 1 g fresh soil (<2 mm), 4 ml of modified universal buffer (pH 6.5) and 1 ml of 100 mM p-NPP were incubated in a sealed 100 ml Erlenmeyer flask at 30˚C for 30 min. The sample was adjusted to 100 ml with deionised water. After incubation the para- nitrophenol (p-NP) was filtered through Whatman-42 and quantified by spectrophotometer at 400 nm by taking p-NP as control.

Urease activity was determined as described by Dash *et al.* [25]. 0.2 ml toluene was added to 5 g soil and 9 ml of Tris-HCl buffer (pH = 9, 0.2 M) followed by 1 ml of 0.2 M urea (substrate solution) to the sample. The samples were subsequently incubated at 37˚C for 2 hrs. After incubation the volume was made up to the mark (50 ml) by adding KCl-$AgSO_4$. The suspension was centrifuged. To 1 ml of supernatant, 1 ml of phenol and 1 ml of alkaline hypo chlorite solution were added. After 5 min, the released ammonium was measured spectrophotometrically at 625 nm.

2.4. Statistical Analysis

The data were statically analyzed using three way analysis of variance (ANOVA) with SPSS 10 statistical software. Pearson's correlation analysis was performed to determine the significant level of correlation. The PCA was done using SX software to discriminate different sites on the basis of the principal components Z_1 and Z_2.

3. Results

3.1. Soil Physico-Chemical Properties

Soil physico-chemical properties of the present study sites have been presented in **Table 1**. Soil moisture content increased gradually along the chronosequence of abandoned field (**Table 1**). ANOVA revealed a significant impact of fallow period on soil moisture content.

Table 1. Soil physico-chemical properties of different abandoned fields at different depths.

Parameters	Depth (cm)	2 yr	4 yr	6 yr	11 yr	15 yr
	0 - 10	5.17 ± 2.39	6.55 ± 3.05	7.96 ± 3.43	9.44 ± 4.32	14 ± 4.91
Moisture (%)	10 - 20	4.70 ± 1.64	5.9 ± 2.34	7.24 ± 2.63	8.76 ± 3.62	12.12 ± 3.95
	20 - 30	3.57 ± 1.18	4.76 ± 2.02	5.94 ± 2.56	7.84 ± 3.15	10.39 ± 3.52
	0 - 10	1.53 ± 0.065	1.48 ± 0.061	1.42 ± 0.06	1.35 ± 0.067	1.25 ± 0.08
Bulk density $g·cm^{-3}$	10 - 20	1.57 ± 0.051	1.52 ± 0.054	1.48 ± 0.063	1.41 ± 0.079	1.33 ± 0.074
	20 - 30	1.60 ± 0.054	1.56 ± 0.049	1.52 ± 0.054	1.48 ± 0.048	1.39 ± 0.098
	0 - 10	24.84 ± 4.03	27.10 ± 3.92	30.16 ± 4.19	32.47 ± 4.50	36.72 ± 4.52
WHC (%)	10 - 20	21.34 ± 2.87	24.00 ± 3.00	27.09 ± 3.25	29.52 ± 3.75	34.61 ± 4.03
	20 - 30	19.72 ± 2.54	22.32 ± 2.16	23.68 ± 2.83	28.16 ± 3.23	33.17 ± 3.72
	0 - 10	5.56 ± 0.08	5.75 ± 0.12	6.00 ± 0.27	6.45 ± 0.11	6.50 ± 0.12
pH	10 - 20	5.35 ± 0.39	5.80 ± 0.07	5.50 ± 0.14	5.64 ± 0.04	5.74 ± 0.03
	20 - 30	5.40 ± 0.18	5.45 ± 0.15	6.07 ± 0.16	6.17 ± 0.12	6.25 ± 0.14
	0 - 10	6.17 ± 1.55	6.34 ± 1.61	8.7 ± 2.11	11.49 ± 2.82	13.43 ± 2.9
SOC $mg·g^{-1}$	10 - 20	5.42 ± 1.23	5.02 ± 1.3	7.61 ± 1.68	9.78 ± 2.55	11.25 ± 2.89
	20 - 30	4.64 ± 0.96	4.24 ± 1.43	6.02 ± 1.81	8.18 ± 2.37	9.57 ± 2.62
	0 - 10	0.66 ± 0.17	0.74 ± 0.18	0.91 ± 0.17	1.09 ± 0.23	1.19 ± 0.22
TN $mg·g^{-1}$	10 - 20	0.06 ± 0.13	0.64 ± 0.12	0.86 ± 0.13	0.98 ± 0.22	1.07 ± 0.23
	20 - 30	0.53 ± 0.1	0.57 ± 0.13	0.72 ± 0.17	0.89 ± 0.2	0.93 ± 0.18
	0 - 10	105.5 ± 32.12	155.36 ± 65.47	274.15 ± 95.26	398.97 ± 114.55	513.78 ± 109.57
MBC $µg·g^{-1}$	10 - 20	81.04 ± 24.49	115.40 ± 42.45	220.35 ± 82.01	290.47 ± 85.44	364.24 ± 92.64
	20 - 30	58.93 ± 17.91	84.97 ± 31.11	158 ± 71	214.39 ± 71.75	280.03 ± 100.19

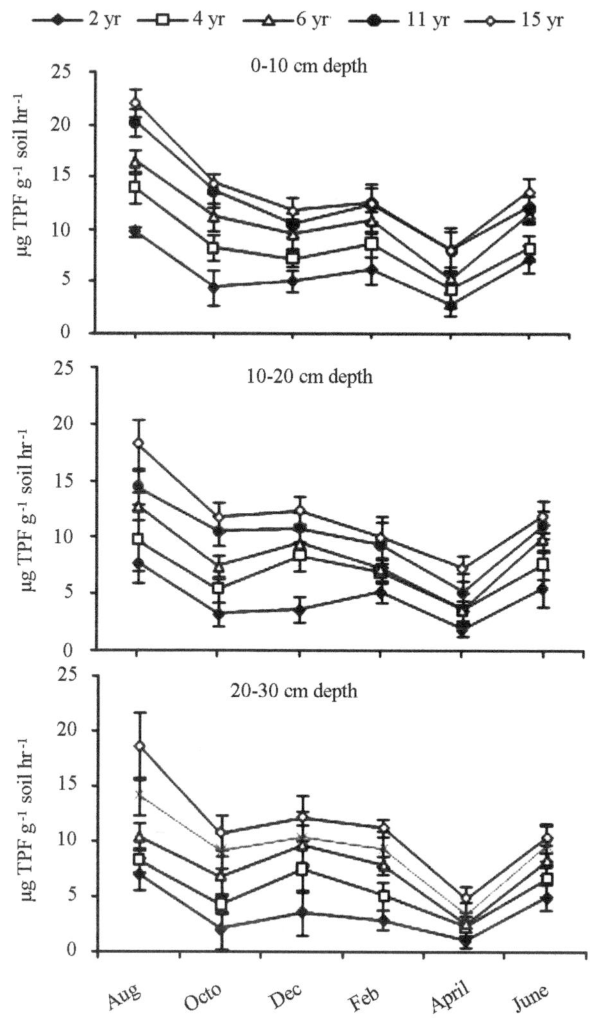

Figure 2. Amylase activity in soil of different abandoned fields during different months and at different depths.

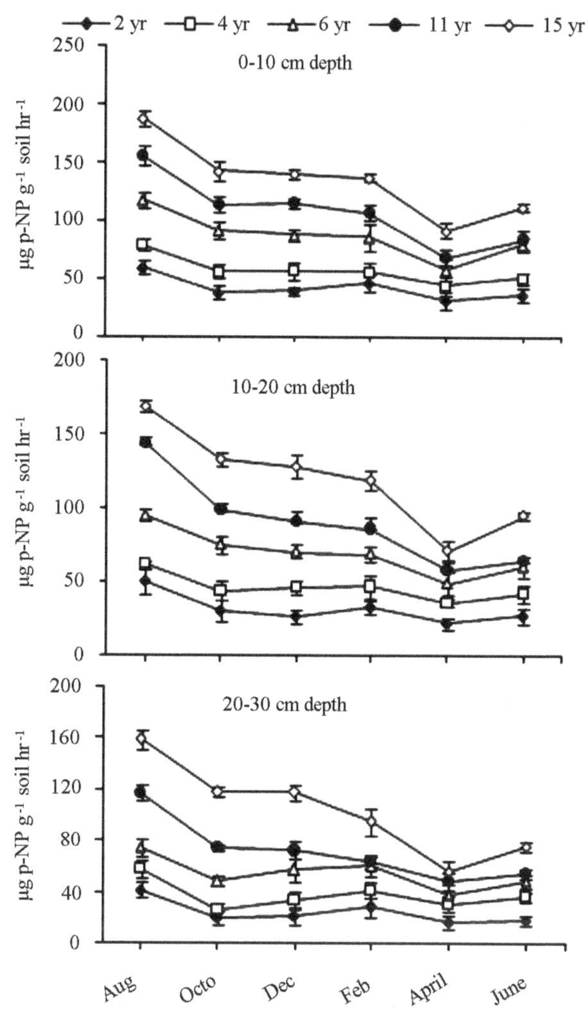

Figure 3. Cellulase activity in soil of different abandoned fields during different months and at different depths.

Decreasing level of soil moisture was found with increase in depth irrespective of fallow period. The bulk density ranged from 1.25 g·cm^{-3} in the surface soil layer (0 - 10 cm) of 15 yr·F land to 1.6 g·cm^{-3} in the subsurface layer (20 - 30 cm) of 2 yr·F land. Lowest water holding capacity (WHC) was recorded at 20 - 30 cm depth of 2 yr abandoned field (19.72%) and highest WHC was recorded in 0 - 10 cm depth of 15 yr abandoned field (36.7%). ANOVA revealed significant difference of WHC between different fallow lands and between depths at $p < 0.001$.

The soil pH was acidic in nature and varied from 5.35 to 6.28. The pH value increased over time since abandonment and highest value was seen in the upper soil layer (0 - 10 cm) of 15 yr·F land. The soil organic matter increased significantly with increasing year of abandonment being highest in 0 - 10 cm depth of 15 yr abandoned field (13.43 mg·g^{-1}) and lowest in 20 - 30 cm depth of 2 yr·F lands (6.17 mg·g^{-1}). Similar trend was observed for total nitrogen in the soil. The Soil microbial biomass carbon ranged from 104.50 to 513 µg·g^{-1} in top soil (0 - 10 cm), 81.04 to 364 µg·g^{-1} of soil in 20 - 30 cm depth and 58.93 to 280 µg·g^{-1} in 20 - 30 cm soil depth.

3.2. Soil Enzyme Activity

Figures 2-6 show amylase, cellulase, dehydrogenase, phosphatase and urease activities respectively in relation to sites, depth and season. Highest activities of all the enzymes were observed in top soil (0 - 10 cm), and were found to be decreasing with increasing depths. Seasonal variations in the enzyme activities in different sites indicated a peak during August (wet season) and lowest enzyme activities was observed in dry season *i.e.* during April. All the enzyme activities increased with the increase in years of abandonment. ANOVA revealed significant difference between sites, depths, and between different seasons at $p < 0.001$. Pearson's correlation analysis indicated significant correlation between en-

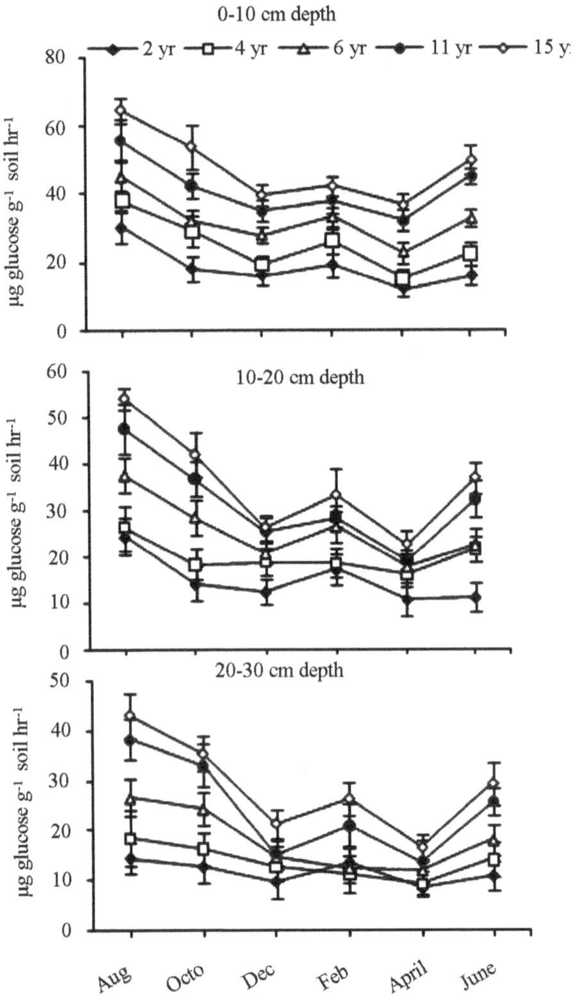

Figure 4. Dehydrogenase activity in soil of different abandoned fields during different months and at different depths.

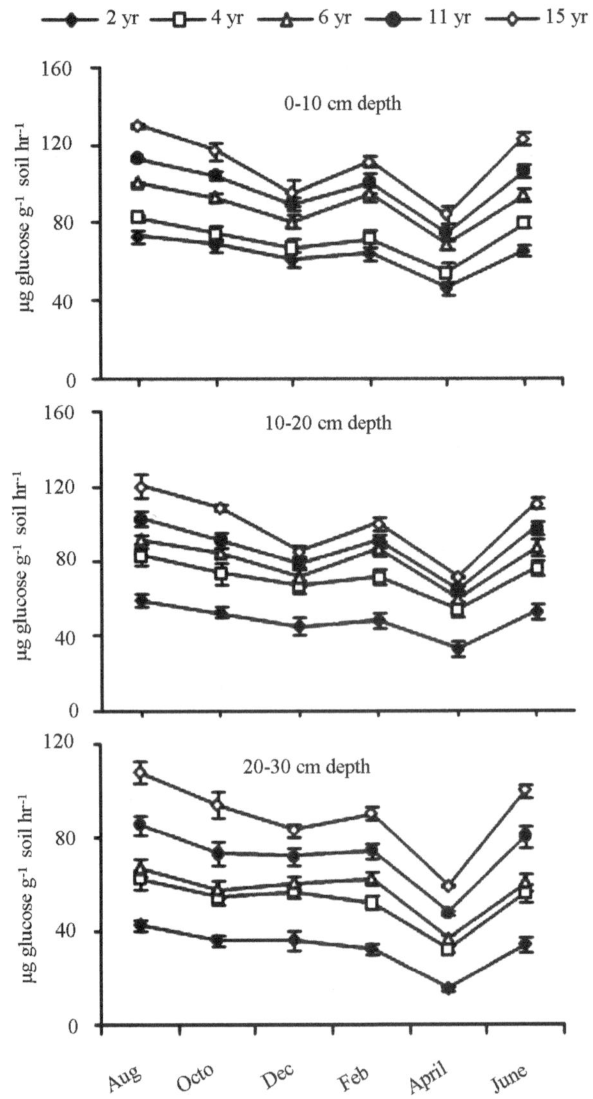

Figure 5. Phosphatase activity in soil of different abandoned fields during different months and at different depths.

zyme activities and soil physico-chemical parameters (**Table 2**).

4. Discussion

4.1. Soil Physico-Chemical Properties

In the present study, soil was slightly acidic ranging from 5.58 to 6.58. Leaching process tends to acidify soils and partially offset by plant growth [26]. Gradual establishment of plants with increasing year of abandonment is supposed to prevent much of the leaching in WHC and decrease the bulk density. As the fallow period gets extended, the inputs of organic residues from colonizing plants increased the carbon and nitrogen contents of soil. During the natural restoration in 3, 7, 10 year abandoned paddy fields an increase in pools of organic matter, total nitrogen, exchangeable K, Ca, Mg and pH was observed with increase of field age [27]. This supports the present findings. Microbial biomass reflects the degree of immobilization of C and N in soil. Gradual increase in microbial biomass carbon (MBC) was recorded in the present study with a maximum amount in the top soil of 15 yr·F land (513.78 $\mu g \cdot g^{-1}$).

This may be attributed to greater input of plant detritus that is ultimately incorporated into the soil and improves the nutrient pool thereby increases the MBC [28]. The MBC dynamic was significantly correlated with soil organic carbon content during the secondary succession as reported by Arunachalam and Pandey [29]. The present study supports the above findings. This result indicated that organic matter played a pivotal role in the building up and development of soil microbial biomass [30].

4.2. Soil Enzyme Activity

Soils enzymes are major indicators of microbial activity,

Table 2. Pearson's Correlation coefficients between soil properties.

	SM	WHC	BD	pH	OC	MBC	TN	AML	CEL	DEH	PHO	URE
SM	1											
WHC	0.45**	1										
BD	−0.34**	−0.71**	1									
pH	0.41**	0.54**	−0.51**	1								
OC	0.85**	0.56**	−0.51**	0.51**	1							
MBC	0.85**	0.6**	−0.54**	0.61**	0.90**	1						
TN	0.86**	0.55**	−0.47**	0.49**	0.97**	0.89**	1					
AML	0.86**	0.55**	−0.47**	0.51**	0.86**	0.88**	0.85**	1				
CEL	0.83**	0.59**	−0.6**	0.48**	0.81**	0.83**	0.81**	0.89**	1			
DEH	0.82**	0.54**	−0.44**	0.4**	0.81**	0.76**	0.81**	0.86**	0.86**	1		
PHO	0.85**	0.64**	−0.59**	0.53**	0.92**	0.88**	0.91**	0.87**	0.86**	0.87**	1	
URE	0.89**	0.57**	−0.52**	0.49**	0.85**	0.88**	0.84**	0.92**	0.93**	0.87**	0.89**	1

**Significant at $p < 0.01$; ^{SM}Soil moisture, ^{WHC} Water holding capacity, ^{BD}Bulk density, ^{OC}Organic carbon, ^{TN}Total nitrogen, ^{AML}Amylase activity, ^{CEL}Cellulase activity, ^{DEH}Dehydrogenase activity, ^{PHO}Phosphatase activity, ^{URE}Urease activity.

and their activities always depend upon soil types, vegetation cover, microbial biomass, and microbial diversity during vegetation succession [31]. Dehydrogenase is an intracellular enzyme that exists only in viable microbial cells and it is considered as an index of microbial activity [32]. In the present study the enzyme activities showed peak value in rainy season (August) and lowest in summer (April). Some hydrolases have a tendency to increase in the rainy season [33]. During summer, due to moisture stress, the microbes are in dormant state and their activities become low, but the entry of rainy season caused spurt in flora and microbial population, and thus increased the enzyme activities. Prolonged precipitation along with enhanced plant growth and rhizoid position result in increased extracellular activities [34,35]. The activity of dehydrogenase, phosphatase and urease was found to decrease in summer [36]. Thus the present finding is in agreement with the above observations.

There was significant increase in extracellular enzymes like amylase, cellulase, phosphatase, urease, and intracellular enzyme *i.e.* dehydrogenase with increasing year of fallow period and significant decrease with increase in soil depth. Significantly positive correlation was found between all the enzyme activities with organic carbon, total nitrogen and microbial biomass carbon (**Table 2**). Increasing age of the land with respect to time caused gradual addition of organic matter to the soil through decomposition of shoot and root biomass of different vegetation developed on the surface layer. Higher organic matter content provides higher substrate which acts as energy source for microorganisms and supports high amount of microorganism, hence higher enzyme activities [30]. Later succession exhibited higher enzyme activity than the early period and upper soil layer showed higher enzyme activity than lower layer of soil. The decrease in these enzyme activities with depth might be attributed to the diminution of biological activity down the soil profile [37].

The data on physico-chemical parameters and enzyme activities of present study and the data of natural forest, natural grassland and crop field obtained in the same topography [38] were subjected to principal component analysis. **Figure 7** illustrates the discrimination of different sites on the basis of Z_1 and Z_2 whose cumulative percent variance was 85.9%. The PC separated the 2 yr, 4yr, 6 yr, 11 yr, 15 yr fallow far away from grassland and crop fields. However the 15 yr F is quite nearer to natural forest which indicates that the chronosequence of abandoned fields showed a general trend of reclamation in terms of some of the soil Physico-chemical and biochemical parameters but still required more time to achieve the soil status of natural forest, crop fields, and grasslands. Self regeneration of vegetation on a degraded land in tropics has relatively better reclamation potential

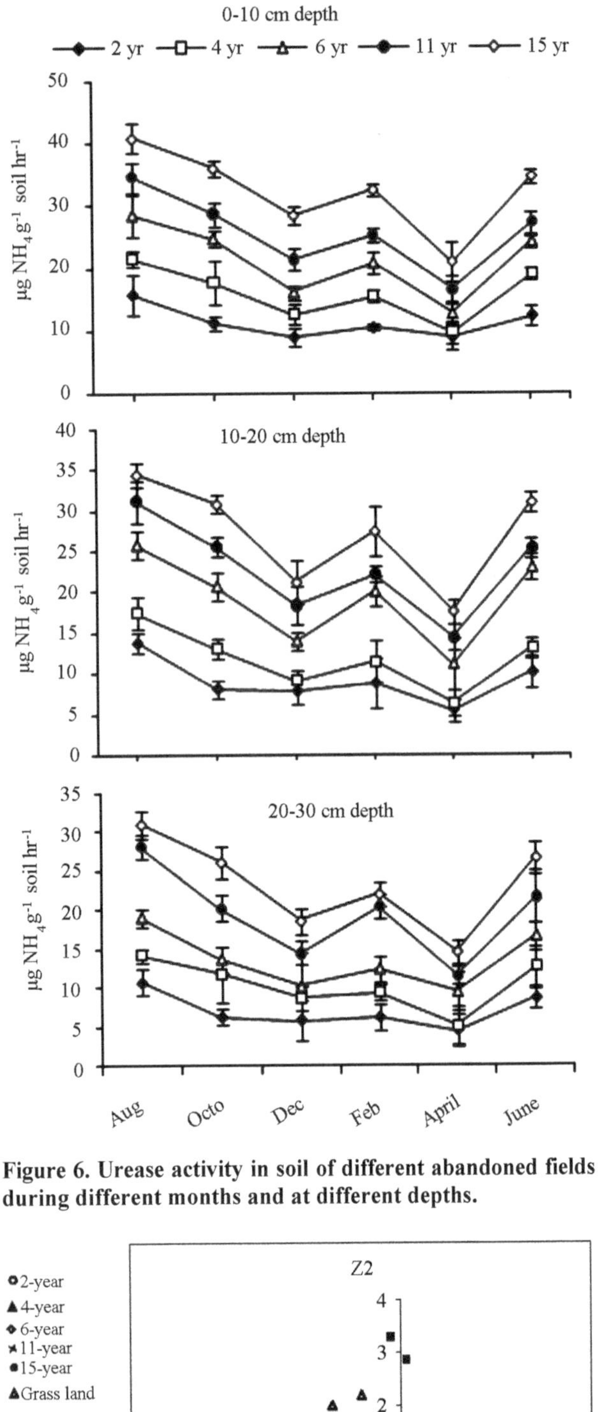

Figure 6. Urease activity in soil of different abandoned fields during different months and at different depths.

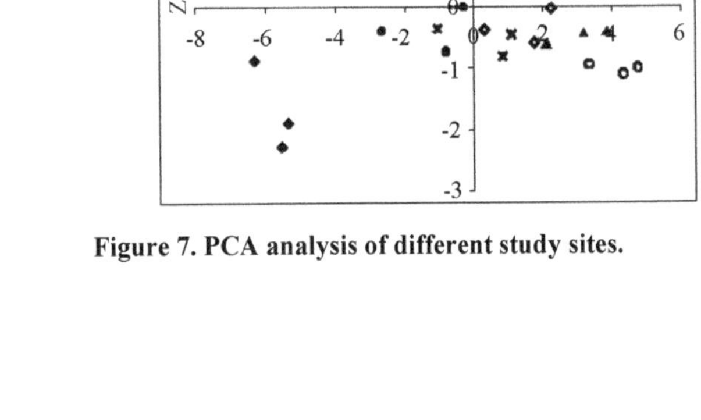

Figure 7. PCA analysis of different study sites.

than that of plantation species [39]. Thus the study justified that the degraded ecosystems derived from natural forest when subjected to abandonment showed the symptoms of reclamation during the course of natural secondary succession.

5. Acknowledgements

The authors are grateful to Prof N. Behera, Head, School of Life Sciences, Sambalpur University for providing necessary facilities to carry out the present work. Also the authors are thankful to Dr. S K Pattanayak, P G Department of Environmental Science, Sambalpur University for some of his important suggestions and comments during the preparation of the manuscript.

REFERENCES

[1] Ministry of Environment and Forests, "Annual Report," Ministry of Environment and Forest, Government of India, New Delhi, 1999.

[2] N. Sanginga, F. Zapata, S. K. A. Danso and G. D. Bowen, "Estimating Nitrogen Fixation in Leucaena and Gliricidia Using Different ^{15}N Labelling Methods," In: K. Mulongoy, M. Gueye and D. C. Spencer, Eds., *Biological Nitrogen Fixation and Sustainability of Tropical Agriculture*, IITA, Wiley-Sayce and AABNF Co-Publication, Chichester, New York, Grisbane, Toronto and Singapore, 1992, pp. 265-275.

[3] K. Maithani, R. S. Tripathi, A. Arunachalam and H. N. Pandey, "Seasonal Dynamics of Microbial Biomass C, N and P during Regrowth of a Subtropical Humid Forest in North-East India," *Applied Soil Ecology*, Vol. 4, No. 1, 1996, pp. 31-37. doi:10.1016/0929-1393(96)00101-1

[4] B. A. Caldwell, "Enzyme Activities as a Component of Soil Biodiversity: A Review," *Pedobiologia*, Vol. 49, No. 6, 2005, pp. 637-644. doi:10.1016/j.pedobi.2005.06.003

[5] M. Scholter, O. Dilly and J. C. Munch, "Indicators for Evaluating Soil Quality," *Agriculture, Ecosystems & Environment*, Vol. 98, No. 1-3, 2003, pp. 255-262. doi:10.1016/S0167-8809(03)00085-9

[6] R. E. J. Boerner, K. L. M. Decker and E. K. Sutherland, "Prescribed Burning Effects on Soil Enzyme Activity in a Southern Ohio Hardwood Forest: A Landscape-Scale Analysis," *Soil Biology & Biochemistry*, Vol. 32, No. 7, 2000, pp. 899-908. doi:10.1016/S0038-0717(99)00208-4

[7] E. I. Ndiaye, J. M. Sandeno, D. McGrath and R. P. Dick, "Integrative Biological Indicators for Detecting Change In Soil Quality," *American journal of Alternative Agriculture*, Vol. 15, No. 1, 2000, pp. 26-36. doi:10.1017/S0889189300008432

[8] V. Acosta-Martinez and M. A. Tabatabai, "Enzyme Activities of a Lime Agricultural Soil," *Biology and Fertility of Soil*, Vol. 31, No. 1, 2001, pp. 85-91. doi:10.1007/s003740050628

[9] M. Ekenler and M. A. Tabatabai, "β-Glucosaminidase Activity of Soils: Effect of Cropping Systems and Its Re-

lationship to Nitrogen Mineralization," *Biology and Fertility of Soils*, Vol. 36, No. 5, 2002, pp. 367-376. doi:10.1007/s00374-002-0541-x

[10] V. Acosta-Martinez, L. Cruz, D. Sotomayor-Ramirez and L. Perez-Alegria, "Enzyme Activities as Affected by Soil Properties and Land Use in a Tropical Watershed," *Applied Soil Ecology*, Vol. 35, No. 1, 2007, pp. 35-45. doi:10.1016/j.apsoil.2006.05.012

[11] V. Acosta-Martinez, D. Acosta-Mercado, D. Sotomayor-Ramirez and L. Cruz-Rodriguez, "Microbial Communities Activities under Different Activities under Different Management in Semi-Arid Soils," *Applied Soil Ecology*, Vol. 38, No. 3, 2008, pp. 249-260. doi:10.1016/j.apsoil.2007.10.012

[12] R. Kizilkaya and O. Dengiz, "Variation of Land Use and Land Cover Effects on Some Soil Physic-Chemical Characteristics and Soil Enzyme Activity," *Zemdirbyste-Agriculture*, Vol. 97, No. 2, 2010, pp. 15-24.

[13] V. Ralte, H. N. Pandey, S. K. Barik, R. S. Tripathy and S. D. Prabhu, "Changes in Microbial Biomass and Activity in Relation to Shifting Cultivation and Horticultural Practices in Sub-Tropical Ever Green Forest Ecosystem of North East India," *Acta Oecologica*, Vol. 28, 2005, pp. 163-172. doi:10.1016/j.actao.2005.04.001

[14] S. K. Joshi, "Production and Decomposition of Litter in a Tropical Mixed Wood Forest from Orissa, India," Ph.D. Thesis, Sambalpur University, Orissa, 1990.

[15] K. L. Mohanty, "Agricultural Guide Book," Goswami Press, Orissa, 1971.

[16] L. T. Benjamin and E. R. Tania, "Stability of Hydrolytic Enzyme Activity and Microbial Phosphorus during Storage of Tropical Rain Forest Soils," *Soil Biology & Biochemistry*, Vol. 42, No. 3, 2010, pp. 459-465. doi:10.1016/j.soilbio.2009.11.029

[17] J. M. Anderson and J. S. I. Ingram, "Tropical Soil Biology and Fertility: A Hand Book of Methods," 2nd Edition, CAB International, Wallingford, 1993.

[18] A. Walkley and I. A. Black, "An Examination of the Degtjareff Method for Determining Organicmatter and A Proposed Modification of the Chromic Acid Titration Method," *Soil Science*, Vol. 37, No. 1, 1934, pp. 251-263. doi:10.1097/00010694-193401000-00003

[19] M. L. Jackson, "Soil Chemical Analysis," Prentice Hall: Englewood Cliffs, 1958.

[20] R. H. Brady and L. T. Kurtz, "Determination of Total Organic and Available Phosphorous in Soils," *Soil Science*, Vol. 59, No. 1, 1945, pp. 39-45. doi:10.1097/00010694-194501000-00006

[21] E. D. Vance, P. C. Brookes and D. S. Jenkinson, "An Extraction Method for Measuring Soil Microbial Biomass C," *Soil Biology & Biochemistry*, Vol. 19, No. 6, 1987, pp. 703-707. doi:10.1016/0038-0717(87)90052-6

[22] P. C. Mishra, R. K. Mohanty and M. C. Dash, "Enzyme Activities in Subtropical Surface Soils under Pasture," *Indian Journal of Agricultural Chemistry*, Vol. 12, No. 1, 1979, pp. 19-24.

[23] L. E. Casida Jr., D. A. Klein and D. Sntoro, "Soil Dehydrogenase Activity," *Soil Science*, Vol. 98, No. 6, 1964, pp. 371-378. doi:10.1097/00010694-196412000-00004

[24] M. A. Tabatabai and J. M. Bremner, "Use of p-Nitrophenyl Phosphatefor Assay of Soil Phosphatase Activity," *Soil Biology and Biochemistry*, Vol. 4, No. 3, 1969, pp. 301-307. doi:10.1016/0038-0717(69)90012-1

[25] M. C. Dash, P. C. Mishra, R. K. Mohanty and N. Bhattm, "Effects of Specific Conductance and Temperature on Urease Activity in Some Indian Soils," *Soil Biology and Bio-Chemistry*, Vol. 13, No. 1, 1981, pp. 73-74. doi:10.1016/0038-0717(81)90105-X

[26] F. R. Troeh and L. M. Thompson, "Soils and Fertility," 5th Edition, Oxford University Press Inc., New York, 1993.

[27] C. Lee, Y. You and R. G. Robinson, "Secondary Succession and Habitat Restoration in Abandoned Rice Fields of Central Korea," *Restoration Ecology*, Vol. 10, No. 2, 2002, pp. 306-314. doi:10.1046/j.1526-100X.2002.00003.x

[28] K. Maithani, A. Arunachalam and R. S. Tripathy, "Influence of Leaf Litter Quality on N-Miniralization in Soils of Sub-Tropical Humid Forest Regrowths," *Biology and Fertility of Soil*, Vol. 27, No. 1, 1998. pp. 44-50. doi:10.1007/s003740050398

[29] A. Arunachalam and H. N. Pandey, "Ecosystem Restoration of Jhum Fallows in Northeast India: Microbial C and N along Altitudinal and Successional Gradients," *Restoration Ecology*, Vol. 11, No. 2, 2003, pp. 1-6. doi:10.1046/j.1526-100X.2003.00013.x

[30] Y. B. Cheng and D.-X. Yue, "Soil Microbial and Enzymatic Activities across a Chronosequence of Chinese Pine Plantation Development on the Loess Plateau of China," *Pedosphere*, Vol. 22, No. 1, 2012. pp. 1-12. doi:10.1016/S1002-0160(11)60186-0

[31] J. P. Taylor, B. Wilson, M. S. Mills and R. G. Burns, "Comparison of Microbial Numbers and Microbial Activities in Surface Soils and Sub Soils Using Various Techniques," *Soil Biology & Biochemistry*, Vol. 34, No. 3, 2002, pp. 387-401. doi:10.1016/S0038-0717(01)00199-7

[32] P. Nannipieri, S. Grego and B. Ceccanti, "Ecological Significance of Biological Activity in Soils," In: J. M. Bollage and G. Stotzky, Eds., *Soil Biochemistry*, Marcel Dekker, New York, 1990, pp. 293-355.

[33] S. Kramer and D. M. Green, "Acid and Alkaline Phosphatase Dynamics and Their Relationship to Soil Microclimate in a Semiarid Woodland," *Soil Biology & Biochemistry*, Vol. 32, No. 2, 2000, pp. 179-188. doi:10.1016/S0038-0717(99)00140-6

[34] T. H. Bell and H. A. I. Henry, "Fine Scale Variability in Soil Extrcellular Enzyme Activity Is Sensitive to Rain Events and Temperature in a Mesic System," *Pedobiologia*, Vol. 54, No. 2, 2011, pp. 141-146. doi:10.1016/j.pedobi.2010.12.003

[35] H. A. I. Henry, "Soil Extracellular Enzyme Dynamics in a Changing Climate," *Soil Biology and Biochemistry*, Vol. 47, 2012, pp. 53-59. doi:10.1016/j.soilbio.2011.12.026

[36] J. F. Dormaar, A. Johnston and S. Smoliak, "Seasonal Changes in Carbon Content, and Dehydrogenazs, Phosphatase, and Urease Activities in Mixed Prairie and Fes-

cue Grassland Ah Horizons," *Journal of Range Management*, Vol. 37, No. 1, 1984, 31-35. doi:10.2307/3898819

[37] M. Matinizadeh, S. A. A. Korori and W. Praznik, "Enzyme Activities in Undisturbed and Disturbed Forest Soil under Oak (*Quercus brantii* var. Persica) as Affected by Soil Depth and Seasonal Variation," *Asian Journal of Plant Sciences*, Vol. 7, No. 4, 2008, pp. 368-374. doi:10.3923/ajps.2008.368.374

[38] P. Prasad, "Soil Microbial Biomass and Activity in Some Tropical Agroecosystems," Ph.D. Thesis, Sambalpur University, Odisha, 1996.

[39] U. Sahani, "Role of Plantation Forestry in Reclamation of Degraded Tropical Forest," Ph.D. Thesis, Sambalpur University, Odisha, 2002.

9

Glomalin and Soil Aggregation under Six Management Systems in the Northern Great Plains, USA

Kristine A. Nichols[1], James Millar[2]

[1]USDA-ARS Northern Great Plains Research Laboratory, Mandan, USA; [2]Precision Soil Management, Redfield, USA.
Email: Kristine.nichols@ars.usda.gov

ABSTRACT

The soil environment is linked to aboveground management including plant species composition, grazing intensity, levels of soil disturbance, residue management, and the length of time of a living plant is growing. Soil samples were collected under rangeland [native grass, rotational grazing (NGRG); tame grass, heavy grazing (TGRG); and tame grass, rotational grazing (TGHG)] and cropland [conventional till (CT); CT plus manure (CTM); and long term no till (NT)] systems. The rangeland systems were hypothesized to have higher glomalin content [measured as Bradford-reactive soil protein (BRSP)] and water stable aggregation (WSA) than the cropland systems. In addition, within both rangeland and cropland systems, BRSP and WSA were expected to decline with increased disturbance due to grazing or tillage and going from native to introduced plant species. Differences were detected for BRSP with NGRG and CTM having the highest values in range and cropland systems, respectively. However, the CTM system had higher BRSP values than one or both of the tame grass systems while the CT and NT systems had similar values. Correlation analysis showed strong relationships between all of the BRSP values and WSA.

Keywords: Glomalin; Arbuscular Mycorrhizal Fungi; Soil Aggregation; Grazing Systems; Crop Production Systems

1. Introduction

The symbiotic relationship between arbuscular mycorrhizal fungi (AMF) and about 90% of all vascular plants provides a direct linkage between above and below-ground activities [1,2]. Soil aggregation is a mechanism used to engineer the soil environment to maximize water and nutrient use efficiency [3]. Glomalin is a glycoprotein produced by AMF which is strongly linked to aggregation [4-6]. The aggregation process and glomalin production are both a carbon cost to AMF and may be hindered by soil disturbance through tillage [4-11] or overgrazing [12-20]. Therefore, the level of mycorrhizal dependency [1] of the host plant as well as photosynthetic activity may impact glomalin content and water stable aggregation (WSA) [2].

In rangeland systems, Liebig *et al.* [21] found that systems under heavy grazing intensity had higher soil C values than systems under moderate intensity. However, Klumpp *et al.* [19] and Ingram *et al.* [20] found that increases in grazing intensity decreased soil C. Unlike the Liebig *et al.* [21] study, these authors also examined microbial community and soil C dynamics under different grazing treatments and found that the rhizosphere environment was the driver behind changes in soil C [19,20]. These results are worth noting because often soil C levels are related to soil aggregation and glomalin concentrations [2, 22].

In cropland systems, no till (NT) treatments had higher soil organic C levels, glomalin concentrations, and more water stable aggregates than conventional till (CT) treatments [5-8,11]. These differences tended to increase with longer periods of time in NT [5,6]. Pikul *et al.* [4] found that particulate organic matter (POM) increased under NT while aggregate wettability decreased. Greater amounts of POM may support higher aggregate stability and water holding capacity while a decrease in aggregate wettability may negatively impact water infiltration and water holding capacity. In addition to C and N levels and aggregate stability, glomalin values also may be related to other AMF parameters such as spore counts, hyphal length or root colonization but these relationships are inconsistent [7,8,13,15,18,23].

Bradford-reactive soil protein (BRSP) and water stable aggregation (WSA) were measured at six sites near Platte, SD, USA. These sites included three rangelands [native grass, rotational grazing (NGRG); tame grass, heavy grazing (TGRG); and tame grass, rotational grazing (TGHG)] and three croplands [conventional till (CT); CT plus manure (CTM); and long term no till (NT)] systems. We hypothesized that the rangeland systems would have higher BRSP and WSA values than the cropland systems and values would increase across these systems in the following order: CT < CTM < NT < TGHG < TGRG < NGRG.

2. Materials and Methods

2.1. Site Description

Soil samples were collected on June 7, 2004 near Platte, SD, USA from under six management systems (NGRG, TGRG, TGHG, CT, CTM, and NT). The sites were selected to show a wide range of management scenarios on the same soil type (an Eakin silty clay loam or fine-silty, mixed, superactive, mesic Typic Argiustolls) and under the same climatic conditions (*i.e.* all six sites within a 10-km radius). The major grass species at the NGRG was big bluestem (*Andropogon gerardii*) with a small amount of switchgrass (*Panicum virgatum* L.). The grass species at the TGRG sites were predominantly bromegrass (*Bromus inermus*) and crested wheatgrass (*Agropyron cristatum*) with small amounts of intermediate wheatgrass (*Thinopyrum intermedium*). The grass species at the TGHG was Kentucky bluegrass (*Poa pratensis*) with some bromegrass. The cropping systems all had a corn-wheat rotation. Two samples were collected from each site at the 0 to 10 cm depth to create a bulk sample. A portion of the soil was sequentially passed through a series of screens to collect 1 to 2 mm aggregates.

2.2. Soil Analyses

Six 2-g subsamples were removed from both bulk soil and 1 to 2 mm aggregate samples and extracted for glomalin-related soil protein (GRSP). Soil was extracted with 50 mM sodium citrate, pH 8.0, for 15 to 26 1-hr cycles at 121˚C [24] followed by extraction with 100 mM sodium pyrophosphate, pH 9.0, for four to nine 1-hr cycles at 121˚C to remove a recalcitrant fraction of glomalin [2]. For both the citrate and the sodium pyrophosphate extractions, samples were centrifuged after each 1-hr extraction cycle to pellet the soil and collect the supernatant. All citrate or pyrophosphate extractions were combined, and the total volume measured.

The Bradford total protein assay was used to measure GRSP as Bradford reactive soil protein (BRSP) [2]. For the samples extracted with pyrophosphate, an aliquot equal in volume to the aliquot used for the unknowns of 100 mM pyrophosphate was added to the BSA standard to remove background cross-reactivity. Total protein values are calculated on a mg Bradford-reactive soil protein (BRSP) g^{-1} soil or 1 to 2 mm aggregates basis. Both the bulk soil and 1 to 2 mm aggregates were corrected for coarse material in the sample using forced water to pass the extracted soil or aggregates through a 1 mm screen leaving the coarse fraction on the screen. Water stable aggregation (WSA) was measured on four subsamples of the 1 to 2 mm aggregates using methodology modified from Kemper and Rosenau [25] and described by Nichols and Wright [2].

2.3. Statistical Analysis

Means and standard error of the mean (SEM) were calculated for the BRSP and WSA. All means comparisons were made at the $\alpha \leq 0.05$ level using PROC MIXED (SAS software, ver. 9.1, SAS Institute, 2004) after the residuals met the assumptions for normality and homogeneity of variance. The correlations between BRSP, for both the citrate and pyrophosphate extractions and WSA were measured using PROC CORR after meeting the assumption for normal data distribution (SAS software, ver. 9.1, SAS Institute, 2004).

3. Results and Discussion

Concentrations of BRSP in both the bulk soil and 1 to 2 mm aggregate samples were similar for each of the individual management scenarios with the exception of the CT site which was lower in the bulk soil than in the 1 to 2 mm aggregates (**Table 1**). The three grazing systems had the highest glomalin concentrations with the NGRG system having the significantly highest values. Because most studies have shown that as grazing intensity increases, soil carbon levels and microbial activities decrease [19], it was unexpected that the TGRG site would have lower BRSP values than the TGHG system. However, a study by Liebig *et al.* [21] showed that a heavily grazed treatment had higher soil carbon levels than a moderately grazed treatment at sites in North Dakota which were in the same semi-arid ecoregion as the sites in this study but were approximately 560 km apart. These studies did indicate that the results were probably dependent upon if the plant species composition was dominated by native or tame grasses which impacts the intensity of the mycorrhizal relationship [1,13] and impacts rooting depth and rhizosphere dynamics [19-21].

In the crop production systems, the CTM site had the highest values while there was no difference between the CT and NT systems in the bulk soil but the CT site had significantly higher BRSP in the 1 to 2 mm aggregates. This was also unexpected since most studies have shown glomalin, or BRSP, increases with longer periods of NT [4-6]. The addition of manure has been shown to increase

Table 1. Values[a] for glomalin extracted with sodium citrate (cit) followed by sodium pyrophosphate (pyro) from either bulk soil or 1 to 2 mm aggregates (aggreg) and measured as Bradford reactive soil protein (BRSP) on a mg protein g^{-1} soil or aggregate weight basis and for water stable aggregation (WSA) at a% of total aggregates. Values are means[b] ± SEs for each management system: native grass, rotational grazing (NGRG); tame grass, heavy grazing (TGRG); tame grass, rotational grazing (TGHG); long term no till (NT); conventional till (CT); and CT plus manure (CTM).

	NGRG	TGRG	TGHG	NT	CT	CTM
BRSP cit soil	9.555 ± 0.187^{A}	6.226 ± 0.249^{B}	6.895 ± 0.090^{B}	4.618 ± 0.152^{D}	4.354 ± 0.222^{D}	5.447 ± 0.128^{C}
BRSP pyro soil	0.700 ± 0.028^{A}	0.357 ± 0.014^{C}	0.379 ± 0.010^{BC}	0.252 ± 0.014^{D}	0.266 ± 0.012^{D}	0.439 ± 0.011^{B}
BRSP total soil	10.255 ± 0.166^{A}	6.583 ± 0.246^{BC}	7.274 ± 0.081^{B}	4.869 ± 0.161^{D}	4.619 ± 0.222^{D}	5.887 ± 0.120^{C}
BRSP cit aggreg	9.291 ± 0.113^{A}	6.423 ± 0.064^{C}	7.165 ± 0.092^{B}	5.093 ± 0.135^{E}	5.664 ± 0.146^{D}	6.128 ± 0.166^{CD}
BRSP pyro aggreg	0.575 ± 0.026^{A}	0.455 ± 0.021^{B}	0.301 ± 0.022^{C}	0.276 ± 0.013^{C}	0.325 ± 0.032^{C}	0.502 ± 0.021^{AB}
BRSP total aggreg	9.865 ± 0.090^{A}	6.878 ± 0.072^{C}	7.466 ± 0.096^{B}	5.369 ± 0.142^{E}	5.989 ± 0.152^{D}	6.630 ± 0.179^{C}
WSA	89.063 ± 1.028^{A}	83.125 ± 0.633^{B}	45.688 ± 0.806^{C}	24.875 ± 0.415^{E}	24.063 ± 0.544^{E}	42.063 ± 1.082^{D}

[a]Values in a row which are followed by a different letter are significantly different at Prob <0.05. [b]N equals six for the BRSP values and four for the WSA.

soil C, which may stimulate aggregation, and BRSP (Wright, personal communication). In addition, other organic compounds, such as humic substances and tannins may be co-extracted with glomalin and measured with the Bradford assay [2,26,27]. This might explain the CTM treatment values, but it was unclear in this study why the long term NT system did not have higher BRSP values than the CT site.

The authors speculate that the crops grown at the cropland sites might have impacted carbon allocation to AMF and AMF allocation to glomalin production and aggregation.

Despite these unexpected results from BRSP and WSA values, the Pearson Correlation Coeffcients showed BRSP was strongly and significantly correlated WSA, r > 0.7 in almost all cases (**Table 2**). This is similar to many other studies where BRSP and WSA are correlated.

Native warm-season grasses, such as big bluestem and switchgrass, are all highly mycorrhizal while cool-season grasses are less dependent upon the mycorrhizal relationship [1,13]. Mycorrhizal fungi and glomalin production are sensitive to carbon exudates from the plant and soil disturbance. Additionally, AMF and glomalin are important in forming and stabilizing aggregates which play significant roles in water infiltration and water retention [5,6,14]. Finally, grazing pressure has been shown to decrease the amount of carbon fixed by photosynthesis and exuded from the roots [19,20]. For the most part, this did occur in this study but given the non-replicated design and the limited number of samples collected at each site, and the limited number of studies on glomalin and WSA in rangeland systems in the literature, more research is needed. Future studies will focus on the biological, chemical, and physical interactions occurring in the soil, particularly in the rhizosphere, which may be used to model above and belowground relationships. These models will help us understand the linkage between levels of soil disturbance and the types and amounts of photosynthetically-derived carbon going belowground, especially as it relates to the plant-mycorrhizal relationship. Ongoing research projects in the northern Great Plains are measuring GRSP and WSA in other range and cropland studies. The range studies include systems with long-term (over 60 years) continuous management different levels of grazing intensity from no grazing to heavy grazing while the cropland studies incorporate perennial and cover crops which may be used as forages.

4. Conclusion

Similar to many other studies, glomalin and aggregate stability were strongly correlated in this study. However, the BRSP and WSA values did not increase from CT < CTM < NT < TGHG < TGRG < NGRG as expected. This was noted in particular in the cropland sites where CTM had the highest values and CT and NT were similar. The complex relationship between AMF and host plant especially as it relates to carbon allocation needs more studies to understand glomalin production and how it relates to aggregation.

5. Abbreviations and Acronyms

BRSP, Bradford reactive soil protein; C, carbon; CT, conventional till; CTM, conventional till plus manure; GRSP, glomalin-related soil protein; IR, infiltration rate; NGRG, native grass rotational grazing; NT, long-term no-till; TGHG, tame grass heavy grazing; TGRG, tame grass rotational grazing; WHC_s, water holding capacity; WSA, water stable aggregation.

6. Acknowledgements

The authors would like to acknowledge Crystal Jundt and Gary Brucker for their assistance with the laboratory

Table 2. Correlation coefficients[a] for glomalin extracted with sodium citrate (cit) followed by sodium pyrophosphate (pyro) from either bulk soil or 1 to 2 mm aggregates (aggreg) and measured as Bradford reactive soil protein (BRSP) on a mg protein g^{-1} soil or aggregate weight basis, for water stable aggregation (WSA) at a% of total aggregates, for the infiltration rate (IR) on a ml per hr basis for the first (1) and second (2) inches of water, and for the water holding capacity at saturation (WHC_s) for the first two inches of water.

	BRSP pyro soil	BRSP total soil	BRSP cit aggreg	BRSP pyro aggreg	BRSP total aggreg	WSA
BRSP cit soil	0.8684**	0.994***	0.9371***	0.5894*	0.9350***	0.7702**
BRSP pyro soil	1.000	0.8855***	0.8740***	0.7811**	0.8895***	0.7700**
BRSP total soil	0.8855***	1.0000	0.9409***	0.6083*	0.9400***	0.7700**
BRSP cit aggreg	0.8740**	0.9409***	1.0000	0.6362*	0.9983***	0.7893**
BRSP pyro aggreg	0.7811**	0.6083*	0.6362*	1.0000	0.6799*	0.7045*
BRSP total aggreg	0.8895***	0.9400***	0.9983***	0.6799*	1.0000	0.8032**

[a]Correlations are significant at the <0.05 (*), <0.01 (**), or <0.001 (***) levels.

analysis, and Drs. Rebecca Phillips and Shannon Osborne for their invaluable comments on earlier drafts of this manuscript.

REFERENCES

[1] M. F. Allen, W. Swenson, J. I. Querejeta, L. M. Egerton-Warburton and K. K. Treseder, "Ecology of Mycorrhizae: A Conceptual Framework for Complex Interactions among Plants and Fungi," *Annual Review of Phytopathology*, Vol. 41, 2003, pp. 271-303. http://dx.doi.org/10.1146/annurev.phyto.41.052002.095518

[2] K. A. Nichols and S. F. Wright, "Contributions of Soil Fungi to Organic Matter in Agricultural Soils," In: F. R. Magdoff and R. R. Weil, Eds., *Soil Organic Matter in Sustainable Agriculture*, CRC Press, New York, 2004, pp. 179-198. http://dx.doi.org/10.1201/9780203496374.ch6

[3] M. C. Rillig and P. D. Steinberg, "Glomalin Production by an Arbuscular Mycorrhizal Fungus: A Mechanism of Habitat Modification?" *Soil Biology and Biochemistry*, Vol. 34, No. 9, 2002, pp. 1371-1374. http://dx.doi.org/10.1016/S0038-0717(02)00060-3

[4] J. Pikul Jr., G. Chilom, J. Rice, A. Eynard, T. E. Schumacher, K. A. Nichols, J. M. F. Johnson, S. Wright, T. C. Caesar-Tonthat and M. Ellsbury, "Organic Matter and Water Stability of Field Aggregates Affected by Tillage in South Dakota," *Soil Science Society of America Journal*, Vol. 73, 2009, pp. 197-206. http://dx.doi.org/10.2136/sssaj2007.0184

[5] S. F. Wright and R. L. Anderson, "Aggregate Stability and Glomalin in Alternative Crop Rotations for the Central Great Plains," *Biology and Fertility of Soils*, Vol. 31, No. 3/4, 2000, pp. 249-253. http://dx.doi.org/10.1007/s003740050653

[6] S. F. Wright, J. L. Starr and I. C. Paltineanu, "Changes in Aggregate Stability and Concentration of Glomalin during Tillage Management Transition," *Soil Science Society of America Journal*, Vol. 63, No. 6, 1999, pp. 1825-1829. http://dx.doi.org/10.2136/sssaj1999.6361825x

[7] F. Borie, R. Rubio, J. L. Rouanet, A. Morales, G. Borie and C. Rojas, "Effects of Tillage Systems on Soil Characteristics, Glomalin and Mycorrhizal Propagules in a Chilean Ultisol," *Soil and Tillage Research*, Vol. 88, No. 1-2, 2006, pp. 253-261. http://dx.doi.org/10.1016/j.still.2005.06.004

[8] T. Caesar-Tonthat, U. M. Sainju, S. F. Wright, W. L. Shelver, R. L. Kolberg and M. West, "Long-Term Tillage and Cropping Effects on Microbiological Properties Associated with Aggregation in a Semi-Arid Soil," *Biology and Fertility of Soils*, Vol. 47, No. 2, 2010, pp. 157-165. http://dx.doi.org/10.1007/s00374-010-0508-2

[9] M. C. Rillig, "Arbuscular Mycorrhizae, Glomalin, and Soil Aggregation," *Canadian Journal of Soil Science*, Vol. 84, No. 4, 2004, pp. 355-363. http://dx.doi.org/10.4141/S04-003

[10] A. Roldán, J. R. Salinas-García, M. M. Alguacil and F. Caravaca, "Soil Sustainability Indicators Following Conservation Tillage Practices under Subtropical Maize and Bean Crops," *Soil and Tillage Research*, Vol. 93, No. 2, 2007, pp. 273-282. http://dx.doi.org/10.1016/j.still.2006.05.001

[11] S. B. Wuest, T. C. Caesar-Tonthat, S. F. Wright and J. D. Williams, "Organic Matter Addition, N, and Residue Burning Effects on Infiltration, Biological, and Physical Properties of an Intensively Tilled Silt-Loam Soil," *Soil and Tillage Research*, Vol. 84, No. 2, 2005, pp. 154-167. http://dx.doi.org/10.1016/j.still.2004.11.008

[12] K. M. Batten, K. M. Scow, K. F. Davies and S. P. Harrison, "Two Invasive Plants Alter Soil Microbial Community Composition in Serpentine Grasslands," *Biological Invasions*, Vol. 8, No. 2, 2006, pp. 217-230. http://dx.doi.org/10.1007/s10530-004-3856-8

[13] M. A. Bingham and M. Biondini, "Mycorrhizal Hyphal Length as a Function of Plant Community Richness and Composition in Restored Northern Tallgrass Prairies (USA)," *Rangeland Ecology and Management*, Vol. 62, No. 1, 2009, pp. 60-67. http://dx.doi.org/10.2111/08-088

[14] S. B. Bird, J. E. Herrick, M. M. Wander and S. F. Wright, "Spatial Heterogeneity of Aggregate Stability and Soil Carbon in Semi-Arid Rangeland," *Environmental Pollution*, Vol. 116, No. 3, 2002, pp. 445-455. http://dx.doi.org/10.1016/S0269-7491(01)00222-6

[15] J. S. Buyer, D. A. Zuberer, K. A. Nichols and A. J. Franzluebbers, "Soil Microbial Community Function, Structure, and Glomalin in Response to Tall Fescue Endophyte Infection," *Plant and Soil*, Vol. 339, No. 1-2, 2011, pp. 401-412. http://dx.doi.org/10.1007/s11104-010-0592-y

[16] A. J. Franzluebbers, S. F. Wright and J. A. Stuedemann, "Soil Aggregation and Glomalin under Pastures in the Southern Piedmont USA," *Soil Science Society of America Journal*, Vol. 64, No. 3, 2000, pp. 1018-1026. http://dx.doi.org/10.2136/sssaj2000.6431018x

[17] E. R. Lutgen, D. Muir-Clairmont, J. Graham and M. C. Rillig, "Seasonality of Arbuscular Mycorrhizal Hyphae and Glomalin in a Western Montana Grassland," *Plant and Soil*, Vol. 257, No. 1, 2003, pp. 71-83. http://dx.doi.org/10.1023/A:1026224209597

[18] M. C. Rillig, F. T. Maestre and L. J. Lamit, "Microsite Differences in Fungal Hyphal Length, Glomalin, and Soil Aggregate Stability in Semiarid Mediterranean Steppes," *Soil Biology and Biochemistry*, Vol. 35, No. 9, 2003, pp. 1257-1260. http://dx.doi.org/10.1016/S0038-0717(03)00185-8

[19] K. Klumpp, S. Fontaine, E. Attard, X. Le Roux, G. Gleixner and J.-F. Soussana, "Grazing Triggers Soil Carbon Loss by Altering Plant Roots and Their Control on Soil Microbial Community," *Journal of Ecology*, Vol. 97, No. 5, 2009, pp. 876-885. http://dx.doi.org/10.1111/j.1365-2745.2009.01549.x

[20] L. J. Ingram, P. D. Stahl, G. E. Schuman, J. S. Buyer, G. F. Vance, G. K. Ganjegunte, J. M. Welker and J. D. Derner, "Grazing Impacts on Soil Carbon and Microbial Communities in a Mixed-Grass Ecosystem," *Soil Science Society of America Journal*, Vol. 72, No. 4, 2008, pp. 939-948. http://dx.doi.org/10.2136/sssaj2007.0038

[21] M. A. Liebig, J. R. Gross, S. L. Kronberg, R. L. Phillips and J. D. Hanson, "Grazing Management Contributions to Net Global Warming Potential: A Long-Term Evaluation in the Northern Great Plains," *Journal of Environmental Quality*, Vol. 39, No. 3, 2010, pp. 799-809. http://dx.doi.org/10.2134/jeq2009.0272

[22] J. Six, S. D. Frey, R. K. Thiet and K. M. Batten, "Bacterial and Fungal Contributions to Carbon Sequestration in Agroecosystems," *Soil Science Society of America Journal*, Vol. 70, No. 2, 2006, pp. 555-569. http://dx.doi.org/10.2136/sssaj2004.0347

[23] X. He, Y. Li and L. Zhao, "Dynamics of Arbuscular Mycorrhizal Fungi and Glomalin in the Rhizosphere Ofartemisia Ordosica Krasch. In Mu Us Sandland, China," *Soil Biology and Biochemistry*, Vol. 42, No. 8, 2010, pp. 1313-1319. http://dx.doi.org/10.1016/j.soilbio.2010.03.022

[24] S. F. Wright, M. Franke-Snyder, J. B. Morton and A. Upadhyaya, "Time-Course Study and Partial Characterization of a Protein on Hyphae of Arbuscular Mycorrhizal Fungi During Active Colonization of Roots," *Plant and Soil*, Vol. 181, No. 2, 1996, pp. 193-203. http://dx.doi.org/10.1007/BF00012053

[25] W. D. Kemper and R. C. Rosenau, "Aggregate Stability and Size Distribution," In: *Methods of Soil Analysis, Part I. Physical and Mineralogical Methods, American Society of Agronomy-Soil Science Society of America*, Madison, WI, 1986, pp. 425-442.

[26] J. J. Halvorson and J. M. Gonzalez, "Tannic Acid Reduces Recovery of Water-Soluble Carbon and Nitrogen from Soil and Affects the Composition of Bradford-Reactive Soil Protein," *Soil Biology and Biochemistry*, Vol. 40, No. 1, 2008, pp. 186-197. http://dx.doi.org/10.1016/j.soilbio.2007.07.022

[27] K. A. Nichols and S. F. Wright, "Comparison of Glomalin and Humic Acid in Eight Native U.S. Soils," *Soil Science*, Vol. 170, No. 12, 2005, pp. 985-997. http://dx.doi.org/10.1097/01.ss.0000198618.06975.3c

10

Developments in Fractionation and Measurement of Soil Organic Carbon: A Review

Yadunath Bajgai[1,2*], Nilantha Hulugalle[3], Paul Kristiansen[1], Melinda McHenry[4]

[1]School of Environmental & Rural Science, University of New England, Armidale, Australia; [2]RNR-Research and Development Centre, Bajo, Department of Agriculture, MoAF, Wangdue Phodrang, Bhutan; [3]NSW Department of Primary Industries, Narrabri, Australia; [4]Centre for Plant and Water Sciences, Central Queensland University, Bundaberg, Australia.
Email: [*]ybajgai@gmail.com

ABSTRACT

Soil organic carbon (SOC) is the percentage measure of carbon (C) derived from living organisms in soil. Stability of soil organic matter (SOM) can be defined in terms of how easily C and nitrogen in the SOM can be decomposed. Due to the implications in the permanence of SOC during sequestration there is scientific interest in fractionation of SOM into different fractions. A large number of SOM fractionation procedures have been developed to distinguish between SOM to study whether it is liable or recalcitrant to activities of soil microbes. There are physical and chemical fractionation techniques. The former is based on particle size and density of soil samples or combination of the two, and the latter on the reaction of chemical on SOM for the separation of stable SOC. Each fraction of SOC in the laboratory can be commonly determined using wet oxidation by Walkley-Black method and dry combustion by LECO CN Analyzer. With the advancement in chemometric statistical techniques; faster, robust, cheaper and non-destructive methods are emerging. The chemometric statistical techniques do not require any reagents for analysis compared with the wet oxidation or dry combustion methods. Thus, these emerging techniques are highly attractive for studies where a large number of analyses are required. For *in situ* measurement of SOC, spectral reflectance technology is developed to facilitate instant measurement in the field using the sensors or by remote sensing.

Keywords: Wet Oxidation; Dry Combustion; Chemometric Statistical Techniques

1. Introduction

Soil organic matter (SOM) is the vast array of carbon (C) compounds in soil and the soil organic carbon (SOC) is the C component of the SOM. Therefore, the SOC is the percentage measure of C deriving from living organisms and it is commonly represented as SOM = SOC × 1.72, with 1.72 as the most commonly used conversion factor [1]. This is based on the assumption that SOM has 58% C content, on average. It is advantageous to report SOC rather SOM for making consistent and reliable comparisons between studies. Recently, a conversion factor of 2.0 was proposed to be more accurate and representative based on the fact that most of the published literature averaged content of 50% C rather than 58% [2]. Slight change in the SOC stock is reported to influence atmospheric C concentration and the fluxes of SOC differ in response to environmental conditions and land management factors [3]. Stability of SOC can be studied using fractionation techniques. This paper summarises the methods of SOC fractionation which helps us understand the characteristic of SOC specially its stability, and provides the summary of main methods used for measurement of SOC in the laboratory and field.

[*]Corresponding author.

2. Fractionation of SOC

2.1. Need for SOC Fractionation

Stability of SOM can be defined in terms of how easily C and nitrogen in the SOM can be decomposed. The identification, isolation and characterisation of SOM fractions have received a great deal of scientific interest because of their implications in the permanence of SOC during sequestration [4,5]. The estimates of different pools are used in mechanistic models (e.g. Roth-C and Century) that predict changes in SOM storage [6,7]. A large number of SOM fractionation procedures have been developed that seek to distinguish between SOM that is more easily decomposed (low stability) and less easily decomposed (high stability) by the soil microbes. The procedures have been recently reviewed and include physical fractionation by size or density, and various chemical fractionation methods that separate SOM by solubility, hydrolysability, or resistance to oxidation [3,8].

2.2. Physical Fraction by Particle Size

Physical fractionation procedures by size is based on the idea that the association of soil particles and their spatial arrangement play a key role in SOM dynamics as bioaccessibility is a prerequisite for decomposition [9,10]. Physical fractionation involves the application of various degrees of disaggregating treatments by dry and wet sieving [11-13], dispersion [14], and density separation and sedimentation [11,15]. The extraction method developed by Cambardella and Elliott [14] is commonly used to isolate particulate (>53 μm) and mineral-associated organic C fractions (<53 μm) after dispersion of soil samples in potassium hexa-meta phosphate solution (5 g L^{-1}) and tumbled overnight. Based on the extraction component of Cambardella and Elliott [14], an automated wet sieving technique was developed in Australia using 50 μm sieve so that >50 μm fraction consists of particulate organic matter (POM) plus sand and <50 μm fraction of mineral-associated organic matter [16].

2.3. Combination of Size and Density Fraction

The general wet sieving process uses sieves of 250 μm and 53 μm to isolate fractions into three aggregate groups. The >250 μm is macroaggregate, 250 - 53 μm is microaggragate and <53 μm is silt + clay size fraction [12,13]. Density fractionation is applied to isolate SOM that is not firmly associated with soil minerals from organo-mineral complexes [11,15]. Associations of SOM to mineral surfaces are most often characterised by a density >1.6 - 2 g cm^{-3}. The fraction of SOM that floats in the >1.6 - 2 g cm^{-3} medium is the light fraction and that settles down is the heavy fraction. Lighter fraction or POM with a density of <1.6 - 2 g cm^{-3} consists mostly of pieces of plant residues and heavier fraction consists of mineral-associated organic C [11,15]. POM also has been separated by a combination of size and density fractionation [17-19] in several steps for each particle size group.

2.4. Chemical Fractionation of SOC

Of the five SOC pools in the Roth-C model, the inert pool [6] is considered to be totally unaffected by microbial attack and thus to undergo no decomposition with time [20]. Several chemical fractionation methods have been used for the isolation of stable SOC; hydrolysis using trifluoroacetic acid (TFA) and hydrochloric acid (HCl) [21,22], oxidative treatments using several oxidants like hydrogen peroxide (H_2O_2) [23], disodium peroxodisulphate ($Na_2S_2O_8$) [24], or sodium hypochlorite (NaOCl) [25]. Hydrolysis removes compounds that are supposed to be potentially biodegradable [21,22], whereas it has been suggested that the treatments using oxidizing reagents to mimic biodegradation to the extent that treatments preferentially remove less protected SOC [23, 25]. Helfrich *et al.* [26] compared five chemical fractionation methods and found that all chemical treatments caused a preferential removal of young, maize-derived SOC, with $Na_2S_2O_8$ and H_2O_2 being most efficient. However, none of the methods were reported to be generally suited for the determination of the inert organic matter pool of the Roth-C model [26].

3. Measurement of SOC

3.1. Wet Oxidation and Dry Combustion Methods

The most commonly used methods for determination of SOC are wet oxidation by Walkley-Black method [27] and dry combustion by LECO CN Analyzer [28]. The latter method estimates SOC more accurately. Due to incomplete oxidation of SOC in a dichromate-sulphuric acid mixture, the Walkley-Black method underestimates SOC concentration when no heating is applied. To overcome the problem of incomplete C recovery, which leads to the underestimation of C concentrations by Walkley-Black method, the analytical results of this technique are adjusted by 1.32 as a correction factor assuming the recovery rate of 76% [29]. However, modified dichromate oxidation techniques that involve extensive heating do not require a correction factor because most of the organic C in the soil is oxidized to CO_2 [27]. Some authors reported that the recovery of 76% is generally too large [30,31] and that it may vary according to land use, soil texture and sampling depth [30-32].

3.2. Chemometric Statistical Methods

With the development and emergence of reflectance spectroscopy techniques, near infrared (NIR) and mid-infrared (MIR) are also used to determine SOC in the laboratory. Using chemometric statistical methods, both NIR and MIR spectroscopy techniques have been used in the past two decades to determine soil properties, including soil organic C and total nitrogen [33-35]. Compared with wet and dry combustion methods, the NIR and MIR spectroscopy techniques are faster, robust, cheaper and non-destructive, and they do not require any reagents for analysis [36,37]. These emerging techniques are, therefore, likely to replace the tedious job of wet and dry combustions in future, and are highly attractive for studies where a large number of analyses are required [34].

3.3. *In Situ* Methods

The techniques used for *in situ* measurement of SOC in the field are by sampling, or by on-the-go detection with sensors mounted on a tractor [38,39] that facilitate instant measurement of SOC. These methods are also non-destructive and relatively cheaper compared to the destructive wet laboratory [27] or dry combustion [28] methods. Remote sensing is another tool for *in situ* measurement of soil properties in the field [40,41] based on spectral laboratory data and this method captures large spatial area unlike the other previously mentioned techniques. The other *in situ* methods include Laser-Induced Breakdown Spectroscopy and Inelastic Neutron Scattering techniques [41]. A recent review covers more detail on SOC dynamics with fractionation and measurement techniques [3].

4. Conclusion

It is advantageous to report SOC rather SOM to consistently and reliably compare between studies. Due to the implications in the permanence of SOC during sequestration, there is a need to fractionate of SOM into different fractions. Further, mechanistic models that predict changes in SOM storage require the estimates of different SOC pools. Thus, a large number of SOM fractionation procedures have been developed to distinguish between SOM to study whether it is liable or recalcitrant. Once fractionated, the most common methods for determination of SOC in laboratory are wet oxidation and dry combustion. More advanced; chemometric statistical techniques are faster, robust, cheaper and non-destructive than wet oxidation and dry combustion methods. With the advancement in spectral reflectance technology, the *in situ* measurement of SOC is done using with sensors or by remote sensing to facilitate instant measurement of SOC in the field.

5. Acknowledgements

The first author, a Bhutanese national, was funded by the Endeavour Postgraduate Award of Australia Awards at the University of New England, Armidale, Australia for his PhD study.

REFERENCES

[1] J. A. Baldock and J. O. Skjemstad, "Soil Organic Carbon/Soil Organic Matter," CSIRO Publishing, Melbourne, 1999.

[2] D. W. Pribyl, "A Critical Review of the Conventional SOC to SOM Conversion Factor," *Geoderma*, Vol. 156, No. 3-4, 2010, pp. 75-83. http://dx.doi.org/10.1016/j.geoderma.2010.02.003

[3] U. Stockmann, M. A. Adams, J. W. Crawford, D. J. Field, N. Henakaarchchi, M. Jenkins, B. Minasny, A. B. McBratney, V. D. R. D. Courcelles, K. Singh, I. Wheeler, L. Abbott, D. A. Angers, J. Baldock, M. Bird, P. C. Brookes, C. Chenu, J. D. Jastrow, R. Lal, J. Lehmann, A. G. O'-Donnell, W. J. Parton, D. Whitehead and M. Zimmermann, "The Knowns, Known Unknowns and Unknowns of Sequestration of Soil Organic Carbon," *Agriculture, Ecosystems & Environment*, Vol. 164, 2013, pp. 80-99. http://dx.doi.org/10.1016/j.agee.2012.10.001

[4] A. F. Plante, J. M. Fernández, M. L. Haddix, J. M. Steinweg and R. T. Conant, "Biological, Chemical and Thermal Indices of Soil Organic Matter Stability in Four Grassland Soils," *Soil Biology and Biochemistry*, Vol. 43, No. 5, 2011, pp. 1051-1058. http://dx.doi.org/10.1016/j.soilbio.2011.01.024

[5] Y. Bajgai, "Effect of Alterative Cropping Management on Soil Organic Carbon," PhD Thesis, University of New England, Armidale, 2013.

[6] K. Coleman, D. S. Jenkinson, G. J. Crocker, P. R. Grace, J. Klír, M. Körschens, P. R. Poulton and D. D. Richter, "Simulating Trends in Soil Organic Carbon in Long-Term Experiments Using RothC-26.3," *Geoderma*, Vol. 81, No. 1-2, 1997, pp. 29-44. http://dx.doi.org/10.1016/S0016-7061(97)00079-7

[7] W. J. Parton, D. S. Schimel, C. V. Cole and D. S. Ojima, "Analysis of Factors Controlling Soil Organic Matter Levels in Great Plains Grasslands," *Soil Science Society of America Journal*, Vol. 51, No. 5, 1987, pp. 1173-1179. http://dx.doi.org/10.2136/sssaj1987.03615995005100050015x

[8] M. von Lützow, I. Kögel-Knabner, K. Ekschmitt, H. Flessa, G. Guggenberger, E. Matzner and B. Marschner, "SOM Fractionation Methods: Relevance to Functional Pools and to Stabilization Mechanisms," *Soil Biology and Biochemistry*, Vol. 39, No. 9, 2007, pp. 2183-2207. http://dx.doi.org/10.1016/j.soilbio.2007.03.007

[9] B. T. Christensen, "Physical Fractionation of Soil and Structural and Functional Complexity in Organic Matter Turnover," *European Journal of Soil Science*, Vol. 52, No. 3, 2001, pp. 345-353. http://dx.doi.org/10.1046/j.1365-2389.2001.00417.x

[10] H. Blanco-Canqui and R. Lal, "Mechanisms of Carbon Sequestration in Soil Aggregates," *Critical Reviews in Plant Sciences*, Vol. 23, No. 6, 2004, pp. 481-504. http://dx.doi.org/10.1080/07352680490886842

[11] B. T. Christensen, "Physical Fractionation of Soil and Organic Matter in Primary Particle Size and Density Separates," *Advances in Soil Science*, Vol. 20, 1992, pp. 1-90.

[12] J. Six, E. T. Elliott and K. Paustian, "Aggregate and soil organic Matter Dynamics under Conventional and No-Tillage Systems," *Soil Science Society of America Journal*, Vol. 63, No. 5, 1999, pp. 1350-1358. http://dx.doi.org/10.2136/sssaj1999.6351350x

[13] J. Six, E. T. Elliott and P. K., "Soil Macroaggregate Turnover and Microaggregate Formation: A Mechanism for C Sequestration under No-Tillage Agriculture," *Soil Biology and Biochemistry*, Vol. 32, No. 14, 2000, pp. 2099-2103. http://dx.doi.org/10.1016/S0038-0717(00)00179-6

[14] C. A. Cambardella and E. T. Elliott, "Particulate Soil Organic-Matter Changes across a Grassland Cultivation Sequence," *Soil Science Society of America Journal*, Vol. 56, No. 3, 1992, pp. 777-783. http://dx.doi.org/10.2136/sssaj1992.03615995005600030017x

[15] A. Golchin, J. M. Oades, J. O. Skjemstad and P. Clarke, "Soil Structure and Carbon Cycling," *Australian Journal of Soil Research*, Vol. 32, No. 5, 1994, pp. 1043-1068. http://dx.doi.org/10.1071/SR9941043

[16] J. Sanderman, J. Baldock, B. Hawke, L. Macdonald, A. Massis-Puccini and S. Szarvas, "National Soil Carbon Research Programme: Field and Laboratory Methodologies," CSIRO, Land and Water, Urrbrae, 2011.

[17] J. Balesdent, E. Besnard, D. Arrouays and C. Chenu, "The Dynamics of Carbon in Particle-Size Fractions of Soil in a Forest-Cultivation Sequence," *Plant and Soil*, Vol. 201, No. 1, 1998, pp. 49-57. http://dx.doi.org/10.1023/A:1004337314970

[18] J. Six, E. T. Elliott, K. Paustian and J. W. Doran, "Aggregation and Soil Organic Matter Accumulation in Cultivated and Native Grassland Soils," *Soil Science Society of America Journal*, Vol. 62, No. 5, 1998, pp. 1367-1377. http://dx.doi.org/10.2136/sssaj1998.03615995006200050032x

[19] J. Six, R. Merckx, K. Kimpe, K. Paustian and E. T. Elliott, "A Re-Evaluation of the Enriched Labile Soil Organic Matter Fraction," *European Journal of Soil Science*, Vol. 51, No. 2, 2000, pp. 283-293. http://dx.doi.org/10.1046/j.1365-2389.2000.00304.x

[20] P. Falloon, P. Smith, K. Coleman and S. Marshall, "How Important Is Inert Organic Matter for Predictive Soil Carbon Modelling Using the Rothamsted Carbon Model?" *Soil Biology and Biochemistry*, Vol. 32, No. 3, 2000, pp. 433-436. http://dx.doi.org/10.1016/S0038-0717(99)00172-8

[21] K. Quénéa, S. Derenne, C. Largeau, C. Rumpel and A. Mariotti, "Influence of Change in Land Use on the Refractory Organic Macromolecular Fraction of a Sandy Spodosol (Landes de Gascogne, France)," *Geoderma*, Vol. 136, No. 1-2, 2006, pp. 136-151. http://dx.doi.org/10.1016/j.geoderma.2006.03.018

[22] N. Poirier, S. Derenne, J. Balesdent, A. Mariotti, D. Massiot and C. Largeau, "Isolation and Analysis of the Non Hydrolysable Fraction of a Forest Soil and an Arable Soil (Lacadee, Southwest France)," *European Journal of Soil Science*, Vol. 54, No. 2, 2003, pp. 243-255. http://dx.doi.org/10.1046/j.1365-2389.2003.00520.x

[23] A. F. Plante, C. Chenu, M. Balabane, A. Mariotti and D. Righi, "Peroxide Oxidation of Clay-Associated Organic Matter in a Cultivation Chronosequence," *European Journal of Soil Science*, Vol. 55, No. 3, 2004, pp. 471-478. http://dx.doi.org/10.1111/j.1365-2389.2004.00626.x

[24] K. Eusterhues, C. Rumpel, M. Kleber and I. Kögel-Knabner, "Stabilisation of Soil Organic Matter by Interactions with Minerals as Revealed by Mineral Dissolution and Oxidative Degradation," *Organic Geochemistry*, Vol. 34, No. 12, 2003, pp. 1591-1600. http://dx.doi.org/10.1016/j.orggeochem.2003.08.007

[25] M. Kleber, R. Mikutta, M. S. Torn and R. Jahn, "Poorly Crystalline Mineral Phases Protect Organic Matter in Acid Subsoil Horizons," *European Journal of Soil Science*, Vol. 56, No. 6, 2005, pp. 717-725. http://dx.doi.org/10.1016/j.geoderma.2004.12.018

[26] M. Helfrich, H. Flessa, R. Mikutta, A. Dreves and B. Ludwig, "Comparison of Chemical Fractionation Methods for Isolating Stable Soil Organic Carbon Pools," *European Journal of Soil Science*, Vol. 58, No. 6, 2007, pp. 1316-1329. http://dx.doi.org/10.1111/j.1365-2389.2007.00926.x

[27] S. J. Kalembasa and D. S. Jenkinson, "A Comparative Study of Titrimetric and Gravimetric Methods for the Determination of Organic Carbon in Soil," *Journal of the Science of Food and Agriculture*, Vol. 24, No. 9, 1973, pp. 1085-1090. http://dx.doi.org/10.1002/jsfa.2740240910

[28] D. W. Nelson and L. E. Sommers, "Total Carbon, Organic Carbon and Organic Matter," In: D. L. Sparks, A. L. Page, P. A. Helmke and R. H. Loeppert, Eds., *Methods of Soil Analysis Part 3—Chemical Methods*, ASA, SSSA, Inc. Madison, 1996, pp. 961-1010.

[29] G. P. Gillman, D. F. Sinclair and T. A. Beech, "Recovery of Organic Carbon by the Walkley and Black Procedure in Highly Weathered Soils," *Communications in Soil Science and Plant Analysis*, Vol. 17, No. 8, 1986, pp. 885-892. http://dx.doi.org/10.1080/00103628609367759

[30] K. R. Brye and N. A. Slaton, "Carbon and Nitrogen Storage in a Typic Albaqualf as Affected by Assessment Method," *Communications in Soil Science and Plant Analysis*, Vol. 34, No. 11-12, 2003, pp. 1637-1655. http://dx.doi.org/10.1081/CSS-120021302

[31] M. Diaz-Zorita, "Soil Organic Carbon Recovery by the Walkley-Black Method in a Typic Hapludoll," *Communications in Soil Science and Plant Analysis*, Vol. 30, No. 5-6, 1999, pp. 739-745. http://dx.doi.org/10.1080/00103629909370242

[32] S. Lettens, B. De Vos, P. Quataert, B. Van Wesemael, B. Muys and J. Van Orshoven, "Variable Carbon Recovery of Walkley-Black Analysis and Implications for National Soil Organic Carbon Accounting," *European Journal of Soil Science*, Vol. 58, No. 6, 2007, pp. 1244-1253. http://dx.doi.org/10.1111/j.1365-2389.2007.00916.x

[33] L. J. Janik, R. H. Merry and J. O. Skjemstad, "Can Mid Infrared Diffuse Reflectance Analysis Replace Soil Extractions?" *Australian Journal of Experimental Agriculture*, Vol. 38, No. 7, 1998, pp. 681-696. http://dx.doi.org/10.1071/EA97144

[34] X. M. Yang, H. T. Xie, C. F. Drury, W. D. Reynolds, J. Y. Yang and X. D. Zhang, "Determination of Organic Carbon and Nitrogen in Particulate Organic Matter and Particle Size Fractions of Brookston Clay Loam Soil Using Infrared Spectroscopy," *European Journal of Soil Science*, Vol. 63, No. 2, 2012, pp. 177-188. http://dx.doi.org/10.1111/j.1365-2389.2011.01421.x

[35] M. Zimmermann, J. Leifeld and J. Fuhrer, "Quantifying Soil Organic Carbon Fractions by Infrared-Spectroscopy," *Soil Biology and Biochemistry*, Vol. 39, No. 1, 2007, pp. 224-231. http://dx.doi.org/10.1016/j.soilbio.2006.07.010

[36] L. J. Janik, J. O. Skjemstad, K. D. Shepherd and L. R. Spouncer, "The Prediction of Soil Carbon Fractions Using Mid-Infrared-Partial Least Square Analysis," *Soil Research*, Vol. 45, No. 2, 2007, pp. 73-81. http://dx.doi.org/10.1071/SR06083

[37] G. M. Vasques, S. Grunwald and J. O. Sickman, "Comparison of Multivariate Methods for Inferential Modeling of Soil Carbon Using Visible/Near-Infrared Spectra," *Geoderma*, Vol. 146, No. 1-2, 2008, pp. 14-25. http://dx.doi.org/10.1016/j.geoderma.2008.04.007

[38] V. Bellon-Maurel and A. McBratney, "Near-Infrared (NIR) and Mid-Infrared (MIR) Spectroscopic Techniques for Assessing the Amount of Carbon Stock in Soils—Critical Review and Research Perspectives," *Soil Biology and Biochemistry*, Vol. 43, No. 7, 2011, pp. 1398-1410. http://dx.doi.org/10.1016/j.soilbio.2011.02.019

[39] L. Cécillon, B. G. Barthès, C. Gomez, D. Ertlen, V. Genot, M. Hedde, A. Stevens and J. J. Brun, "Assessment and Monitoring of Soil Quality Using Near-Infrared Reflectance Spectroscopy (NIRS)," *European Journal of Soil Science*, Vol. 60, No. 5, 2009, pp. 770-784. http://dx.doi.org/10.1111/j.1365-2389.2009.01178.x

[40] E. Ben-Dor, K. Patkin, A. Banin and A. Karnieli, "Mapping of Several Soil Properties Using DAIS-7915 Hyperspectral Scanner Data—A Case Study over Clayey Soils in Israel," *International Journal of Remote Sensing*, Vol. 23, No. 6, 2002, pp. 1043-1062. http://dx.doi.org/10.1080/01431160010006962

[41] A. Chatterjee, R. Lal, L. Wielopolski, M. Z. Martin and M. H. Ebinger, "Evaluation of Different Soil Carbon Determination Methods," *Critical Reviews in Plant Sciences*, Vol. 28, No. 3, 2009, pp. 164-178. http://dx.doi.org/10.1080/07352680902776556

11

Changes in Soil Composition and Floral Coverage on a Glacial Foreland Chronosequence in Southern Iceland

Lawrence H. Tanner[1], Ann E. Walker[2], Morgan Nivison[1], David L. Smith[1]

[1]Department of Biological Sciences, Le Moyne College, Syracuse, USA; [2]Department of Earth & Planetary Sciences, and Science Alliance Center of Excellence, University of Tennessee, Knoxville, USA.
Email: tannerlh@lemoyne.edu

ABSTRACT

The land surface in front of the Skaftafellsjökull in southern Iceland, exposed by ice recession commencing about the start of the twentieth century, constitutes a foreland with a maximum age of about 100 years and a more distal outwash plain. The ages of different surfaces within this sequence are constrained by moraines of known or estimated ages. Across this chronosequence, we measured at various sites the extent of floral coverage of the surface, the soil carbon and nitrogen contents of the substrate and the soil CO_2 flux rate. All measured parameters exhibit values increasing with distance from the ice front, which correlates approximately with age. The strongest correlations are seen between distance and the carbon and nitrogen concentrations of the soil. Marked horizonation of the soil is observed only on the oldest surfaces (100+ years).

Keywords: Skaftafellsjökull; Glacial Foreland; Chronosequence; Soil Carbon; Soil CO_2 Flux

1. Introduction

Current concerns with climate change often focus on the carbon cycle, and more specifically, the various sources and sinks of carbon exchangeable on short (decadal) time scales. To this end, various studies have attempted to quantify the reservoirs and fluxes of carbon from both marine and continental reservoirs [1]. At a broad approximation, the biomass of primary producers stores c. 600 Gt of elemental carbon, somewhat less than the atmospheric reservoir of c. 760 Gt·C and the estimated 1600 Gt·C stored in soils and marine sediments [2]. The carbon in these reservoirs is rapidly exchangeable, with an annual flux of roughly 60 Gt from the atmosphere to the biosphere (by photosynthesis) and approximately the same amount returned via respiration, decay and fermentation. Not surprisingly, given their immense bioproductivity, tropical ecosystems account for approximately 20% of the terrestrial organic carbon inventory, yet tundras also contain a not inconsiderable 8% of carbon stocks [3].

Accurate tracking of changes in the global carbon cycle will require a thorough understanding of the fluxes of carbon in all environments. Moreover, knowledge of how these fluxes will vary as a response to on-going and anticipated climate change is crucial to predicting changes in the size of these carbon reservoirs in the future. Currently, the ability of high-latitude lands to store excess atmospheric carbon is much lower than for tropical lands [4]. But warming at high latitudes, which likely will be of substantially greater magnitude than it will be in the tropics, is likely to modify the role of high latitude soils by changing both carbon storage, as bioproductivity increases, and carbon emission, as respiration is stimulated. As described in Callaghan and others [5], the response of arctic and subarctic ecosystems to climate change (e.g., changes in primary production) will depend on the response of microbial decomposers to temperature change and litter and water availability.

Glacial foreland chronosequences have long been considered optimal settings for the study of rates and processes of landscape modification through floral colonization and pedogenesis. For example, Messer [6] found that across glacial foreland chronosequences in southern Norway soil depth and organic content increased, and pH decreased with surface age. Significantly, this study found that although some soil property chronofunctions, e.g., cation exchange capacity, were variable with local climate conditions, others, such as pH and organic content, were not. Similarly, Matthews [7] well-documented the importance of organic processes in development and horizonation of the solum. However, the effects of climate warming in these environments are likely to be limited by nutrient availability. The importance of nutrient availability, specifically nitrogen fixation, was first investigated by Crittendon [8], who studied lichen colonization of glacial drift in southern Iceland. Later, Hodkinson and others [9] studied glacial foreland chronosequences on Svalbard and found that cyanobacterial colonization was important for raising nutrient status prior to vascular plant colonization. This study also found that vascular plant species richness increased for the first c. 100 years, beyond which bryophytes increased in dominance.

This study examines the rates and interactions of the processes of floral colonization and soil formation in the sub-arctic environment of southern Iceland. The study was conducted on a chronosequence, *i.e.*, a sediment surface on which the age of exposure is fairly well known. We conducted our study on the deposits of the Skaftafellsjökull that are currently being exposed by glacial recession.

2. Study Location

Vatnajökull, the largest ice cap in Iceland (and Europe), measures 8100 km^2 (from Landsat in 1991) and is the source of nearly 30 individual outlet glaciers, most of which occupy valleys or broad, subglacial channels. For the most part, these glaciers reached their Little Ice Age maximum extent near the end of the nineteenth century, and have been in recession much of the time since. Comprehensive monitoring of the location of the ice front for most of these glaciers was initiated in the early 1930s, although earlier data exist for some glaciers. Data on the movements of these glaciers have been and continue to be collected by a combination of scientists and local residents. Measurements were made initially by tape measure and later by theodolite, and are archived by the Icelandic Glaciological Society. Tabulations of these data are published regularly in the journal *Jökull*. Most of these glaciers were in retreat through the 1930s and into the 1940s, responding to a warming climate, but following cooling early in the latter half of the twentieth century, recession rates slowed through the 1950s and 1960s, and some glaciers re-advanced both in the early 1950s and in the late 1960s through 1970s. Consistent warming since the mid-1980s has led to retreat of almost all of these outlet glaciers over the last two decades.

Retreat of these glaciers over the past century or more has exposed foreland areas with varying topography. Most of the non-surging glaciers have left recessional moraines that can be dated reasonably accurately. The foreland of the Skaftafellsjökull was chosen for this study for two reasons: 1) the age of the land surface at various distances from the glacier terminus can be constrained by the observations of ice recession and readvance described above; and 2) the location within Vatnajökull National Park (or Vatnajökulsþjóðgarður, formerly Skaftafell National Park) has protected a portion of the land from grazing by large herbivores (sheep and reindeer) for several decades, allowing a natural progression of ecological succession. The most distal moraine of the Skaftafellsjökull that is clearly identifiable is located about 2 km from the current ice front and is approximately dated to the position of the terminus in 1904. A much more pronounced topography is formed by a set of nested moraines that may be associated with re-advance in the early 1950s [10]. A similar interval of re-advance occurred in the 1970s [10], and the most proximal set of moraines dates to the position of the glacial front as late as 1980. Between the moraines are flatter areas formed by alluvial outwash, incised channels and terraces. The ages of these inter-moraine surfaces are not as well constrained, other than by the moraines that form their boundaries, due to the potential of reactivation of the outwash channels over the course of many years. Therefore, the ages of the surfaces can best be characterized as the pre-1904 outwash plain, the 1904 moraine, the 1904-1950 outwash surface, the 1950s moraine, the 1950s-1970 outwash surface, the 1980 moraine, and the post-1980 surface. Thus, the outwash and foreland of the Skaftafellsjökull forms a chronosequence, more specifically a post-incisive chronosequence, *sensu* Vreeken [11] and Hugget [12], in which distance from the ice-front equates (approximately) to total age of the surface.

The study site is located at 64°N latitude, and despite the proximity to the Arctic Circle, is subject to a cool maritime climate. The annual mean temperature along the south coast is 4°C to 6°C, with a winter (January) mean of near 0°C, and a summer (July) mean of 10°C, although air temperatures in proximity of the glacier are greatly variable and often significantly cooler. The land surface closest to the glacier is subject to katabatic winds of variable strength. Mean annual precipitation on this area of the south coast is approximately 1500 mm.

3. Methods

We established 13 study sites between the current front of the glacier and the outwash plain distal to the position of the glacier at the start of the twentieth century, about 2 km from the present position of the terminus (**Figure 1**). At all of the sites, the substrate is stony and well drained.

Ten of these sites lie within an area that is fenced to prevent the entry of grazing animals, with the remaining three falling outside this boundary. At each site, we marked out a 50 m line oriented parallel to the glacial front and set sample locations every 10 m along this line.

The position of each site was recorded by GPS. At each of the five sample locations, vegetation (if any) was

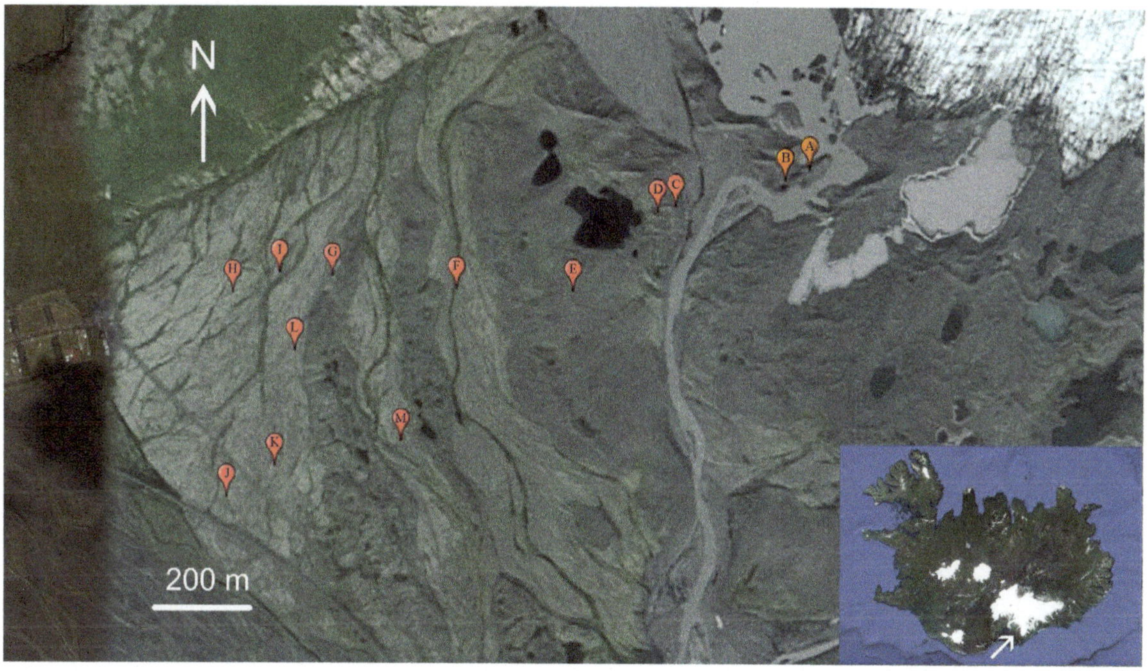

Figure 1. Aerial view of the foreland and outwash plain of the Skaftafellsjökull (ice visible to upper right). Sites B-D are in the ridges from the 1970s-1980s. The 1950s moraines occur on either side of Site M. The 1904 ridge is subdued, but is approximated by the position of Site I. Image adadpted from Google Earth®, acquisition date 10/2012.

cleared and two 10 cm soil cores were drawn with a 2.5 cm diameter hand auger. The vegetation coverage was determined by visual estimation within a 0.25 m^2 quadrat that was flipped four times to provide complete coverage for one square meter. The soil CO_2 flux was measured at each sample location with a LI-COR LI-8100A® soil gas chamber system equipped with soil moisture and temperature probes. The collection chamber was set on a soil collar with 8 cm vertical offset. CO_2 concentrations were measured continuously for 90 seconds following a 45 second purge cycle and the soil CO_2 flux calculated from the regression line in units of μmol $CO_2/sec/m^2$.

Soil carbon and nitrogen content were measured from the soil cores. Each core was dried, homogenized, sieved with a 2 mm screen to remove larger rocks and root fragments, and the sub-2 mm fraction pulverized. From each processed sample, 0.1 to 0.125 g was drawn for analysis with a Leco TruSpec CN® by combustion in a pure O_2 atmosphere at 950°C. The weight percent carbon and nitrogen were calculated from the composition of the evolved gases as measured by infrared and thermal conductivity cells, respectively.

In summary, the data collected at each of the 13 study sites included five soil CO_2 flux measurements, 10 soil samples and 20 measurements of the floral coverage.

4. Results

The goal of this project was to characterize the processes and rates of change of the land surface of this chronosequence following glacial retreat. Our investigation demonstrates a general trend of increasing plant cover, soil carbon, soil nitrogen and soil CO_2 flux with increasing distance from the glacier, which approximates the age of the chronosequence surface, although with substantial variations among sample locations along each sample line (**Table 1**, **Figure 2**). Given that the range of ages of the land surfaces encompassed by this chronosequence spans approximately just over one century only, pedogenesis is more difficult to quantify. Distinctly visible horizonation is visible only on the oldest surfaces, as discussed below.

In addition to measuring the extent of vegetation cover at each sample location, we also made note of prominent plant species; a partial list of common species includes: red fescue grass (*Festuca rubra*), moss campion (*Silene acaulis*), woolly willow (*Salix lanata*), dwarf birch (*Betula nana*), downy birch (*B. pubescens*), crowberry (*Empetrum nigrum*), bearberry (*Arctostaphylos uva-ursi*), fleabane (*Erigeron* sp.), wild thyme (*Thymus praecox*), lady's bedstraw (*Galium verum*), heather (*Calluna vulgaris*), lady's mantle (*Alchemilla* sp.), purple saxifrage (*Saxifraga oppositifolia*), rough horsetail (*Equisetum hiemale*), biting stonecrop (*Sedum acre*), hairy stonecrop (*S. villosum*), various mosses, and foliose and crustose lichens.

Plant cover is sparse to nonexistent on the youngest surfaces in proximity to the glacier. Site A is located on gravelly till proximal to a proglacial lake, on a slope facing the glacier (**Figure 3(A)**) and therefore has an expo-

Table 1. Summary data from all study sites. Cover = % surface area with vegetation (mean of 20 measurements); C% and N% represent mean of 10 samples each with standard deviation in parentheses; Soil CO_2 flux represents mean of five measurements (reported in μmol/sec/m^2).

Site	Surface	Age	Distance (m)	Cover%	C% (σ)	N% (σ)	Soil CO_2 flux
A	recent till	post-1980	236	<5	0.075 (0.006)	0.027 (0.033)	0.07
B	moraine	1980	278	35	0.134 (0.061)	0.032 (0.030)	0.11
C	moraine	1970-1980	564	65	0.19 (0.11)	0.163 (0.186)	1.97
D	moraine	1970-1980	620	80	0.508 (0.207)	0.171 (0.145)	0.18
E	terrace	1950s-1970	970	60	0.108 (0.067)	0.091 (0.038)	0.4
F	terrace	1950s-1970	1209	75	0.181 (0.046)	0.18 (0.047)	1.82
G	outwash	1950	1405	65	0.645 (0.374)	0.165 (0.029)	2.08
H	outwash	pre-1904	1672	90	0.614 (0.296)	0.12 (0.056)	0.38
I	boulder moraine	1904	1527	93	1.101 (1.096)	0.143 (0.132)	0.66
J	outwash-grazed	pre-1904	1935	43	0.199 (0.047)	0.156 (0.01)	0.5
K	moraine-grazed	1904	1785	58	0.454 (0.396)	0.176 (0.059)	0.45
L	outwash	1904-1950	1589	85	0.598 (0.304)	0.26 (0.223)	1.08
M	terrace-grazed	1950s	1485	38	0.463 (0.074)	0.142 (0.089)	0.77

sure age of several decades at most. Vegetation is nearly absent with the exception mainly of widely separated tufts of grass (red fescue). The values of soil carbon, nitrogen and soil CO_2 flux measured here are the lowest among all studied sites. Site B is directly in the lee of the most proximal ridge of the 1980 moraine. Floral cover is substantially higher, consisting mainly of moss, but also including abundant grass, wooly willow and crowberry (**Figure 3(B)**). Measurements of the carbon and nitrogen content of the soil are accordingly higher (**Table 1**). Sites C and D are also located on till within the nested moraines associated with the position of the ice front in the 1970s to 1980s (**Figure 1**). The floral cover is higher at these sites, and the measured soil carbon and nitrogen are successively higher at both of these sites. Soil CO_2 flux at Site C was an order of magnitude higher than at Sites B or D.

Site E was situated on an elevated terrace located between the most recent moraine ridges and those formed in the early 1950s and the 1970s. The surface is covered by gravel to boulder-size clasts, although soil cores demonstrate a sandy-pebbly substrate beneath this coarse veneer, suggesting that the surface has been modified by deflation. Floral cover is lower and more discontinuous than at Sites C and D, consisting of patches of moss, willow and birch, and values of soil carbon, nitrogen and CO_2 flux are all lower than at Sites C and D. Site F was situated on the terrace of a stream incised proxi higher on these alluvial gravels, dominated by moss, willow, northern crowberry and bearberry. Measurements of soil carbon, nitrogen and CO_2 flux are all elevated with respect to Site F, and are similar to the values measured at Site C. Sites G and L are on the outwash flat immediately distal to the 1950s boulder moraine. Although floral cover at Site G is lower than at Sites L and F, with moss providing the dominant coverage, soil carbon and soil CO_2 flux are elevated with respect to both sites.

Sites H and I are distal to the glacier, with Site H situated on the outwash plain beyond the most distal moraine, an area of ridge and swale topography (**Figure 3(D)**). The oldest moraine is a subdued boulder ridge dated to 1904. Site I was located on the flat of this moraine. These two sites displayed the most complete floral coverage of all sites in the study, in excess of 90% of the surface area. Most of this coverage consists of thick moss, with subordinate willow and birch. Soil carbon and nitrogen are elevated at both sites, and carbon at Site I is the highest among the study sites, with a value of just over 3% representing the highest individual measurement in the study. Soil CO_2 flux is not similarly elevated at these sites, however. Site H, representing the oldest part of the chronosequence, demonstrates the clearest horizonation of the substrate, with a 10 cm silty-sandy, pale brown (5 YR 5/2) A horizon overlying a greenish-gray (5 GY 5/1), sandy-pebbly C horizon (**Figure 4**). Sites J, K and M all are located outside of the fence that protects the park land from grazing. Although the ages of the surfaces differ at these sites (M is 1950s, K is 1904, J is pre-1904), the floral cover and soil carbon content for these sites is significantly reduced with respect to comparably aged sites that are not grazed. Soil nitrogen and soil CO_2 flux are similar, however.

5. Discussion

5.1. Carbon Storage and Soil CO_2 Flux

As shown in **Figure 2**, the carbon and nitrogen content of the soil increases with distance from the ice front, dem-

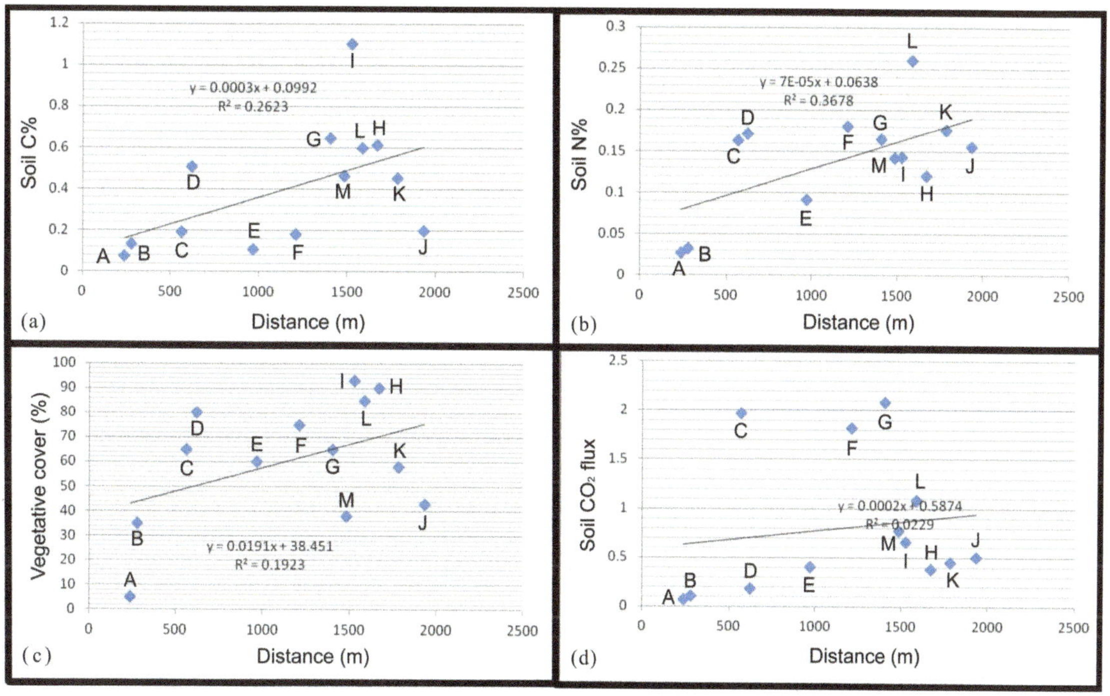

Figure 2. Measured parameters of (a) soil carbon (weight%), (b) soil nitrogen (weight%), (c) floral coverage (% surface area) and (d) soil CO_2 flux (μmol/sec/m$_2$) plotted against distance from the ice front. Letters reference sample locations from Figure 1.

Figure 3. Sample sites on the Skaftafellsjökull chronosequence. (A) Site A is located on an ice contact till surface inclined toward the glacier. The vegetation consists solely of scattered tufts of red fescue. (B) Site B is near Site A, but behind an ice contact ridge with aspect away from the glacier. The wood stake is in a patch of crowberry. (C) Site M is situated between the two ridges formed by re-advance in the early 1950s. The surface is vegetated by moss and low patches of birch. (D) Site I is located on a low boulder ridge formed in 1904. Most of the shrubbery consists of birch, with a small tuft of willow.

onstrating clearly that elemental carbon and nitrogen storage on the chronosequence are at least in part a function of time. The rate of increase appears as roughly a linear function. Thus, a carbon content of ~1%, as at Site I (the maximum site value), where the surface is approximately 100 years old, equates to a mass of elemental carbon of 2.4 kg/m^2, or 2.4 × 10^4 kg/ha, assuming a uniform carbon concentration through a 10 cm solum in a soil with a mean dry density of 2.4 g/cm^3. Across the chronosequence, therefore, elemental carbon is sequestered in the subsurface at a rate of c. 240 kg/ha/yr. The ratio of carbon to nitrogen in these soils is extremely variable, but averages about 3. Thus, the rate of nitrogen storage is roughly one-third that of carbon. These rates accord well with those from similar studies of chronosequences in the Swiss Alps [13].

The relationship between soil CO_2 flux and distance is not nearly so clear. Soil CO_2 is generated primarily by root respiration and aerobic decomposition of soil organic matter. Both factors vary directly with temperature, and the lower soil temperatures in proximity to the ice will have an inhibiting effect, particularly on rates of decomposition. Root respiration will be a greater source of soil CO_2 in the presence of large vascular plants.

Thus, surfaces dominated by bryophytes are likely to produce a lower CO_2 flux than those vegetated surfaces with a more diverse plant assemblage. For example, sites H and I have the highest values of floral coverage, but relatively low soil CO_2 flux rates, likely due to the fact that most of the surface is covered by mosses.

5.2. Soil Formation

Multiple studies have documented the dominance of physical processes in the initial stages of soil formation on glacial forelands. For example, Romans and others [14] described initial formation of a subsurface silty layer in the Breiðamerkurjökull chronosequence within a several years of exposure, and development of a weak silt-droplet fabric within thirty years. Similarly, Frenot and others [15] determined, through study of Antarctic proglacial chronosequences, that mechanical translocation, and mechanical processes in general, is much more important than biological modification of the soil in the initial stages of development. These findings mirrored those of Matthews [7]. Furthermore, Arnalds [16] found that plant succession is intimately related to physical substrate modification. Pioneer species are characterized by very shallow root systems that permit them to colonize the unmodified substrate. Modification of the substrate by cryogenesis decreases over time with the development of surface lag, and eolian processes continue in importance through formation of silt caps. Deeper-rooted plant species colonize only following stabilization of the surface (ending cryogenesis) and deeper translocation of the nutrient-bearing and water-retaining silty layer.

The primary difference in the parent material at our different sample sites is in particle size. The glacial foreland deposits are proximal to the glacial front, generally within 1.5 km of the current ice position, and thus are

Figure 4. Soil profile at Site H (pre-1904). (A) The surface is covered mainly by a mat of moss, with secondary red fescue (f), crowberry (c) and foliose lichen (l). (B) Detail of (A) illustrates organic enrichment of silty A Horizon over C Horizon.

particularly coarse-grained. Our preliminary investigations show that the ice-contact deposits (moraines) are largely matrix to clast-supported gravels and boulder deposits. The proximal outwash deposits are more variable, ranging from gravelly clast-supported to matrix-supported gravelly sands. Distal outwash deposits (beyond the foreland) are uniformly matrix-supported gravelly sands. In the course of our investigation, we observed no clear evidence of the separation of distinct soil horizons until we examined the oldest surfaces on the outwash plain. Here we found a distinct A Horizon, in which the color demonstrates organic enrichment, and the grain size indicates mechanical translocation of fine-grained sediment, undoubtedly of eolian origin.

Almost all soils in Iceland are derived from volcanic materials [17] and thus may be related to the soil order Andisols. In addition to tephra, these may form on glacial tills, glacio-fluvial and glacio-lacustrine deposits, and igneous rock surfaces [18]. The sample sites are located on either glacial forelands or glacial outwash plains. Thus the parent material for both sets of sample sites consists of glacial debris, deposited either as outwash or as ice-contact deposits (moraines), and thus can be related to the Selsund pedon of Arnalds and Kimble [19]. All of the parent material is derived ultimately from volcanic activity, which is primarily basaltic. There is no significant difference in parent material composition between sample sites. Hence all soils in this study can be regarded as andic, either cryic andisols (or Cryands), primarily on the moraines, or andic cryofluvents on the outwash deposits [20].

5.3. Floral Cover

The development of the floral cover is controlled by substrate, as described above, and also aspect (slope orientation), particularly in proximity to the ice front, where the influence of katabatic winds is strongest. At Sites A and B, for example, difference in floral cover relates primarily to aspect toward the ice at A and away from the ice at B. Similarly, Site E is located on an exposed terrace on an older surface than Site D, but has lower floral coverage. Human intervention is also a factor in controlling vegetation. A portion of the surface in the park boundaries at Skaftafell has been protected from grazing since the late 1970s. The decreased floral coverage at sites J, K and M demonstrates the significance of the effect of grazing on the plant cover.

6. Conclusion

Measurements of the extent of vegetative cover, soil carbon and nitrogen content and soil CO_2 flux on the chronosequence formed by the twentieth century retreat of the Skaftafellsjölull demonstrate that all four parameters increase in value approximately as a function of distance from the current ice front, which serves as a proxy for time. In addition to distance (time) from the ice, aspect is an important control on floral cover of the surface through shelter, or lack thereof, from katabatic winds. Additionally, the type of vegetation is a control on soil CO_2 flux as surfaces dominated entirely by bryophytes undergo lower rates of root respiration than do areas dominated by vascular plants. Well-developed soils, with clearly differentiated A and C horizons form through organic carbon accumulation and translocation of fine-grained sediment on the oldest surfaces studied.

7. Acknowledgements

Dr. Guðrun Gisladottir, of the University of Iceland, provided encouragement and advice at the initiation of this project. We also thank Jessica Curry, Corinne Murray, Marcia Ruwe and Ben Thivierge for assistance in the field and laboratory. Finally, we thank the management of Vatnajökull National Park for granting access for this study.

REFERENCES

[1] P. Bousquet, P. Peylin, P. Ciais, C. Le Quere, P. Friedlingstein and P. P. Tansa, "Regional Changes in Carbon Dioxide Fluxes of Land and Oceans Since 1980," *Science*, Vol. 290, 2000, pp. 1342-1346. doi:10.1126/science.290.5495.1342

[2] L. R. Kump, J. F. Kasting and R. G. Crane, "The Earth System," 2nd Edition, Pearson-Prentice Hall, New York, 2004.

[3] Food and Agriculture Organization of the United Nations, "State of the World's Forests," 2001. http://www.fao.org/docrep/003/y0900e/y0900e00.htm

[4] B. B. Stephens, K. R. Gurney, P. P. Tans, C. Sweeney, W. Peters, L. Bruhwiler, P. Ciais, M. Ramonet, P. Bousquet, T. Nakazawa, S. Aoki, T. Machida, G. Inoue, N. Vinnichenko, J. Lloyd, A. A. Jordan, M. Heimann, O. Shibistova, R. L. Langenfelds, L. P. Steele, R. J. Francey and A. S. Denning, "Weak Northern and Strong Tropical Land Carbon Uptake From Vertical Profiles of Atmospheric CO_2," *Science*, Vol. 316, No. 5832, 2007, pp. 1732-1735. doi:10.1126/science.1137004

[5] T. V. Callaghan, S. Jonasson, H. Nichols, R. B. Hetwood and P. A. Wookey, "Arctic Terrestrial Ecosystems and Environmental Change," *Philosophical Transactions in the Physical Sciences: Engineering*, Vol. 352, No. 1699, 1995, pp. 259-276. doi:10.1098/rsta.1995.0069

[6] A. C. Messer, "Regional Variations in Rates of Pedogenesis and the Influence of Climatic Factors on Moraine Chronosequences, Southern Norway," *Arctic and Alpine Research*, Vol. 20, No. 1, 1988, pp. 31-39. doi:10.2307/1551696

[7] J. A. Matthews, "The Ecology of Recently Deglaciated Terrain: A Geoecological Approach to Glacier Forelands and Primary Succession," Cambridge University Press, New York, 1992.

[8] P. D. Crittendon, "Nitrogen Fixation by Lichens on Glacial Drift in Iceland," *New Physiologist*, Vol. 74, No. 1, 1975, pp. 41-49. doi:10.1111/j.1469-8137.1975.tb01337.x

[9] I. D. Hodkinson, S. J. Coulson and N. R. Webb, "Community Assembly along Proglacial Chronosequences in the High Arctic: Vegetation and Soil Development in North-West Svalbard," *Journal of Ecology*, Vol. 91, No. 4, 2003, pp. 651-663. doi:10.1046/j.1365-2745.2003.00786.x

[10] O. Sigurðsson, "Glacier Variations in Iceland 1930-1995," *Jökull*, Vol. 45, 1998, pp. 3-26.

[11] W. J. Vreeken, "Principle Kinds of Chronosequences and Their Significance in Soil History," *Journal of Soil Science*, Vol. 26, No. 4, 1975, pp. 378-394. doi:10.1111/j.1365-2389.1975.tb01962.x

[12] R. J. Huggett, "Soil Chronosequences, Soil Development, and Soil Evolution: A Critical Review," *Catena*, Vol. 32, No. 3, 1998, pp. 155-172. doi:10.1016/S0341-8162(98)00053-8

[13] F. Conen, M. V. Yakutin, T. Zumbrunn and J. Leifeld, "Organic Carbon and Microbial Biomass in Two Soil Development Chronosequences Following Glacial Retreat," *European Journal of Soil Science*, Vol. 58, No. 3, 2007, pp. 758-762. doi:10.1111/j.1365-2389.2006.00864.x

[14] J. C. C. Romans, L. Robertson and D. L. Dent, "The Micromorphology of Young Soils from South-East Iceland," *Geografiska Annaler: Series A, Physical Geography*, Vol. 62, No. 1-2, 1980, pp. 93-103. doi:10.2307/520456

[15] Y. Frenot, B. V. Vliet-Lanoe and J.-C. Gloaguen, "Particle Translocation and Initial Soil Development on a Glacier Foreland, Kerguelen Islands, Subantarctic," *Arctic and Alpine Reearch*, Vol. 27, No. 2, 1995, pp. 107-115. doi:10.2307/1551892

[16] Y. Frenot, J. C. Gloaguen, M. Cannavacciuolo and A. Bellido, "Primary Succession on Glacial Forelands in the Subarctic Kerguelen Islands," *Journal of Vegetative Science*, Vol. 9, No. 1, 1998, pp. 75-84. doi:10.2307/3237225

[17] O. Arnalds, "Volcanic Soils of Iceland," *Catena*, Vol. 56, No. 1, 2004, pp. 3-20. doi:10.1016/j.catena.2003.10.002

[18] O. Arnalds, C. T. Hallmark and L. P. Wilding, "Andisols

from Four Different Regions of Iceland," *Soil Science Society of America Journal*, Vol. 59, No. 1, 1995, pp. 161-169. doi:10.2136/sssaj1995.03615995005900010025x

[19] O. Arnalds and J. Kimble, "Andisols of Deserts in Iceland," *Soil Science Society of America Journal*, Vol. 65, No. 6, 2001, pp. 1778-1786. doi:10.2136/sssaj2001.1778

[20] A. Arnalds, "Carbon Sequestration and the Restoration of Land Health," *Climatic Change*, Vol. 65, No. 3, 2004, pp 333-346. doi:10.1023/B:CLIM.0000038204.60219.0a

12

Changes in Bacterial Density, CO_2 Evolution and Enzyme Activities in Poultry Dung Amended Soil

Lakshmikanti Bhoi, Pramod Chandra Mishra*

Department of Environmental Sciences, Sambalpur University, Orissa, India.
Email: *profpcmishra@gmail.com

ABSTRACT

The utilization of cattle and poultry manure as organic fertilizer improves soil productivity, but arsenic contaminated poultry dung may interfere in soil metabolism and soil fertility. The study was conducted to assess the effects of poultry dung as well as arsenic contamination on soil properties in 1%, 3% and 5% poultry dung amended soil and 1, 5 and 10 ppm sodium arsenite contaminated soil. pH and conductivity were found to be increasing with increase in poultry dung in soil. Other chemical parameters like nitrate, phosphate and organic carbon were found higher in poultry dung amended soil than that of arsenic contaminated soil. Soil bacteria, CO_2 evolution and enzymatic activities like amylase, invertase and dehydrogenase were also found higher in poultry dung amended soil suggesting the effectiveness of poultry dung in enhancing soil productivity, even if it was contaminated by As through feed additive.

Keywords: Soil; Poultry Dung; Arsenic; Soil Enzyme; CO_2 Evolution; Bacteria

1. Introduction

The increasing demand of chicken meat has prompted more poultry farming with consequent effects on increased utilization of organic wastes (e.g. chicken dung manure) as fertilizers. Organic wastes contain varying amounts of water, mineral nutrients, organic matter [1,2]. While the use of organic wastes as manure has been in practice for centuries world-wide [3,4], there still exists a need to assess the potential impacts of poultry manure on soil chemical properties and crop yield and in particular evaluating the critical application levels. Moreover, the need and utilization of poultry manure has overtaken the use of other animal manure (e.g. pig manure, kraal manure) because of its high content of nitrogen, phosphorus and potassium [5,6]. Escalating prices of inorganic fertilizers due to the increase in the fuel prices has also prompted the use of poultry manure [7]. Similarly, organic wastes are also being advocated for by different environmental organizations world-wide to preserve the sustainability of agricultural systems. Recent studies have shown a host of nutrient management practices undertaken by small scale African farmers [8]. While the relative adoption rates between organic and mineral nutrients vary by location, the incidence of organic practices is often more than the use of mineral fertilizers. Increase in nitrogen levels from 40% - 60% and 17% - 38% with respect to control for Norfolk sandy soils and Cecil sandy loam soils, respectively following application of manure has been reported [9]. In addition, application of poultry manure to soil enhances concentration of water soluble salts. Plants absorb plant nutrients in the form of soluble salts, but excessive accumulation of soluble salts (or soil salinity) suppresses plant growth.

Roxarsone is added to poultry feed at the rate of 22.7 to 45.4 grams per ton, or 0.0025 to 0.005 per cent as feed additive [10]. Most of the roxarsone passes through the birds and is excreted unchanged. Each broiler excretes about 150 milligrams of roxarsone during the 42-day growth period in which it is administered [11]. Chemical and microbial reactions readily transform roxarsone into inorganic forms of arsenic. These inorganic forms are then subject to a variety of chemical and biological reactions in the soil. Soil mineralogy, soil moisture, soil pH, and microbial reactions all determine arsenic mobility and its toxicity.

2. Materials and Methods

The physico-chemical analysis like pH, conductivity, organic carbon, nitrate, phosphate and arsenic content of soil samples were done following standard methods. pH and conductivity were measured using digital pH meter and conductivity meter with automatic temperature compensation and calibrated with calibration solutions [12]. Organic Carbon was determined by Walkey-Black titration method [13], Nitrate was estimated by phenol disul-

*Corresponding author.

phonic acid method [14] and phosphate by stannous chloride method [15]. Estimation of arsenic was done by silver diethyldithiocarbamate method [16].

Amylase and invertase activities in soil were measured employing Sorenson's buffer (0.06 M) with 1% soluble starch for amylase activity and 5% soluble starch for invertase activity [17]. The dehydrogenase activity was estimated by 2,3,5-tetrazolium chloride reduction technique [18]. The CO_2 evolution from soil was measured by alkali absorption method [19,20]. Dilution plate technique was employed for the enumeration of soil bacteria [21].

Laboratory experiment was conducted taking arsenic contaminated soil and poultry dung amended soil to study the soil parameters. Arsenic contaminated soil (1 ppm, 5 ppm and 10 ppm), poultry dung amended soil (1%, 3% and 5%) and control with only soil (3 replicates) were selected. The soil biochemical parameters were measured on 0th, 15th, 30th, 45th and 60th day.

3. Results and Discussion

3.1. Physico-Chemical Characteristics

pH: The pH of soil samples was found to be alkaline in nature. pH in arsenic contaminated soil varied from 8.22 ± 0.31 (0th day) to 8.26 ± 0.29 (60th day) ($F_1 = 470.846^*$, $F_2 = 7.384^{**}$, $p < 0.05$), where as pH of poultry dung amended soil varied from 8.42 ± 0.35 (0th day) and 8.5 ± 0.31 (60th day) ($F_1 = 309.611^*$, $F_2 = 7.567^{**}$, $p < 0.05$). However, the pH of control sample was found to be less than that of arsenic contaminated and poultry dung amended soils (**Table 1**). The pH levels were expected to rise with the addition of the poultry dung due to release of ammonia from the decomposing manure.

Conductivity: Conductivity was found to be increasing with increase in arsenic concentration and poultry dung in soil. In arsenic contaminated soils, conductivity ranged from 49.64 ± 2.74 µS/cm on 0th day to 52.54 ± 1.34 µS/cm on 60th day ($F_1 = 3.607^{**}$, $F_2 = 9.393^{**}$, $p < 0.05$) where as it was highest (280.7 ± 121.77 µS/cm) on 60th day and lowest (277.46 ± 121.21 µS/cm) on 0th day ($F_1 = 113703.2^*$, $F_2 = 40.573^{**}$, $p < 0.05$) in poultry dung amended soil. Like pH, conductivity of control was found less than that of arsenic contaminated and poultry dung amended soils (**Table 1**). Soil EC is the indication of the salinity status of the soil. The high EC in poultry dung is attributable to higher salt levels of N and P nutrients which are proportionally high.

Organic carbon: The organic carbon content in poultry dung amended soils showed an increase with 43.4 ± 1.04 mg/g on 0th day to 47.26 ± 0.6 mg/g on 60th day ($F_1 = 33.24^*$, $F_2 = 0.342^{**}$, $p < 0.05$). But in case of arsenic contaminated soil it was found reverse as 32.3 ± 4.15 mg/g on 0th day to 31.2 ± 4.68 mg/g on 60th day ($F_1 = 58.17^*$, $F_2 = 28.532^*$, $p < 0.05$) which is less than that of control (**Table 1**).

Nitrate: The nitrate content in arsenic contaminated soil was lower than control and was found to be 0.211 ± 0.022 mg/g on 0th day which gradually decreased with increase in days to 0.129 ± 0.058 mg/g on 60th day ($F_1 = 5.344^{**}$, $F_2 = 16.361^*$, $p < 0.05$). Poultry dung amended soil showed increased nitrate content with increase in days as 0.271 ± 0.022 mg/g on 0th day to 0.298 ± 0.012 mg/g on 60th day ($F_1 = 3.877^{**}$, $F_2 = 0.277^{**}$, $p < 0.05$) (**Table 1**).

Phosphate: The phosphate content in arsenic contaminated soil was found to be much lower than that of control and poultry dung amended soil. In control the phosphate content was found to be 4.16 mg/g on 0th day and 4.39 mg/g on 60th day. Both Arsenic contaminated and poultry dung amended soil showed decreased phosphate content in soil with increase in days *i.e.* 2.95 ± 0.15

Table 1. pH, EC, OC, nitrate and phosphate content of arsenic contaminated and poultry dung amended soil.

parameters↓	days↓	control	Arsenic contaminated soil				poultry dung amended soil			
			1 ppm As	5 ppm As	10 ppm As	Avg.	1% PD	3% PD	5% PD	Avg.
pH*	0	7.82	7.92	8.2	8.54	8.22 ± 0.31	8.02	8.56	8.69	8.42 ± 0.35
	60	7.85	8	8.21	8.58	8.26 ± 0.29	8.15	8.59	8.76	8.5 ± 0.31
EC*	0	49.46	48.09	48.03	52.81	49.64 ± 2.74	143.1	310.7	378.6	277.46 ± 121.21
	60	51.52	51.07	52.85	53.71	52.54 ± 1.34	145.8	313.8	382.5	280.7 ± 121.77
Organic Carbon**	0	35.2	34.8	34.6	27.5	32.3 ± 4.15	42.9	42.7	44.6	43.4 ± 1.04
	60	36.8	33.6	34.2	25.8	31.2 ± 4.68	46.7	47.2	47.9	47.26 ± 0.6
Nitrate*	0	0.241	0.236	0.205	0.193	0.211 ± 0.022	0.258	0.259	0.297	0.271 ± 0.022
	60	0.198	0.195	0.113	0.081	0.129 ± 0.058	0.288	0.295	0.312	0.298 ± 0.012
Phosphate*	0	4.16	3.12	2.93	2.81	2.95 ± 0.15	4.86	4.89	4.92	4.89 ± 0.03
	60	4.39	2.28	2.18	2.14	2.2 ± 0.07	3.66	3.38	3.31	3.45 ± 0.18

*Significant; **Not significant (Two-way ANOVA, $p < 0.05$).

mg/g to 2.2 ± 0.07mg/g (F_1 = 11.839*, F_2 = 4.178**, $p < 0.05$) and 4.89 ± 0.03 mg/g to 3.45 ± 0.18 mg/g (F_1 = 0.037**, F_2 = 5.746**, $p < 0.05$) respectively (**Table 1**).

Mitchell and Donald reported that repeated application of poultry litter can increase the soil carbon content along with nitrogen and phosphorus [22]. With addition of poultry dung, there was increased phosphate content [2, 23] where in case of arsenic contaminated soil, the organic carbon, nitrate and phosphate content was found lower than control which may be due to their utilization by microbial components in soil.

3.2. Soil Bacteria

The total soil bacterial ($\times 10^3$ CFU/g soil) population was found to be higher in poultry dung amended soil than that of control and As contaminated soil. In arsenic contaminated soil, it was found decreasing with increasing days *i.e.* 20 ± 2 on 0th day to 13 ± 2 on 60th day. Whereas, control and poultry dung amended soil showed increased soil bacterial density with increasing days *i.e.* 24 on 0th day to 31 on 60th day and 36 ± 4 on 0th day to 40 ± 2 on 60th day respectively (**Figure 1**).

Many authors have reported toxic effects of various metals on microorganisms and found heavy metals as known inhibitors of bacterial population [24-30]. They are of the view that microorganisms are the first group of soil organisms to be affected because of their ubiquity, abundance, shape and small size and consequently extremely large surface area to volume ratio. It is known that sufficient metal exposure would normally results in immediate death of cells due to changes in their viability or competitive ability.In case of Poultry dung contaminated with As, there was no detrimental effect on soil bacteria.

3.3. CO_2 Evolution

The rate of CO_2 evolution ($g/m^2/hr$) was found to be higher in poultry dung amended soils than the control and arsenic contaminated soils (**Figure 2**). Control soil showed CO_2 evolution of 0.083 on 0th day and 0.095 on 60th day. Arsenic contaminated soil showed decreased CO_2 evolution of 0.053 ± 0.014 on 0th day to 0.031 ± 0.009 on 60th day. Whereas poultry dung amended soil showed an increased CO_2 evolution of 0.226 ± 0.038 on 0th day to 0.341 ± 0.018 on 60th day. Such significant increase may be due to the fact that the poultry dung amendments in soil provide the nutritive elements for mineralization by micro flora [31-33].

3.4. Soil Enzymes

Amylase Activity: Amylase activity (mg glucose/g soil/24 hr) varied from 0.689 on 0th day to 0.744 on 60th day in control and 0.723 ± 0.077 on 0th day to 0.786 ± 0.043 on 60th day in poultry dung amended soil. Unlike control and poultry dung amended soil, arsenic contaminated soil showed a decreased amylase activity with increase in days *i.e.* 0.577 ± 0.095 on 0th day to 0.383 ± 0.118 on 60th day (**Figure 3**).

Invertase Activity: Invertase activity (mg glucose/g soil/24 hr) also showed the same trend as that of amylase activity. It varied from 1.595 on 0th day to 1.662 on 60th day in control and 0.812 ± 0.482 on 0th day to 2.108 ± 0.091 on 60th day in poultry dung amended soil. Unlike control and poultry dung amended soil, arsenic contaminated soil showed a decreased invertase activity with increase in days *i.e.* 0.812 ± 0.482 on 0th day to 0.717 ± 0.453 on 60th day (**Figure 4**).

Dehydrogenase Activity: Like amylase and invertase activity, dehydrogenase activity (µg triphenyl formazan /g soil/24 hr) showed an increased value in respect to days with 2.436 on 0th day to 2.603 on 60th day in control and 2.562 ± 0.472 on 0th day to 2.885 ± 0.226 on 60th day in poultry dung amended soil. Arsenic contaminated soil

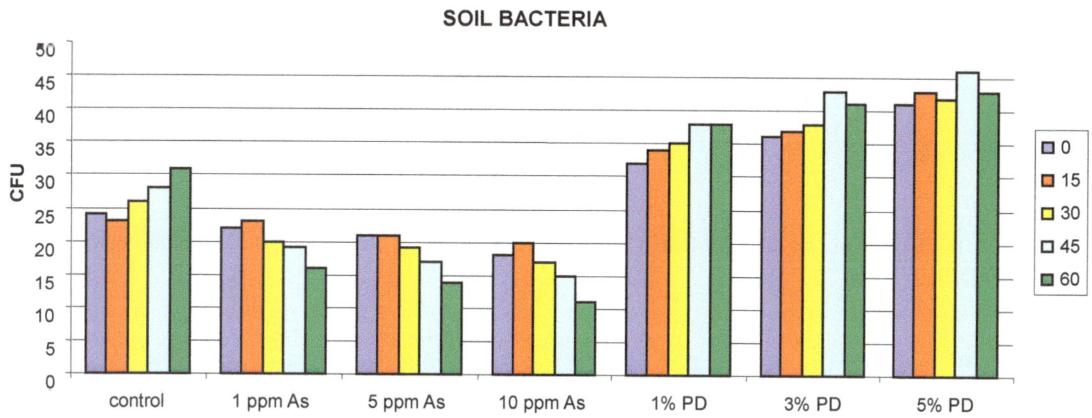

Figure 1. Soil bacteria in contaminated soil and poultry dung amended soil. Arsenic contaminated soil: (F_1 = 9.723*, F_2 = 0.871, $p < 0.05$); Poultry dung amended soil: (F_1 = 141.99*, F_2 = 13.471*, $p < 0.05$); *Significant; **Not Significant (Two-way ANOVA).**

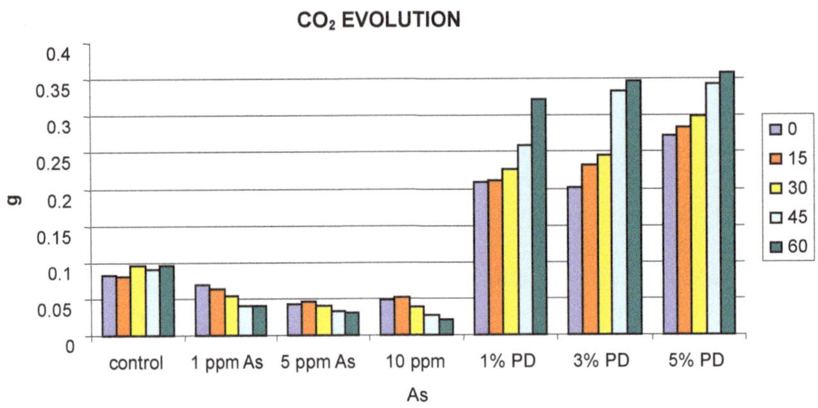

Figure 2. CO_2 **evolution in arsenic contaminated soil and poultry dung amended soil. Arsenic contaminated soil:** $F_1 = 13.108^*$, $F_2 = 3.36^*$, $p < 0.05$**); Poultry dung amended soil: (**$F_1 = 48.797^*$, $F_2 = 4.83^*$, $p < 0.05$**); *Significant (Two-way ANOVA).**

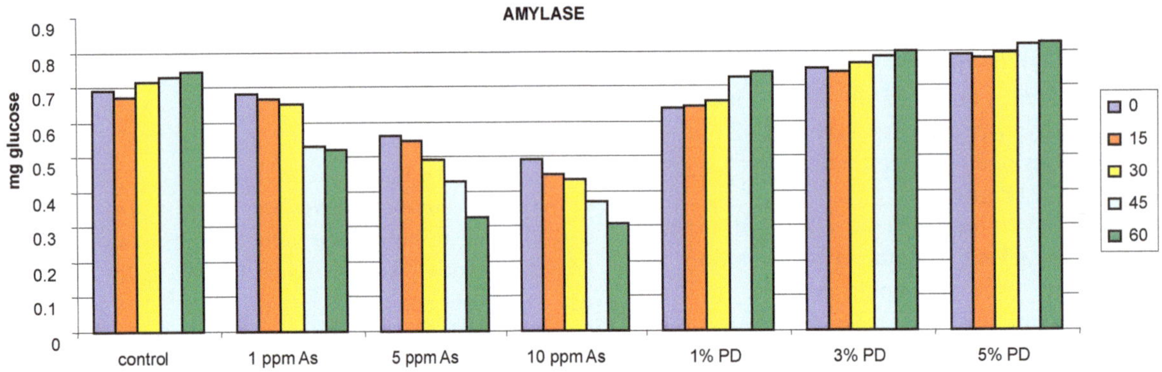

Figure 3. Amylase activity in arsenic contaminated soil and poultry dung amended soil. Arsenic contaminated soil: ($F_1 = 28.042^*$, $F_2 = 3.535^*$, $p < 0.05$**); Poultry dung amended soil: (**$F_1 = 69.542^*$, $F_2 = 17.899^*$, $p < 0.05$**); *Significant (Two-way ANOVA).**

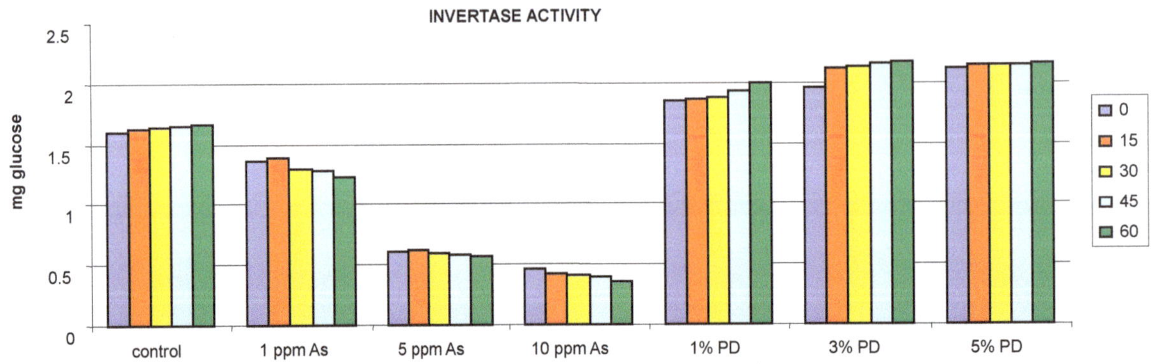

Figure 4. Invertase Activity in arsenic contaminated soil and poultry dung amended soil. Arsenic contaminated soil: ($F_1 = 1069.054^*$, $F_2 = 1.587^{**}$, $p < 0.05$**); Poultry dung amended soil: (**$F_1 = 199.081^*$, $F_2 = 5.577^*$, $p < 0.05$**); *Significant; **Not Significant (Two-way ANOVA).**

showed decreased enzyme activity with increase in days with 1.891 ± 0.219 on 0th day to 1.619 ± 0.42 on 60th day (**Figure 5**).

4. Conclusion

In poultry dung amended soils, bacterial density, CO_2 evolution and the enzymatic activities significantly increased ($P < 0.05$) with increase in dung amendments indicative of favourable soil condition. It is known that sufficient metal exposure would normally results in immediate death of cells due to changes in their viability or competitive ability [29] which resulted in decreased

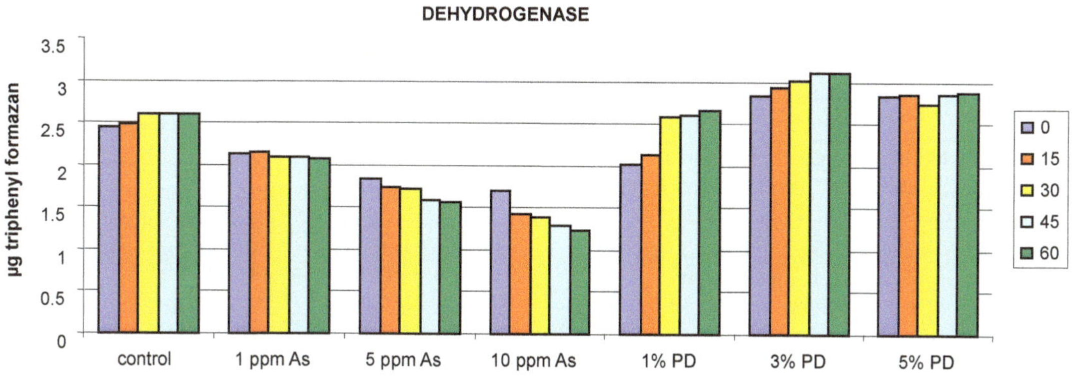

Figure 5. Dehydrogenase activity in arsenic contaminated soil and poultry dung amended soil. Arsenic contaminated soil: (F_1 = 97.384*, F_2 = 1.252, $p < 0.05$); Poultry dung amended soil: (F_1 = 20.97*, F_2 = 3.394*, $p < 0.05$); *Significant; **Not Significant (Two-way ANOVA).**

microbial population along with lower CO_2 evolution in arsenic contaminated soil. The reduction in the level of activities of enzymes in arsenic contaminated soils may be due to 1) masking of active groups 2) by protein demodulation 3) by other effects on enzyme configuration 4) the decreased level of contribution from microorganisms and v) failure of the resistant organisms to elaborate the enzymes [34]. Therefore, poultry dung amendment favoured the soil fertility status.

5. Acknowledgements

The authors are thankful to the University Grants Commission, New Delhi for providing financial support in the form of a major research project (No.34-68/2008(SR) dt 24. 12. 2008) to PCM.

REFERENCES

[1] D. R. Edwards and T. C. Daniel, "Environmental Impacts of On-Farm Poultry Waste Disposal: A Review," *Bioresource Technology*, Vol. 41, No. 1, 1992, pp. 9-33. doi:10.1016/0960-8524(92)90094-E

[2] N. C. Brady and R. R. Weil, "The Nature and Properties of Soils," 11th Edition, Prentice Hall International, Inc., 1996.

[3] D. C. Clay, V. Kelly, E. Mpyisi and T. Reardon, "Input Use and Conservation Investments among Farm Households in Rwanda: Patterns and Determinants," In: Barrett, Place and Aboud, Eds., *Natural Resources Management in African Agriculture*: *Understanding and Improving Current Practices*, CABI Publishing, Oxon and New York, 2002, pp. 103-114.

[4] M. E. López- Masquera, F. Cabaleiro, M. S. Sainz, A. López-Fabal and E. Carral, "Fertilizing Value of Broiler Litter: Effects of Drying and Pelletizing," *Bioresource Technology*, Vol. 99, No. 13, 2008, pp. 5626-5633. doi:10.1016/j.biortech.2007.10.034

[5] P. R. Warman, "The Effect of Fertilizer, Chicken Manure and Dairy Manure on Timothy Yield, Tissue Composition and Soil Fertility," *Agricultural Wastes*, Vol. 18, No. 4, 1986, pp. 289-298. doi:10.1016/0141-4607(86)90074-0

[6] A. J. Schjegel, "Effect of Composted Manure on Soil Chemical Properties and Nitrogen Use by Grain Sorghum," *Journal of Production Agriculture*, Vol. 5, No. 1, 1992, pp. 153-157.

[7] F. Place, C. B. Barrett, F. H. Ade, J. J. Ramisch and B. Vanlauwe, "Prospects for Integrated Soil Fertility Management Using Organic and Inorganic Inputs: Evidence from Smallholder African Agricultural Systems," *Food Policy*, Vol. 28, No. 4, 2003, pp. 365-378. doi:10.1016/j.foodpol.2003.08.009

[8] M. Uddin, A. Kashem and T. Osman, "Effect of Organic and Inorganic Amendments on the Phytoavailability of Phosphorous to Corn (*Zea mays*)," *Open Journal of Soil Science*, Vol. 2, No. 1, 2012, pp. 50-54. doi:10.4236/ojss.2012.21008

[9] P. W. Chescheir, L. M. Westserman and L. M. Safley, "Laboratory Methods for Estimating Available Nitrogen in Manures and Sludge," *Agricultural Wastes*, Vol. 18, No. 3, 1986, pp. 175-195. doi:10.1016/0141-4607(86)90112-5

[10] C. V. Miller, T. C. Hancock and J. M. Denver, "Environmental Fate and Transport of Arsenical Feed Amendments for Animal Agriculture," American Geophysical Union, 2000 Spring Meeting: Integrative Geoscience Solutions—A Start for the New Millennium, Washington DC, 30 May-3 June 2000.

[11] A. K. Kpomblekou, R. O. Ankumah, and H. A. Ajwa, "Trace and Non Trace Element Contents of Broiler Litter," *Communications in Soil Science and Plant Analysis*, Vol. 33, No. 11-12, 2002, pp. 1799-1811.

[12] M. L. Jackson, "Soil Chemical Analysis," Prentice Hall India, 1973, p. 498.

[13] A. Walkley and I. A. Black, "Determination of Organic Carbon in Soil," *Soil Science*, Vol. 37, No. 1, 1934, pp. 29-38. doi:10.1097/00010694-193401000-00003

[14] H. J. Harper, "The Accurate Determination of Nitrates in Soils. Phenoldisulfonic Acid Method," *Industrial and Engineering Chemistry*, Vol. 16, 1924, pp. 180-183. doi:10.1021/ie50170a037

[15] S. R. Olsen, C. V. Cole, F. S. Watanabe and L. A. Dean, "Estimation of Available Phosphorus in Soils by Extraction with Sodium Bicarbonate," USDA Cire, US Gov. Print. Office, Washington DC, Vol. 939, 1954, pp. 1-19.

[16] D. C. Ballinger, R. J. Lishaka and M. E. Gales, "Application of Silver Diethyldithiocarbamate Method to Determination of Arsenic," *Journal of the American Water Works Association*, Vol. 54, 1962, p. 1424.

[17] P. C. Mishra, R. K. Mohanty and M. C. Dash, "Enzyme activity in Subtropical Surface Soils under Pasture," *Indian Journal of Agricultural Chemistry*, Vol. 12, 1979, pp. 9-24.

[18] B. F. Casida, "Microbial Metabolic Activity in Soil as Measured by Dehydrogenase Determination," *Applied and Environmental Microbiology*, Vol. 34, 1977, pp. 630-636.

[19] M. Witkamp, "Rate of Carbon Dioxide Evolution from the Forest Floor," *Ecology*, Vol. 47, 1966, pp. 492-494. doi:10.2307/1932992

[20] J. P. E. Anderson, "Soil Respiration," In: A. L. Page, R. H. Miller and D. R. Kenney, Eds., *Method of Soil Analysis*, Part 2, Chemical and Microbial Properties, 2nd Edition, 1982, pp. 837-871.

[21] D. K. Jha, G. D. Sharma and R. R. Mishra, "Soil Microbial Population Numbers and Enzyme Activities in Relation to Altitude and Forest Degradation," *Soil Biology and Biochemistry*, Vol. 24, No. 8, 1992, pp. 761-767. doi:10.1016/0038-0717(92)90250-2

[22] P. Mitchell and B. Donald, "The Nature and Significance of Public Exposure to Arsenic: A Review of Its Relevance to Southwest England," *Environmental Geochemistry and Health*, Vol. 17, No. 2, 1995, pp. 57-82. doi:10.1007/BF00146709

[23] T. M. Agbede, S. O. Ojeniyi and A. J. Adeyemo, "Effect of Poultry Manure on Soil Physical and Chemical Properties, Growth and Grain Yield of Sorghum in Southwest Nigeria," *American-Eurasian Journal of Sustainable Agriculture*, Vol. 2, No. 1, 2008, pp. 72-77.

[24] M. J. Jordan and M. P. Lechevalier, "Effect of Zinc Smelter Emissions on Forest Soil Microflora," *Canadian Journal of Microbiology*, Vol. 21, 1975, pp. 1855-1865. doi:10.1139/m75-269

[25] S. T. Williams, T. McNeilly and E. M. H. Wellington, "The Decomposition of Vegetation Growing on Metal Mine Waste," *Soil Biology and Biochemistry*, Vol. 9, 1997, pp. 271-275. doi:10.1016/0038-0717(77)90034-7

[26] M. Wood, "A Mechanism of Aluminum Toxicity to Soil Bacteria and Ecological Implications," In: R. A. Date, *et al.*, Eds., *Plant-Soil Interactions at Low pH*, Kluwer Academic Publishers, Netherlands, 1995, pp. 173-179.

[27] S. Dahlin, E. Witter, A. M. Martensson, A. Turner and E. Baath, "Where's the Limit? Changes in the Microbiological Properties of Agricultural Soils at Low Levels of Metal Contamination," *Soil Biology and Biochemistry*, Vol. 29, 1997, pp. 1405-1415. doi:10.1016/S0038-0717(97)00048-5

[28] R. G. Kuperman and M. M. Carreiro, "Soil Heavy Metal Concentrations, Microbial Biomass and Enzyme Activities in a Contaminated Grassland Ecosystem," *Soil Biology and Biochemistry*, Vol. 29, 1997, pp. 179-190. doi:10.1016/S0038-0717(96)00297-0

[29] K. E. Giller, E. Witter and S. P. McGrath, "Toxicity of Heavy Metals to Microorganisms and Microbial Processes in Agricultural Soils: A Review," *Soil Biology and Biochemistry*, Vol. 30, No. 10-11, 1998, pp. 1389-1414. doi:10.1016/S0038-0717(97)00270-8

[30] B. Stenberg, "Monitoring Soil Quality of Arable Land: Microbiological Indicators," *Acta Agriculturae Scandinavica Section B. Soil and Plant Science*, Vol. 49, 1999, pp. 1-24.

[31] T. Fan, B.A. Stewart, W. Yong, L. Junjie and Z. Guangye, "Long-Term Fertilization Effects on Grain Yield, Water-Use Efficiency and Soil Fertility in the Dryland of Loess Plateau in China," *Agriculture, Ecosystems & Environment*, Vol. 106, No. 4, 2005, pp. 313-329. doi:10.1016/j.agee.2004.09.003

[32] S. B. Wuest, T. C. Caesar-Ton That, S. F. Wright and J. D. Williams, "Organic Matter Addition, N, and Residue Burning Effects on Infiltration, Biological, and Physical Properties of an Intensively Tilled Silt-Loam Soil," *Soil & Tillage Research*, Vol. 84, No. 2, 2005, pp. 154-167. doi:10.1016/j.still.2004.11.008

[33] P. Wang, J. T. Durkalski, W. Yu, H. A. J. Hoitink and W. A. Dick, "Agronomic and Soil Responses to Compost and Manure Amendments under Different Tillage System," *Soil Science*, Vol. 171, No. 6, 2006, pp. 456-467. doi:10.1097/01.ss.0000227824.77488.ac

[34] G. Tyler, "Heavy Metals in Soil Biology and Biochemistry," In: E. A. Paul and J. N. Ladd, Eds., *Soil Biochemistry*, Moral Dekker, New York, 1991, pp. 371-414.

13

Acacia trotilis and *Calotropis procera*: Do They Substantially Promote Soil Carbon Sequestration?

Taoufik Saleh Ksiksi

Department of Biology, Faculty of Science, UAE University, Al-Ain, UAE.
Email: tksiksi@uaeu.ac.ae

ABSTRACT

Very little is known about the type and mix of desert plant species and their management to optimize carbon sequestration in desert ecosystems. Overgrazing is one important practice that affects soil carbon cycling and therefore sequestration. Improving soil carbon in desert ecosystems may be best through the use of native trees and shrubs. Acacia tortilis and calotropis procera are two important species in the United Arab Emirates (UAE). The former is a native species that improves biodiversity and the latter is not native and has been reported to be an indicator of overgrazing. The average soil organic matter (SOM) content was higher in soils dominated by *A. tortilis* when compared to those dominated by *C. procera*; 2.98 and 1.34; respectively ($P < 0.05$). Moreover, *A. tortilis* leaves had a higher OM content than *C. procera* leaves (94.1% and 90.6%; respectively). The higher OM content of *A. tortilis* leaves explains the higher contribution of this species to the overall soil organic matter inputs. There was also a significant effect of shrub species on total SOC ($P < 0.05$). A total of about 14.7 tons of SOC were added per hectare in the areas dominated by *A. tortilis*. While only about 6.6 tons of SOC were added to the areas dominated by *C. procera*. In short, it is believed that both species substantially promote soil carbon sequestration. Some significant superiority of the native *A. tortilis* has been shown. But much has to be done to investigate the mix of plant species that promote the best soil carbon sequestration in the desert areas. Further studies are required in order to assess temporal as well as spatial variations in soil carbon sequestration in the UAE deserts. This will certainly help, in addition to other practices, in mitigating CO_2 emission.

Keywords: Desert Soils; Nutrient Cycling; Soil Carbon

1. Introduction

Soil restoration and woodland regeneration are sound strategies to increase soil carbon pool [1]. Increasing soil carbon improves soil quality, productivity and long-term sustainability. Equally important are the growing concerns about global greenhouse gas emission issues, which call for proper management of the terrestrial carbon pool. Specifically, this calls for a better understanding of carbon sequestration and ways to optimize it. Especially that desert lands are believed to contain small amounts of soil carbon. In general terms, carbon sequestration in terrestrial ecosystems can be defined as the net removal of carbon dioxide from the atmosphere into long-lived pools of carbon. These pools can be living above-ground biomass, wood products, living biomass in soils such as roots and micro-organisms or recalcitrant organic and inorganic carbon in soils [2]. In the UAE, for example, soil carbon pools become more important and relevant if we consider the vast areas covered by desert ecosystems, which occupy at least two-third of the country's land area.

Fortunately, terrestrial carbon pools could be significantly enhanced by adopting sound management practices in desert ecosystems. As in other agricultural systems, the potential of desert ecosystems to store carbon is dependent on how adequately the soil-plant resources are managed. For instance, promoting healthy perennial plant species is a management option that could improve the terrestrial carbon pool, through increased rooting depth. This will be even more pertinent if native shrub and tree species were used.

Unfortunately, very little is known about the type and mix of desert plant species that optimize carbon stocks in desert ecosystems. Additionally management principles and practices which can maintain carbon stocks through time are not well defined and adopted. *Panicum*, for instance, offers an excellent carbon sequestration option because of its deep rooting system and perenniality [3]. Understandably, plant species differ markedly in their impact on soil carbon concentration and distribution, mainly because of differences in their root systems. For example, the mean carbon concentration in the top 10 cm of soil in areas dominated by *Panicum maximum* was 3.31% compared to 1.89% and 0.74% in areas dominated by *Themeda triandra* and *Aristida jerichoensis*, respectively [4]. Furthermore, the restoration of some types of soils with *Astrebla* species has been attributed to a substantial increase in organic carbon concentration in the top 5 cm of soil [5]. Such differences in soil carbon under different species can be attributed to root systems' characteristics, specifically root turnover, which is a central component of ecosystem carbon and nutrient cycling [6]. In areas where *Prosopis* and Acacia are adapted, 6.2 × 109 Mg of carbon would be sequestered [7]. These types of carbon sequestration could offset CO_2 emission due to fossil fuel burning.

In the UAE, desired and undesired species are becoming part of the desert ecosystems. An undesired species that is prevalent in many parts of the UAE deserts is *colotropis procera* (Aiton) W.T. Aiton. It is common in many parts of the UAE desert as it is an indicator of overgrazing [8]. Desired species—such as *acacia toritillis*—are also an integral part of the UAE deserts. *Acacia toritillis* promotes floral diversity as well as provides feed sources for the majority of wildlife as well as livestock species in the country [9]. But do *C. procera* and *A. toritillis* have a potential to substantially improve soil carbon sequestration? What is the extent of such sequestration in the UAE soils? Those are some of the questions that the present endeavor will try to address.

2. Materials and Methods

2.1. Study Site

The study was conducted in the surrounding area of Al-Ain city in the (UAE). The average minimum temperature in Al-Ain is 22˚C while the average maximum is 35.8˚C (**Table 1**). The annual average long term rainfall is 119.7 mm (**Table 2**) and the humidity is 58% (**Table 3**). Three locations were selected where the two plant species grow. The aim was to collect pair samples with similar soil characteristics. The soils were characterized as sandy to sandy loam.

2.2. Sample Collection

Soil samples—about 150 grams—were collected from the various locations from the top soil layer (5 cm) and at 10 cm deep. A total of 72 samples (2 Species × 2 Positions × 2 Depths × 9 Replicates) were collected during Spring and during Winter of 2009-2010. Samples were collected from underneath the shrub canopy and away from the shrub canopy; referred to as in and out; respectively. Leaf samples were collected from each tree to assess percent OM and OC. Percent dry matter loss and percent moisture losses were also assessed. All samples were then transported to the UAE University labs for analyses.

2.3. Sample Analyses

Soil and leaf samples were first air dried for 48 hours. Soil samples were sieved to remove coarse material. Crucibles were then used to oven-dry each sample at 105 degrees C for 72 hours. Moisture content was calculated for the soils samples following this step (formula a). Combustion was performed for 3 hours on the soil and leaf samples to estimate organic matter (formula b). Percent organic carbon was calculated as a fraction of OM (formula c).

a) Moisture content: Sample loss/Dry Weight of Sample.

b) Percent Soil Organic Matter (SOM): Sample loss in combustion/Dry Weight of Sample.

c) Percent Soil Organic Carbon (SOC): Organic matter × 0.58.

In order to assess the total bulk SOC that the two spe-

Table 1. Monthly variation in the air temperature (˚C) (1965-2001) of ten meteorological stations in Al-Ain UAE. Minimum (top row) and maximum (bottom row) (Ministry of Agriculture and fisheries UAE, 1965-2001).

Station	Jan.	Feb.	Mar.	Apr.	May	Jun.	Jul.	Aug.	Sep.	Oct.	Nov.	Dec.	Mean
Al-Ain	14.8	14.9	18.5	20.4	23.9	27.1	30	30.2	26.6	23	18.9	15.8	22.0
	23.3	26.9	30.8	35.4	40.4	43.5	43.5	44.9	42.0	37.6	32.4	28.8	35.8

Table 2. Monthly variation in the annual rainfall (mm) during (1965-2001) of ten meteorological stations in Al-Ain UAE (Ministry of Agriculture and fisheries UAE, 1965-2001).

Station	Jan.	Feb.	Mar.	Apr.	May.	Jun.	Jul.	Aug.	Sep.	Oct.	Nov.	Dec.	Total
Al-Ain	68.1	45.7	2.7	Trace	0.0	0.0	0.0	0.0	0.0	0.0	0.0	3.2	119.7

Table 3. Monthly variation in the relative humidity (%) during (1965-2001) of ten meteorological stations in Al-Ain UAE (Ministry of Agriculture and fisheries UAE, 1965-2001).

Station	Jan.	Feb.	Mar.	Apr.	May.	Jun.	Jul.	Aug.	Sep.	Oct.	Nov.	Dec.	Mean
Al-Ain	66	64	59	53	50	53	53	54	56	58	63	66	58

cies add to the soils, it has been decided to estimate bulk SOC based on an approximate number of shrubs growing in the study sites. It has been assumed that an average 20 shrubs and *A. tortilis* and 5 shrubs of *C. procera* were growing in the site. It was also estimated that the age of shrubs was 15 and 10 years for *A. tortilis* and *C. procera*; respectively. The total amount of SOC per square meter was calculated following the method (formula d) reported by [10].

d) Soil C ($g \cdot m^{-2}$) = $z \times p_b \times c \times 10$.

where z = thickness of each sample depth (cm), p_b = bulk density (1.7 $g \cdot cm^{-3}$) of each sample depth and c = carbon concentration ($g \cdot C \cdot kg^{-1}$ soil) of each sample depth. The results will be reported in tons per hectare ($tons \cdot Ha^{-1}$).

2.4. Statistical Analyses

ANOVA analyses were performed to compare main effects (season, shrub species, soil depth and position in relation to shrub canopy) and all interactions. SYSTAT11 was used to perform all analyses [11].

3. Results

3.1. Soil Organic Matter

3.1.1. The Effects of Tree Species

Soil organic matter (SOM) was different in soils dominated with *A. tortilis* than in those dominated with *C. procera* ($P < 0.05$; **Table 4**; **Table 5**). The average SOM content was higher in soils dominated by *A. tortilis* when compared to those dominated by *C. procera*; 2.98 and 1.34; respectively (**Table 4**). When comparing canopy positions (*i.e.* underneath or away from the shrub) in areas dominated by either *A. tortilis* or *C. procera*, no differences were detected ($P > 0.05$). The above findings do not agree with much of what was reported elsewhere. Results from the sonorant desert suggested higher soil fertility underneath live plants, regardless of species and phenology [12]. Except for cactus, which was suggested to use nutrients and remove fertility areas around them. There was an improvement in soil characteristics underneath *A. tortilis* when compared to open grassland area, away from the shrub [13]. Their findings restrict these differences, however, to lightly grazed sites. Additionally, significant differences in soil characteristics, underneath vs away from the tree/shrub canopies, in lightly grazed areas were found [14].

While in our study, the grazing history is unknown but heavy camel grazing has been a wide spread practice in the area. The increase in camel populations within the UAE during the past few decades is also an indicator of the pressure on the desert ecosystems [15]. It is strongly believed, therefore, that while the dominant shrub species contribute to SOM, but a substantial part of this contribution may be the result of the associated species that grow in soils where *A. tortilis* or *C. procera* dominate. The associated species contributed to soil fertility in soils dominated with *A. tortilis* in Ethiopia [14]. Acacia tortilis has also been reported to improve associated species in

Table 4. Summary averages for percent soil organic matter (OM) and soil organic carbon (OC) for the two seasons at two different soil depths under *Acacia tortilis* and *Calotropis procera* species growing in the deserts of the UAE.

			Soil Depths					
			5		10			
	Shrub Species	under/out of Shrub	% OM	% OC	% OM	% OC	Average % OM	Average % OC
Spring			2.13	1.23	2.02	1.17	2.07	1.20
	Acacia		2.77	1.61	2.70	1.56	2.73	1.59
		In	2.94	1.71	2.85	1.65	2.90	1.68
		Out	2.60	1.51	2.54	1.47	2.57	1.49
	Calotropis		1.49	0.86	1.34	0.78	1.41	0.82
		In	1.38	0.80	1.36	0.79	1.37	0.79
		Out	1.60	0.92	1.31	0.76	1.45	0.84
Winter			2.46	1.43	2.04	1.19	2.25	1.31
	Acacia		3.52	2.04	2.93	1.70	3.23	1.87
		In	3.66	2.12	2.90	1.68	3.28	1.90
		Out	3.37	1.96	2.96	1.72	3.17	1.84
	Calotropis		1.40	0.81	1.15	0.67	1.28	0.74
		In	1.34	0.78	1.24	0.72	1.29	0.75
		Out	1.46	0.85	1.06	0.62	1.26	0.73
grand total			2.29	1.33	2.03	1.18	2.16	1.25

Table 5. ANOVA analysis percent soil organic matter (SOM) for the two seasons at two different soil depths under *Acacia tortilis* and *Calotropis procera* species growing in the deserts of the UAE.

Source	Sum-of-Squares	df	Mean-Square	F-ratio	P
Season	1.165	1	1.165	1.747	0.189
Taxa	96.154	1	96.154	144.220	0.000
Depth	2.499	1	2.499	3.748	0.055
Season*Taxa	3.525	1	3.525	5.287	0.023
Season*Depth	0.827	1	0.827	1.240	0.267
Taxa*Depth	0.149	1	0.149	0.223	0.637
Season*Taxa*Depth	0.381	1	0.381	0.572	0.451
Error	90.674	136	0.667		

desert ecosystems [9,13]. Some other desert species were also reported to improve soil characteristics beyond their canopy [16].

3.1.2. The Effects of Season

SOM was highest during winter collection than during spring ($P < 0.05$). The difference between the two shrub species is more pronounced during winter. During which *A. tortilis* had an average SOM of 3.23% while *C. procera* had an average SOM of 1.28%. During spring the average SOM for *A. tortilis* and *C. procera* was 2.73% and 1.41%; respectively.

3.1.3. The Effects of Soil Depth

The average SOM was highest in the top 5 cm ($P = 0.055$). SOM was 2.29% and 2.03% at 5 cm and 10 cm soil depth; respectively. For *A. tortilis*, SOM was 3.14% and 2.81% for 5 cm and 10 cm soil depths; respectively. While for *C. procera* SOM was 1.82% and 1.63% for the two depths, respectively. Marked differences in the top 5 cm of the soil profile were reported by [14].

3.1.4. Leaf Content

Overall *A. tortilis* leaves had a higher OM content than *C. procera* leaves (94.1% and 90.6%; respectively) at $P < 0.05$ (**Figure 1**). But little fluctuations were observed during the 28-day period (data not shown). The higher OM content of *A. tortilis* leaves explains the higher contribution of this species to the overall soil organic matter inputs. This is another reason to encourage the plantations of such native species in the UAE deserts.

As for the leaf moisture content, *A. tortilis* contained a slightly lower level (52.2% and 84.6%; respectively) and lost moisture at a relatively faster rate than *C. procera* (**Figure 2**). But toward the end of the 28-day period, both species had comparable moisture contents (9.7% and 11.3%; respectively).

3.2. Soil Organic Carbon

Percent SOC was transformed into bulk tons of SOC per hectare. Please see the methodology section for more details. There was significant effect of shrub species on SOC ($P < 0.05$). An estimated total of about 14.7 tons of SOC were added per hectare in the areas dominated by *A. tortilis* (**Table 6**). While about 6.6 tons of SOC were added to the areas dominated by *C. procera*.

Variations between soil depths was also detected ($P < 0.05$). The average SOC at 5 and 10 cm depths was 15.5 tons·Ha^{-1} and 13.9 tons·Ha^{-1} for *A. tortilis*; while SOC was 7.1 tons·Ha^{-1} and 6.1 tons·Ha^{-1} for *C. procera*; respectively.

Many other studies also report positive SOC sequestration but many disagree on the estimated amount per hectare; mainly because of ecosystem differences and variations in the adopted experimental protocols. An average of about 26 tons·Ha^{-1} of SOC in the grazing lands of Ethiopia was reported [17]. While others [10] reported about 14.7 tons·Ha^{-1} of SOC in the top 10 cm of the soil profile semiarid acacia woodland. The accumulation of SOC at 0 - 10 cm depth was estimated to be a staggering 61.2 tons·Ha^{-1} [18]. Some of the changes to SOC were attributed, and rightly so, to land management practices [19] such as overgrazing [20]. In the UAE, and across much of the region, overgrazing has been reported as one of the main threats facing desert environments [21].

For *A. tortilis*, we can estimate an annual SOC addition of about 0.98 tons·Ha^{-1} would be added to the desert soils of the UAE. While an estimated annual addition of 0.66 tons·Ha^{-1} of SOC in soils dominated by *C. procera*. Restoring grasslands to woody grasslands, where *A. tortilis* was growing in the sahel, would add about 0.8 tons·Ha^{-1} of SOC annually [22]. The huge differences between the estimates in our study and those in many other studies may be attributed to floral understory. In the UAE, the understory of grass species is much less when compared to the Ethiopian grasslands, for instance.

4. Discussion

Acacia tortilis is an important native species to the UAE and needs to be grown in large areas as part of the current attempts to re-vegetate the desert. Assuming an average success rate of 20 individuals per hectare, we can

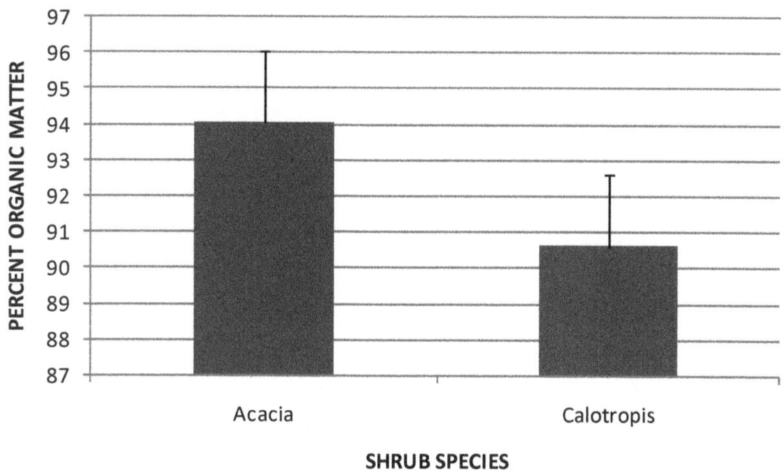

Figure 1. Percent leaf organic matter content (%OM) for both *Acacia tortilis* and *Calotropis procera* species growing in the deserts of the UAE.

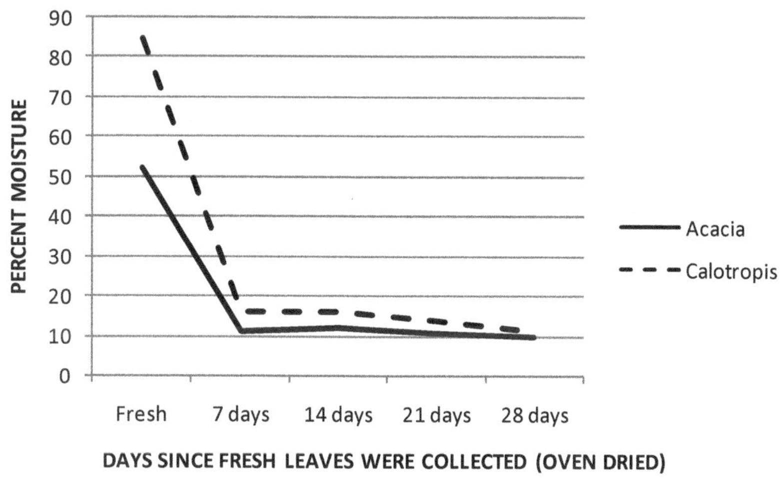

Figure 2. Percent moisture loss for both *Acacia tortilis* and *Calotropis procera* species growing in the deserts of the UAE.

Table 6. Average soil organic carbon (SOC) at two different soil depths under and away from the canopies of *Acacia tortilis* and *Calotropis procera* species growing in the deserts of the UAE.

Soil Organic Carbon		Canopy Position		
Taxa	Soil Depth	Under	Away	Average
Acacia	5	16282.22	14728.61	15505.42
	10	14180.83	13571.67	13876.25
Acacia Average		15231.53	14150.14	14690.83
Calotropis	5	6696.11	7536.67	7116.39
	10	6417.50	5855.55	6136.53
Calotropis Average		6556.80	6696.11	6626.46
Grand Average		10894.17	10423.12	10658.64

sequester 19.6 Mt in the next 10 years if we plant 2000 hectares of *A. tortilis*. The other equally important benefits of improved species diversity and soil improvement are to be taken into consideration.

For *A. tortilis*, based on the above numbers, we can estimate an annual SOC addition of about 0.98 tons·Ha^{-1} to be added to the UAE desert ecosystem. While an estimated annual addition of 0.66 tons·Ha^{-1} of SOC in soils dominated by *C. procera*. This highlights the importance of re-vegetating our desert ecosystems using species that promote carbon sequestration. The direct benefits are the greening of these ecosystems, while indirect benefits may include the creation of islands of fertility underneath these shrubs and the improvement of the vegetative cover in the floral understory. Islands of fertility in the Sonoran Desert underneath mesquite canopies played important ecosystem functions [22].

Furthermore, SOC inputs enrich soil characteristics. An increase of 1 ton of soil carbon of degraded cropland soils may increase crop yield by 20 to 40 $kg \cdot Ha^{-1}$ for wheat, 10 to 20 $kg \cdot Ha^{-1}$ for maize, and 0.5 to 1 $kg \cdot Ha^{-1}$ for cowpeas. In addition to enhancing food security, carbon sequestration has may possibly offset fossil fuel emissions by 0.4 to 1.2 gigatons of carbon per year, or 5 to 15% of the global fossil-fuel emissions [1].

Finally and to address the question stated in the project title, it is believed that both species substantially promote soil carbon sequestration. It is important to note that one species is a native wanted species (*A. tortilis*) and the other is an introduced unwanted plant species (*C. procera*). Some significant superiority of the native *A. tortilis* has been shown. As for the extent of soil carbon sequestration, some evidence of vertical as well as horizontal variability was shown. Much has to be done, however, to investigate the mix of plant species that promote the best soil carbon sequestration in desert areas. Besides enhancing food security, soil carbon sequestration offsets global fossil fuel emission by up to 15% [1]. The extent of such quantitative estimates within the region and more specifically within the UAE is unknown. More detailed studies are to be initiated in order to assess temporal as well as spatial variations in soil carbon sequestration in the UAE deserts.

5. Acknowledgements

The investigator would like to express his sincere appreciation to the Research Affairs at the United Arab Emirates University for the financial support of this project under fund Grant # 01-04-2-11/09. The support from the Biology Department and the Faculty of Science is also much appreciated. The investigator would also like to express his gratitude to Drs Mohamed T. Moussa and Nael Fawzi for their assistance in data collection.

REFERENCES

[1] R. Lal, "Soil Carbon Sequestration Impacts on Global Climate Change and Food Security," *Science*, Vol. 304, No. 5677, 2004, pp. 1623-1627. doi:10.1126/science.1097396

[2] R. C. Dahlman and G. K. Jacobs, "Research Challenges for Carbon Sequestration in Terrestrial Ecosystems," Anl. Gov., USA, 2001, pp. 718-720.

[3] Z. Ma, C. Wood and D. Bransby, "Soil Management Impacts on Soil Carbon Sequestration by Switchgrass," *Biomass and Bioenergy*, Vol. 18, No. 6, 2000, pp. 469-477. doi:10.1016/S0961-9534(00)00013-1

[4] J. O. Carter, K. A. Day, W. B. Hall and S. M. Howden, "Evaluation of the Impact of Climate Change on Northern Australian Grazing Industries," A Final Report for the Rural Industries Research and Development Corporation (RIRDC), 1998, p. 291.

[5] T. H. Chen, A. Henderson-Sellers, P. C. D. Milly, A. J. Pitman, A. C. M. Beljaars, J. Polcher, F. Abramopoulos, A. Boone, S. Chang, F. Chen, Y. Dai, C. E. Desborough, R. E. Dickinson, L. Dümenil, M. Ek, J. R. Garratt, N. Gedney, Y. M. Gusev, J. Kim, R. Koster, E. A. Kowalczyk, K. Laval, J. Lean, D. Lettenmaier, X. Liang, J.-F. Mahfouf, H.-T. Mengelkamp, K. Mitchell, O. N. Nasonova, J. Noilhan, A. Robock, C. Rosenzweig, J. Schaake, C. A. Schlosser, J.-P. Schulz, Y. Shao, A. B. Shmakin, D. L. Verseghy, P. Wetzel, E. F. Wood, Y. Xue, Z.-L. Yang and Q. Zeng, "Cabauw Experimental Results from the Project for Intercomparison of Land-Surface Parameterization Schemes," *Journal of Climate*, Vol. 10, No. 6, 1997, pp. 1194-1215. doi:10.1175/1520-0442(1997)010<1194:CERFTP>2.0.CO;2

[6] R. A. Gill and R. B. Jackson, "Global Patterns of Root Turnover for Terrestrial Ecosystems," *New Phytologist*, Vol. 147, No. 1, 2000, pp. 13-31. doi:10.1046/j.1469-8137.2000.00681.x

[7] D. Geesing, P. Felker and R. L. Bingham, "Influence of Mesquite (Prosopis glandulosa) on Soil Nitrogen and Carbon Development: Implications for Global Carbon Sequestration," *Journal of Arid Environments*, Vol. 46, No. 2, 2000, pp. 157-180. doi:10.1006/jare.2000.0661

[8] A. Wezel and S. Bender, "Degradation of Agro-Pastoral Village Land in Semi-Arid Southeastern Cuba," *Journal of Arid Environments*, Vol. 59, No. 2, 2004, pp. 299-311. doi:10.1016/j.jaridenv.2003.12.010

[9] A. El-Keblawy and T. Ksiksi, "Artificial Forests as Conservation Sites for the Native Flora of the UAE," *Forest ecology and management*, Vol. 213, No. 1-3, 2005, pp. 288-296. doi:10.1016/j.foreco.2005.03.058

[10] M. Lemenih and F. Itanna, "Soil Carbon Stocks and Turnovers in Various Vegetation Types and Arable Lands along an Elevation Gradient in Southern Ethiopia," *Geoderma*, Vol. 123, No. 1-2, 2004, pp. 177-188. doi:10.1016/j.geoderma.2004.02.004

[11] K. Kroeger, "SYSTAT 11. Getting Started," Manual. Richmond, CA, SYSTAT Software Inc., USA, 2004.

[12] B. J. Butterfield and J. M. Briggs, "Patch Dynamics of Soil Biotic Feedbacks in the Sonoran Desert," *Journal of Arid Environments*, Vol. 73, No. 1, 2009, pp. 96-102. doi:10.1016/j.jaridenv.2008.09.012

[13] F. Abdallah, Z. Noumi, B. Touzard, A. O. Belgacem, M. Neffati and M. Chaieb, "The Influence of Acacia Tortilis (Forssk.) Subsp. Raddiana (Savi) and Livestock Grazing on Grass Species Composition, Yield and Soil Nutrients in arid Environments of South Tunisia," *Flora-Morphology, Distribution, Functional Ecology of Plants*, Vol. 203, No. 2, 2008, pp. 116-125.

doi:10.1016/j.flora.2007.02.002

[14] E. Abule, G. N. Smit and H. A. Snyman, "The Influence of Woody Plants and Livestock Grazing on Grass Species Composition, Yield and Soil Nutrients in the Middle Awash Valley of Ethiopia," *Journal of Arid Environments*, Vol. 60, No. 2, 2005, pp. 343-358. doi:10.1016/j.jaridenv.2004.04.006

[15] D. J. Gallacher and J. P. Hill, "Effects of Camel Grazing on the Ecology of Small Perennial Plants in the Dubai (UAE) Inland Desert," *Journal of Arid Environments*, Vol. 66, No. 4, 2006, pp. 738-750. doi:10.1016/j.jaridenv.2005.12.007

[16] J. Li, C. Zhao, H. Zhu, Y. Li and F. Wang, "Effect of Plant Species on Shrub Fertile Island at an Oasis-Desert Ecotone in the South Junggar Basin," *China Journal of Arid Environments*, Vol. 71, No. 4, 2007, pp. 350-361. doi:10.1016/j.jaridenv.2007.03.015

[17] J. Nyssen, H. Temesgen, M. Lemenih, A. Zenebe, N. Haregeweyn and M. Haile, "Spatial and Temporal Variation of Soil Organic Carbon Stocks in a Lake Retreat Area of the Ethiopian Rift Valley," *Geoderma*, Vol. 146, No. 1-2, 2008, pp. 261-268. doi:10.1016/j.geoderma.2008.06.007

[18] M. P. McClaran, J. Moore-Kucera, D. A. Martens, J. Van Haren and S. E. Marsh, "Soil Carbon and Nitrogen in Relation to Shrub Size and Death in a Semi-Arid Grassland," *Geoderma*, Vol. 145, No. 1-2, 2008, pp. 60-68. doi:10.1016/j.geoderma.2008.02.006

[19] B. A. El Tahir, D. M. Ahmed, J. Ardö, A. M. Gaafar and A. A. Salih, "Changes in Soil Properties Following Conversion of Acacia Senegal Plantation to Other Land Management Systems in North Kordofan State, Sudan," *Journal of Arid Environments*, Vol. 73, No. 4-5, 2009, pp. 499-505. doi:10.1016/j.jaridenv.2008.11.007

[20] A. El-Keblawy, T. Ksiksi and H. El Alqamy, "Camel Grazing Affects Species Diversity and Community Structure in the Deserts of the UAE," *Journal of Arid Environments*, Vol. 73, No. 3, 2009, pp. 347-354. doi:10.1016/j.jaridenv.2008.10.004

[21] P. L. Woomer, A. Toure and M. Sall, "Carbon Stocks in Senegal's Sahel Transition Zone," *Journal of arid environments*, Vol. 59, No. 3, 2004, pp. 499-510. doi:10.1016/j.jaridenv.2004.03.027

[22] J. D. Schade and S. E. Hobbie, "Spatial and Temporal Variation in Islands of Fertility in the Sonoran Desert," *Biogeochemistry*, Vol. 73, No. 3, 2005, pp. 541-553. doi:10.1007/s10533-004-1718-1

14

Spatial Variation in Soil Chemistry on a Sub-Antarctic Island

Everhard Christiaan Conradie, Valdon R. Smith

Department of Botany & Zoology, Stellenbosch University, Stellenbosch, South Africa.
Email: vs2@sun.ac.za

ABSTRACT

On both west and east sides of sub-Antarctic Marion Island (47°S, 38°E), total Na and exchangeable Na, Mg and K concentrations in the soil decline with increasing distance inland and altitude, related to a decrease in the intensity of seaspray deposition. On the east side, the coastal plain is wide and slopes gently up to the mountainous interior and total C, total N and soil moisture content all decrease significantly, whereas bulk density increases significantly, as one moves away from the sea, reflecting a gradual change from organic, wet, low bulk density peats characteristic of lowland coastal regions to mineral, dry, high bulk density volcanic soils characteristic of inland areas. On the west side, the narrow coastal plain is bounded by an escarpment that rises up very steeply to the highland interior. There, sampling was largely restricted to the coastal plain (soils are rare on the escarpment and interior) and did not cover the same transition from organic to mineral soils as on the east side. Hence, total C, total N and bulk density did not change significantly with increasing distance inland on the west side. Most total Mg is in the mineral fraction of the soil, with a lesser contribution by organic, exchangeable and soil solution forms of Mg. On the east side the gradual transition from highly organic peats to very mineral soils results in an increase in total Mg going inland, but on the west, where there was not this change in soil minerality, total Mg decreased with increasing distance inland, reflecting the decreasing intensity of seaspray. Once the between-side differences in the influence of altitude and distance from the sea are accounted for, there are significant differences in soil chemical composition between the two sides of the island. Overall, west side soils are more influenced by both seaspray and the parent volcanic basalts than are east side soils.

Keywords: Sub-Antarctic; Soil Nutrients; Altitudinal Variation; Seaspray; Mineral-Organic Gradient; Soil Organic Matter

1. Introduction

Marion Island (47°S, 38°E, 2100 km southeast of Cape Town) has a hyperoceanic climate typical of the sub-Antarctic region [1]. It is cool (annual mean air temperature is 6°C, the mean temperatures of the coldest and the warmest month differ by only 4°C), wet (mean annual rainfall is 2200 mm·y^{-1}, mean relative humidity 80%) and windy (mean wind velocity is 25 km·h^{-1}, gales occur on 110 days per year). The island is volcanic, 290 km^2 in area and consists of a central highland (highest peak 1230 m above sea level) that on the eastern and northern sides slopes down gently to a coastal plain 3 to 5 km wide. On the island's western and southern sides the coastal plain is narrow (400 to 900 m wide) and meets an escarpment rising up very steeply to the mountainous interior. The most prevalent (and strongest) winds are from the west [2] so that haline plant communities resistant to saltspray extend much further inland (hundreds of meters) on the west side of the island than they do on its east side (rarely more than a few score meters). The island's ecosystem is classified as sub- Antarctic tundra [3].

An early aim of the ecological research program on the island was to draw up a nutrient budget for the whole terrestrial ecosystem [4], which involved quantifying nutrient inputs to, and losses from, the island and assessing the spatial patterns of nutrient concentrations in its soils and plants. This reflected the influence of the Tundra Biome Project of the International Biological Programme (1967-1974), a major objective of which was to elucidate patterns of distribution of nutrients within tundra ecosystems [5]. Research at the island has subsequently provided much information on the magnitudes of nutrient transfers from the ocean to the island through saltspray, precipitation and dryfallout [6] and also by seabirds and seals that feed in the ocean and deposit excreta, shells, moulted feathers and fur on the island [7]. In contrast, soil nutrient concentrations have been measured only at a few localities, and detailed nutrient budgets have been drawn up for only eight of the island's 41 plant communities [8,9], all within a 1.5 ha area about 35 m a.s.l. and 600 m from the coast on the island's eastern side. Some soil nutrient information exists for other areas on the east side but there is none for other parts of the island. In fact, prior to this study there was no information on the spatial variation in soil chemical composition at the island.

One component of a nutrient budget is the *soil nutrient standing stock*, which is the mass of a particular nutrient per square meter of soil down to a specified depth. This depends on the concentration of that nutrient in, and the bulk density (mass per volume) of, the soil. Hence, to construct a nutrient budget on a whole-island basis it is important to know how soil nutrient concentration and soil bulk density change with increasing distance inland and elevation, and if the changes are different on the various sides of the island. Here, we report the results of an investigation aimed at providing such information for the island's eastern and western sides.

2. Material and Methods

2.1. Sampling Localities and Protocol

In April/May 2010, soils were sampled from 77 localities on the east side, and 55 localities on the west side, of the island (**Figure 1**). Each locality's latitude and longitude were recorded using a GPS and used with a digital elevation model in ArcGIS 8 (ESRI, California) to estimate its altitude and distance to the sea, and also to construct **Figure 1**. A 5 cm diameter core, 25 cm deep where the soil was deep enough to allow it, was taken from the soil and the surface vegetation, litter and conspicuous roots removed. If the core contained conspicuous horizons, these were separated. The volume of the whole core or horizon section was calculated from its diameter and length.

2.2. Soil Moisture and Chemical Analyses

The fresh core or horizon section was weighed, air dried and weighed again. A weighed subsample of the air dried soil was dried in an oven at 105˚C for 12 hours and reweighed. From the fresh, air-dried and oven-dried masses, moisture content was calculated on an oven-dried basis. From the oven-dried mass and volume of the soil core or horizon section, bulk density was calculated.

All chemical analyses were performed on the air-dried subsamples and the concentrations expressed on an oven-dried mass basis using the air-dry:oven-dry mass ratios. For total P, Ca, Mg, K and Na, a soil subsample was extracted with boiling $HCl:HNO_3$ mixture (3:1) and the concentrations of P and the cations in the extract determined by Inductively Coupled Plasma—Optical Emission Spectrometry (Varian Vista-MPX; Varian, Inc. California). Total N was measured with a Leco FP-528

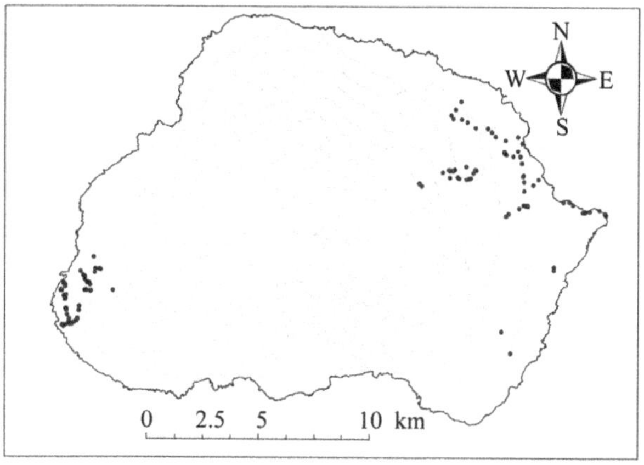

Figure 1. Map of Marion island with 100 m contour lines and showing localities where soils were sampled.

nitrogen analyser (LECO Corporation, Michigan). Total carbon was determined by the Walkley-Black method [10]. Exchangeable cations were determined by extracting soil subsamples with 1 M ammonium acetate solution and measuring the concentrations of Ca, Mg, Na and K in the extract by ICP-OES.

2.3. Data Analyses

Linear regression analysis was used to test if the magnitudes of the soil chemistry variables were significantly ($P \leq 0.05$) related to altitude or distance to the sea. For total Na and exchangeable Na and Mg, the relationships were approximately negative exponential ones, so the data were log-transformed. Two regression analyses were carried out, one on the east side data and the other on the west side data. If one or both of the regressions were significant, a homogeneity-of-slopes analysis was used to test if the slopes differed significantly between east and west sides. If they did not, the between-sides difference in concentrations was tested by analysis of covariance; if they did, a separate-slopes test was used (in both instances altitude or distance to the sea was the covariate). Where the values of a chemical variable were unrelated to altitude or distance to sea, the east-west difference in its mean values was tested by analysis of variance. The statistical analyses were done using Statistica 9 software [11].

3. Results and Discussion

Table 1 compares the relationships (slopes) of the soil chemistry values with altitude and with increasing distance from the sea between the eastern and western sides of the island, and also gives the mean values for each side (or the means adjusted for the effect of altitude and distance from the sea).

On both the eastern and the western sides of the island, concentrations of exchangeable Na, Mg and K declined

Table 1. Relationships between soil chemistry variables and altitude or nearest distance to sea. Only slopes of significant regressions are reported. Where slopes or means differed significantly ($P \leq 0.05$) between the east and west side of the island, one of the pair is marked with an asterisk*. Where a variable was significantly influenced by altitude or distance to sea, the mean value is the mean (±standard error) adjusted for that influence and is given in italics. Otherwise, actual means (±standard error) are reported. For total Na and exchangeable Na and Mg, the influence of altitude or distance to sea was approximately a negative exponential one so the data were log-transformed for the analysis. In that case the logs of the adjusted means and standard errors are reported and the exponential value of the (log) mean given in brackets. The reported slope values for total C, total N, CEC, exch. K and exch. Ca are the actual slope values × 10^4. N is the number of samples.

	N	Altitude				Distance to the Sea			
		Slope		Means ± S.E. (Adj. Mean ± S.E.)		Slope		Means ± S.E. (Adj. Mean ± S.E.)	
	(East/West)	East	West	East	West	East	West	East	West
Range (m)		7 - 409	27 - 130			20 - 4958	200 - 1268		
Moisture content (%)	77/55	−1.69	−8.59	834 ± 73.7	640 ± 79.1	−0.101		829 ± 77.9	641 ± 80.1
Bulk density (g/cm^3)	77/55	12.18		0.17 ± 0.025	0.25 ± 0.026	0.771		0.18 ± 0.027	0.25 ± 0.028
Total C (g/kg)	76/55	−417.43		205 ± 11.4	155 ± 12.1*	−29.89		207 ± 12.1	155 ± 12.3*
Total N (g/kg)	77/55	−28.54		13.6 ± 0.66	9.9 ± 0.71*	−2.13		12.8 ± 0.63	9.0 ± 0.75*
Total P (mg/kg)	74/54			667 ± 35.2	654 ± 68.5			667 ± 35.2	654 ± 68.5
Total Na (mg/kg)	74/54		−1.05	6.5 ± 0.07 (644)	7.5 ± 0.09* (1778)	−0.14	−1.01*	6.46 ± 0.066 (636)	7.46 ± 0.0775* (1745)
Total K (mg/kg)	74/54			262 ± 14.8	679 ± 87.4*		−0.995	255 ± 51.8	682 ± 52.6*
Total Ca (mg/kg)	74/54			2036 ± 144	5966 ± 634.7*		−7.25	1942 ± 385.1	5999 ± 390.7*
Total Mg (mg/kg)	74/54	14.07		2770 ± 596.3	6782 ± 630.2*	0.87	−8.91*	2761 ± 584.9	6818 ± 593.4*
Exchangeable Na (meq/100g)	77/55	−0.07	−1.97*	0.29 ± 0.086 (1.33)	1.15 ± 0.101* (3.17)	−0.54	−1.35*	0.21 ± 0.081 (1.23)	1.12 ± 0.096* (3.08)
Exchangeable K (meq/100g)	77/55	−21.13	−54.62	0.84 ± 0.057	0.56 ± 0.062*	−1.54	−4.49	0.86 ± 0.06	0.56 ± 0.06*
Exchangeable Ca (meq/100g)	77/55		−549.93	3.5 ± 0.44	5.4 ± 0.469*		−35.37	3.7 ± 0.46	5.4 ± 0.48
Exchangeable Mg (meq/100g)	77/55	−0.58	−1.01	1.3 ± 0.09 (3.50)	2.1 ± 0.109* (7.91)	−0.433	−0.643	1.2 ± 0.09 (3.30)	2.1 ± 0.12* (7.84)

significantly with increasing altitude or distance inland. Saltspray and aerosols of seawater (together, these are henceforth termed seaspray) is the sole source of Na, and the main sources of Mg and K, at the island [12], so this is unsurprising. It also explains the decrease in total Na with increasing distance inland.

On the east side, total C, total N and soil moisture content declined significantly, while bulk density increased significantly, with increasing altitude and distance inland. The four variables varied in a similar pattern on the west side but the effect of the both covariates was significant only for moisture content. That total C, total N and bulk density did not change significantly with increasing distance inland on the west side can be explained as follows.

Soil moisture content and total N concentration are positively, and bulk density negatively, strongly correlated with total C concentration [13]. Organic soils (peats) are wet, have low bulk density and high total C and N concentrations. Mineral soils are dry, with a high bulk density and low total C and N concentrations. Mineral soils also tend to have high total Ca and Mg concentrations, since much of the total complement of these two elements is found in the parent volcanic rock and ash. On the eastern side, total C, total N and soil moisture all decline significantly as one moves away from the sea and up in elevation, reflecting the change from organic, wet, low bulk density peats characteristic of lowland coastal regions to the mineral, dry, high bulk density volcanic soils characteristic of inland areas. On the island's west side, a narrow (400 to 900 m wide) coastal plain meets an escarpment rising up very steeply to the mountainous interior. Soils are rare on the escarpment and absent in the interior, the substrate being lava or volcanic ash.

Most samples on the western side were of the coastal plain soils; sampling did not extend as far inland or as high up as it did on the eastern side. There was thus not the same strong gradation from organic to mineral soils with increasing distance inland as there was on the eastern side. This accounts for the lack of a significant relationship of total C, total N and bulk density with altitude or distance to the sea on the western side.

On the eastern side, total Mg concentration increased significantly, whereas on the western side it decreased significantly, with increasing distance from the sea. This is also attributable to the fact that on the west the sampling did not extend over as a wide range of organic versus mineral soil types as on the east coast. Most total Mg is in the mineral fraction of the soil (the basaltic parent lava and scoria). The organic, exchangeable and dissolved forms contribute little to total Mg, except perhaps in peats close to the sea. On the east side, the gradual transition from highly organic peats to very mineral soils represented by the samples results in an increase in total Mg with increasing distance inland. On the west side there was not this range of soil minerality and the change (decrease) in total Mg with increasing distance inland reflects the decreasing intensity of seaspray. This is not to say that west side soils are not more mineral, on the whole, than east side soils—it is the change in minerality that is important, and that is greater on the east side. West side soils are actually intrinsically more mineral than east side soils, as is shown by the significantly lower total C concentrations of the west side soils when adjusted for the effect of altitude or distance from the sea. Mineral soils have a higher bulk density and are drier than organic soils, and the altitude- or distance to sea-adjusted differences in bulk density and moisture content between the west and east sides of the island, although not significant at $P \leq 0.05$, also point toward west side soils being intrinsically more mineral and less organic than the east side soils.

Because of their greater minerality, west side soils have higher concentrations of total Ca, Mg and K than east side soils. Because the wind (especially strong wind) is predominantly westerly, west side soils also have higher concentrations of total Na and exchangeable forms of Na, Ca, and Mg than do east side soils. In contrast, soil exchangeable K concentrations are lower on the west side than the on east side. The reason for this is uncertain. K does possess cation exchange properties not shown by the other exchangeable cations; an important one is that it can convert from an easily exchangeable to a fixed form in some types of clay [14]. The identities of clay minerals in Marion Island soils are a matter of some controversy [3]; indeed, most of the controversy concerns whether there are any clays at all [15]. However, recent Raman spectroscopic analysis reveals that crystalline minerals such as biotite and muscovite, both capable of weathering into K-fixing clays, are common on the island [16]. Possibly, K-fixation is more intense in the mineral-rich west coast soils than in the organic east coast ones. Another possible explanation for the lower exchangeable K concentration in the west side than east side soils is that the high amount of Na reaching the soil surface through seaspray on the western side displaces exchangeable K as it percolates through the soil. On the east coast, replacement of K by Na on the cation exchange complex would be much less intense because of the much less intense deposition of seaspray.

4. Conclusion

Altitude and nearest distance to the sea are strongly interrelated, so it is not surprising that soil chemical composition is influenced by both. The influence is through a decreasing influence of seaspray with increasing distance inland and a change from organic peats characteristic of the coastal plain to mineral soils more influenced by parent basalts further inland and higher up. Furthermore, the dependence of soil chemical composition on altitude or distance from the sea is a function of aspect—here we showed that it differs markedly between the west and east sides of the island. There are intrinsic differences (*i.e.* once this between-side difference in the influence of altitude and distance from the sea on soil chemical composition has been accounted for) in soil chemical composition between east and west sides. These findings are crucial for modeling nutrient cycling budgets on a whole island basis.

5. Acknowledgements

This study was funded by the South African National Research Foundation (grant SNA2008050700003) and the fieldwork was supported logistically by the Antarctica and Islands Directorate of the South African Department of Environmental Affairs and Tourism.

REFERENCES

[1] B. R. Schulze, "The Climate of Marion Island," In: E. M. Van Zinderen Bakker, J. M. Winterbottom and R. A. Dyer, Eds., *Marion and Prince Edward Islands*, A.A. Balkema, Cape Town, 1971, pp. 16-31.

[2] P. C. Le Roux, "Climate and Climate Change," In: Chown, S. L and Froneman, P. W. Eds., *The Prince Edward Islands: Land-Sea Interactions in a Changing Ecosystem*, African SunMedia, Stellenbosch, 2008, pp. 39-64.

[3] V. R. Smith and L. Mucina, "Vegetation of Subantarctic Marion and Prince Edward Islands," In: L. Mucina and M. C. Rutherford, Eds., *The Vegetation of South Africa, Lesotho and Swaziland*, Strelitzia, Vol. 19, South African National Biodiversity Institute, Pretoria, 2006, pp. 698-

723.

[4] E. M. Van Zinderen Bakker, "The Second South African Biological Expedition to Marion Island 1971-72," *South African Journal of Antarctic Research*, Vol. 13, 1973, pp. 60-63.

[5] O. W. Heal, "Introduction," In: L. C. Bliss, O. W. Heal and J. J. Moore, Eds., *Tundra Ecosystems: A Comparative Analysis*, Cambridge University Press, Cambridge, 1981, pp. xxvii-xxxvii.

[6] V. R. Smith, "Chemical Composition of Precipitation at Marion Island (Sub-Antarctic)," *Atmospheric Environment*, Vol. 21, No. 5, 1987, pp. 1159-1165. doi:10.1016/0004-6981(87)90243-5

[7] V. R. Smith and P. W. Froneman, "Nutrient Dynamics in the Vicinity of the Prince Edward Islands," In: S. L. Chown and P. W. Froneman, Eds., *The Prince Edward Islands: Land-Sea Interactions in a Changing Ecosystem*, African SunMedia, Stellenbosch, 2008, pp. 165-179.

[8] V. R. Smith, "Production and Nutrient Dynamics of Plant Communities on a Sub-Antarctic Island. 5. Nutrient Budgets and Turnover times for Mire-grasslands, Fjaeldmark and Fernbrakes," *Polar Biology*, Vol. 8, 1988, pp. 255-269. doi:10.1007/BF00263174

[9] V. R. Smith, "Terrestrial and Freshwater Primary Production and Nutrient Cycling," In: S. L. Chown and P. W. Froneman, Eds., *The Prince Edward Islands: Land-Sea Interactions in a Changing Ecosystem*, African SunMedia, Stellenbosch, 2008, pp. 181-214.

[10] L. E. Allison, "Organic Carbon," In: C. A. Black, Ed., *Methods of Soil Analysis, Part* 2, *Chemical and Microbiological Properties*, American Society of Agronomy, Madison, 1965, pp. 1367-1378.

[11] StatSoft Inc., STATISTICA Version 9.0., 2009. www.statsoft.com

[12] V. R. Smith, "Soil Chemistry of Marion Island (Sub-Antarctic)," *South African Journal of Science*, Vol. 74, 1978, pp. 174-175.

[13] V. R. Smith and M. Steenkamp, "Classification of the Terrestrial Habitats on Sub-Antarctic Marion Island based on Vegetation and Soil Chemistry," *Journal of Vegetation Science*, Vol. 12, No. 2, 2001, pp. 181-198. doi:10.2307/3236603

[14] E. W. Russell, "Soil Conditions and Plant Growth," Ninth Edition, Longman, London, 1960.

[15] K.-H. Gribnitz, L. E., Kent and R. R. Dixon, "Volcanic Ash, Ash Soils and the Inferred Quaternary Climate of Sub-Antarctic Marion Island," *South African Journal of Science*, Vol. 82, 1986, pp. 629-635.

[16] L. C. Prinsloo, P. Colomban, J. D. Brink and K. I. Meiklejohn, "A Raman Spectroscopic Study of the Igneous Rocks on Marion Island: A Possible Terrestrial Analogue for the Geology on Mars," *Journal of Raman Spectroscopy*, Vol. 42, No. 4, 2011, pp. 626-632. doi:10.1002/jrs.2756

15

Effect of Industrial Sludge Application on Soil Nitrogen and Wheat Plant Response

Sutapa Bose[1,2*], **Asim K. Bhattacharyya**[1]

[1]School of Environmental Sciences, Jawaharlal Nehru University, New Delhi, India; [2]Department of Earth Sciences, Indian Institute of Science Education and Research (IISER) Kolkata PO: BCKV Main Campus, Mohanpur Nadia, India.
Email: *sutaparai@gmail.com

ABSTRACT

Wazirpur industrial area of Delhi generates a huge quantity of sludge per day, which is highly acidic in nature (pH 2.7 to 4.4) and contains macronutrients, micronutrients as well as toxic metals. A pot-culture experiment was conducted by taking the two soils (JNU and Chhattarpur) amended with sludge (0%, 10%, 20%, 30%), pretreated with lime (0%, 0.5% and 1%). Two wheat seedlings were planted per pot containing 3 kg sludge amended or control soil and the experiment was carried out till harvesting (four months) in a glass house. Lime treatments enhanced the N content in wheat plant in almost all cases. Sludge and lime treatments enhanced dry weight in wheat plants grown in Chhattarpur soil and dry weight increased with time. Maximum growth was observed in 0.5 lime treated and 20% sludge amended soils. But we have to take an account about any kind of metal toxicity before disposal of this waste to land.

Keywords: Industrial Sludge; Lime Treatment; Land Disposal; Soil Nitrogen; Wheat Plant

1. Introduction

Like many countries in South Asia, India is regarded as an agricultural country. Indian agricultural soils have been used extensively for crop production for extended period of time and are usually low in soil organic matter and essential plant nutrients for crop. Addition of chemical fertilizers is most commonly used, which increased the cost of crops. To reduce such expenditure and to manage the soil quality of land, sludge application (which is considered an organic fertilizer) is an alternative source of plant nutrients for crops.

Land application of industrial sludge is an option for safe and economic methods for waste disposal. The application of urban industrial sludge mixed with sewage sludge to agricultural soil is generally most economic outlet for waste disposal, in this way it is possible to recycled the plant nutrients such as N, P and organic matter and many more macro and micro nutrients. On the other hand it can affect the soil fertility, food safety and natural ecosystem in general. Many researches have been carried out on the effect of applying sewage to agricultural soil as regards both the potential effects due to sewage toxic components [1-11]. Sewage sludge provides labile organic matter in sufficient quantities to stimulate soil microorganisms. This kind of organic residue improves the soil physical characteristics: increasing soil's water holding capacity and percentage of stable aggregates and enhanced the nutritional quality of soils [12]. It can improve the physical properties of soils [13] by increasing soil porosity, and the stability of soil aggregation [14]. So, sludge can be used as soil fertilizer provided that pretreatments have done for its metal contents. Liming to soil offers a mean of minimizing the risk of food chain contamination by reducing the plant uptake of sludge-born heavy metals [15-17]. Urban sludge application on land increased the dry matter yield of maize over control [18]. According to several studies the effectiveness of liming varies depending on the soil, metal, pH value of the limed soils and crop species [19,20]. Matula and Pechová [21] studied that application of lime in lower pH value increased nitrification in soil. Nitrogen supply in plants mainly depends on soil available nitrogen so before cultivation it is necessary to estimate N release from mineralization. In a field experiment the shoot dry weight of lowland rice (*Oryza sativa* L.) increased with the increase in shoot N uptake up to flowering stage of plant growth. At harvest, N uptake in the shoot decreased due to translocation to the grain.

The aim of the present study is to investigate the behavior of sludge amended soil in respect to available nitrogen in soils and response of wheat plants in context of root shoot length, dry weight and nitrogen content in it. The industrial sludge, which was used for the experiment,

*Corresponding author.

was acidic in nature, so that, sludge was treated with different dosages of lime and then the sludge was mixed with soils collected from two different parts of Delhi. A pot culture experiment was conducted in a glass house to see the response of wheat plants grown in lime treated or untreated industrial sludge amended soils.

2. Material and Methods

2.1. Samples Collection

Samples were collected from different location of of Wazirpur industrial area, Delhi (**Figure 1**) for each season (summer, monsoon and winter) from road-side dumps by following the standard methods. Soil samples were collected from 5 different spots from JNU as well as from Chhattarpur and were homogenized separately. Soil sampling was done once during the study period. Soon after the collection of sludge and soils, the pH Electrical Conductivity (EC), Water Holding Capacity (WHC) and Moisture Content (MC) were measured [22] and remaining samples were kept for drying. Then air-dried sludge samples were grinded and pass through 2 mm sieve and homogenized to make a representative sample. Soil samples were also processed by following the same methods.

2.2. Preparation for Glass House Experiment

The homogenized sludges were treated with three different dosages of lime (0%, 0.5% and 1%) separately. Then the lime treated and untreated sludges were mixed with two soils (Chhattarpur and JNU) separately at the rate of 0%, 10% 20% and 30% (dry weight basis). Then the sludge amended and control soils were put in new earthen pots (3 kg/pot).

The wheat grains (HD1553) were collected from National Seed Corporation (NSC), Pusa, New Delhi. The plants were grown in two trays with two different soils. Two wheat seedlings of 10 days old were transferred per pots containing 3 kg sludge amended soil or control soil.

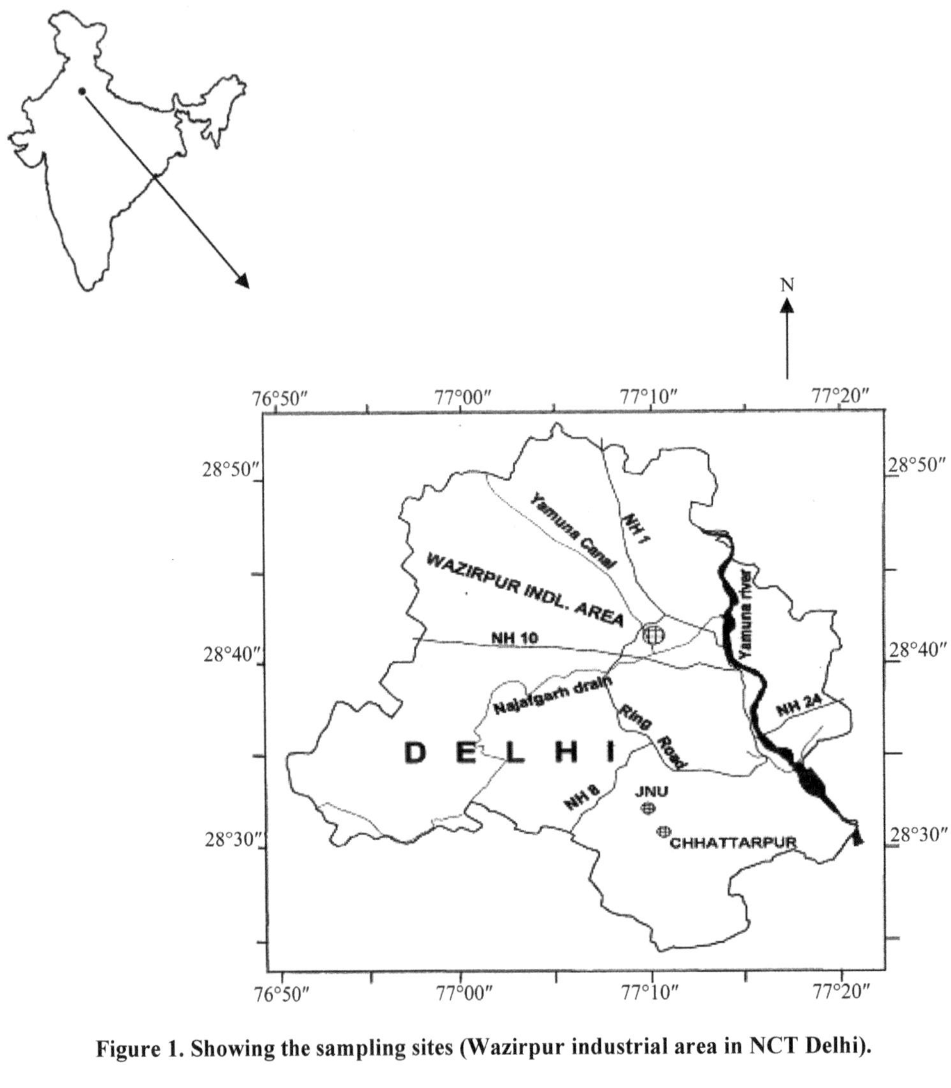

Figure 1. Showing the sampling sites (Wazirpur industrial area in NCT Delhi).

Distilled water was given to plants to maintained 50% moisture of WHC in sludge amended and control soils throughout the experiment. Whole experiment was carried out in a glass house maintaining the temperature at 20˚C - 30˚C. The experiment was performed with three replicates and was continued from January to April 2004.

2.3. Collection and Preservation of Soil and Plant Samples

Plants and soils samples were taken out from the pots in different stages of plant growth, viz. planting, tillering, flowering, grain formation and harvesting respectively. At each harvest plants were removed carefully and washed first with tap water followed by de-ionized water. Root and shoot were separated and lengths were measured. Then folded it in brown paper and kept in oven at 80˚C for 48 hrs. After drying root and shoot weights were taken separately. Wheat grains were also measured after harvesting stage of wheat plant growth. The soil samples were air-dried at ambient room temperature, and passed through 2 mm sieve for chemical analysis. The available (mineralizable) nitrogen analyzed by following Subbian and Asija [23] method. The soil samples were stored in an airtight polythene bags and kept at 4˚C to prevent any kind of microbial degradation. The total nitrogen of plant samples was analyzed through acid digestion followed by Kjeldahl method [24]. The parameters were carried out in triplicate for better results.

2.4. Data Analysis

In the present study, the plant total nitrogen and soil available (mineralizable) nitrogen (amended with different proportion of lime treated sludge) were monitored on five growth stages of plants (days). Similarly, plant's nitrogen was also monitored on four growth stages of plants. These parameters were studied for root and shoot separately. Each set of data was arranged to accomplish two sets of 2-factor ANOVA as follows:

1) Keeping sludge treatment constant, growth period and lime treatment were taken as two factors.

2) Keeping lime treatment constant, growth period and sludge amendment were taken as two factors.

3. Results and Discussion

3.1. Available (Mineralizable) Nitrogen in Soils

The available nitrogen of Chhattarpur control soil (49 ± 2.9 ppm) was higher than JNU control soil (42 ± 2.2 ppm). Sludge application increased the available nitrogen in both soils. Sludge amendment enhanced the available nitrogen in alkaline soil than acidic soil [25]. Lime treatment (0.5%) decreased the available nitrogen in Chhattarpur soil, whereas 1% lime application enhanced the available nitrogen in sludge amended Chhattarpur soil. In JNU soil lime treatment gradually and slowly increased and got stabilized the available nitrogen concentration in sludge amended JNU soil, in most of the cases. ANOVA results show the variation in available nitrogen, were highly significant. With the successive stages of plant growth the changes in available (mineralizable) nitrogen in soils were also highly significant. It is clear that higher amount of lime application had neutralized the acidity of amended soil. Hence, there was an increase amount of available nitrogen in those soils. Lime treatment gradually, but slowly increased (or stabilized in some cases) the available (mineralizable) nitrogen concentration in waste amended JNU soil. Both mineralization and immobilization take place simultaneously and may not be quantified separately. Gaseous N loss would also depend on pH and temperature regimes of soil. The pH inhibits mineralization only at the levels whose pH is below 5 and higher than 8. Nitrogen mineralization dynamics are predominantly determined by the soil properties such as total C, total N, soil texture and water holdidg capacity [26]. A sufficiently long latent time (8 weeks) was needed for tolerant microorganisms to release or to start nitrogen minerilazation. The nitrogen mineralization was positive in the presence of metals in most of the cases [27].

Figures 2 and **3** showing that available (mineralizable) nitrogen in soils increased till flowering stage and the gradually decreased with the successive stages of wheat plant growth which indicated that the rate of release of available (mineralizable) N was higher than the uptake or immobilization on nitrogen in waste amended soil till flowering stage of wheat plant growth. The changes in available (mineralizable) nitrogen were significant with lime treatments, sludge amendment as well as with growth periods ($P < 0.05$).

3.2. Dry Weight of Wheat Plant's Root-Shoot-Grain

Sludge amendments increased plant growth significantly as compared to the control soil and the growth was higher in the plant grown in Chhattarpur pure soil than JNU soil. Lime amendment showed a positive effect on plant growth at each sludge amendment, which is likely due to neutralizing pH, reducing the availability of heavy metals. The dry wt. of crops increased with increasing soil pH [4] which was also observed in the present study. The control soils were alkaline in nature *viz.* pH of Chhattarpur soil was 8.66 and JNU soil was 8.37 whereas, the sludge was very acidic (pH 3.05) in nature [17,28]. The dry weights of harvested grains were higher in wheat plants grown in JNU soil than Chhattarpur soil. Sludge application increased the dry weight of grains in all cases.

Lime treatment increased the dry weight of grain. The changes in shoot dry weight were significant with lime treatments, sludge amendment as well as with growth periods ($P < 0.05$). (**Table 1**) The changes in root dry weight were significant with lime treatments as well as with growth periods while these were insignificant with sludge amendment ($P < 0.05$).

3.3. Lengths of Wheat Plant's Root-Shoot

The length of wheat plant's shoot was higher in plants grown in JNU control soil than Chhattarpur control soil (**Table 2**). 10% sludge application reduced the shoot length as compared with control one. But 20% and 30% sludge application increased the shoot length. The changes in root-shoot length were almost insignificant with lime treatments, sludge amendment as well as with growth periods ($P < 0.05$). Lime treatment increased the shoot length slightly. Length of wheat plant's root increased with increasing percent of sludge (10% and 20%) but decreased at 30% sludge amendment, which showed that application of sludge had some, limiting factors, which prevents the growth of plants. Lime treatments increased the root length. Over the time (*i.e.* stages of plant growth) the root length increased significantly.

3.4. Total Nitrogen in Wheat Plant

The total nitrogen in wheat plant's shoot and root were 0.44% ± 0.02% and 0.15% ± 0% of dry weight at the day of plantation respectively. The nitrogen concentration in wheat plant's shoot increased with the successive stages of plants growths in most of the cases. It was observed that the available nitrogen in sludge amended soil decreased during grain formation to harvesting, which may be due to higher uptake of nitrogen in wheat plants (**Figures 2, 3**). Sludge application (20% and 30%) decreased the N content in wheat plants, which might be due to toxic effects of heavy metals. But, the total nitrogen increased in wheat plants shoots grown in 10% sludge amended Chhattarpur soil. The changes in total nitrogen in wheat shoot were almost insignificant with lime treatments, sludge amendment as well as growth periods ($P < 0.05$). The changes in total nitrogen in wheat root were largely significant with lime treatments as well as sludge amendment while insignificant with growth periods ($P < 0.05$). The sludge contained higher concentration of heavy metals, which caused the stress of nutrients to the plants [17,28,29]. Sludge application in Chhattarpur soil enhanced the total nitrogen in wheat plant's root, except 30% sludge amended Chhattarpur soil (**Figure 4**). The

Table 1. Dry weight (g/Pot) of wheat plant's shoot and root grown in sludge amended soils in different stages of plant growth.

Sample code	Stages of plant growth							
	Tillering		Flowering		Grainform		Harvesting	
	Shoot	Root	Shoot	Root	Shoot	Root	Shoot	Root
C-0-0	1.22 ± 0.02	0.34 ± 0.02	3.1 ± 0.06	0.52 ± 0.04	3.85 ± 0.04	0.92 ± 0.05	5.21 ± 0.02	1.25 ± 0.03
C-10-0	1.04 ± 0.03	0.31 ± 0.01	2.8 ± 0.05	0.48 ± 0.04	3.42 ± 0.05	0.68 ± 0.05	5.34 ± 0.07	0.95 ± 0.05
C-10-0.5	1.15 ± 0.06	0.35 ± 0.02	3 ± 0.04	0.53 ± 0.03	3.66 ± 0.05	0.72 ± 0.06	5.56 ± 0.07	1.08 ± 0.05
C-10-1.0	1.21 ± 0.05	0.33 ± 0.03	3.2 ± 0.02	0.55 ± 0.04	3.79 ± 0.02	0.75 ± 0.04	5.84 ± 0.05	1.15 ± 0.05
C-20-0	1.54 ± 0.06	0.34 ± 0.04	3.6 ± 0.03	0.42 ± 0.05	3.91 ± 0.03	0.75 ± 0.06	5.92 ± 0.04	0.95 ± 0.04
C-20-0.5	1.68 ± 0.03	0.38 ± 0.04	3.8 ± 0.04	0.46 ± 0.04	4.16 ± 0.04	0.81 ± 0.06	6.13 ± 0.06	1.02 ± 0.05
C-20-1.0	1.98 ± 0.02	0.41 ± 0.02	3.86 ± 0.02	0.51 ± 0.03	4.28 ± 0.02	0.85 ± 0.07	6.28 ± 0.05	1.05 ± 0.06
C-30-0	1.86 ± 0.06	0.44 ± 0.05	4.3 ± 0.05	0.5 ± 0.01	4.76 ± 0.05	0.71 ± 0.07	6.11 ± 0.06	0.89 ± 0.06
C-30-0.5	2.05 ± 0.05	0.49 ± 0.06	4.1 ± 0.07	0.53 ± 0.02	4.55 ± 0.08	0.82 ± 0.07	6.42 ± 0.07	0.95 ± 0.07
C-30-1.0	2.21 ± 0.08	0.51 ± 0.05	4.5 ± 0.04	0.57 ± 0.04	4.62 ± 0.09	0.88 ± 0.07	6.44 ± 0.08	0.99 ± 0.03
J-0-0	1.14 ± 0.05	0.27 ± 0.05	2.54 ± 0.05	0.33 ± 0.02	2.75 ± 0.05	0.73 ± 0.07	4.88 ± 0.04	1.11 ± 0.03
J-10-0	1.08 ± 0.02	0.24 ± 0.02	2.33 ± 0.03	0.28 ± 0.03	2.55 ± 0.04	0.57 ± 0.07	4.73 ± 0.05	0.86 ± 0.05
J-10-0.5	1.28 ± 0.02	0.28 ± 0.03	2.62 ± 0.02	0.31 ± 0.02	2.75 ± 0.03	0.59 ± 0.07	4.95 ± 0.03	0.97 ± 0.05
J-10-1.0	1.34 ± 0.03	0.27 ± 0.04	2.6 ± 0.02	0.33 ± 0.01	2.76 ± 0.07	0.63 ± 0.08	5.06 ± 0.02	0.99 ± 0.07
J-20-0	1.35 ± 0.04	0.3 ± 0.03	2.91 ± 0.03	0.35 ± 0.03	2.99 ± 0.07	0.68 ± 0.05	5.16 ± 0.07	0.79 ± 0.07
J-20-0.5	1.45 ± 0.05	0.34 ± 0.05	3.12 ± 0.04	0.37 ± 0.03	3.22 ± 0.03	0.73 ± 0.05	5.37 ± 0.08	0.82 ± 0.06
J-20-1.0	1.49 ± 0.02	0.35 ± 0.04	3.05 ± 0.04	0.39 ± 0.04	3.29 ± 0.05	0.77 ± 0.05	5.52 ± 0.09	0.88 ± 0.04
J-30-0	1.38 ± 0.02	0.38 ± 0.05	3.15 ± 0.02	0.43 ± 0.04	3.29 ± 0.08	0.56 ± 0.06	5.11 ± 0.08	0.77 ± 0.07
J-30-0.5	1.55 ± 0.03	0.4 ± 0.03	3.36 ± 0.03	0.46 ± 0.04	3.54 ± 0.08	0.64 ± 0.06	5.38 ± 0.06	0.81 ± 0.08
J-30-1.0	1.67 ± 0.06	0.42 ± 0.04	3.45 ± 0.04	0.5 ± 0.02	3.62 ± 0.06	0.68 ± 0.05	5.49 ± 0.08	0.85 ± 0.05

[C = Chhattarpur soil, J = JNU soil; 0, 10, 20, 30 (% of sludge amendments); 0, 0.5, 1 (% of Lime treatments). All the values are mean of three replicates].

Table 2. Length (cm) of wheat plant's shoot and root grown in sludge amended soils in different stages of plant growths.

Sample code	Stages of plant growths							
	Tillering		Flowering		Grainform		Harvesting	
	Shoot	Root	Shoot	Root	Shoot	Root	Shoot	Root
C-0-0	33.2 ± 1.2	5.4 ± 0.3	48.35 ± 3.2	7.1 ± 0.5	64.2 ± 4.5	8.2 ± 0.5	72.9 ± 5.5	8.5 ± 0.8
C-10-0	28.3 ± 2.3	6.3 ± 0.5	46.2 ± 2.5	8.5 ± 0.6	61.5 ± 4.6	9.1 ± 0.4	70.5 ± 5.4	9.4 ± 0.5
C-10-0.5	29.7 ± 2.1	6.8 ± 0.4	49.3 ± 3.2	8.9 ± 0.8	65.5 ± 4.2	9.8 ± 0.6	75.3 ± 5.1	10.5 ± 0.6
C-10-1.0	30.1 ± 2.5	7.2 ± 0.5	51.8 ± 4.2	9.1 ± 0.7	67.9 ± 5.7	10.5 ± 0.6	77.5 ± 5.2	11.6 ± 0.7
C-20-0	35.8 ± 2.6	6.1 ± 0.6	50.2 ± 4.2	7.3 ± 0.5	64.3 ± 5.2	8.9 ± 0.6	75.6 ± 4.2	9.3 ± 0.9
C-20-0.5	37.3 ± 3.3	6.4 ± 0.6	55.6 ± 4.5	7.8 ± 0.6	68.9 ± 5.4	8.5 ± 0.8	79.2 ± 4.1	9.5 ± 0.9
C-20-1.0	39.5 ± 3.4	6.9 ± 0.6	58.3 ± 3.8	8.4 ± 0.4	72 ± 5.8	9.3 ± 0.7	80.5 ± 5.7	9.9 ± 0.8
C-30-0	34.2 ± 3.5	4.9 ± 0.2	48.1 ± 3.7	6.9 ± 0.5	55.2 ± 5.4	7.8 ± 0.4	62.9 ± 4.1	8.1 ± 0.8
C-30-0.5	37.2 ± 2.8	5.4 ± 0.3	50.6 ± 4.5	7.5 ± 0.7	59.2 ± 5.2	8.4 ± 0.8	66.2 ± 4.2	8.7 ± 0.8
C-30-1.0	38.1 ± 4.2	5.9 ± 0.5	52.3 ± 4.6	7.8 ± 0.5	62.1 ± 4.2	8.2 ± 0.7	69.2 ± 3.8	8.9 ± 0.7
J-0-0	35.2 ± 2.5	6.2 ± 0.4	52.9 ± 4.2	8.2 ± 0.6	67.3 ± 4.3	8.9 ± 0.8	74.6 ± 3.3	9.7 ± 0.8
J-10-0	28.3 ± 3.4	6.8 ± 0.4	51.3 ± 4.1	8.9 ± 0.7	64.8 ± 4.3	9.7 ± 0.8	74.3 ± 4.2	10 ± 0.7
J-10-0.5	30.5 ± 3.1	7.3 ± 0.6	55.2 ± 4	9.3 ± 0.8	68.2 ± 4.2	10.5 ± 0.8	76.3 ± 3.8	11.5 ± 1.0
J-10-1.0	30.8 ± 3.2	7.1 ± 0.5	57.8 ± 3.8	9.5 ± 0.8	70.2 ± 4.3	10.6 ± 0.5	78.4 ± 3.3	11.4 ± 1.0
J-20-0	35.8 ± 3.2	7.5 ± 0.5	52.3 ± 3.9	8.9 ± 0.9	65.3 ± 5.7	9.7 ± 1.0	74.6 ± 4.2	9.9 ± 0.9
J-20-0.5	39.1 ± 2.5	7.9 ± 0.4	55.2 ± 4	9.5 ± 0.9	70.2 ± 5.1	10.5 ± 0.8	78.5 ± 4.1	10.9 ± 0.5
J-20-1.0	38.5 ± 3.6	8.1 ± 0.4	57.4 ± 4.7	9.9 ± 0.7	74.3 ± 4.9	11.2 ± 1.0	81.6 ± 5.0	11.6 ± 0.8
J-30-0	36.8 ± 3.5	5.8 ± 0.6	45.3 ± 4.1	7.8 ± 0.5	58.3 ± 5.5	8.6 ± 0.5	62.7 ± 3.6	8.8 ± 0.7
J-30-0.5	39.2 ± 3.7	6.4 ± 0.3	49.6 ± 3.8	8.4 ± 0.6	62.1 ± 5.8	9.5 ± 0.6	66.3 ± 3.3	9.8 ± 0.8
J-30-1.0	40.1 ± 3.6	6.6 ± 0.2	49.5 ± 4.5	8.9 ± 0.6	63.5 ± 5.5	9.9 ± 0.6	68.9 ± 3.1	10.1 ± 0.8

[C = Chhattarpur soil, J = JNU soil; 0, 10, 20, 30 (% of sludge amendments); 0, 0.5, 1 (% of Lime treatments). All the values are mean of three replicates].

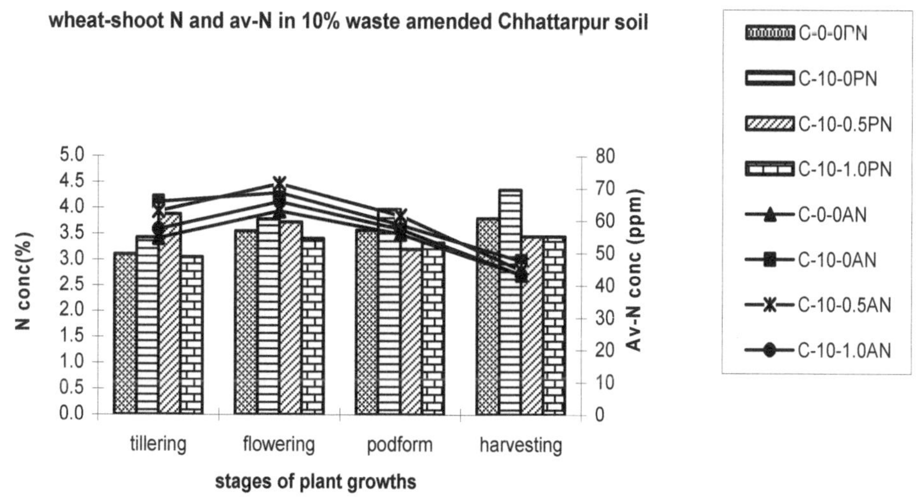

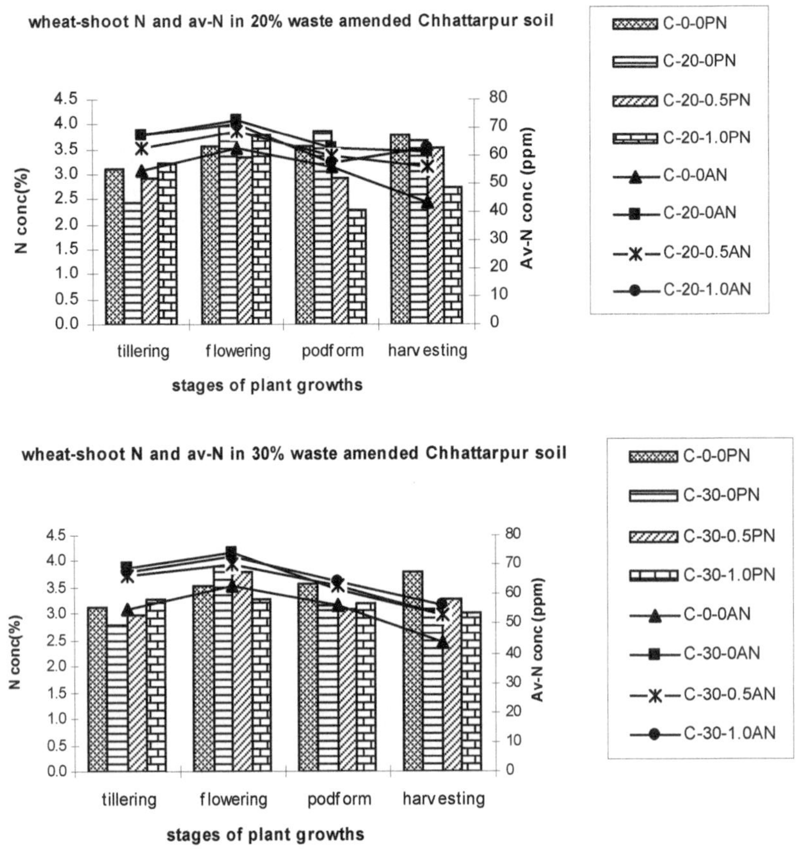

Figure 2. Available N in 10%, 20% and 30% waste amended Chhattarpur soil in different stages of wheat plant growth.

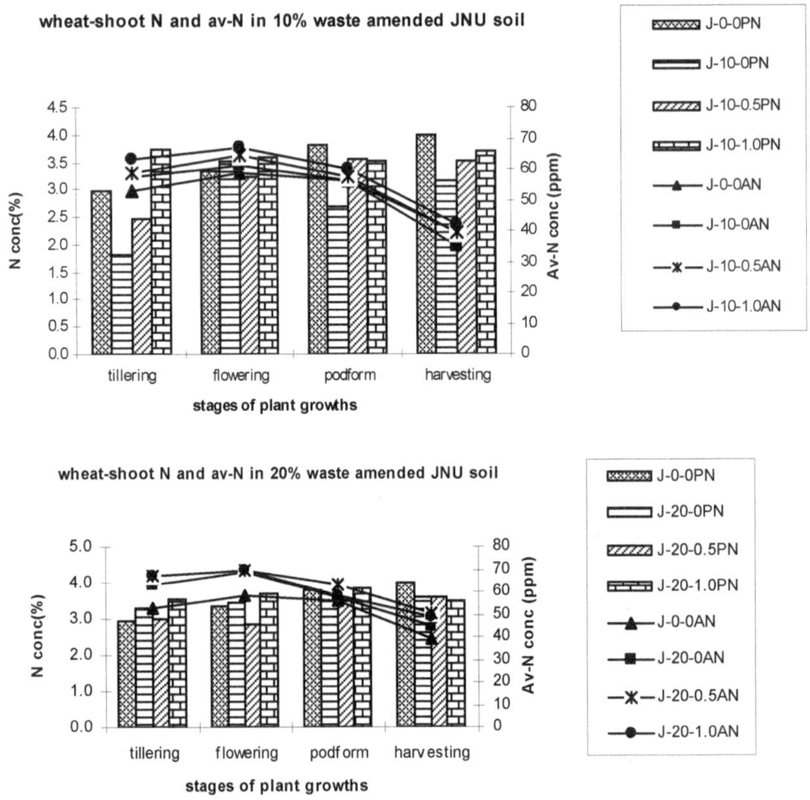

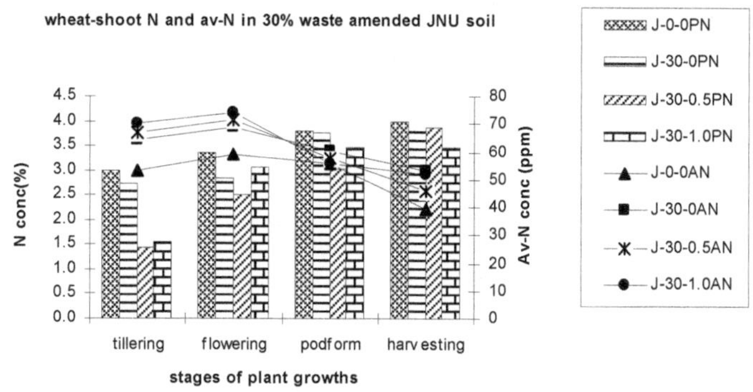

Figure 3. Available N in 10%, 20% and 30% waste amended JNU soil in different stages of wheat plant growth.

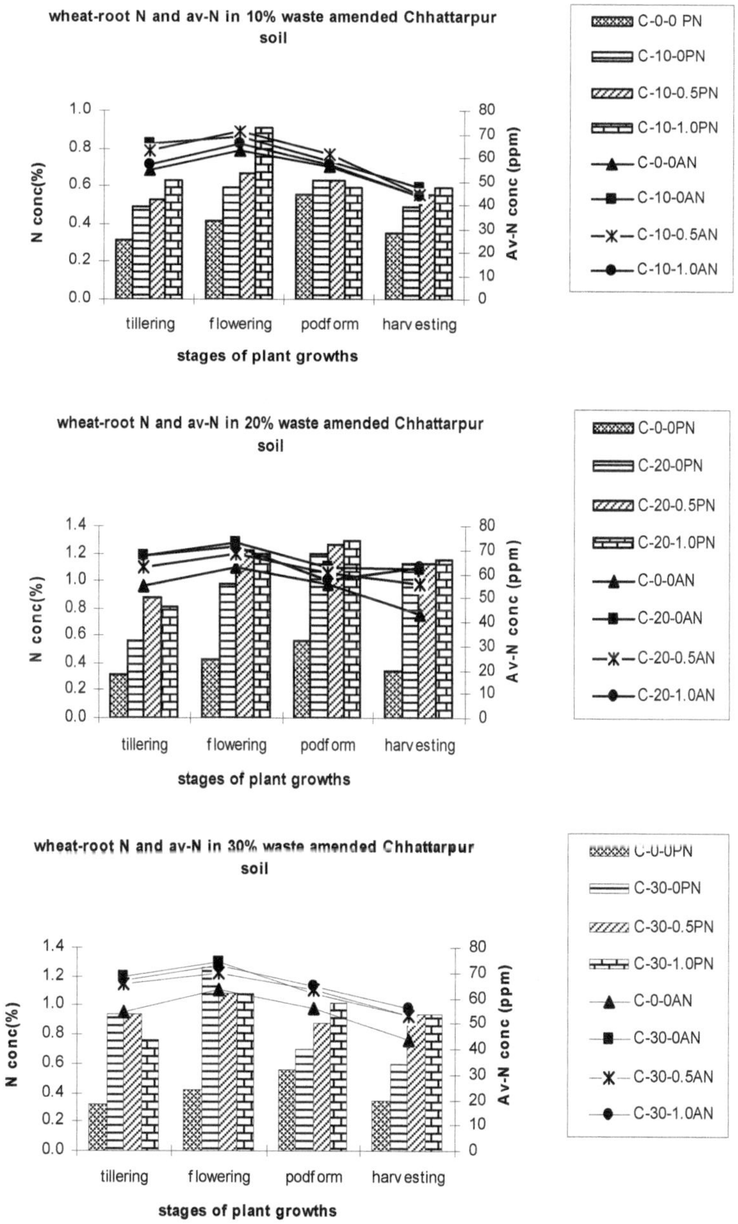

Figure 4. Wheat root Nitrogen and available N in 10%, 20% and 30% waste amended Chhattarpur soil in different stages of wheat plant growth.

nitrogen content of root in wheat plants grown in sludge amended JNU soil increased than control soils (**Figure 5**).

Lime treatment did not make any significant changes in nitrogen content in wheat plant's shoots. The total nitrogen in root increased till grain formation stage, after that reduced slightly. The available nitrogen in sludge amended soil had decreased continuously from grain formation to harvesting stage, where nitrogen concentrations in root increased in grain formation stage and slightly decreased in harvesting stage (**Figure 6**). At harvesting stage the uptake of available nitrogen may be same or more but the nitrogen were used in grain maturation. So, there was a slight decrease of total nitrogen in wheat plants at harvesting stage. The nitrogen concentration in wheat grains increased in the plants grown in 10% sludge amended soil, but it decreased in other sludge amendments (20% and 30%). Lime treatment did not make any significant changes in grain nitrogen concentration. The nitrogen in shoot was always >80% of the total nitrogen concentration in wheat plants in all cases At harvesting stage the nitrogen was almost 90% in above ground part of wheat plants as compared to the total nitrogen concentration in plants (**Figure 6**).

4. Conclusion

Available nitrogen in both the soils increased with sludge

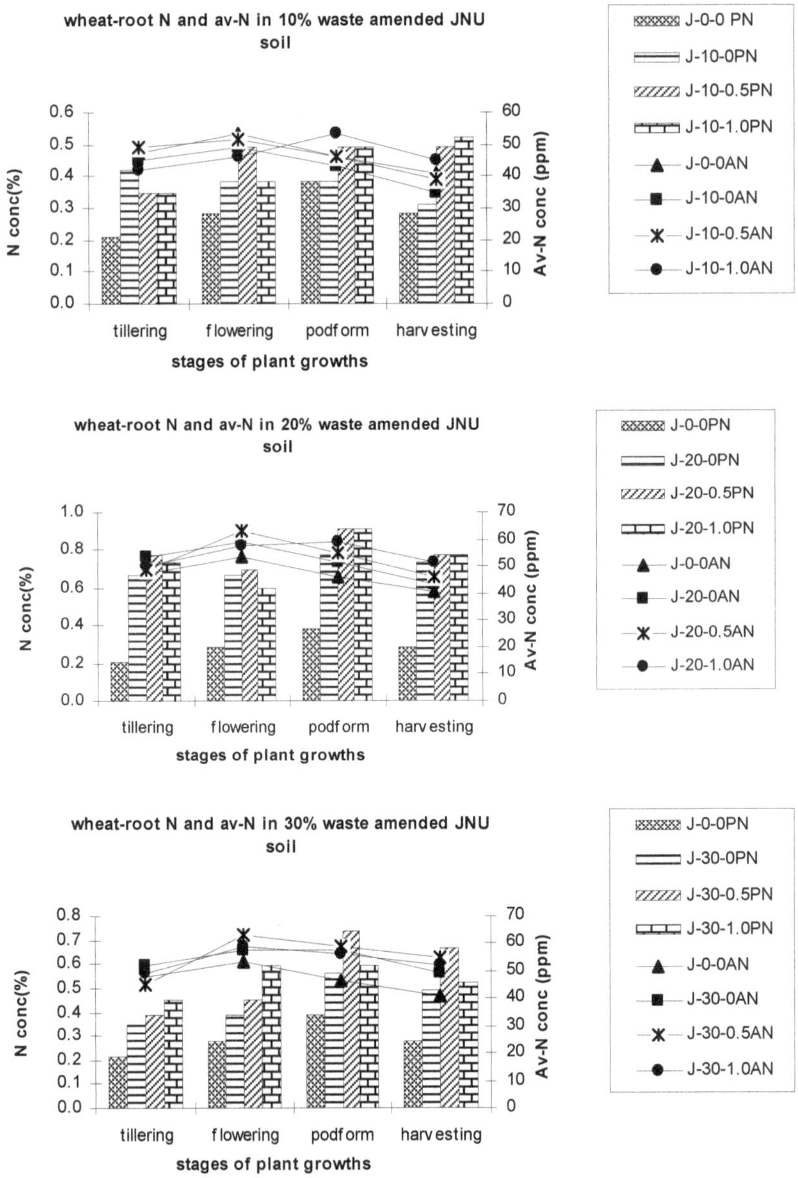

Figure 5. Wheat root Nitrogen and available N in 10%, 20% and 30% waste amended JNU soil in different stages of wheat plant growth.

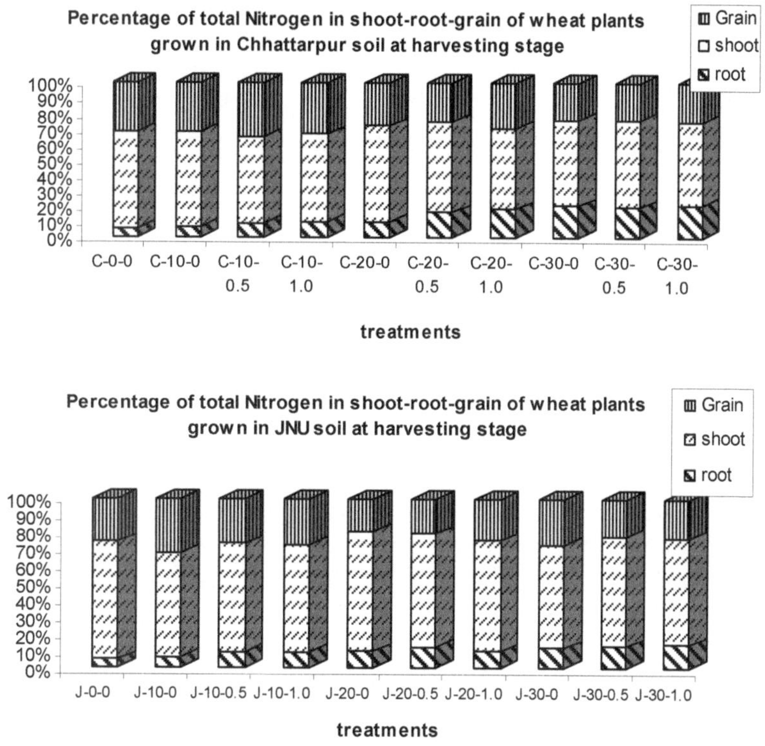

Figure 6. The percentage of wheat root-shoot-grain Nitrogen concentration in Chattarpur and JNU soil at harvesting stage.

addition and lime treatment. Wheat plants observed increase in plant biomass as well as grain yields with the both sludge and lime amendments except 30% sludge amendment. The nitrogen content in both root and shoot increased with 10% sludge amendment but continuously decreased with other amendments *i.e.*, 20% and 30% sludge amendments. Total nitrogen in root increased till grain formation and onward decreased which indicated that during grain filling and maturation the nitrogen content was utilized by upper part of the wheat plants. Available nitrogen was also continuously decreased during grain formation and harvesting stage. At harvesting stage the total nitrogen in shoot was 90% of the whole plant nitrogen. So, from this study, it seems that 20% sludge can be disposed off the land after 0.5 lime treatment. But, before make any final suggestion the heavy metal content in the consumable part of the wheat plant should be taken into an account pursued by field trial by following USEPA land application of sewage sludge [30].

5. Acknowledgements

Authors are thankful to Dean of School of Life Sciences, Jawaharlal Nehru University, New Delhi, India, for providing glass house during the pot-culture experiment and, Sutapa Bose is thankful to University Grant Commission (UGC), India to provide fund in the form of Junior Research Fellowship (JRF) followed by Senior Research Fellowship (SRF) during period of her work.

REFERENCES

[1] A. C. Chang, A. L. Page and F. T. Bingham, "Heavy Metal Absorption by Winter Wheat Following Termination of Cropland Sludge Application," *Journal of Environmental Quality*, Vol. 11, No. 4, 1982, pp. 705-708. doi:10.2134/jeq1982.00472425001100040030x

[2] E. Witter and J. Lopez-Real, "Nitrogen Losses During Composting of Sewage Sludge and the Effectiveness of the Clay Soil, Zeolite and Composting the Volatilized Ammonia," *Biological Wastes*, Vol. 23, No. 4, 1988, pp. 279-294. doi:10.1016/0269-7483(88)90016-X

[3] S. R. Wild, M. L. Berrow and K. C. Jones, "The Persistence of Polynuclear Aromatic Hydrocarbons (Pahs) in Sewage Sludge Amended Agricultural Soils," *Environmental Pollution*, Vol. 72, No. 2, 1991, pp. 141-157. doi:10.1016/0269-7491(91)90064-4

[4] S. R. Smith, "Effects of Sewage Sludge Application on Soil Microbial Processes and Soil Fertility," In: B. A. Stewart, Ed., *Advances in Soil Science*, Springer-Verlag, New York, 1991, pp. 191-212. doi:10.1007/978-1-4612-3144-8_4

[5] B. J. Alloway and A. P. Jackson, "The Behavior of Heavy Metals in Sewage Sludge-Amended Soils," *Science of the Total Environment*, Vol. 100, 1991, pp. 151-176. doi:10.1016/0048-9697(91)90377-Q

[6] M. E. Mesquita, J.-M. Vieira, E. Silva and H. Domingues, "Copper Adsorption by a Schistic Soil. Application of

Sewage Sludge," *Environmental Technology*, Vol. 15, No. 11, 1994, pp. 1089-1094. doi:10.1080/09593339409385518

[7] H. Hani, A. Siegenthaler and T. Cardinas, "Soil Effects Due to Sewage Sludge Application in Agriculture," *Nutrient Cycling in Agroecosystems*, Vol. 43, No. 1-3, 1996, pp. 149-156.

[8] M. Pietrantonio, A. Liberatori, B. M. Petronio and M. Pietrolett, "Change in Soil's Metal Carrying Capacity as a Result of Urban Sludge Addition," *Water, Air, and Soil Pollution*, Vol. 132, No. 1-2, 2000, pp. 141-151. doi:10.1023/A:1012044202829

[9] J. J. Kelly, M. Haggblom and R. L. Tate, "Effects of the Land Application of Sewage Sludge on Soil Heavy Metal Concentrations and Soil Microbial Communities," *Soil Biology and Biochemistry*, Vol. 31, No. 10, 1999, pp. 1467-1470. doi:10.1016/S0038-0717(99)00060-7

[10] I. Walter and G. Cuevas, "Chemical Fractionation of Heavy Metals in a Soil Amended with Repeated Sewage Sludge Application," *Science of the Total Environment*, Vol. 226, No. 2-3, 1999, pp. 113-119. doi:10.1016/S0048-9697(98)00374-X

[11] P. Parkpain, K. Klanrong, R. DeLaure and A. Jugsujinda, "Metal Leachability from Sewage Sludge Amended Thai Soils," *Journal of Environmental Science and Health*, Vol. 37, No. 5, 2002, pp. 765-791. doi:10.1081/ESE-120003588

[12] I. Marcote, T. Hernandez, C. Garcia and A. Polo, "Influence of One or Two Successive Annual Application Organic Fertilizers on the Enzyme Activity of a Soil under Barley Cultivation," *Biores Tecnnology*, Vol. 79, No. 2, 2001, pp. 147-154. doi:10.1016/S0960-8524(01)00048-7

[13] P. Parkpian, K. Klankrong, R. DeLaune and A. Jugsujinda, "Metal Leachability from Sewage Sludge-Amended Thai Soils," *Journal of Environmental Science and Health*, Vol. 37, No. 5, 2002, pp. 765-791. doi:10.1081/ESE-120003588

[14] M. Pagliai, G. Guidi, M. La Marca, M. Giachelti and G. Lucamante, "Effects of Sewage Sludges and Compost on Soil Porosity and Aggregation," *Journal of Environmental Quality*, Vol. 10, No. 4, 1981, pp. 556-561. doi:10.2134/jeq1981.00472425001000040028x

[15] I. C. Smith and B. L. Carson, "Trace Metals in the Environment," Vol. I, Thallium, Ann Arbor Science Publishers, Ann Arbor, 1977, p. 394.

[16] D. E. William, J. Vlamis, J. Pukite and J. E. Corey, "Metal Movement in Sludge-Amended Soil: A Nine-Year Study," *Soil Science*, Vol. 143, No. 2, 1987, pp. 124-131. doi:10.1097/00010694-198702000-00007

[17] S. Bose and A. K. Bhattacharyya, "Heavy Metal Accumulation in Wheat Plant Grown in Soil Amended with Industrial Sludge," *Chemosphere*, Vol. 70, No. 7, 2008, pp. 1264-1272. doi:10.1016/j.chemosphere.2007.07.062

[18] K. J. Rao and M. V. Santaram, "Effects of Urban Solid Wastes on Dry Matter Yield, Uptake of Micronutrients and Heavy Metal by Maize Plant," *Journal of Environmental Biology*, Vol. 17, No. 1, 1996, pp. 25-32.

[19] T. D. Hinsley, D. E. Alexander, H. E. Redborg and E. L. Zielger, "Differential Accumulations of Cadmium and Zinc by Corn Hybrids Grown on Soil Amended with Sewage Sludge," *Journal of Agronomy*, Vol. 74, No. 3, 1982, pp. 469-474. doi:10.2134/agronj1982.00021962007400030017x

[20] I. L. Pepper, D. F. Bezedicek, A. S. Baker and J. M. Sims, "Silage Corn Uptake of Sludge-Applied Zn and Cd as Affected by Soil pH," *Journal of Environmental Quality*, Vol. 12, No. 2, 1983, pp. 270-275. doi:10.2134/jeq1983.00472425001200020024x

[21] M. Jiríand and M. Pechová, "A Simplified Approach to Liming and Its Evaluation," *Communications in Soil Science and Plant Analysis*, Vol. 33. No. 15-18, 2002, pp. 15-18.

[22] P. R. Hess, "A Text Book of Soil Chemical Analysis," Chemical Publishing Co., Inc., New York, 1972, pp. 197-299.

[23] B. V. Subbiah and G. L. Asija, "A Rapid Procedure for the Determination of Available Nitrogen in Soils," *Current Science*, Vol. 25, No. 8, 1956, pp. 259-260.

[24] J. M. Bremner, "Détermination of Nitrogen in Soil by the Kjeldahl Method," *Journal of Agricultural Science*, Vol. 55, No. 1, 1960, pp. 11-33. doi:10.1017/S0021859600021572

[25] D. R. Morris and D. J. Lathwell, "Anaerobically Digested Dairy Manure as Fertilizer for Maize in Acid and Alkaline Soils," *Communications in Soil Science and Plant Analysis*, Vol. 35, No. 11-12, 2004, pp. 1757-1771.

[26] F. Amlinger, B. Gotaz, P. Dreher, J. Geszti and C. Weissteiner, "Nitrogen in Biowaste and Yard Waste Compost: Dynamics of Mobilization and Availability—A Review," *European Journal of Soil Biology*, Vol. 39, No. 3, 2003, pp. 107-116. doi:10.1016/S1164-5563(03)00026-8

[27] H. Abdennaceur, N. Jedidi, A. Mohamed Cherit, A. M'Hiri, A. Boudabous and O. van Cleemput, "Mineralization of Nitrogen in a Clayey Loamy Soil Amended with Organic Wastes Enriched with Zn, Cu And Cd," *Bioresource Technology*, Vol. 64, No. 1, 1998, pp. 39-45. doi:10.1016/S0960-8524(97)00153-3

[28] S. Bose, S. Chandrayan, V. Rai, A. K. Bhattacharyya and A. L. Ramanathan, "Translocation of Metals in Pea Plants Grown on Various Amendment of Electroplating Industrial Sludge," *Bioresource Technology*, Vol. 99, No. 10, 2008, pp. 4467-4475. doi:10.1016/j.biortech.2007.08.020

[29] S. Bose, A. Jain, V. Rai and A. L. Ramanathan, "Chemical Fractionation and Translocation of Heavy Metals in *Canna indica* L., Grown on Industrial Waste Amended Soil," *Journal of Hazardous Materials*, Vol. 160, No. 1, 2008, pp. 187-193.

[30] United States Environmental Protection Agency (USEPA), "Process Design Manual: Land Application of Sewage Sludge and Domestic Septage," EPA/625/R-95/001, Washington DC, 1995.

Appendix

The ANOVA results showed that there was no significant variation with shoot length with lime treatments but significant with waste treatments with growth periods (**Tables 3-4**). There were significant variations of root length with growth periods vs. lime treatments (**Tables 5-6**). There were significant variations in shoot and root dry weight with lime treatment as well as waste treatments with time (**Tables 7-10**).

Table 3. Two-factor ANOVA to test significance of difference in shoot length of wheat plants due to growth period vs. lime-treatment (0.05 level of significance).

Sl	Sample Type		Diff. due to rows (lime treat.)			Diff. due to columns (growth period)		
No.	W.T.	Types of soil	F-value	S/I	P-value	F-value	S/I	P-value
			3.862548358			3.862548358		
1	10	Chhattarpur	2.80089534	I	0.10085409	1.36952767	I	0.31327613
2	20	Chhattarpur	1.04325427	I	0.4194115	1.77487315	I	0.22174935
3	30	Chhattarpur	1.08555646	I	0.40359894	3.28148057	I	0.07247967
4	10	JNU	3.59780665	I	0.05900359	3.42628045	I	0.06589362
5	20	JNU	2.34824843	I	0.14068319	5.09836706	S	0.02473357
6	30	JNU	4.45204426	S	0.03526918	16.2355516	S	0.00056451

[S = Significant, I = Insignificant, L.T = Lime Treatment, W.T = Waste Treatment].

Table 4. Two-factor ANOVA to test significance of difference in shoot length of wheat plants due to growth period vs. waste treatment (0.05 level of significance).

Sl	Sample Type		Diff. due to rows (waste treat.)			Diff. due to columns (growth period)		
No.	L.T.	Types of soil	F-value	S/I	P-value	F-value	S/I	P-value
			5.14325285			4.757062664		
1	0	Chhattarpur	2.90301471	I	0.13126295	4.06873986	S	0.06787549
2	0.5	Chhattarpur	2.23234569	I	0.18848399	2.15438156	I	0.19459993
3	1.0	Chhattarpur	0.64899014	I	0.55570678	1.1870683	I	0.39087041
4	0	JNU	2.40100834	I	0.17137175	2.20585797	I	0.188207
5	0.5	JNU	1.97041866	I	0.21987956	6.06722174	S	0.03006147
6	1.0	JNU	2.66796403	I	0.14827998	0.9644555	I	0.468265

Table 5. Two-factor ANOVA to test significance of difference in root length of wheat plants due to growth period vs. lime-treatment (0.05 level of significance).

Sl	Sample Type		Diff. due to rows (lime treat.)			Diff. due to columns (growth period)		
No.	W.T.	Types of soil	F-value	S/I	P-value	F-value	S/I	P-value
			3.862548358			3.862548358		
1	10	Chhattarpur	7.68367347	S	0.00747719	3.64285714	I	0.05734072
2	20	Chhattarpur	11.1152263	S	0.0022114	2.04526749	I	0.17805747
3	30	Chhattarpur	46.6595745	S	8.2566E–06	14.3829787	S	0.00088341
4	10	JNU	5.97297297	S	0.01592064	2.02702703	I	0.18066299
5	20	JNU	15	S	0.00075721	8.3553719	S	0.00573319
6	30	JNU	92.2173913	S	4.4661E–07	10.826087	S	0.00242189

Table 6. Two-factor ANOVA to test significance of difference in root length of wheat plants due to growth period vs. waste treatment (0.05 level of significance).

Sl	Sample Type		Diff. due to rows (waste treat.)			Diff. due to columns (growth period)		
No.	L.T.	Types of soil	F-value	S/I	P-value	F-value	S/I	P-value
			5.14325285			4.757062664		
1	0	Chhattarpur	2.95184591	I	0.12805859	0.70786517	I	0.58156322
2	0.5	Chhattarpur	30.9130435	S	0.00069225	2.47826087	I	0.15858281
3	1.0	Chhattarpur	12.6521739	S	0.00704109	3.83524027	I	0.07584365
4	0	JNU	30.1	S	0.00074453	1.425	I	0.3249824
5	0.5	JNU	16.6470588	S	0.00356017	3.82352941	I	0.07627518
6	1.0	JNU	16.0851064	S	0.00388401	2.40425532	I	0.16598793

Table 7. Two-factor ANOVA to test significance of difference in dry wt of wheat plant's shoot due to growth period vs. lime-treatment (0.05 level of significance).

Sl	Sample Type		Diff. due to rows (lime treat.)			Diff. due to columns (growth period)		
No.	W.T.	Types of soil	F-value	S/I	P-value	F-value	S/I	P-value
			3.8625387			3.8625387		
1	10	Chhattarpur	4.20874228	S	0.04060626	621.723262	S	9.5143E–11
2	20	Chhattarpur	21.8091038	S	0.00018257	632.987198	S	8.7795E–11
3	30	Chhattarpur	36.2245354	S	2.3701E–05	440.704479	S	4.433E–10
4	10	JNU	27.8083333	S	6.9627E–05	4693.825	S	1.0879E–14
5	20	JNU	45.5981364	S	9.0945E–06	2163.11448	S	3.5401E–13
6	30	JNU	41.6209744	S	1.3323E–05	947.602808	S	1.4397E–11

Table 8. Two-factor ANOVA to test significance of difference in dry wt of wheat plant's shoot due to growth period vs. waste treatment (0.05 level of significance).

Sl	Sample Type		Diff. due to rows (waste treat.)			Diff. due to columns (growth period)		
No.	L.T.	Types of soil	F-value	S/I	P-value	F-value	S/I	P-value
			5.1432494			4.7570552		
1	0	Chhattarpur	29.2108183	S	0.0008079	224.023777	S	1.5105E–06
2	0.5	Chhattarpur	163.399597	S	5.8601E–06	1783.92552	S	3.0709E–09
3	1.0	Chhattarpur	34.4543499	S	0.00051387	375.318959	S	3.2512E–07
4	0	JNU	18.6513872	S	0.00266015	382.00769	S	3.0844E–07
5	0.5	JNU	18.8861997	S	0.00257545	421.615483	S	2.298E–07
6	1.0	JNU	16.6040024	S	0.00358368	319.740553	S	5.2421E–07

Table 9. Two-factor ANOVA to test significance of difference in dry wt of wheat plant's root due to growth period vs. lime-treatment (0.05 level of significance).

Sl	Sample Type		Diff. due to rows (lime treat.)			Diff. due to columns (growth period)		
No.	W.T.	Types of soil	F-value	S/I	P-value	F-value	S/I	P-value
			3.8625387			3.8625387		
1	10	Chhattarpur	3.88824826	S	0.04921775	109.643717	S	2.1007E–07
2	20	Chhattarpur	4.13899921	S	0.04231368	118.920572	S	1.4725E–07
3	30	Chhattarpur	1.22792023	I	0.35508087	29.9721674	S	5.1446E–05
4	10	JNU	4.94354839	S	0.02686225	213.58871	S	1.1143E–08
5	20	JNU	0.80088735	I	0.52403263	50.4266854	S	5.9525E–06
6	30	JNU	0.47305089	I	0.70860701	19.9606418	S	0.00025753

Table 10. Two-factor ANOVA to test significance of difference in dry wt of wheat plant's root due to growth period vs. waste treatment (0.05 level of significance).

Sl	Sample Type		Diff. due to rows (waste treat.)			Diff. due to columns (growth period)		
No.	L.T.	Types of soil	F-value	S/I	P-value	F-value	S/I	P-value
			5.1432494			4.7570552		
1	0	Chhattarpur	0.3559322	I	0.71437168	74.0550847	S	3.9385E–05
2	0.5	Chhattarpur	0.25141777	I	0.78549864	51.8941399	S	0.00011023
3	1.0	Chhattarpur	0.32450936	I	0.7348202	41.4472843	S	0.00020966
4	0	JNU	0.56346927	I	0.59668444	33.5778288	S	0.00038031
5	0.5	JNU	0.24285139	I	0.79174007	26.6129682	S	0.00072719
6	1.0	JNU	0.53441802	I	0.6115191	30.6032541	S	0.0004932

16

Recovery of Soil Test Phosphorus from an Acidic Soil Amended with Organic and Inorganic Phosphorus

Farzana Rumi, Md. Abul Kashem*, Khan Towhid Osman

Department of Soil Science, University of Chittagong, Chittagong, Bangladesh.
Email: *kashem00@yahoo.com

ABSTRACT

Information on soil test phosphorus (P) in soil treated with organic amendments is important to a sound management of manure additions to agricultural fields. This study compared the recovery of cow manure, chicken manure, city compost P relative to triple super phosphate P (TSP) for an acidic soil with different antecedent soil test P (STP). Phosphorus was added at rates of 0, 100, 200, 400 and 800 mg P·kg^{-1} soil based on total P. The soil was incubated at field capacity for 1, 4, 8 and 16 weeks (wk) after which they were extracted using $NaHCO_3$ (Olsen) Mehlich-3, Kelowna and Bray & Kurtz-1 extractants. Regardless of extractants, after 1 wk incubation, the highest STP source was the TSP and the least was the city compost. Soil Test P increased with the addition of amendments from different P sources. Among the amendments, soil test P in TSP amended soil gradually decreased but in the city compost amended soil slightly increased with incubation time, whereas the changes of soil test P with time in the cow and chicken manures amended soil was very negligible. Across the amendments and rates of P additions, the value of extractable P with Olsen was of 55 mg·kg^{-1} (16%), with Mehlich-3 was of 112 mg·kg^{-1} (32%), with Kelowna was of 88 mg·kg^{-1} (24%) and with Bray & Kurtz was of 104 mg kg^{-1} (29% of total added P). The P extraction efficiency was in the order: $NaHCO_3$ < Kelowna < Bray & Kurtz-1 < Mehlich-3. This study indicates that P in organic amendments reflects plant available P through the entire incubation period but P in the TSP are likely to under estimate after 8 wk of incubation.

Keywords: Extraction Methods; Organic Amendments; Phosphorus; Incubation

1. Introduction

Phosphorus is generally added to agricultural soils as inorganic fertilizer or as organic manures. As there are generally no obvious adverse effects of high soil P concentrations on plant growth, farmers have frequently added P in amounts which exceed crop removal. This is particularly in region with intensive pig, poultry or cattle production systems. As a consequence, P has been accumulated in many intensively farmed agricultural soils [1].

Organic fertilizer such as, animal manure, green manure, compost and sewage sludge may be added to cultivated soil [2]. The increasing demand of chicken meat has prompted more poultry farming with consequent effects on increased utilization of organic waste (e.g. chicken manure) as fertilizers. Organic wastes contain varying amounts of water, mineral nutrients, organic matter [3]). While the use of organic wastes as manure has been in practice for centuries world-wide and in the recent times [4-5], there still exists a need to assess the potential impacts of chicken manure on soil chemical properties and crop yield and in particular evaluating the critical application levels. Furthermore, chicken manure is preferred amongst other animal wastes because of its high concentration of macro-nutrients [6,7].

Several soil test phosphorus (STP) methods are used to estimate extractable soil P. The higher the level of STP, the higher is the risk of P exports from the soil. STP methods can be divided into two broad categories: agronomic P tests and environmental P tests. Agronomic tests such as Olsen, Kelowna and Mehlich-3 have been developed to estimate the amount of P that will be available to a crop throughout the growing season [8-10]. However environmental soil P tests extract a portion of soil P that is easily lost through surface runoff or subsurface flow and therefore use mild extractrants, for example, distilled water [11], $CaCl_2$ [12] and NH_4Cl [13]. Recently, agronomic soil P tests have been used to derive indices of environmental risk such as the degree of P saturation [14] and P index [15]. Several studies have shown that agronomic soil P tests are often correlated with soil P measured by environmental P tests as well as P in the runoff and drainage waters [16,17].

*Corresponding author.

Application of city waste compost and livestock manures can benefit crop production as a valuable source of nitrogen and phosphorus. The efficiency of added mineral fertilizer P for increasing soil test P levels is typically <20%. Furthermore, the availability of city waste, chicken-and cattle-manure P to crops and it impact on soil P pools may differ from that of inorganic fertilizer. Biosolids P was 4 to 7 times less available to plants than triple super phosphate P [18]. Phosphorus in pig slurry was at least 90% as effective as P in the fertilizer because it contains 70% to 90% inorganic P [19]. However some studies suggested that manure P may be equally or more available than fertilizer P [20]. These results indicate that P availability with organic amendments is variable and not understood. A sustainable amount of effort has been put into understanding soil P chemistry specifically as it relates to P fixation P availability from inorganic fertilizer applied to soil [21]. By comparison, far less research has been devoted to understanding P availability from organic sources of P from manures and compost. Understanding city waste compost P and animal manure P chemistry in soil and trying to predict P availability is an important component for making nutrient management plans that maximize economic benefits and minimize environmental risks. Different organic amendments contain varying amounts of P. The tendency is to manage these amendments based on total P in the amendment or total loading of P. This strategy will be appropriate only if the P in these amendments behaves similarly in the soil. Therefore, the objectives of this study were

1) to compare the soil test P in soil treated with cow manure, chicken manure, city compost and TSP fertilizer P sources;

2) to determine the changes in soil test P with time treated with either organic or inorganic P sources and

3) to investigate the relationship among the soil test P extraction methods.

2. Materials and Methods

2.1. Characteristics of Soil and Organic Amendments

An acidic mineral soil, Parhartoli Sandy Loam (Aeric Endoaquepts) was used in this study. A composite soil sample from a crop producing field collected from the surface layer (0 - 15 cm) was air dried and crushed to pass through 2-mm sieve. Soil properties and some characteristics of organic amendments are presented in **Table 1**. The soil texture was measured by hydrometer method [22]. Soil pH was measured using 1:2.5 soil to water ratio and pH of organic amendments at 1:10 ratio. Organic carbon was measured by Walkley and Black [23] and CEC was measured by extraction with 1 M NH_4OAc (pH 7.0) [24]. Cow manure was collected from Chittagong University Campus, city compost from Halishore of Chittagong City and chicken manure was from the Veterinary and Animal Sciences University of Chittagong, Bangladesh. Total P in soil and organic amendments was determined by H_2O_2-H_2SO_4 wet oxidation methods [25].

2.2. Incubation Experiment

The experiment was carried out using three organic amendments: cow manure, chicken manure, and city compost in comparison with TSP to measure soil test P in different incubation time. Cow manure was collected from the Chittagong University Campus, chicken manure from the Veterinary and Animal Sciences University of Chittagong and city compost was collected from the composting plant of Chittagong City Corporation, Halishahar, Chittagong, Bangladesh. After collection, these composts were air dried, ground and sieved to analysis and incubation study.

One hundred gram of soil was mixed with cow manure, chicken manure, city compost and TSP at rates equivalent to 0, 100, 200, 400 and 800 mg P kg^{-1} soil based on total P on oven dry weight basis. Each treatment was replicated three times. The soil was incubated at field capacity for 1, 4, 8 and 16 weeks at room temperature in a 250 ml plastic jar with perforated lids to allow gas exchange. Total number of jars was 240 [4 (amend) × 5 (rate) × 4 (time) × 3 (rep)]. Samples were weighed weekly and additional water was added as required. After each incubation time, soils of 60 jars were collected, air-dried, ground and passed through a 2 mm sieve for analysis. Soil P was extracted using four different extractants [8-10,26] (**Table 2**). Phosphorus in the various extracts and digests was measured by ascorbic acid blue color method [27]. A UV/Vis Double Beam Spectrophotometer (T80) was used to measure absorbance at wave length of 882 nm.

Table 1. Characteristics of the soil and organic amendments.

Characteristics	Soil	Cow manure	Chicken manure	City compost
Field capacity (%)	37	-	-	-
Sand (%)	57	-	-	-
Silt (%)	30	-	-	-
Clay (%)	13	-	-	-
Textural class	Sandy loam	-	-	-
Organic matter (%)	1.77	-	-	-
Cation exchange capacity (cmol kg^{-1})	5.07	-	-	-
pH	5.1	7.9	8.3	7.1
Total P (%)	0.0595	0.78	1.79	0.68

Table 2. Different methods used for extracting P from soil.

Methods	Solution	Soil to solution ratio (W/V)	Extraction time	Reference
Olsen	0.5 M $NaHCO_3$, pH 8.5	1:20	30 min at 120 strokes per minute	[8]
Kelowna	0.2 N CH_3COOH + 0.013 NH_4F	1:10	15 min at 180 strokes per minute	[9]
Mehlich-3	0.2 N CH_3COOH + 0.013 N HNO_3 + 0.015 N NH_4F + 0.25 N NH_4NO_3 + 0.001 M EDTA	1:10	5 min at 180 strokes per minute	[10]
Bray and kurtz P-1	0.03 N NH_4F + 0.025 N HCl	1:10	5 min 120 strokes per minute	[26]

2.3. Net Phosphorus Concentration and Extraction Efficiency

To remove the effect of changes in extractable P in the control soil, we calculated the net P concentration as the difference between P in the amended soil and the control soil at each time of incubation (1, 4, 8 and 16 wk).

2.4. Statistical Analyses

The experiment was statistically analyzed as a completely randomized factorial design with three factors, amendments, rate of P additions and incubation time using analysis of variance (ANOVA). Data from each extractant was analyzed independently. Minitab statistical software, Released 11 [28] was used for analysis.

3. Results and Discussion

The results are discussed on the basis of net P concentration in amended soils. The analysis of variance (ANOVA) of the data was computed to evaluate the effect of different amendments, incubation time and rate on soil test P. The analysis of variance showed extractable P was significantly influenced by amendments, incubation time, rates of P addition and their interactions. Positive values of net P indicate that additions from organic and inorganic P sources increased extractable P (**Table 3**).

3.1. Effects of Amendment on Soil Test P

There were significant differences in the soil test P among the amendments regardless of soil test P extraction methods. The differences obtained among the amendments on soil test P were rate and time dependent (**Table 3**). The results obtained after 1 wk incubation, as extracted by $NaHCO_3$ (Olsen-P) are used to illustrate the amendment effect of soil test P. The trend in the data obtained with other three extractants was similar to that of the $NaHCO_3$-extractable P (Olsen-P) (**Figure 1**).

After 1 wk incubation, the highest soil test P source was the TSP and the least was the city compost amended soil. The regression with $NaHCO_3$, Mehlich-3, Kelowna and Bray & Kurtz-1 extractable P against the rate of P additions were significant ($p < 0.001$), considering amendment and incubation time combination (**Table 3**). Slope of the regression reflects the increase in extractable soil P with each increment of P added by the amendments (**Table 4**). The slope of $NaHCO_3$-extractable P was 0.08 for cow manure, 0.11 for chicken manure, 0.06 for city compost and 0.25 for TSP after 1 wk incubation. The corresponding slopes after 16 wk incubation were 0.09 for cow manure, 0.12 for chicken manure, 0.14 for city compost and 0.15 for TSP. The slopes of STP after 1, 4 and 8 wk incubation for cow and chicken manures were similar to those after 16 wk incubation remained relatively unchanged. However, the slopes for city compost gradually increased with incubation time, on the other hand, the slopes of extractable P decreased in the TSP amended soil (**Table 4**). The slope of Olsen P after 1 wk incubation was in the order of: TSP > chicken manure > cow manure > city compost. A similar order of slopes of extractable P in soils was found for the other three extractants (**Table 4**). Slopes indicate that city compost had the smallest STP compared to other two organic amendments. The highest extractable P obtained by TSP after 1 wk incubation is not surprising due to the high solubility of the inorganic P fertilizer. These results indicate that P extractability does not depend on the total amounts of added P, but on the characteristics of P sources. The smaller slopes of extractable P in the organic amended soils compared to inorganic fertilizer is consistent with the results of other investigators [18,29]. McCoy *et al.* [18] reported that city waste compost (Biosolids) P was four to seven times less available to plants than TSP fertilizer P source. These results indicate that there would be less environmental risk with cow manure and city compost P compared to other amendments at equal loading of total P.

3.2. Effects of Incubation Time on Soil Test P

The effects of incubation time on the soil test P in the amended soils depend on the nature of amendment and application rate (**Tables 1** and **3**). Olsen P is taken into account to represent the time effect on soil test P. In general, regardless of extractants and rates of P addition, soil test P in the TSP amended soils decreased and city compost amended soils increased with incubation time. For example, Olsen P at the lowest rate of P addition (100 $mg \cdot kg^{-1}$) varied from 9.0 to 10, 23 to 26, 8 to 15 and 39 to 8 $mg \cdot kg^{-1}$ from 1 to 16 wk incubation in the cow manure

Table 3. The F-values of the analysis of variance of the effects of amendments, rate and time on the net extractable P.

Treatment	Olsen-P	Melich3-P	Kelowna-P	Bray & Kurtz-1
Amendment	1215*	2054*	1518*	3064*
Rate	144*	550*	168*	648*
Time	1525*	7765*	6652*	11382*
Amendment × rate	54*	561*	326*	757*
Amendment × time	147*	442*	314*	543*
Time × rate	14*	169*	89*	120*
Amendment × rate × time	7*	112*	43*	75*

*Significant at $p < 0.0001$.

Table 4. Slope and coefficient of determination (R^2) of the regression of soil extractable P at 1, 4, 8 and 16 weeks of incubation.

Amend	1 week		4 week		8 week		16 week	
	Slope	R^2	Slope	R^2	Slope	R^2	Slope*	R^2
				Olsen P				
COM	0.08	0.98	0.09	0.98	0.08	0.93	0.09	0.94
CHM	0.11	0.94	0.15	0.99	0.12	0.88	0.12	0.88
CFC	0.06	0.94	0.08	0.87	0.12	0.92	0.14	0.88
TSP	0.25	0.98	0.24	0.98	0.19	0.94	0.15	0.92
				Mehlich-3 extractable P				
COM	0.20	0.99	0.21	0.99	0.18	0.98	0.21	0.97
CHM	0.25	0.97	0.28	0.99	0.25	0.91	0.21	0.89
CFC	0.19	0.99	0.21	0.98	0.26	0.97	0.27	0.89
TSP	0.82	0.99	0.36	0.98	0.33	0.97	0.28	0.97
				Kelowna extractable P				
COM	0.19	0.99	0.18	0.90	0.19	0.98	0.21	0.97
CHM	0.30	0.99	0.28	0.94	0.30	0.96	0.31	0.99
CFC	0.13	0.98	0.18	0.99	0.21	0.98	0.23	0.83
TSP	0.49	0.99	0.37	0.98	0.26	0.98	0.22	0.98
				Bray & Kurtz-1 extractable P				
COM	0.22	0.97	0.20	0.95	0.23	0.99	0.22	0.97
CHM	0.36	0.99	0.34	0.94	0.37	0.94	0.38	0.93
CFC	0.10	0.95	0.15	0.93	0.18	0.98	0.31	0.84
TSP	0.57	0.98	0.36	0.96	0.35	0.97	0.30	0.97

COM = Cow manure; CHM = Chicken manure; CFC = City compost; TSP = Triple super phosphate; *Slopes have unit of mg extractable P·kg^{-1} added P.

chicken manure, city compost and in the soil test P amended soils, respectively. The corresponding values in the same amended soils at the highest rate of P addition (800 mg·kg^{-1}) varied from 70 to 77, 106 to 103, 55 to 88 and 220 to 122 mg·kg^{-1}, respectively (**Figure 1**). The trend of soil test P changes was similar with other extractants in different amended soils but the magnitude of extractability was higher than the 0.5 M $NaHCO_3$ extractant.

The differences in magnitude of declined in soil test P between TSP and manure suggest the P retention mechanisms. The fertilizer P was mixed with the soil in the form of granules, high P concentration around the granules would favour the precipitation of P probably as dicalcium phosphate dehydrate [30]. In contrast, with chicken and cow manures, low concentration of P will favor sorption and desorption rather than precipitation. The increase of city compost P with incubation time indicating net mineralization of city compost P. The behavior of city compost P was comparable with biosolids P. Extractable and labile (H_2O plus $NaHCO_3$ extractable P) biosolids P increased with incubation time [29,31].

3.3. Comparison of the Phosphorus Extractability in Soil

Tukey's test for multiple means comparison showed that the means (n = 192) of Olsen, Mehlich-3, Kelowna and Bray & Kurtz-1 extractable P were significantly different ($p < 0.05$; **Figure 2**). The average net value of extractable P with Olsen was of 55 mg·kg^{-1} (16% of total added P), with Mehlich-3 was of 112 mg·kg^{-1} (32% of added total P), with Kelowna was of 88 mg·kg^{-1} (24% of total added P) and with Bray & Kurtz was of 104 mg·kg^{-1} (29% of total added P). The amount of P extracted from amended soils increased in the order: $NaHCO_3$ < Kelowna < Bray & Kurtz-1 < Mehlich-3 (**Figure 2**). Sims [16] noted that Olsen extractant has less ability to remove P from the soil than the acidic extractants. In order to ascertain whether or not the several soil extracting methods considered could provide a comparable and consistent evaluation of extractable P pool, values of soil extractable P as determined by different procedures were related to each other using regression analysis. The differences among the P extraction methods probably arose from the fact that plant available P in the soil is not from a discreet fraction but from a continuum of fractions: extracting agents preferentially extract from different fractions depending on their reactions with soil components involved in P sorption [32]. Further more, each extracting solution has a different ability to extract varying portions of soil P because they were targeted at different pools of soil P [33]. The mean of Olsen P is significantly small than the means of other three extractants. This is an indication that other extracting agents extracted some forms of labile P that are not immediately available to 0.5 M $NaHCO_3$ extractant. Though, $NaHCO_3$ is used for calcareous soil, however, it has been commonly used as a soil P extractant in acid and neutral soils [8]. The coefficient of determination (R^2) varied from 0.74 to 0.91 indicating that any of these extractants can be used to estimate P

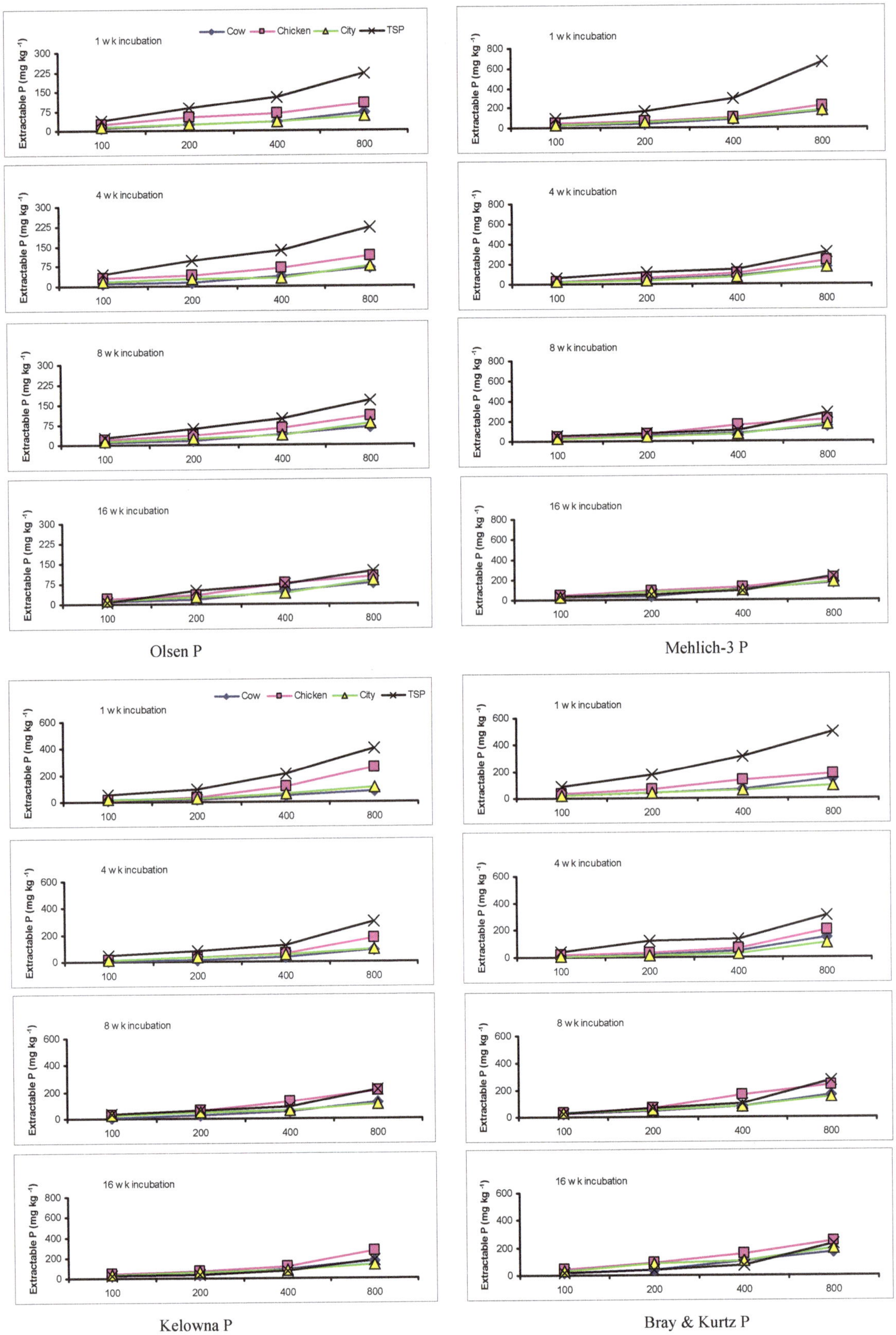

Figure 1. Net extractable P concentration (mg·kg^{-1}) following incubation of amended soil extracted with Olsen, Mehlich-3, Kelowna and Bray & Kurtz-1.

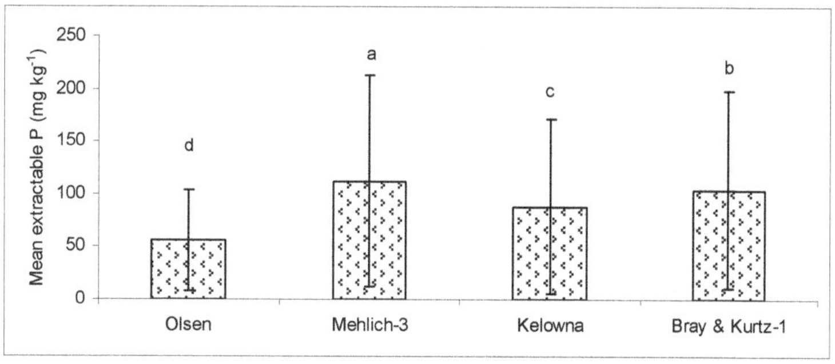

Figure 2. Mean of P extracted by the different extractable methods. Means indicated with the same letter were not significantly different ($p < 0.05$).

extractability (available P) in soils. It is possible to predict Mehlich-3, Kelowna and Bray & Kurtz extractable using the following regression equation.

1) Mehlich 3 extractable P = 1.817 × Olsen P + 10.94;

2) Kelowna extractable P = 1.494 × Olsen P + 4.86;

3) Bray & Kurtz extractable P = 1.702 × Olsen P + 9.54.

4. Conclusion

Soil test P increased in soils amended with organic and inorganic P sources, but the increase varied with P sources of different amendments. Extractable soil was the least with city compost and the highest with the TSP while cow and chicken manures P was intermediate between city compost and TSP at 1 wk of incubation. Among the amendments, soil test P in TSP amended soil decreased and increased in city compost amended soil with increasing incubation time. The high correlation coefficient between the various extractants, shows that all of them will provide a good index of P labiality in this soil. Our results indicate that the P in cow and chicken manures behaved almost similarly but city and TSP behaved differently upon addition to the soil and hence should be take into consideration in managing their land application for optimum plant growth and environmental point of view.

REFERENCES

[1] H. Tunney, "A Note on a Balance Sheet Approach to Estimating the Phosphorus Fertilizer Needs of Agriculture," *Irish Journal of Agricultural Research*, Vol. 29, No. 2, 1990, pp. 149-154.

[2] W. E. Splittstosser, "Vegetable Growing Handbook: Organic and Traditional Method," 3rd Edition, New York, USA, 1998.

[3] D. R. Edwards and T. C. Daniel, "Environmental Impacts on Farm Poultry Waste Disposal: A Review," *Bioresource Technology*, Vol. 41, No. 1, 1992, pp. 9-33. doi:10.1016/0960-8524(92)90094-E

[4] D. C. Clay, V. Kelly, E. Mpyisi and T. Readon, "Input Use and Conservation Investments and Determinants," In: C. B. Barrett, F. Place and A. A. Aboud, Eds., *Natural Resource Management in African Agriculture: Understanding and Improving Current Practices*, CABI Publishing, Oxon and New York, 2002, pp. 103-114.

[5] M. E. Lopez-Masquera, F. Cabaleiro, M. S. Sainz, A. Lopez-Fabal and E. Carral, "Fertilizing Value of Boiler Litter: Effect of Drying and Palletizing," *Bioresource Technology*, Vol. 99, No. 13, 2008, pp. 5626-5633. doi:10.1016/j.biortech.2007.10.034

[6] P. R. Warman, "The Effect of Fertilizer, Chicken Manure on Timothy Yield, Tissue, Composition and Soil Fertility," *Agric Wastes*, Vol. 18, No. 4, 1986, pp. 289-298. doi:10.1016/0141-4607(86)90074-0

[7] J. Duncan, "Composting Chicken Manure," WSU Cooperative Extension, King County Master Gardener and Cooperative Extension Livestock Advisor, Washington State University, Pullman, 2005.

[8] S. R. Olsen, C. V. Cole, F. S. Watanabe and L. A. Dean, "Estimation of Available Phosphorus in Soils," USDA, Washington, 1954.

[9] A. Mehlich, "Mehlich 3 Soil Extractant: A Modification of Mehlich 2 Extractant," *Communications in Soil Science and Plant Analysis*, Vol. 15, No. 12, 1984, pp. 1409-1416. doi:10.1080/00103628409367568

[10] W. Van Lierop, "Determination of Available Phosphorus in Acid and Calcareous Soils with Kelowna Multiple-Element Extractant," *Soil Science*, Vol. 146, No. 4, 1988, pp. 284-291. doi:10.1097/00010694-198810000-00009

[11] F. Van der Paaw, "An Effective Water Extraction Method for the Determination of Plant Available Phosphorus," *Plant and Soil*, Vol. 34, No. 1, 1971, pp. 467-481. doi:10.1007/BF01372799

[12] M. L. Self-Davis, P. A. Moore and B. C. Joern, "Determination of Water or Dilute Salt-Extractable Phosphorus in Soils," In: G. M. Pierzynski, Ed., *Methods of Phosphorus Analysis for Soils, Sediments, Residuals, and Waters*, Kansas State University, Manhattan, 2000, p. 24.

[13] G. J. Racz, "Release of Phosphorus in Organic Soils under Aerobic and Anaerobic Conditions," *Canadian Journal of Soil Science*, Vol. 59, No. 3, 1979, pp. 337-339. doi:10.4141/cjss79-038

[14] M. Zhou and Y. Li, "Phosphorus Sorption Characteristics of Calcareous Soils and Limestone from the Southern Everglades and Adjacent Farmlands," *Soil Science Society of America Journal*, Vol. 65, No. 5, 2001, pp. 1404-1412. doi:10.2136/sssaj2001.6551404x

[15] A. N. Sharpley, J. L. Weld, D. B. Beegle, P. J. A. Kleinman, W. J. Gburek, P. A. Moor Jr. and G. Mullins, "Development of Phosphorus Indices for Nutrient Management Planning Strategies in the United States," *Journal of Soil and Water Conservation*, Vol. 58, No. 3, 2003, pp. 137-152.

[16] J. T. Sims, "Soil Test Phosphorus: Olsen P," In: G. M. Pierzynski, Ed., *Methods for Phosphorus Analysis for Soils, Sediments, Residuals and Water*, Kansas State University, Manhttan, 2000, pp. 20-21.

[17] R. W. McDowell and A. N. Sharpley, "Phosphorus Losses in Subsurface Flow Bfore and after Manure Application to Intensively Farmed Land," *Science of the Total Environment*, Vol. 278, No. 1-3, 2001, pp. 113-125. doi:10.1016/S0048-9697(00)00891-3

[18] J. L. McCoy, L. J. Sikora and R. R. Weil, "Plant Availability of Phosphorus in Sewage Sludge Compost," *Journal of Environmental Quality*, Vol. 15, No. 4, 1986, pp. 403-409. doi:10.2134/jeq1986.00472425001500040016x

[19] H. Tunney and B. Pommel, "Phosphorus Uptake by Ryegrass from Monocalcium Phosphate and Pig Manure on Two Soils in Pots," *Irish Journal of Agricultural & Food Research*, Vol. 26, No. 2-3, 1987, pp. 189-198.

[20] P. M. Gale, M. D. Muller, C. Cieslik, D. D. Tyle, B. N. Duck, M. Kirchner and J. McClure, "Phosphorus distribution and Availability in Response to Diary Manure Applications," *Communications in Soil Science and Plant Analysis*, Vol. 31, No. 5-6, 2000, pp. 553-565. doi:10.1080/00103620009370459

[21] E. F. Khasawneh, E. C. Sample and E. J. Kamprath, "The Role of Phosphorus in Agriculture," ASA, CSSA and SSSA, Madison, 1980.

[22] G. J. Bouyoucos, "Hydrometer Method Improved for Making Particle Size Analysis of Soils," *Agronomy Journal*, Vol. 54, No. 5, 1962, pp. 464-465. doi:10.2134/agronj1962.00021962005400050028x

[23] A. Walkley and I. A. Black, "An Examination of Degtjareff Method for Determining Soil Organic Matter and a Proposed Modification of the Chromic Acid Titration Method," *Soil Science*, Vol. 37, No. 1, 1934, pp. 29-38. doi:10.1097/00010694-193401000-00003

[24] Soil Survey Laboratory Staff, "Soil Survey Laboratory Methods Manual," USDA-SCS, Washington, 1992.

[25] O. O. Akinremi, N. Armisen, M. A. Kashem and H. H. Janzen, "Evaluation of Analytical Methods for Total Phosphorus in Organic Amendments," *Communications in Soil Science and Plant Analysis*, Vol. 34, No. 19-20, 2003, pp. 2981-2991.doi:10.1081/CSS-120025220

[26] R. H. Bray and L. T. Kurtz, "Determination of Total, Organic and Available form of Phosphorus in Soil," *Soil Science*, Vol. 59, No. 1, 1945, pp. 39-45. doi:10.1097/00010694-194501000-00006

[27] J. Murphy and J. P. Riley, "A Modified Single Solution Methods for the Determination of Available Phosphate in Natural Water," *Acta Analytica*, Vol. 27, 1962, pp. 31-36. doi:10.1016/S0003-2670(00)88444-5

[28] Minitab Inc., "Minitab User Guide Release II," Minitab, State College, Pennsylvania, 1996.

[29] M. A. Kashem, O. O. Akinremi and O. J. Racz, "Phosphorus Fractions in Soil Amended with Organic and Inorganic Phosphorus Sources," *Canadian Journal of Soil Science*, Vol. 84, No. 1, 2004, pp. 83-90. doi:10.4141/S03-018

[30] G. J. Racz and R. J. Soper, "Reaction Products of Orthophosphates in Soils Containing Varying Amounts of Calcium and Magnesium," *Canadian Journal of Soil Science*, Vol. 47, No. 3, 1967, pp. 223-230. doi:10.4141/cjss67-035

[31] M. A. Kashem, O. O. Akinremi and G. J. Racz, "Extractable Phosphorus in Soil Amended with High Rates of Organic Amendments," *Canadian Journal of Soil Science*, Vol. 84, No. 4, 2004, pp. 459-467. doi:10.4141/S03-085

[32] Council for Agricultural Science and Technology, "Relevance of Soil Testing to Agriculture and the Environment," CAST, Ames, 2000, pp. 1-12.

[33] M. Zhang, R. Wright, D. Heaney and D. Vanderwel, "Comparison of Different Phosphorus Extraction and Determination Methods Using Manured Soils," *Canadian Journal of Soil Science*, Vol. 84, No. 4, 2004, pp. 469-475. doi:10.4141/S02-023

17

Compost of Different Stability Affects the Molecular Composition and Mineralization of Soil Organic Matter

Bekele Eshetu, Christel Baum, Peter Leinweber

Faculty of Agricultural and Environmental Sciences, Soil Science, University of Rostock, Rostock, Germany.
Email: eshetubekele@yahoo.com, christel.baum@uni-rostock.de, peter.leinweber@uni-rostock.de

ABSTRACT

This study investigated the C mineralization and chemical modification of a typical tropical soil amended with regional compost of different stability. Compost samples were produced from coffee pulp and fruit and vegetable waste in a method of small heap composting and the samples were collected in three different phases of composting. Both the fresh waste and compost samples were analyzed for their phytotoxicity. These samples were added to a tropical Nitisol at the rate of 48 $t \cdot ha^{-1}$ and a control was set up without amendment. The CO_2-C respired was determined during 98 days of incubation and the incubated samples were taken at the start and end of incubation for molecular-chemical analysis by Pyrolysis-Field Ionization Mass Spectrometry (Py-FIMS). The fresh waste yielded a germination index (GI) < 26% indicating phytotoxicity but this disappeared in all the composts (GI > 100%). The CO_2-C respired was best explained by a first order plus linear model. A soil amended with a compost taken at the thermophilic phase attained the lowest overall organic C loss. In general, the Py-FIMS revealed a significant enrichment of stable N-compounds during the incubation in all amended soils compared to the control. Furthermore, among the compost-soil mixtures Py-FIMS indicated significantly higher increases in the proportions of carbohydrates, peptides and phenols/lignin monomers at the expense of fatty acids and sterols in soil amended with composts from the thermophilic phase. Thermal volatilization curves of Py-FIMS indicated enrichments of stable N-compounds and peptides in compost amended soil. This was a result of enhanced decomposition and stabilization of decomposition products by physical protection through association with clay and soil aggregates. In summary, application of compost shortly after reaching the high temperature phase was shown to be more efficient in organic C sequestration in a clay-rich tropical agricultural soil than mature composts.

Keywords: Compost; Soil Organic Matter; Soil Respiration; Soil Carbon Sequestration; Pyrolysis-Field Ionization Mass Spectrometry; Molecular Composition

1. Introduction

Organic waste recycling in agriculture through composting is increasing as an environmentally sustainable waste management strategy [1]. It improves the soil fertility as compost increases the soil organic matter (SOM) content [2,3] and modifies the chemical, physical and biological properties of SOM [4-6]. Moreover, the added compost can lead to significant changes in the composition of SOM and minimize carbon (C) mineralization from the soil [3,7-9]. This will have a global implication as it is in line with the Kyoto protocol (1997) that aims at reducing greenhouse gas emissions through C sequestration in soils. Some authors suggested a significant impact of land application of organic residues to meet the emission reduction target agreed under the Kyoto protocol [10,11]. The strategy has more implications in tropical regions like Ethiopia where the degradation of SOM has been accelerated by the high temperature and a wide set of soil fertility issues [12]. Moreover, tropical soils have lost a significant part of their original SOM pool, and therefore, have a capacity to sequester C by adopting recommended management practices. For these soils the estimated annual C sequestration potential is 0.34 to 0.52 Pg [13].

Coffee (*Coffea arabica* L.) is one of the abundant agricultural products generates expressive amounts of agricultural residues during processing [14]. Ethiopia is the third coffee producer country in the world with a production accounting for 61% of the annual commodity exports [15]. At present considerable amounts of coffee processing by-products like coffee pulp (40% of the wet weight) are generated and mainly damped to water bodies, unsanitary landfills and partly applied to agricultural fields without pre-treatment. This causes freshwater eutrophication, health problems, adverse effects on soil properties such as induced anaerobic conditions and release of phytotoxic compounds impairing plant growth as well as offensive odour and greenhouse gases generation when collected, transported and disposed [15,16]. Moreover, a further increase of the amount of agro-industrial by-products can be assumed by the current development plan of Ethiopia. This triggers the need to find a sink for this growing amount of waste and at the same time opens up the chance for recycling the waste as compost for the improvement of soil fertility and the sequestration of C in soil. However, composting of organic waste has to be optimized to avoid adverse effects to plant growth and minimize overall C losses in the transformation process.

Aerobic incubation experiments were used to measure organic matter decomposition or soil C sequestration potential in compost-fertilized, inorganic fertilized or non-amended soils [7,17]. Moreover, better understanding of the decomposition dynamics need fitting the resulting carbon dioxide (CO_2) release data to different kinetic models used to describe the mineralization of organic waste. Incubation of different waste mixtures (prepared with sewage sludges, manures, city refuse, cotton waste, olive-mill wastewater, and sweet sorghum bagasse) after they had been composted for various periods with soil revealed decreases in CO_2 evolution with longer composting duration [17]. Moreover, references [5,7,8] evaluated the significance of adding compost of diverse origin in improving the soil C sequestration. All reported compost application increased the amount of stable compound retained in the soil and minimize the overall C loss. However, the majority of this information has been generated in temperate soils and little information is available on C mineralization of typical tropical soils amended with regional composts of different stability. The molecular-chemical composition of those compost-soil-mixtures has never been studied.

Pyrolysis-field ionization mass spectrometry (Py-FIMS) was used to characterize organic matter in sewage farm soils [18], in different stages of biowaste composting [19], and to study long-term compost effects on the molecular composition of SOM [3]. Recently Py-FIMS was used to characterize the organic matter composition of fresh agro-industrial wastes (including coffee pulp) and significant correlations between the organic composition and soil biochemical properties after application were reported [16]. However, neither Py-FIMS nor another highly sensitive speciation method has been applied so far to disclose effects of differently stabilized composts on C transformations in tropical soils.

Therefore, the objectives of the study were (1) to investigate how the CO_2-respiration from a tropical Nitisol was altered following the application of differently stabilized composts, and (2) to evaluate effects of the differently stabilized composts on the molecular composition of SOM and its short-term modification during an incubation experiment.

The overall objective was to find out which compost is favorable to improve the soil fertility by sequestration of C in the soil based on a low CO_2-release to the atmosphere.

2. Materials and Methods

2.1 Soil and Composting Samples

The soil was sampled from a coffee plantation farm located in the southwestern highlands of Ethiopia and beloged to the major Soil Unit "Nitosol" (FAO, 1998). The sampling area is situated 36°36'E longitude and 7°56'N latitude (**Figure 1**). The elevation varies between 1533 - 1548 m a.s.l. Mean maximum and minimum temperature range from 26°C to 30°C and 11°C to 14°C, respectively, and the mean of precipitation ranges from 1131 to 1150 mm. Thus, the climate may be described as warm tropical. A total of 50 subsamples were taken from a total area of 5.91 h, traversed in a zig-zag pattern, from soil depth of 0 - 30 cm using a GPS and an Auger. The subsamples were well mixed in a plastic container and then a representative sample of 3 kg was placed in a labeled plastic bag. This soil sample was air-dried and milled to pass through a 2-mm-sieve. Its main characteristics were 84% clay, 14% silt and 1% sand, pH ($CaCl_2$) 5.2, electrical conductivity 2.39 mS $(cm)^{-1}$, 3.9% C_{org}, 0.3% N_t, C/N ratio of 12.0, and 0.1% S.

Composts were produced from 8 m^3 fruit and vegetable waste (dominated by fruit), 1600 kg wet coffee pulp waste and 300 kg garden trimmings (dry leaves, and young tree branches) in a method of small heap composting. Heaps were piled in a bamboo box of dimensions 1.5 m (l) × 1.0 m (w) × 0.8 m (h). The platform used has been considered as pilot compost production unit by the Addis Ababa city council environmental protection authority, Ethiopia.

The regional climate is tropical, with average temperatures between 15°C and 25°C. Three piles were constructed for experimental purposes with three replicates each and monitored as treatment COM1 (composed of coffee pulp waste), COM2 (fruit and vegetable waste), and COM4 (coffee pulp waste with fruit and vegetable waste in which the ratio was 50:50 by volume). The same quantity of soil was added as additional source of microbial colonization and garden trimmings to improve the structure. The experiment lasted from November 2010 to February 2011. The piles were manually turned each week during the active phase which lasts approximately 13 to 17 days, and then in a 15-day-interval during the maturation phase. Depending on the situation the piles

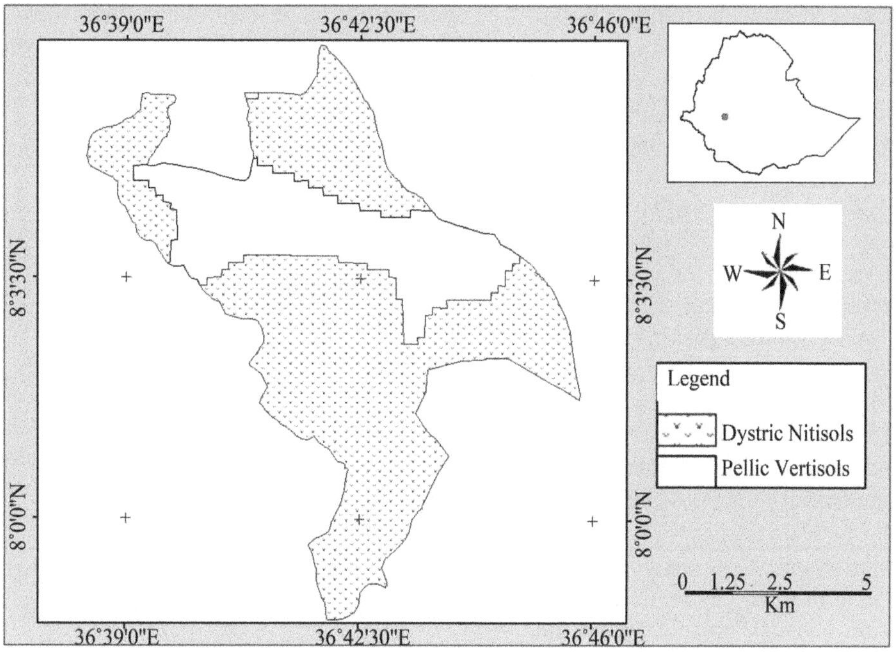

Figure 1. Map of Ethiopia showing area of study and soil type.

were watered so as to maintain the moisture above 40%. Temperatures were measured daily for the first two week and in a 3-day-interval during the next period always in the early morning using digital thermometers (0.5 and 1.0 m in length) at 2 different points of the heaps (25 and 60 cm depth). The composting was considered to be finished when the temperature of the mixture remained stable and near ambient (about 21°C).

Representative compost samples (about 1 kg) were taken by mixing nine subsamples from different levels of the section in the pile along the whole profile at different phases of the composting, day 8 (thermophilic phase = T), day 24/36 (mesophilic phase = M, depending on the duration each treatment lasted in this phase), and 89/119 (final compost = F, depending on the duration each treatment lasted in this phase). In this way composts of different degree of stability were obtained from three different composting piles. The samples were air dried and ground to pass through a 0.5-mm-sieve. The main properties of the composts are shown in **Table 1**.

2.2 Phytotoxicity Test of Compost Samples

Raw coffee pulp waste (RCPW) and composts sampled at the end of the thermophilic phase (COM1-T, COM2-T, COM4-T), mesophilic phase (COM1-M, COM2-T, COM4-M) and final compost (COM1-F, COM2-F, COM4-F) were investigated for possible phytotoxic effects. A germination test was carried out using garden cress (*Lepidium sativum*) and radish (*Raphanus sativus*) seeds as suggested by [20]. In brief, Whatman filter papers were placed in a 145/20 mm disposable petri dish. The filter paper was wetted with 5 ml of filtered compost-water-extract. The extract was prepared by putting 10 g of compost sample (dry weight) in a 100 ml flask filled with 100 ml of distilled water and shaking at 125 rpm for 16 hours at room temperature using an electric shaker. Then, 50 seeds of radish and cress were placed on the wetted paper. Distilled water was used as a control. The petri dish was covered and then kept in the oven at 25°C. At day 3 and 6 germinated seeds were counted and roots length was measured for cress and radish, respectively. Germination percentages (G% = (No of seed germinated in each extract/No of seed germinated in control) × 100) and root growth percentage (RG% = (mean root length in each extract/mean root length in control) × 100) was determined. The germination index (GI%), which is sensitive for phytotoxicity, was then calculated as (GI% = (G% × RG%)/100) [20]. Each treatment was replicated four times.

2.3. Aerobic Incubation and Respiration Measurement

10 g dry weight of sieved soil samples (<2 mm) were thoroughly mixed with organic amendment at desired application rates (48 $t \cdot ha^{-1}$) and placed in 100 ml incubation vessels in five replicates each. They were monitored as treatments S+RCPW, S+COM1-T, S+COM1-M, S+COM1-F, S+COM2-T, S+COM2-M, S+COM2-F, S+COM4-T, S+COM4-M and S+COM4-F. Soil controls were run without any amendment. Distilled water (1 to 5 ml) was added to the soil-compost-mixtures to keep the moisture at 60% of water-holding capacity. The in-

cubation vessels were placed in a glass container containing 25 ml of 0.05 M NaOH and made air tight with a film plaster. To maintain sufficient O_2 in the vessels they were opened several times in the first week and for two hours per day during the following weeks. Empty vessels were used as blanks. The CO_2 evolved was measured by titration of the NaOH solution with 0.05 M HCl after the carbonate was precipitated by adding excess 0.05 M $BaCl_2$ in 24-hour-intervals. The incubation was carried out in a temperature-controlled incubator at 25°C for 98 days. Subsamples (2 g) were taken destructively from each treatment at the start (day 1) and end (day 98) of incubation for chemical and mass spectrometric analyses. Percent increase and decrease in soil organic C as a result of added composts and subsequent mineralization were computed using a mathematical formula:% increase $C_{org} = (C_t/C_s) \times 100) - 100)$, % decrease $C_{org} = (100 - (C_a/C_b) \times 100)$ where C_t is total organic C of the compost-soil mixture, C_s is organic C of the control soil, C_b is organic C at the end of incubation, and C_a is organic C at the start of incubation.

2.4. Chemical Analyses: Basic Data and Molecular-Chemical Characterization

The soil samples were analyzed for pH, EC in 0.01 M $CaCl_2$ suspensions (1:2.5 w/v) and composting samples were analyzed for pH, EC in H_2O suspensions (1:10 w/v). Whereas the concentrations of total organic carbon (C_{org}), total nitrogen (N_t) and total sulfur (S_t) were determined for both soil and compost samples using a CNS analyzer (Vario EL III; Elementar Analysensysteme, Hanau, Germany).

For Py-FIMS analyses composting pile COM4 was selected because this treatment reached stability in a shorter time period than the other treatments and high analyses costs restricted the number of samples that could be measured. About 3 mg of the air dried, ground and homogenized soil samples amended with differently stabilized COM4 were thermally degraded in the ion source (emitter: 4.7 kV, counter electrode −5.5 kV) of a double-focusing Finnigan MAT 95 mass spectrometer. All samples were heated in a vacuum of 10^{-4} Pa from 50°C to 700°C in temperature steps of 10°C over a time period of 18 minutes. Between magnetic scans the emitter was flash-heated to avoid residues of pyrolysis products. About 65 magnetic scans were recorded for the mass range m/z 15 to 900.

Ion intensities were referred to 1 mg of the sample. For each of the single scans, the absolute and relative ion intensities of ten classes of compounds in the OM were calculated by summation of the ion intensities of indicator signals to obtain thermograms of their volatilization and averaged Py-FI mass spectra. This procedure was done for each three replicate measurements per soil sample and the results were averaged for statistical analyses.

2.5. Statistical Analyses

The C-losses during the incubation experiment were fitted to mathematical models using the non-linear regression procedure. Means and standard errors were calculated for chemical and phytotoxicity parameters, CO_2-C volatilization data and ion intensities from Py-FIMS. Comparisons between means of ion intensities of compound classes of different treatments were done by applying One Way ANOVA test (LSD mean comparison method were used). All statistics were computed using data analysis and graphic software (Origin 8.1 G).

3. Results

3.1. Germination Index (GI)

In the experiment with garden cress the GI varied between 3% and 112% (**Table 1**). All the compost samples yielded GI greater than 100% whereas the RCPW yielded 3%. Moreover, a germination delay was observed in the treatments with fresh coffee pulp waste. In the test with radish the GI varied between 26% and 132%. Similar to the results with cress all the compost samples yielded GI greater than 100% where as the RCPW yielded 26%. The phytotoxicity of RCPW and the sanitation effect of composting were not reflected by the Germination percentages (GP) as all samples had GP above 80% (not shown).

3.2. Carbon Mineralization

In all treatments the cumulative respiration curves showed two distinct phases for CO_2-C evolution. An initial most intensive biological transformation phase was followed by a slower second phase (**Figure 2**). The maximum of CO_2-C release was achieved in the first three weeks of incubation in all treatments. The equation first order plus linear model $C_t = C_1\left(1 - e^{-k_1 * t}\right) + K_2 t$ gave the best fit for the cumulative CO_2-C respired. The different parameters in the equation have a biological meaning, *i.e.*, C_t is the amount of organic C mineralized, C_1 is the amount of the labile C ($mg \cdot g^{-1}$), K_1 (day^{-1}) and K_2 (day^{-1}) are the rate constants for the mineralization of labile and recalcitrant C, respectively, and t is the incubation time in days. The CO_2-C release followed the order S+RCPW > S+COM-M > S+COM-T > S+COM-F > CONTROL in all treatments. However, the cumulative CO_2-C release curves of treatments S+COM4-T and S+COM4-F were not significantly different (**Figure 2**).

In the short time of three weeks between 35% and 69% of the total C had been evolved. The cumulative

Table 1. Germination index (GI) and chemical properties of row coffee pulp waste and composts of different degree of stability taken at different composting time (t). Different letters indicate that samples are significantly different ($p < 0.05$) according to Fisher LSD with in a column and values followed by the same letters with in a column were not significantly different at 5% probability level.

Compost samples	Composting time (days)	pH_{H_2O}	C_{org} ($g \cdot kg^{-1}$)	N_{tot} ($g \cdot kg^{-1}$)	S ($g \cdot kg^{-1}$)	C/N	EC ($ms \cdot cm^{-1}$)	GI (%)	
								Cress	Radish
COM1-T	8	8.81(0.03)ab	151.0(2.0)a	10.5(0.0)a	1.70(0.0)ad	14.37	1.99(1.3)a	106(8.9)a	121(8.5)a
COM1-M	36	9.29(0.03)a	149.0(4.0)a	12.7(0.5)b	2.22(0.0)b	11.72	2.13(0.0)a	109(4.6)a	120(8.4)a
COM1-F	119	9.07(0.02)ab	139.2(3.0)b	12.7(0.3)b	2.40(0.0)b	10.98	2.07(0.5)a	102(16.9a	125(4.8)a
COM2-T	8	8.70(0.2)ab	77.0(2.5)c	6.8(0.1)c	1.35(0.0)c	11.40	1.67(0.0)b	112(8.0)a	118(5.6)a
COM2-M	24	9.15(0.08)ab	73.2(2.0)c	7.0(0.1)c	1.80(0.0)d	10.48	1.09(0.0)c	113(9.0)a	120(7.5)a
COM2-F	89	8.96(0.27)ab	73.1(1.0)c	7.3(0.1)c	1.80(0.0)d	9.98	1.33(0.0)f	115(3.5)a	132(3.9)a
COM4-T	8	8.05(0.01)c	107.9(3.0)d	8.74(0.0)d	1.40(0.0)c	12.35	1.36(0.0)f	107(9.4)a	111(12.1)a
COM4-M	24	8.65(0.02)b	93.3(1.0)e	9.32(0.1)d	1.70(0.0)d	10.01	1.07(1.0)c	107(4.7)a	108(8.4)a
COM4-F	89	8.55(0.05)bc	91.49(3.0)e	8.99(0.0)d	1.88(0.0)d	10.18	1.11(0.0)c	108(8.8)a	115(9.3)a
RCPW	0	5.81(0.01)d	397.0(1.6)f	21.1(0.3)e	3.10(0.2)e	18.86	6.73(0.0)d	3(1.1)b	26(3.0)b

COM: compost; T: thermophilic phase; M: mesophilic phase; F: final stage; RCPW: row (fresh) coffee pulp waste; C_{org}: total organic carbon; N_{tot}: total nitrogen; C/N: carbon to nitrogen ratio; EC: electrical conductivity and standard errors in brackets.

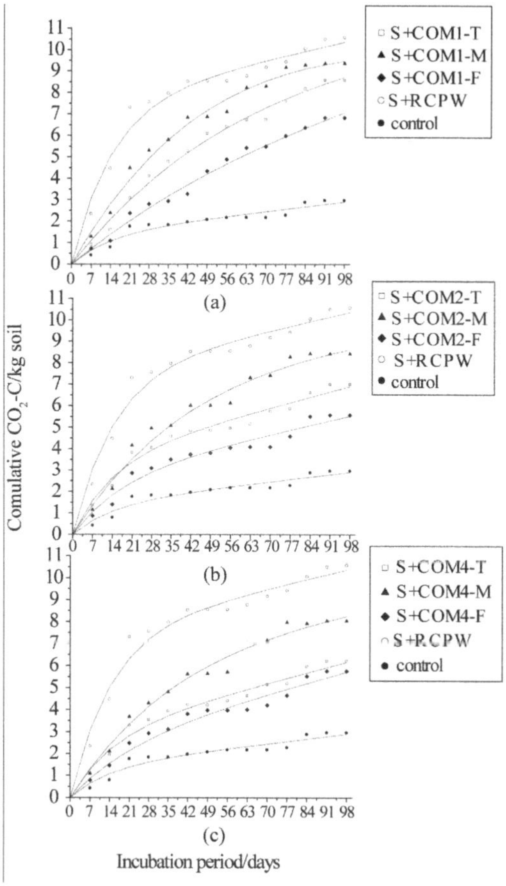

Figure 2. Comulative CO_2-C release from the control, soil amended with raw coffee pulp waste (S+RCPW), and composts of different degree of stability made up of mainly composting of: (a) coffee pulp waste (S+COM1-T, S+COM1-M, S+COM1-F); (b) fruit and vegetable waste (S+COM2-T, S+COM2-M, S+COM2-F); and (c) equal proportion of coffee pulp waste with fruit and vegetable waste (S+COM4-T, S+COM4-M, S+COM4-F).

CO_2-C respired from the amended soils were varied between 5.6 $g \cdot C \cdot (kg \cdot soil)^{-1}$ and 10.6 $g \cdot C \cdot (kg \cdot soil)^{-1}$ at the end of the incubation (**Table 2**). The rate constants K_1 were higher in treatment S+RCPW and lower in the control whereas the higher rate constants K_2 were observed in treatment S+COM-F. In all treatments the K_1 values exceeded the K_2 values (**Table 2**). During the whole incubation period the CO_2-C release from S+RCPW (10.6 $g \cdot C \cdot (kg \cdot soil)^{-1}$) was significantly higher than in soils amended with any of the compost materials.

The application of differently stabilized composts significantly affected the CO_2-C release from soil (**Table 2**). Addition of COM4-F to soil resulted in the lowest CO_2-C release (5.76 g C ($kg \cdot soil^{-1}$)) followed by COM4-T (6.22 $g \cdot C \cdot (kg \cdot soil)^{-1}$) and COM4-M (8.03 $g \cdot C \cdot (kg \cdot soil)^{-1}$). Respiration of soils amended with differently stabilized COM1 and COM2 followed the same order. Considering the loss in percent of C_{org} derived from the C_{org} contents at the start and end of the incubation showed the smallest overall C loss (ranges from 1.42% to 2.43%) from the soil amended with COM-T during the incubation (**Table 2**).

3.3. Pyrolysis-Field Ionization Mass Spectrometry (Py-FIMS)

In the Py-FIMS analyses of the control and soils amended with differently stabilized COM4 the thermograms of total ion intensity (TII) showed a reduction in intensity and a shift towards higher pyrolysis temperature during incubation. For instance the thermal evolution of molecules reached maximum intensities in the temperature 460°C and 480°C at start of the incubation and shifted to 490°C and 510°C at the end of incubation (see upper right inserts in **Figure 3**).

Table 2. Carbon mineralized (g·C·kg^{-1} soil), total amount (g·kg^{-1}) of selected elements, C/N ratio, percent increase in soil organic carbon % inc. C_{org}, percent decrease of organic carbon (% dec. C_{org}), the labile pool rate constant K_1 (day^{-1}) and recalcitrant pool rate constant K_2 (day^{-1}) of the control and amended soil during the incubation. Values followed by the same letters with in a column were not significantly different at 5% probability level (S+COM: soil-compost mixture; T: thermophilic phase; M: mesophilic phase; F: final stage; S+RCPW: soil-row coffee pulp waste mixture). Numbers in brackets represent standard errors.

Treatments	K_1	K_2	CO_2-C	At day 1					At day 98				
				C_{org}	N_{tot}	S	C/N	% inc. C_{org}	C_{org}	N_{tot}	S	C/N	% dec. C_{org}
S+COM1-T	0.06	0.00	8.57(1.43)a	42.03	3.5	0.7	11.89	5.60	41.2	3.4	0.8	11.90	2.04
S+COM1-M	0.08	0.01	9.35(2.6)a	42.01	3.6	0.7	11.60	5.53	40.3	3.4	0.7	11.94	4.08
S+COM1-F	0.04	0.01	6.83(1.43)b	41.77	3.6	0.8	11.72	4.93	40.7	3.4	0.7	12.00	2.55
S+COM2-T	0.07	0.00	6.99(2.02)b	40.58	3.3	0.7	12.28	1.96	40.0	3.3	0.7	12.00	1.42
S+COM2-M	0.08	0.00	8.41(1.83)a	40.50	3.4	0.7	11.97	1.75	39.3	3.4	0.8	11.54	2.90
S+COM 2-F	0.06	0.01	5.57(1.67)c	40.46	3.4	1.3	11.78	1.66	39.2	3.3	0.7	11.91	3.05
S+COM4-T	0.07	0.00	6.22(1.2)c	41.20	3.4	0.8	12.17	3.51	39.9	3.3	0.7	12.03	3.26
S+COM4-M	0.07	0.00	8.03(2.08)a	40.93	3.4	0.7	11.96	2.82	39.3	3.4	0.8	11.63	4.09
S+COM4-F	0.05	0.00	5.76(0.85)c	40.77	3.5	0.7	11.74	2.44	39.8	3.4	0.7	11.71	2.50
S+RCPW	0.12	0.00	10.57(1.7)d	47.60	3.7	0.8	12.89	19.60	42.6	3.7	0.8	11.61	10.49
Control	0.04	0.00	2.97(1.41)e	39.80	3.3	0.7	12.15	0.00	38.3	3.2	0.7	11.82	3.66

In the Py-FI mass spectra, a particular enrichment of N-compounds during the incubation in compost treated soil was reflected by m/z signals 67, 110 and 117 being more intense in **Figures 3(b)**, **(d)** and **(f)** than in **Figures 3(a)**, **(c)** and **(e)**. This difference, however, was not shown in the control (**Figures 3(g)** and **(h)**). Comparison of mass spectra at the start and end incubation indicated increased proportions of carbohydrates (e.g. more intense m/z signals 84, 96, 110 in **Figures 3(b)**, **(d)**, **(f)** and **(h)** than in **Figures 3(a)**, **(c)**, **(e)** and **(g)**), lignin building blocks (e.g. more intense m/z signals 156, 168, 178, 192, 194, 202, 216 and 218 in **Figures 3(b)**, **(d)**, **(f)** and **(h)** than in **Figures 3(a)**, **(c)**, **(e)** and **(g)**). A decreased proportion of homologous series of free fatty acids and alkenes was indicated by m/z signals at 252, 266, 280, 294, 308 322, 336, and 350 (e.g. less intensive in **Figures 3(a)**, **(c)**, **(e)** and **(g)** than in **Figures 3(b)**, **(d)**, **(f)** and **(h)**) in all incubated treatments, but most pronounced in S+COM4-T.

The TII values were significantly larger in all amended soils than in the control at start of the incubation (**Table 3**). However, significantly higher TII's ($p < 0.05$) were recorded only in the treatment S+COM4-T than the control, both at the start and end of incubation. This well agreed with the C_{org} concentrations at the start and end of the incubation experiment (**Table 2**).

The relative ion intensities (% TII) of compound classes indicated a significant ($p < 0.05$) enrichment of N-compounds during the incubation in all amended soils (by 39.1% in S+COM4-T, by 28.6% in S+COM4-M and by 40.9% in S+COM4-F). However, in the control the proportions of N-compounds did not significantly change between the start and end of incubation (**Table 3**). Among compost amended treatments significantly higher increases in the proportions of carbohydrates (by 65.3%), phenols/lignin monomers (by 12.9%) and peptides (by 43.5%) at the expense of fatty acids and sterols were observed in the treatment S+COM4-T than in any other treatment. The compound classes phenols/lignin monomers and alkylaromatics are the major constituents of the SOM both at the start and end of incubation. The proportions of fatty acids and sterols were the smallest but an active component in the compost amended soils in which their proportion decreased largely during the incubation (**Table 3**). The proportions of alkylaromatics, lignin dimers and lipids did not differ among treatments, neither at the start nor at the end of incubation.

The temperature-resolved volatilization curves for N-compounds and peptides showed a clear difference in thermal stability between S+COM4-T and the control during the incubation (**Figure 4**). At start of the incubation the thermograms for these two compound classes indicated no change in the thermal stability between treatments S+COM4-T and the control. At the end of incubation, however, clear shifts of peaks to higher pyrolysis temperature were observed in S+COM4-T when compared with the control.

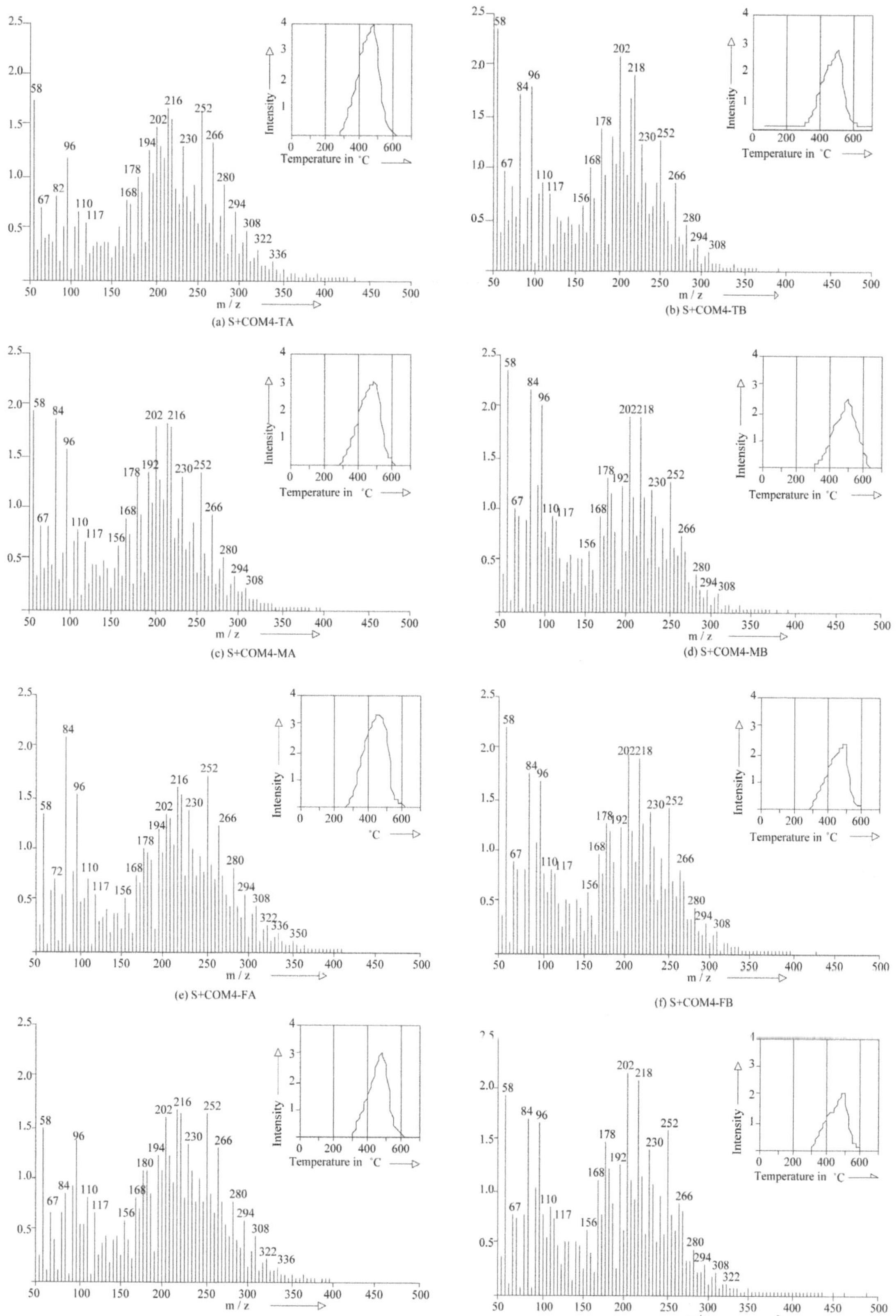

Figure 3. Summed and averaged pyrolysis-field ionization mass spectra and thermograms of total ion intensity (TII) (upper right inserts) of the non-treated control soil and soil amended with different degree of stability composts (COM: compost; T: thermophilic phase; M: mesophilic phase; F: final stage); at the start of incubation (---A) and end of incubation(---B).

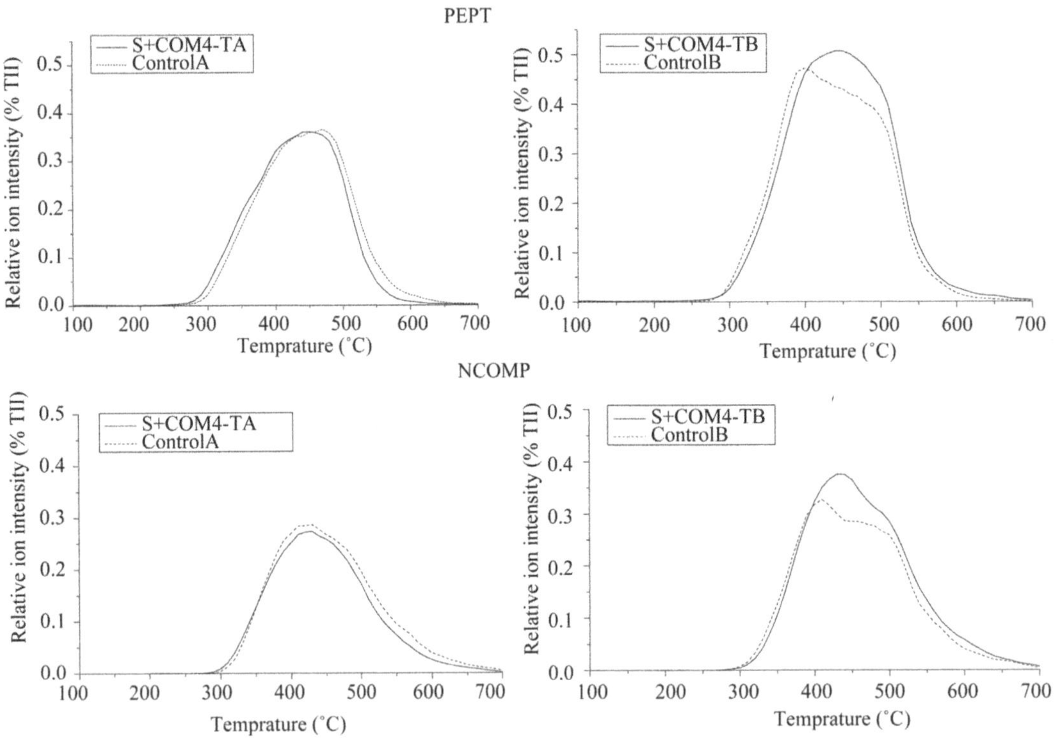

Figure 4. Thermograms for the evolution of peptides (PEPT) and mainly heterocyclic N-compounds (NCOMP) from non-treated control soil (Control) and soil amended with thermophilic compost (S+COMP4-T) at the start of incubation (Control A, S+COMP4-TA) and end of incubation (Control B, S+COMP4-TB).

Table 3. Total ion intensity (TII), volatile matters (VM %) and relative abundance (% TII) of principal classes of compounds from incubated samples at the start of incubation (---A) and end of incubation (----B). Values followed by the same letters with in a column and column without letters were not significantly different at 5% probability level.

Treatment	TII	CARB	PHLM	LDIM	LIPID	ALKY	NCOMP	STER	PEPT	SUBE	FATTY	VM (%)
S+COM4-T A	60.0(1.5)a	4.9(0.1)a	12.4(0.3)e	7.7(0.3)	3.1(0.3)	15.6(0.2)	4.6(0.1)a	0.3(0.1)a	6.2(0.1)a	0.0(0.0)	0.3(0.1)a	21.7(1.3)
S+COM4-T B	41.5(2.2)b	8.1(0.1)b	14.0(0.1)f	8.1(0.1)	2.8(0.1)	15.1(0.1)	6.4(0.1)b	0.1(0.0)b	8.9(0.1)bc	0.0(0.0)	0.0(0.0)b	23.7(0.5)
S+COM4-M A	49.7(4.3)b	7.8(0.1)b	13.5(0.3)f	8.0(0.1)	3.1(0.1)	15.5(0.3)	5.6(0.1)c	0.2(0.0)ab	8.3(0.1)d	0.0(0.0)	0.1(0.0)b	20.0(1.9)
S+COM4-M B	41.2(7.2)b	9.2(0.5)c	13.6(0.5)f	8.3(0.3)	2.8(0.4)	14.9(0.4)	7.2(0.2)d	0.1(0.0)b	9.4(0.4)c	0.0(0.0)	0.0(0.0)b	19.8(1.2)
S+COM4-F A	57.2(2.1)a	7.5(0.1)b	12.2(0.3)e	8.0(0.0)	3.5(0.1)	15.5(0.2)	4.4(0.1)a	0.3(0.0)a	7.4(0.0)e	0.0(0.0)	0.4(0.0)a	24.0(2.1)
S+COM4-F B	35.1(0.7)bc	7.8(0.2)b	13.5(0.2)f	8.3(0.0)	3.2(0.2)	15.1(0.3)	6.2(0.1)b	0.1(0.0)b	8.7(0.1)bd	0.0(0.0)	0.1(0.0)b	23.4(0.5)
Control A	42.9(0.5)b	5.4(0.1)d	13.1(0.1)f	8.3(0.0)	3.8(0.2)	15.6(0.1)	5.1(0.2)c	0.2(0.0)a	6.4(0.2)a	0.0(0.0)	0.1(0.0)b	18.8(0.1)
Control B	30.6(1.0)c	7.7(0.3)b	14.0(0.4)f	8.5(0.1)	3.4(0.1)	15.0(0.3)	5.6(0.1)c	0.2(0.1)ab	8.3(0.2)d	0.0(0.0)	0.1(0.0)b	23.3(0.7)

CARB: carbohydrates; PHLM: phenols/lignin monomers; LIPID: lipids; NCOMP: N-compounds; ALKY: alkylaromatics; LDIM: lignindimers; STER: sterols; PEPT: peptides; SUBE: suberin; FATTY: *n*-Fatty Acids (*n*-C_{16} to *n*-C_{34}) and standard errors in brackets; S: soil; COM: compost; T: thermophilic phase; M: mesophilic phase; F: final stage; RCPW: row (fresh) coffee pulp waste.

4. Discussion

4.1. Phytotoxicity of Starting Material and Composts

The low GI of 3% with cress and 26% with radish measured for the RCPW indicates its high phytotoxcity [20,21]. Moreover, the germination delay observed in this treatments can be attributed to the high salt content (electrical conductivity = 6.73 mS·cm^{-1}) of the sample [18]. A GI < 50% means high phytotoxicity, values between 50 and 80% a moderate phytotoxicity and values > 80% no phytotoxicity [20]. Therefore, the GI > 80% in all compost samples indicated that phytotoxicity of the raw material disappeared during the composting. The observed values of GIs greater than 100% in all composting samples agreed with [21] who found a range of GIs between 25% and 151% in a study with 28 composts and cress seeds. This indicates the promotion of the germination by compost samples [22]. A GI of >80% for radish in all samples including the RCPW indicates a lower sensitivity of this seedling against the RCPW.

4.2. Organic Matter Mineralization

The observed pattern in CO_2-C release agreed with the model used by [9]. They compared four different models to describe the mineralization dynamics from soil amended with alkaline treated municipal biosolids. The first rapid phase corresponded to the fast microbial decomposition of labile compounds [2,17]. The second slower phase corresponded to the slow decomposition of resistant organic compounds. The CO_2-C release from RCPW and compost-amended soils above that from the untreated control soil can be explained by the addition of labile C sources. This agrees with [7,23] who reported increased respiration of soils after amendment with organic waste of diverse origins and stability from 7th to 30th day of incubation. Similarly, the observed difference in the CO_2-C release between the RCPW and compost amended soils resulted mostly from the first three weeks of decomposition. In the later incubation period a faster degradation of organic compounds was observed in soil amended with COMP-F ($K_2 = 0.008$) than in RCPW and soil amended with COM-T ($K_2 = 0.001$) (**Table 2**).

The variation in the CO_2-C release among the soils amended with differently stabilized composts despite of similar initial C_{org} contents and C/N ratios indicated a C immobilization. This immobilization was stronger in the soil amended with COM-T than in the soil amended with COM-M. This confirms reference [8] who observed that the degree of stabilization of added compost materials must not necessarily agree with the efficiency in C conservation. Moreover, some authors found an inverse relationship between the CO_2-C release and the degree of transformation/stabilization of compost materials added [7,17,23]. One explanation for the partial disagreement of our data for respiration and stabilization with the above three references might be the quantity of added labile components with COM-T. This quantity might have been to low to promote the microbial decomposition significantly in the initial phase of incubation [19]. Another explanation is the stabilization of C by clay-organic matter associations [24]. Moreover, the lowest C_{org} loss (ranges from 1.4% and 2.4%) in the soil amended with COM-T indicated the suitability of composts from an early composting phase. This confirms that composts with a short composting period can be optimal for C sequestration in soil [7,9].

4.3. Changes in SOM Composition

The larger TII in the compost-amended treatments than in the control at the start of incubation coincided with the C_{org} concentrations in the treatments. It is explained by the additional organic matter from the composts [25,26]. However, the larger TII in S+COM4-T than in any other treatment at the end of incubation proved selective preservation of compounds that were possibly mineralized in samples with longer composting time. The best explanation for such a selective protection is the binding of organic matter at clay surfaces, and a consecutive aggregation of clay-organic matter complexes [24,27,28].

The signal patterns of Py-FI mass spectra indicated the clearest impact of compost application by increasing proportions of N-compounds at the expense of fatty acids and sterols (**Figure 2**, **Table 3**). The stabilization of N-compounds in compost amended soils compared to the controls was also reflected by changes in the thermal volatilization of this compound class after 98 days of incubation (**Figure 3**). This suggests that the organic matter in compost-amended soils was relatively enriched in heterocyclic N-compounds which obviously resisted microbial decomposition. This indicates a residual relative enrichment of rather stabile compounds during the decomposition [28-30]. This explanation agrees with reference [31], who described enrichments of heterocyclic N-compounds in SOM also beneath forest trees. In the same line, enrichments of N-compounds were detected at the end of composting by Py-FIMS and FT-IR spectroscopy [19]. However, these N-compounds were completely decomposed in re-circulated leachates in larger scale lysimeters [32,33], likely caused by the lack of sorptive surfaces. This indirectly points to the importance of the sorptive surfaces of the clay minerals in the present experiment.

The temperature-resolved Py-FIMS showed that thermally stable carbohydrates, phenols/lignin monomers, and peptides were more enriched during the incubation of S+COM4-T than in any other compost amended treatment. This indicates that these compound classes were microbially synthesized during incubation or selectively preserved. The pronounced enrichment of phenols/lignin monomers agreed with [34] who stated that composition of SOM after compost amendment changed mainly by increases in the proportion of lignins. The stronger decrease in the lignin content in a soil amended with final compost than with early compost (treatment S+COM4-T) underlined the advantage of the early compost. Moreover, significantly increased carbohydrate proportions in the variant S+COM4-T during the incubation proved a selective carbohydrate preservation. This agrees with findings of [35,36] that progressive compost maturity reflected a decrease of the carbohydrate content.

In agreement with references [19,37], we found a progressive association of bio-labile peptides during composting.

4.4. Explanation of the Organic Matter Mineralization by Changes in the Molecular Composition of SOM

Computing the percentage decrease of ion intensities in the temperature range < 400˚C (**Figure 3**) as indicator of thermally labile compounds during incubation revealed the largest decrease by 49% in S+COM4-M, followed by 35% in S+COM4-T, 28% in S+COM4-F whereas no decrease was observed in the control. This result is consistent with the order in CO_2-C release during the incubation experiment (**Figure 2**). Moreover, the changes in TII values (10^6 counts·mg^{-1}) of the compound classes (not shown) confirmed the respired CO_2-C mainly resulted from reduced ion intensities of phenols/lignin monomers, lignin dimers, lipids, alkylaromatics, fatty acids and sterols during incubation for 98 days. On the other hand thermally stable compounds (> 400˚C according to [38]) relatively increased by 12% in S+COM4-M, 16% in S+COM4-T and S+COM4-F but only by 1% in the control. Moreover, computing the relative increase in thermally stable carbohydrates (volatilized >400˚C), N-compounds (volatilized > 380˚C) and peptides (volatilized > 350˚C - 400˚C) confirmed that addition of composts particularly from the end of the thermophilic phase to soil significantly enriched the SOM during incubation by stabilized compounds of microbial and plant origin.

The decrease in TII mainly in the temperature range below 400˚C (thermally labile compounds) in the compost amended soils only implied that a large portion of labile components of the compost material itself was transformed or degraded during the incubation process. This supports the idea that decomposition of the added composts in soil goes along with humification in the composts themselves [34]. Here we show that this decomposition process is accompanied by the pronounced development of thermally stable carbohydrates, phenols/lignin monomers, and peptides of plant and microbial origin if the composts from the early phase of composting were applied. This was really a positive contribution of adding compost to SOM composition in the clay dominated Nitisol because this was not observed following NPK fertilization of soil [36] and matured compost application in sandy soil [39].

5. Conclusions

1) The combination of laboratory incubation and Py-FIMS for the first time provided compelling evidence for effects of the compost stability on C mineralization and the molecular composition SOM when composts were applied to a typical clay-rich tropical soil.

2) The combined results revealed that it would be sufficient to end up the composting of coffee pulp and fruit waste immediately after a short thermophilc phase as its application better conserved the C (a) during composting and (b) during transformations in the soil.

3) Application of compost from the early phase of composting may result in a quicker mineralization of biologically labile organic matter from the compost and a better enrichment of SOM with stable compounds of plant and microbial origin such as carbohydrates, phenols/lignin monomers and peptides at the expense of free fatty acids and sterols compared to mature compost. This may have considerable economic implications as composting of coffee pulp in heaps may take 6 to 8 months to reach the mature stage.

4) Therefore, the application of compost from the early composting phase can be recommended not only as a soil amendment but also as a measure to mitigate CO_2-enrichment in the atmosphere. Forthcoming studies will be directed to disclose in more detail the stabilization mechanism of biologically labile, compost-derived organic molecules in tropical and other soils.

6. Acknowledgments

The ECPB-DAAD program provided a scholarship for B. Eshetu. The Mass Spectrometry Laboratory of the Rostock Soil Science group was supported by the "Exzellenzförderprogramm" of the Ministerium für Bildung, Wissenschaft und Kultur Mecklenburg-Western Pomerania, project UR 07 079. Furthermore, we thank Dr. R. Beese and Dipl.-Chem. Kai-Uwe Eckhardt, University of Rostock, for technical support and data analyses.

REFERENCES

[1] E. Favoino and D. Hogg, "The Potential Role of Compost in Reducing Greenhouse Gases," *Waste Management & Research*, Vol. 26, No. 1, 2008, pp. 61-69. doi:10.1177/0734242X08088584

[2] B. Ceccanti, G. Masciandaro and C. Macci, "Pyrolysis-gas Chromatography to Evaluate the Organic Matter Quality of a Mulched Soil," *Soil Tillage Research*, Vol. 97, No. 1, 2007, pp. 71-78. doi:10.1016/j.still.2007.08.011

[3] B. Eshetu, G. Jandl and P. Leinweber, "Compost Changed Soil Organic Matter Molecular Composition: A Study by Py-GC/MS and Py-FIMS," *Compost Science and Utilization*, Vol. 20, No. 4, 2012, pp. 230-238.

[4] S. Nardi, F. Morari, A. Berti, M. Tosoni and L. Giardini, "Soil Organic Matter Properties after 40 Years of Different Use of Organic and Mineral Fertilizers," *European Journal of Agronomy*, Vol. 21, No. 3, 2004, pp. 357-367. doi:10.1016/j.eja.2003.10.006

[5] A. Fabrizio, F. Tambone and P. Genevini, "Effect of

Compost Application Rate on Carbon Degradation and Retention in Soils," *Waste Management*, Vol. 29, No. 1, 2009, pp. 174-179.

[6] L. D. Lima, M. S. Santos, W. H. Scherer, J. R. Schneider, C. A. Duarte, B. H. E. Santos and I. V. Esteves, "Ef-Fects Of Organic And Inorganic Amendments On Soil Organic Matter Properties," *Geoderma*, Vol. 150, No. 1-2, 2009. pp. 38-45. doi:10.1016/j.geoderma.2009.01.009

[7] C. Mondini, M. L. Cayuela, T. Sinicco, F. Cordaro, A. Roig and M. A. Sánchez-Monedero, "Greenhouse Gas Emissions and Carbon Sink Capacity of Amended Soils Evaluated under Laboratory Conditions," *Soil Biology and Biochemistry*, Vol. 39, No. 6, 2007, pp. 1366-1374. doi:10.1016/j.soilbio.2006.12.013

[8] M. A. Sánchez-Monedero, M. L. Cayuela, C. Mondini, N. Serramia and A. Roig, "Potential of Olive Mill Wastes for Soil Carbon Sequestration," *Waste Management*, Vol. 28, No. 4, 2008, pp. 767-773.

[9] J. D. Gillis and G. W. Price, "Comparison if Novel Model to Three Conventional Models Describing Carbon Mineralization from Soil Amended with Organic Residues," *Geoderma*, Vol. 160, No. 3-4, 2011, pp. 304-310. doi:10.1016/j.geoderma.2010.09.025

[10] P. Smith, D. S. Powlson, J. U. Smith, P. Falloon and K. Coleman, "Meeting Europe's Climate Change Commitments: Quantitative Estimates of the Potential for Carbon Mitigation by Agriculture," *Global Change Biology*, Vol. 6, No. 5, 2001, pp. 525-539. doi:10.1046/j.1365-2486.2000.00331.x

[11] R. Lal, "Soil Management and Restoration for C Sequestration to Mitigate the Accelerated Greenhouse Effect," *Progress in Environmental Science*, Vol. 1, No. 4, 1999, pp. 307-326.

[12] O. Davidson, K. Halsnaes, S. Huq, M. Kok, B. Metz, Y. Sokona and J. Verhagen, "The Development and Climate Nexus: The Case of Sub-Saharan Africa," *Climate Policy*, 31, Supl. 1, 2003, pp. 97-113. doi:10.1016/j.clipol.2003.10.007

[13] R. Lal, "The Potential of Soils of the Tropics to Sequester Carbon and Mitigate the Greenhouse Effect," *Advances in Agronomy*, Vol. 74, 2002, pp. 155-192. doi:10.1016/S0065-2113(01)74033-3

[14] P. S. Murthy, M. R. Manjunatha, G. Sulochannama and M. M. Naidu, "Extraction, Characterization and Bioactivity of Coffee Anthocyanins," *European Journal of Biological Sciences*, Vol. 4, No. 1, 2012, pp. 13-19.

[15] B. Gezahegne, L. Fikre and M. Wakjira, "Exploring the Suitability of Coffee Pulp Compost as Growth Media Substitute in Greenhouse Production," *International Journal of Agricultural. Research*, Vol. 6, No. 3, 2011, pp. 255-267. doi:10.3923/ijar.2011.255.267

[16] W. Negassa, C. Baum and P. Leinweber, "Soil Amendement with Agro-Industrial Products: Moluclar-Chemical Compostions and Effects on Soil Biochemical Activites and Phosphrous Fractions," *Journal of Plant Nutrition and Soil Science*, Vol. 174, No. 1, 2011, pp. 113-120. doi:10.1002/jpln.201000034

[17] M. P. Bernal, M. A. Sanchez-Monedero, C. Paredes and A. Roig, "Carbon Mineralization from Organic Wastes at Different Composting Stages during Incubation with Soil. Agriculture," *Ecosystems & Environment*, Vol. 69, No. 3, 1998, pp. 175-189. doi:10.1016/S0167-8809(98)00106-6

[18] P. Leinweber, O. Blumensteins and H.-R. Schulten, "Organic Matter Composition in Sewage Farm Soils: Investigations by ^{13}C-NMR and Pyrolysis-Field Ionization Mass Spectrometry," *European Journal of Soil Science*, Vol. 47, No. 1, 1996, pp. 71-80. doi:10.1111/j.1365-2389.1996.tb01373.x

[19] E. Smidt, K.-U. Eckhardt, P. Lechner, H.-R. Schulten and P. Leinweber, "Characterization of Different Decomposition Stages of Biowaste Using FT-IR Spectroscopy and Pyrolysis-Field Ionization Mass Spectrometry," *Biodegradation*, Vol. 16, No. 1, 2005, pp. 67-79. doi:10.1007/s10531-004-0430-8

[20] F. Zucconi, A. Monaco and M. Forte, "Phytotoxins during the Stabilization of Organic Matter," In: J. K. R. Gasser, Ed., *Composting of Agricultural and other Wastes*, Elsevier Applied Science Publication, New York, 1985, pp. 73-86.

[21] K. Lasaridi, I. Protopapa, M. Kotsou, G. Pilidis, T. Manios and A. Kyriakou, "Quality Assessment of Composts in the Greek Market: The Need for Standards and Quality Assurance," *Journal of Environmental Management*, Vol. 80, No. 1, 2006, pp. 58-65. doi:10.1016/j.jenvman.2005.08.011

[22] R. Paradelo, A. B. Moldes, M. Rodríguez and M. T. Barral, "Relationship between Heavy Metals and Phytotoxicity in Composts" *Ciencia y Tecnologia Alimentaria*, Vol. 6, No. 2, 2008, pp. 143-151. doi:10.1080/11358120809487639

[23] J. A. Pascual, C. Hernandez, C. García and M. Ayuso, "Carbon Mineralization in Arid Soil Amended with Organic Wastes of Varying Degrees of Stability," *Communications in Soil Science and Plant Analysis*, Vol. 29, No. 7-8, 1998, pp. 835-846. doi:10.1080/00103629809369989

[24] N. S. Bolan, A. Kunhikrishnan, G. K. Choppala, R. Thangarajan and J. W. Chung, "Stabilization of Carbon in Composts and Biochars in Relation to Carbon Sequestration and Soil Fertility," *Science of Total Environment*, Vol. 424, No. 1, 2012, pp. 264-270. doi:10.1016/j.scitotenv.2012.02.061

[25] P. Leinweber and H.-R. Schulten, "Dynamics of Soil Organic Matter Studied by Pyrolysis-Field Ionization Mass Spectrometry," *Journal of Analytical and Applied Pyrolysis*, Vol. 25, No. 1, 1993, pp. 123-136. doi:10.1016/0165-2370(93)80036-Y

[26] C. Sorge, R. Miiller, P. Leinweber and H. -R. Schulten, "Pyrolysis-Mass Spectrometry of Whole Soils, Soil Particle-Size Fractions, Litter Materials and Humic Substances: Statistical Evaluation of Sample Weight, Residue, Volatile Matter and Total Ion Intensity," *Fresenius Journal of Analytical Chemistry*, Vol. 346, No. 6-9, 1993, pp. 687-703. doi:10.1007/BF00321275

[27] H.-R. Schulten, P. Leinweber and C. Sorge, "Composition of Organic Matter in Particle Size Fractions of an Agricultural Soil," *Journal of Soil Science*, Vol. 44, No. 4, 1993, pp. 677-691.

doi:10.1111/j.1365-2389.1993.tb02332.x

[28] M. A. Bustamante, D. Said-Pullicino, C. Paredes, J. A. Cecilia and R. Moral, "Influences of Winery-Distillery Waste Compost Stability and Soil Type on Soil Carbon Dynamics in Amended Soils," *Waste Manage*ment, Vol. 30, No. 10, 2010, pp. 1966-1975. doi:10.1016/j.wasman.2010.03.012

[29] B. H. Janssen, "Nitrogen Mineralization in Relation to C:N Ratio and Decomposability of Organic Materials," *Plant and Soil*, Vol. 181, No. 1, 1996, pp. 39-45. doi:10.1007/BF00011290

[30] H.-R. Schulten and M. Schnitzer, "The Chemistry of Soil Organic Nitrogen—A Review," *Biology and Fertility of Soils*, Vol. 26, No. 1, 1997, pp. 1-15. doi:10.1007/s003740050335

[31] C. E. Stewart, J. C. Neff, K. L. Amatangelo and P. M. Vitousek, "Vegetation Effects on Soil Organic Matter Chemistry of Aggregate Fractions in a Hawaiian Forest," *Ecosystems*, Vol. 14, No. 3, 2011, pp. 382-397. doi:10.1007/s10021-011-9417-y

[32] M. Franke, G. Jandl and P. Leinweber, "Organic Compounds in Re-Circulated Leachates of Aerobic Biological Treated Municipal Solid Waste," *Biodegradation*, Vol. 17, No. 5, 2006, pp. 473-485. doi:10.1007/s10532-005-9019-5

[33] M. Franke, G. Jandl and P. Leinweber, "Analytical Pyrolysis of Re-Circulated Leachates: Towards an Improved Municipal Waste Treatment," *Journal of Analytical and Applied Pyrolysis*, Vol. 79, No. 1-2, 2007, pp. 16-23. doi:10.1016/j.jaap.2006.11.011

[34] J. Leifeld, S. Siebert and I. Kögel-Knaber, "Changes in the Chemical Composition of Soil Organic Matter after Application of Compost," *European Journal of Soil Science*, Vol. 53, No. 2, 2002, pp. 299-309. doi:10.1046/j.1351-0754.2002.00453.x

[35] B. Chefetz, P. G. Hatcher, Y. Hadar and Y. Chen, "Chemical and Biological Characterization of Organic Matter during Composting of Municipal Solid Waste," *Journal of Environmental Quality*, Vol. 25, 1996, pp. 776-785. doi:10.2134/jeq1996.00472425002500040018x

[36] P. Leinweber, G. Jandl, C. Baum, K.-U Eckhardt and E. Kandeler, "Stability and Composition of Soil Organic Matter Control Respiration and Soil Enzyme Activities," *Soil Biology and Biochemistry*, Vol. 40, No. 6, 2008, pp. 1496-1505. doi:10.1016/j.soilbio.2008.01.003

[37] R. Spaccini and A. Piccolo, "Molecular Characteristics of Humic Acids Extracted from Compost at Increasing Maturity Stages," *Soil Biology & Biochemistry*, Vol. 41, No. 6, 2009, pp. 1164-1172. doi:10.1016/j.soilbio.2009.02.026

[38] K. Kalbitz D. Schwesig, J. Schmerwitz, K. Kaiser, L. Haumeier, B. Glaser, R. Ellerbrock and P. Leinweber, "Changes in Properties of Soil-Derived Dissolved Organic Matter Induced by Biodegradation," *Soil Biology and Biochemistry*, Vol. 35, No. 8, 2003, pp. 1129-1142. doi:10.1016/S0038-0717(03)00165-2

[39] F. J. González-Vila, G. Almendros and F. Madrid, "Molecular Alterations of Organic Fractions from Urban Waste in the Course of Composting and Their Further Transformation in Amended Soil," *Science of Total Environment*, Vol. 236, No. 1-3, 1999, pp. 215-229. doi:10.1016/S0048-9697(99)00284-3

18

Effects of Tillage Practices on Soil Properties under Maize Cultivation on Oxic Paleustalf in South Western Nigeria

Bola A. Senjobi[1*], Olufunmilayo T. Ande[2], A. E. Okulaja[3]

[1]Federal University of Agriculture, Abeokuta, Nigeria; [2]Institute of Agric Research and Training, Obafemi Awolowo University, Apata, Ibadan; [3]Department of Crop Production, Olabisi Onabanjo University, Ago-Iwoye, Nigeria.
Email: *bolasenjobi@gmail.com, *senjobiba@funaab.org.ng, *bolasenjobi@yahoo.com, funmiande@yahoo.com

ABSTRACT

A study was conducted to determine the effect of tillage practices on soil properties under maize cultivation in savanna ecosystem of South Western Nigeria. The tillage systems evaluated were zero, convectional and traditional tillage systems. The results showed that there is a significant difference in soil parameters and crop morphologies at $p < 0.05$ among the tillage systems. The traditional tillage system resulted to the most favorable soil environment, for crop growth and best performance of crop followed by conventional and no-tillage system in the area are studied respectively. The significant difference in yields adduced to lower bulk density, higher water holding capacity and porosity which increased plant root proliferation and optimal utilization of soil nutrients under tilled methods. Hence tillage methods have the capability to increase production while no-tillage is better under long term production for sustainable land use.

Keywords: Tillage Practices; Soil Properties; Maize; Sustainable Production

1. Introduction

The ever-increasing population scenario of Nigeria, makes it a necessity for more land to be opened up and efficiently managed for increased food production. The Nigerian traditional tillage method involves the use of hoes to make mounds size of which depends on the crop to be grown on. This method is gradually being replaced by conventional method of plough and harrow. Considering the fragile nature of Nigeria savannah soils, this method has further impoverished the soils due to erosion, increased loss of soil moisture by evaporation and compaction. Tillage, according to [1], is a terminology that is applied to the creation of enabling environment for the germination and growth of crops. Technically according to [2], tillage refers to the mechanical stirring of soil to provide a suitable soil environment for growth of crops. Ashaye [3] affirmed that the productivity of any soil depends on the way and how it is managed through cultivation practices imposed on it. Appropriate tillage practices are those that avoid the degradation of soil properties but maintain crop yields as well as ecosystem stability [4], [5,6]. In recent years, the interest in conservation tillage systems has increased in response to the need to limit erosion and promote water conservation [7,8]. This has been considered to provide the best opportunity for halting degradation and for restoring and improving soil productivity [9]. Conservation tillage is being adopted by local farmers in order to protect the soils from adverse effects of climate. However, the effects of this method need to be examined further in savanna soils of Nigeria characterized with limited rainfall, high temperature and fragile soils prone to erosion.

*Corresponding author.

The major stable crop grown widely in south western savanna is maize which cultivation and utilization have increased due to greater demands for consumption and for industrial purposes. Hence more land is being opened up yearly to sustain increase demands. An understanding of the climate and soil environment is necessary in order to select appropriate management practices such as tillage practices, fertilization, irrigation etc., as well as modification of the soil nutrients to optimize maize yield. This has necessitated the use of modern implements for tillage practices to facilitate farming operations. However, poor choice of implements and soil management practices, inappropriate tillage systems, seed bed types have often been identified as major constraints to cereal crop production in the savannah agro-ecological zones of South-Western, Nigeria [6]. Lal [6], therefore advocated that in tropical countries, soil tillage practices that have significant effects compared to zero tillage should be encouraged since mechanized tillage appeared to be indispensable in large scale farming. This is necessary, since a granular structure does not only encourage rapid infiltration and good retention of rainfall but also provides adequate moisture, air exchange capacity within the soil and minimizes resistance to root penetration.

This study was carried out to examine the influence of tillage practices on properties of soils under a maize cultivation and to determine the effective and appropriate tillage practice(s) for sustainable maize production.

2. Materials and Methods

2.1. Field Work

Composite surface samples (0 - 15 cm) and sub-surface samples (15 - 30 cm) were taken at the lower slope of the toposequence selected for the research. In addition to the above, a representative profile pit was dug and samples were taken from it. These were later air-dried, crushed and sieved before taken to the laboratory for physical and chemical analyses of the soils.

Three tillage operations were imposed on the soil. These were traditional, conventional and zero tillage operations. The experiments were laid out in a randomized complete block design with three replicates. Maize was used as test crop and morphological properties such as stem girth, number of internodes, number of cob/plant, fresh weight of cobs with bract. Fresh weight of grains cob, fresh weight of 10 grains and grain yield/ha were taken.

2.2. Laboratory Analyses

Soil analysis were carried out later to observe the variations in soil properties due to applied different tillage operations. The soil particle size analysis was done mechanically by using the hydrometer method of Bouyoucos [10] with sodium hexa-metaphosphate as the dispersing agent. Soil pH was determined potentiometrically in both water and 0.01 M Potassium Chloride solution (1.1). Exchangeable bases were extracted with neutral ammonium acetate solution; Ca and Mg were determined by atomic absorption spectrophotometer, and K and Na by flame photometer. Exchangeable acidity was determined by the KCL extraction method [11]. Total nitrogen was determined by the macro-kjeldah method [12]. Available Phosphorus was determined by the Bray and Kurtz [13], method and organic carbon by the method of Walkley and Black [14]. ECEC was determined as the sum of exchangeable bases and exchangeable acidity.

All the parameters collected were subjected to analysis of variance and the means were compared using Duncan Multiple Range Test.

3. Results and Discussion

Table 1 shows the result of the physical and chemical properties of the profile pit dug on the experimental site. The soil textures ranged from clay loam to sandy clay. The bulk density ranged from 1.24 to 1.59 g/cm^3. Gravel concentration and total porosity around the rooting zone were 28.7% to 66.8% and 40.0% to 53.3% respectively.

Hydraulic conductivity, permeability and water holding capacity (WHC) fluctuated with depth down the profile. This may probably be due to variation in soil texture down the profile pit. The organic carbon, pH and exchangeable bases with exception of potassium and sodium decreases with depth down the pit. The soil was classified as Oxic Paleustalf because of its Ustic moisture regime, percent CEC greater than 35% in the sub soil and Oxic diagnostic horizon.

Table 2 shows the soil properties of the studied site before tillage at 0 - 15 cm and 15 - 30 cm depths. It was observed that the bulk density and porosity decreased with depth while the pH varied between 4.82 and 4.85. The organic carbon and exchangeable bases with the exception of phosphorus were low. The low nutrient status of the soil might be due to continuous cultivation without corresponding return of organic materials to the soil. The clay content range from 28.8% to 30.4% while

Table 1. Physical and chemical properties of the soil profile of studied site.

Depth cm	Hydraulic conductivity cm/hr	Permeability cm/hr	Bulk Density g/cm^3	Porosity %	Texture	WHC %	% Clay	% Silt	% Sand	Gravel Content %	% N	pH H_2O	H^+ cmol/kg	% O.C.	P µg/g	K^+ cmol/kg	Na^+ cmol/kg	Ca^{2+} cmol/kg	Mg^{2+} cmol/kg	CEC cmol/kg	% BS
0 - 32	370.8	258.6	1.12	53.5	CL	42.2	31.4	16.0	52.6	29.7	0.23	6.03	0.09	2.30	9.78	0.75	0.92	3.96	3.05	8.68	98.97
32- 61	210.6	147.0	1.59	40.0	SC	27.9	49.4	18.0	32.6	66.8	0.04	5.01	0.08	0.55	2.13	0.72	0.87	3.34	2.55	7.48	98.94
61 - 93	390.0	271.8	1.58	40.4	CL	29.7	35.4	26.0	38.6	67.8	0.05	4.47	0.06	0.48	0.00	0.85	1.13	4.21	3.41	9.60	98.76
93 - 150	291.0	154.8	1.48	44.2	SC	36.3	41.4	20.0	38.6	19.6	0.00	3.48	0.03	0.01	0.00	0.62	0.85	4.07	3.28	8.82	99.66

Table 2. Physical and chemical properties of the surface samples of the studied site before tillage.

Depth cm	Bulk Density g/cm^3	Porosity %	% WHC	% Clay	% Silt	% Sand	% N	pH H_2O	H^+ cmol/kg	% O.C.	P μg/g	K^+ cmol/kg	Na^+ cmol/kg	Ca^{2+} cmol/kg	Mg^{2+} cmol/kg	CCEC cmol/kg	% BS
0 - 15	1.14	50.5	42.5	28.8	19.4	57.8	0.08	4.82	0.11	0.83	2.98	0.73	0.94	2.12	1.89	5.68	98.10
15 - 30	1.12	48.2	44.6	30.4	16.7	52.9	0.09	4.85	0.09	0.74	2.74	0.70	0.80	2.33	2.04	5.87	98.49

silt content ranged from 16.7% to 19.4% and the percentage sand varied from 52.9 to 57.8.

Table 3 shows the physical and chemical properties of the surface soil of the studied site under zero tillage. The water holding capacity decreased down the soil depth while the porosity increased down the depth. The organic carbon and exchangeable base contents decreased down the depth, while the bulk density increased with depth.

Under conventional tillage system, the water holding capacity (WHC) and porosity decreased down the depth while the bulk density increased down the depth (**Table 4**). The organic carbon and exchangeable cations contents decreased more compare to zero tillage method.

The chemical properties of the surface soil under traditional tillage showed that the organic carbon content, exchangeable bases and water holding capacity values decreased with depth while the bulk density and porosity increased with depth (**Table 5**).

Generally, the values of the physical properties under the traditional and conventional tillage systems were significantly higher ($p < 0.05$) than those of the zero tillage (**Figure 1**). **Figure 1** showed that tillage practices affected the bulk density, porosity and water holding capacity of the soil because of higher bulk density and lower infiltration rate of the soils under traditional and conventional tillage systems. Variation in the exchangeable bases across the different tillage methods, with consistent decreased in their values down the depth, was an indication that the planted crop, maize, (a surface feeder) obtained its nutrient from the topsoil. It further proved the nutrient depleting ability of the crop as shown in **Figure 1**. The water holding capacity of the tilled soil was higher than the untilled soil. This was because the bulk density of the tilled soil was lower than that of the untilled soil and the lower the bulk density, the higher the water holding capacity (**Figure 1**). The tillage of the soil therefore created a better soil physical environment for root growth and development. This was evident in the higher values of root volume in the tilled soil than in the untilled soils, with the conventional tillage greater than the traditional and zero tillage (**Figure 2**). There was no difference between the soil properties before tillage and under zero tillage (**Tables 2** and **3**). This may probably be due to the slashing of the bush regrowth at the site before the application of the herbicide (Round-up glyphosphate) for the effective weed control. It was observed from **Figure 2** that most of the plant parameters were significantly different ($p < 0.05$) except fresh weight of 10 grains under the three tillage systems. It was observed that traditional tillage system performed better than conventional and zero tillage systems because of the highest stem height, leaf number, number of untilled soil. This was because the bulk density of the tilled soil was lower than that of the untilled soil and the lower the bulk density, the higher the water holding capacity (**Figure 1**). The tillage of the soil therefore created a better soil physical environment for root growth and development. This was evident in the higher values of root volume in the tilled soil than in the untilled soils, with the conventional tillage greater than the traditional and zero tillage (**Figure 2**). However, there was no difference between the soil properties before tillage and under zero tillage (**Tables 2** and **3**). This may probably be due to the slashing of the bush regrowth at the site before the application of the herbicide (Round-up glyphosphate) for the effective weed control. It was observed from **Figure 2** that most of the plant parameters were significantly different ($p < 0.05$) except fresh weight of 10 grains under the three tillage systems. It was observed that traditional tillage system performed better than conventional and zero tillage systems because of the highest stem height, leaf number, and number of Cobs/plant, fresh weight of cob with bracts, and fresh weight of bract and fresh weight of 10 grains were obtained under the traditional tillage system. This may be due to the lower bulk density, high porosity and water drainage as shown in **Tables 3-5**. This is in accordance with the findings of [9], who noted that ridging system of field layout, in general improves soil drainage and crop performance compared with that planted on the no tilled soils.

The soils bulk density under traditional tillage system was lower than that of zero tillage which could have been as a result of pulverization of the soil which enhances plant root proliferation, leading to more of the soil being optimally used [4].

Tillage alters the soil environment by reducing the strength of the soil matrix and increasing the porosity thereby improving the permeability of the soil for the flow of both water and gas.

Osuji [15] findings on soils classified as Ferric Luvisol showed that nutrient accumulated under traditional tillage was relatively high after deeper ploughing; thereby

Table 3. Physical and chemical properties of the surface samples of the studied site under zero tillage.

Depth cm	Bulk Density g/cm^3	Porosity %	% WHC	% Clay	% Silt	% Sand	% N	pH H_2O	H^+ cmol/kg	% O.C.	P µg/g	K^+ cmol/kg	Na^+ cmol/kg	Ca^{2+} cmol/kg	Mg^{2+} cmol/kg	CEC cmol/kg	% BS
0 - 15	1.26	48.6	37.2	25.4	14.0	60.6	0.04	5.64	0.08	1.31	1.57	0.63	0.72	2.42	2.23	6.00	98.68
15 - 30	1.31	53.7	35.1	31.4	16.0	52.6	0.03	4.58	0.08	1.02	1.38	0.58	0.62	2.33	2.03	5.58	98.59

Table 4. Physical and chemical properties of the surface samples of the studied site under conventional tillage.

Depth cm	Bulk Density g/cm^3	Porosity %	% WHC	% Clay	% Silt	% Sand	% N	pH H_2O	H^+	% O.C.	P µg/g	K^+ cmol/kg	Na^+ cmol/kg	Ca^{2+} cmol/kg	Mg^{2+} cmol/kg	CEC cmol/kg	% BS
0 - 15	1.09	62.1	48.4	17.4	4.0	78.6	0.03	6.07	0.16	3.92	1.28	0.71	0.87	2.34	2.11	6.03	97.42
15 - 30	1.21	58.5	44.6	21.4	10.0	68.6	0.02	5.89	0.10	2.57	0.97	0.70	0.72	1.84	1.67	4.93	98.01

Table 5. Physical and chemical properties of the surface samples of the studied site under traditional tillage.

Depth cm	Bulk Density g/cm^3	Porosity %	% WHC	% Clay	% Silt	% Sand	% N	pH H_2O	H^+	% O.C.	P µg/g	K cmol/kg	Na cmol/kg	Ca^{2+} cmol/kg	Mg^{2+} cmol/kg	CEC cmol/kg	% BS
0 - 15	1.02	56.20	51.7	14.4	6.0	79.6	0.02	5.92	0.11	3.14	0.97	0.68	0.74	1.93	1.63	4.98	97.84
15 - 30	1.07	59.10	47.8	19.4	8.0	72.6	0.01	4.87	0.07	2.35	0.68	0.51	0.70	1.77	1.70	4.68	98.53

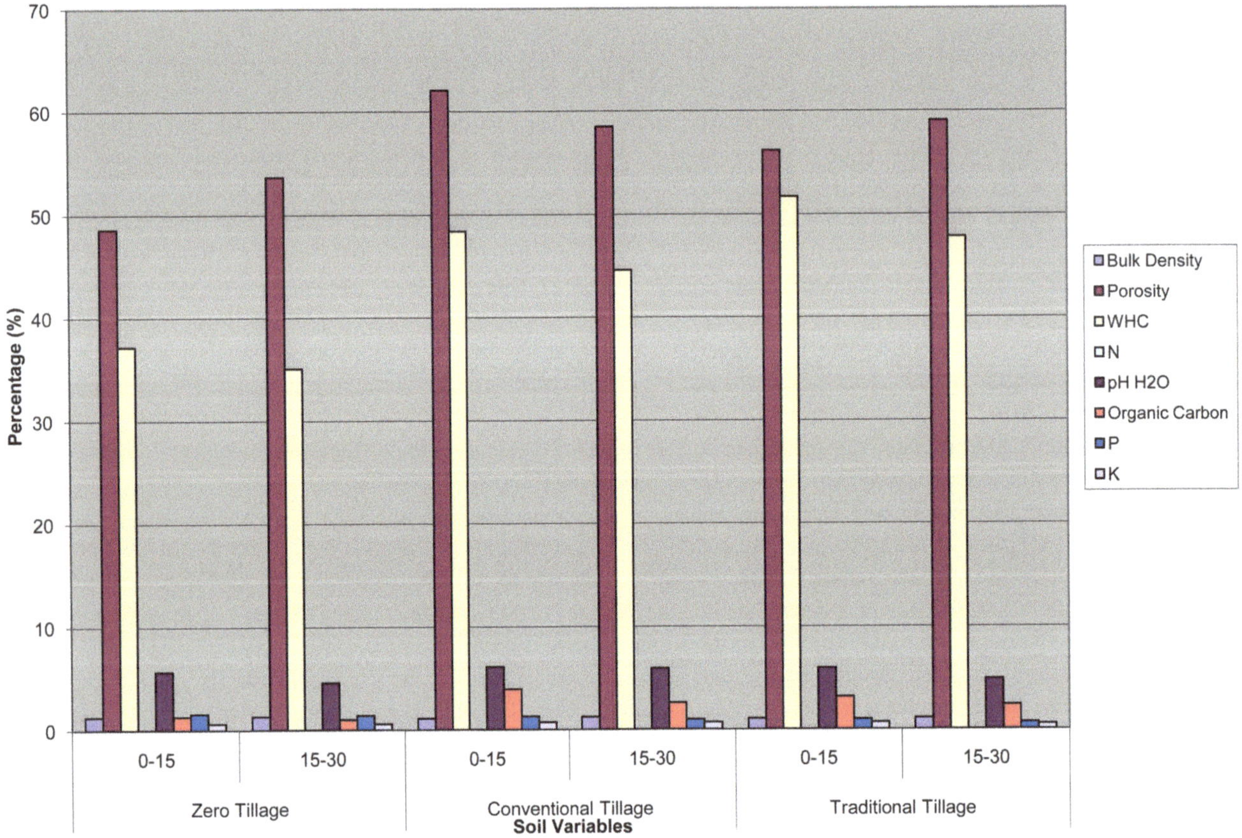

Figure 1. Soil properties under the three tillage systems at 0 - 15 cm and 15 - 30 cm depths P > 0.05.

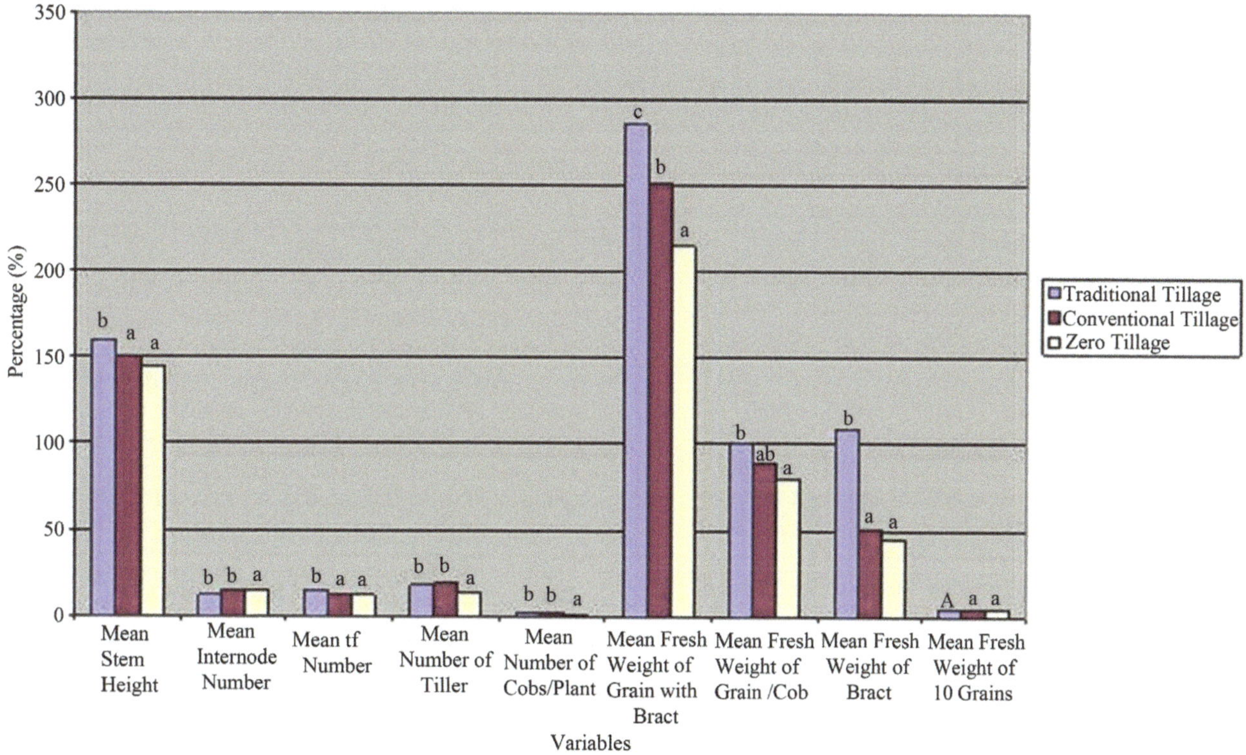

Figure 2. The agronomic variables of planted crop under the three tillage systems at 0 - 15 cm and 15 - 30 cm depths P > 0.05. Note: Variables with the same letters are not significantly different from each other.

concentrating plant nutrients for better plant growth and development. Changes in chemical properties are dependent mainly on the organic matter content of the soil which was observed to be low in the studied soil under traditional tillage system. This must have been due to the soil manipulation, which optimized condition for seed germination and seedling emergence. Traditional tillage system generally improves soil drainage and crop performance compared with that planting on heavy soils and sandy loam soil [16]. (Similar results were also reported by [17]. From the aforementioned yield and yield components of maize, traditional tillage performed better compared with conventional and zero tillage systems. Unger [8], Singh and Rachiee [18], and Nitant and Singh [19], reported that deeper ploughing prior to ridging allows maximum absorption of rainwater and reduces weed population at the initial stage of crop growth which ultimately increases crop yield attributes. Ajuwon [20], observed that in relatively low rainfall areas in Nigeria, zero tillage system was found to produce lower yields of maize on topsoil with penetration resistance of greater than 0.05 MPA or with minimal earthworm activity.

The three tillage systems showed significant difference in the morphological properties of maize except in fresh weight of 10 grains at $p < 0.05$.

Furthermore, traditional tillage is better than zero tillage. In terms of cost of herbicides, conventional tillage is better than zero tillage. This possibly was due to poor permeability [15], poor aeration and compaction of the zero tillage.

The poor growth in zero tillage could also be a result of reduced infiltration rate according to. However, with high soil temperature it can be shown that zero tillage has potential for all year round maize production especially when the rainfall regime is not high. According to [21], the higher the soil moisture under zero tillage is, the lower the rate of maize production is in the wet season. It was noted that weed development was more pronounced on zero tillage.

4. Summary and Conclusion

Inappropriate tillage systems have been identified as major constraints to major staple crop especially maize production in the savannah agro-ecological zones in South Western Nigeria. The traditional tillage system provides a more favorable soil environment for crop growth and yield than those which are planted using conventional and zero tillage systems.

The difference in yield was as a result of the lower bulk density which gave room for the pulverization of the soil causing increased plant root proliferation and optimal utilization of soil nutrients. Zero tillage system in the savannah vegetation will require an effective system of weed control. For the land to be effectively utilized under these three tillage system, the use of fertilizers is necessary due to the low nutrients observed in the soils. Statistically, significant differences among the three tillage systems were more pronounced in the traditional tillage except in fresh weight of ten grains. Based on the observed results, the traditional tillage should be given the highest priority for maize production in the savannah vegetation for optimum performance, proper soil management and economic benefits.

Traditional tillage (the ridge method) has been found to be an effective soil conservation practice, which can reduce soil degradation in the study area considering the soil types.

REFERENCES

[1] J. H. Davies, "Tillage Research and Development in Nigeria," *Proceedings of the First National Tillage Symposium of the NSAE*, 1983, pp. 15-20.

[2] G. A. Makanjuola, "Appropriate Machines for Tillage in Nigeria," *Proceedings of the First National Tillage Symposium of the NSAE*, 1983, pp. 21-34.

[3] T. I. Ashaye, "Suitability of Nigerian Soils to Mechanical Cultivation," *Proceedings of the First National Tillage Symposium of the NSAE*, 1983, pp. 57-69.

[4] R. Lal, "Importance of Tillage System in Soil and Water Management in the Tropics, in Soil Tillage and Crop Production," In: R. Lal, Ed., *IITA Proc. Series*, No. 2, 1979, pp. 25-32.

[5] R. Lal, "Soil Tillage and Crop Production," IITA, Ibadan, 1979.

[6] R. Lal, "Mechanized Tillage System, Effects on Properties of a Tropical Alfisoil in the Watershed Cropped to Maize," *Soil Tillage Research*, Vol. 6, 1985, pp. 149-162.

[7] N. R. Hulugalle, R. Lal and O. A. Opara-Nadi, "Effect of Spatial Orientation of Mulch on Soil Properties and Growth of Yam (*Dioscorea rotundata*) and Cocoyam (*Xanthosoma sagitifolium*) on an Ultisol," *J. Root Crops*, Vol. 12, 1986, pp. 37-45.

[8] P. Unger, "Conservation Tillage Systems," *Advances in Soil Sciences*, Vol. 13, 1990, pp. 27-68.

[9] R. Lal, "Effect of Slope Length on Erosion of an Alfisol in Western Nigeria," *Geoderma*, Vol. 31, No. 3, 1983, pp. 185-193. doi:10.1016/0016-7061(83)90012-5

[10] G. H. Bouyoucous, "A Recalibration of the Hydrometer Method for Making Mechanical Analysis," *Agronomy Journal*, Vol. 43, No. 9, 1951, pp. 434-438. doi:10.2134/agronj1951.00021962004300090005x

[11] E. O. Mclean, "Aluminium: In Methods of Soil Analysis," In: C. A. Black, Ed., 1965.

[12] M. L. Jackson, "Soil Chemical Analysis," Prentice Hall, New York, Agronomy No. 9, Part 2, American Society of Agronomy, Madison, 1962, pp. 978-998.

[13] R. H. Bray and I. T. Kurtz, "Determination of Total Organic and Available Forms of Phosphorus in Soils," *Soil Science*, Vol. 59, No. 1, 1945, pp. 39-46.

[14] A. Walkley and C. A. Black, "Determination of Organic Matter in the Soil Science," Vol. 37, 1934, pp. 540-556.

[15] G. Osuji, "Zero-Tillage for Erosion Control and Maize Production in the Humid Tropics," *Journal of Environmental Management*, Vol. 23, 1990, pp. 193-201.

[16] J. O. Akinyemi, O. E. Akinpelu and A. O. Olaleye, "Performance of Cowpea under Three Tillage Systems, on an Oxic Pareustaff in South Western Nigeria," 2003.

[17] D. Nangju, "Effect of Density, Plant Type and Season on Growth and Yield of Cowpea," *Journal of the American Society for Horticultural Science*, Vol. 104, No. 4, 1979, pp. 466-470.

[18] Singh and Rachiee, *International Symposium on Pests of Grain Legumes*, Ibadan, 8-13 November 1976.

[19] H. S. Nitant and P. Singh, "Effect of Deep Tillage on Dry Land Production of Red Grain (*Cajanus cajan* L.) in Central India," *Soil and Tillage Research*, Vol. 34, No. 1, 1995, pp. 17-26. doi:10.1016/0167-1987(94)00454-M

[20] S. O. Ajuwon, "Intrinsic Deficiencies of Zero Tillage," *Proceedings of the 1st National Tillage Symposium*, Ilorin, 22-25 November 1983, 25 p.

[21] W. L. Lindsay, P. L. G. Vlek and S. H. Chien, "Phosphate Minerals," In: J. B. Dixon and S. B. Weed, Eds., *Minerals in Soil Environment*, 2nd Edition, Soil Science Society of America, Madison, 1989, pp. 1089-1130.

19

Assessment of Soil C and N Stocks and Fractions across 11 European Soils under Varying Land Uses

Karolien Denef[1*], Ilaria Del Galdo[2], Andrea Venturi[2], M. Francesca Cotrufo[1,3]

[1]Natural Resource Ecology Laboratory, Colorado State University, Fort Collins, Colorado, USA; [2]Environmental Science Department, Second University of Naples, Caserta, Italia; [3]Department of Soil and Crop Sciences, Colorado State University, Fort Collins, Colorado, USA.
Email: *Karolien.Denef@Colostate.edu

ABSTRACT

In this study, we measured the stocks and pool sizes of soil organic carbon (SOC) and total soil nitrogen (TN), and their natural ^{13}C and ^{15}N abundance across a wide range of temperate European ecosystems. The objectives were to examine any distinct isotope patterns with land use or climate, and how C and N in these different ecosystems are distributed among soil organic matter (SOM) fractions to better predict soil C and N dynamics and longer term persistence. Soils were sampled to 30 cm depth at 11 sites of the Nitro Europe (NEU) network and included four forests, three grasslands and four croplands. Surface soil samples were fractionated using a combined size-density fractionation protocol separating light (LF) from heavy particulate organic matter (hPOM) by density and silt-from-clay-associated SOM by size. Down-profile natural abundance ^{15}N patterns pointed towards a closed N cycle in the forest sites, while ^{13}C patterns suggested differences in plant water use efficiency across the C3 grassland sites. The forests and grassland sites stored the majority of surface SOC and TN in the LF and hPOM pools. Sustained sequestration of C and N in these rather labile pools will rely on management practices that minimize soil disturbance and increase C input. We also found that the mineral fraction (silt and clay) in the cropland soils stored less C and N per unit of fraction mass compared to the forests and grasslands, which points towards a lower mineral-OM stabilization efficiency of cropland soils. Finally, our study revealed total POM (LF plus hPOM) as a strong predictor of SOC and TN differences, particularly among the non-cropped sites. This study shows that these sites, independent of soil type and climate, store a large fraction of C and N in POM pools that are particularly vulnerable to soil disturbance such as caused by land use change.

Keywords: Soil Organic Carbon; Soil Nitrogen; Soil Organic Matter Fractions; Natural ^{15}N and ^{13}C Abundance

1. Introduction

Globally, soils store a tremendous amount of organic carbon (~2300 Pg C) and nitrogen (~140 Pg N) [1], of which the majority is contained in the soil organic matter (SOM). The benefits of SOM for soil fertility are widely recognized due to its contribution to soil structure and aeration, nutrient and water holding capacity. The organic matter content of a soil reflects the balance between plant, animal, microbial and erosional inputs, and losses due to mineralization, leaching and erosion. It is therefore a dynamic property, greatly vulnerable to land use and climate [2] and with important feedbacks to the atmospheric green house gas (GHG) balance and the rate of climate change [3]. The increased demand for accurate soil C and N stock assessments and predictions of C and N changes as a result of land use/cover and climate change has triggered large-scale and long-term measurements of soil C and N stocks and pools globally [4-6]. Such empirical studies provide critical information to quantify the response of soil properties to management and changing climate across regions, ultimately supporting climate change and farm policy, food security, and overall ecosystem health. Monitoring soil C and N stocks is imperative not only for understanding how soils change in response to land use and shifting climate patterns, but also for validating and reducing the uncertainty around estimates of biogeochemical prediction models as Century [7], DayCent [8], Roth C [9] and DNDC [10] commonly used in regional and national GHG inventories. Model outputs typically come with large uncertainties [11,12], which is at least

*Corresponding author.

partly due to the still limited knowledge about ecosystem processes and properties involved in C and N cycling [13]. Consequently, direct measurements by repeated soil inventories are urgently needed to improve our understanding of C and N dynamics and further constrain model estimates [6,12].

Whole soil C and N stock changes are difficult to detect in the short term as the bulk of SOM is stabilized and has turnover times measured in hundreds to thousands of years [14]. This would require long term measurements to capture modest changes in a robust manner [15]. Soil organic matter fractionations can be helpful in studying long-term C and N dynamics, as they can reveal early indications of changes in SOC and TN stocks [16-19] hence predict how ecosystems will respond to a particular change. A number of physical fractionation methods have been successful in terms of isolating more active dissolved (DOM) and particulate (POM) organic matter pools that are likely to participate in short term soil processes (e.g. nutrient cycling) and respond more rapidly to land use or management changes, from more passive mineral-associated SOM pools that are relatively recalcitrant and more resistant to disturbance [20-22]. Structural analyses have revealed unique molecular characteristics of these SOM fractions based on their degree of decomposition, the dominance of different chemical classes, and the relative contributions from plant and microbially-derived C linking their physical characteristics (*i.e.* how they are operationally defined) to their ecological function [23]. Particulate organic matter fractions, obtained either by particle size (often >53 μm) or density (usually 1.3 - 1.8 g cm^{-3}) separation, are composed primarily of identifiable and young plant litter in various stages of decomposition that is chemically similar to its source material, whereas SOM associated with silt and clay minerals (usually < 53 μm), occluded within aggregates, or associated in higher density organo-mineral complexes, is generally further decomposed and more microbially altered [24,25], containing a greater proportion of microbial products and more resistant C as compared with larger or lighter fractions [26]. Assessing how the bulk of SOM in different ecosystems is distributed among these ecologically meaningful fractions can provide insights into SOM behavior upon management or land use change. This will ultimately help decision-making on best management practices to maintain or increase soil C and N. For example, Stewart *et al.* [21] stressed the importance of conservation management (reduced disturbance and increased C input) in the Virginia Coastal Plains to avoid rapid losses of SOC, as most of the change in SOC upon changes in management was observed in the active POM pools.

Soil ^{13}C and ^{15}N natural abundance measurements are another useful tool forinferring changes in C and Ndynamics and reducing uncertainty in estimates of soil C storage. The natural abundance of ^{13}C and ^{15}N in SOM has been shown to be an integrative measure of the ecosystem factors and processes that produced it, and its measurement has yielded insight into the dominant processes guiding the biological and physical dynamics in a soil system [e.g., 27]. Soil ^{13}C values are closely related to vegetation cover (C3 versus C4 plants) and moisture availability. They can inform about historic vegetation or climatic shifts, and can be used to measure SOC turnover [28]. Mineralization and processes associated with SOM formation induce additional variations in the soil^{13}C abundance, contributing to the often observed ^{13}C enrichment with soil depth [29,30]. Soil ^{15}N values reflect soil N sources and their fractionation during N transformation [31,32]. Generally, closed N cycles (*i.e.* limited N losses) are expressed by ^{15}N patterns with lower ^{15}N at the soil surface and relatively higher ^{15}N with depth, as has been found in temperate forests not subject to significant N inputs, whereas ^{15}N enriched soils suggest open N cycles with greater N losses via fractionating pathways such as nitrification and denitrification [31,32]. Nitrogen fertilizer or atmospheric deposition inputs may complicate the interpretation of natural abundance ^{15}N gradients in soil [33].

The aim of our investigation was to assess how C and N in different ecosystems are distributed among SOM fractions to predict longer term persistence and SOM behavior when subject to disturbance, such as changes in climate, management or land use. To do this we 1) assessed SOC and TN stocks as well as ^{13}C and ^{15}N natural abundance of 11 soil profiles (0 - 0 cm) across Europe, and 2) evaluated the distribution and variability of SOC and TN among SOM fractions in these temperate soils of varying land uses. Land use affects SOC and TN stocks, their distribution among SOM fractions and their ^{13}C and ^{15}N values across the soil profile. Our hypothesis was that forests and grasslands would store more C and N in the more labile POM fractions compared to croplands. We also expected a more uniform ^{13}C and ^{15}N pattern across the soil profile of the cropland sites due to tillage, and for the forest sites, a ^{15}N pattern indicative of a closed N cycle, with depleted ^{15}N values in the surface soils as compared to ^{15}N values at deeper depths.

2. Materials and Methods

2.1. Site Description

Soil cores were collected from 11 sites across Europe during spring and summer of 2007 representing a variety of climates, land-use and soil properties. The sites cover different climatological zones (**Figure 1**) from Finland in the north to Italy in the south, and from the UK in the west to Hungary in the east [34,35] and are part of the Nitro Europe Level-3 "Super Sites" network. Addition-

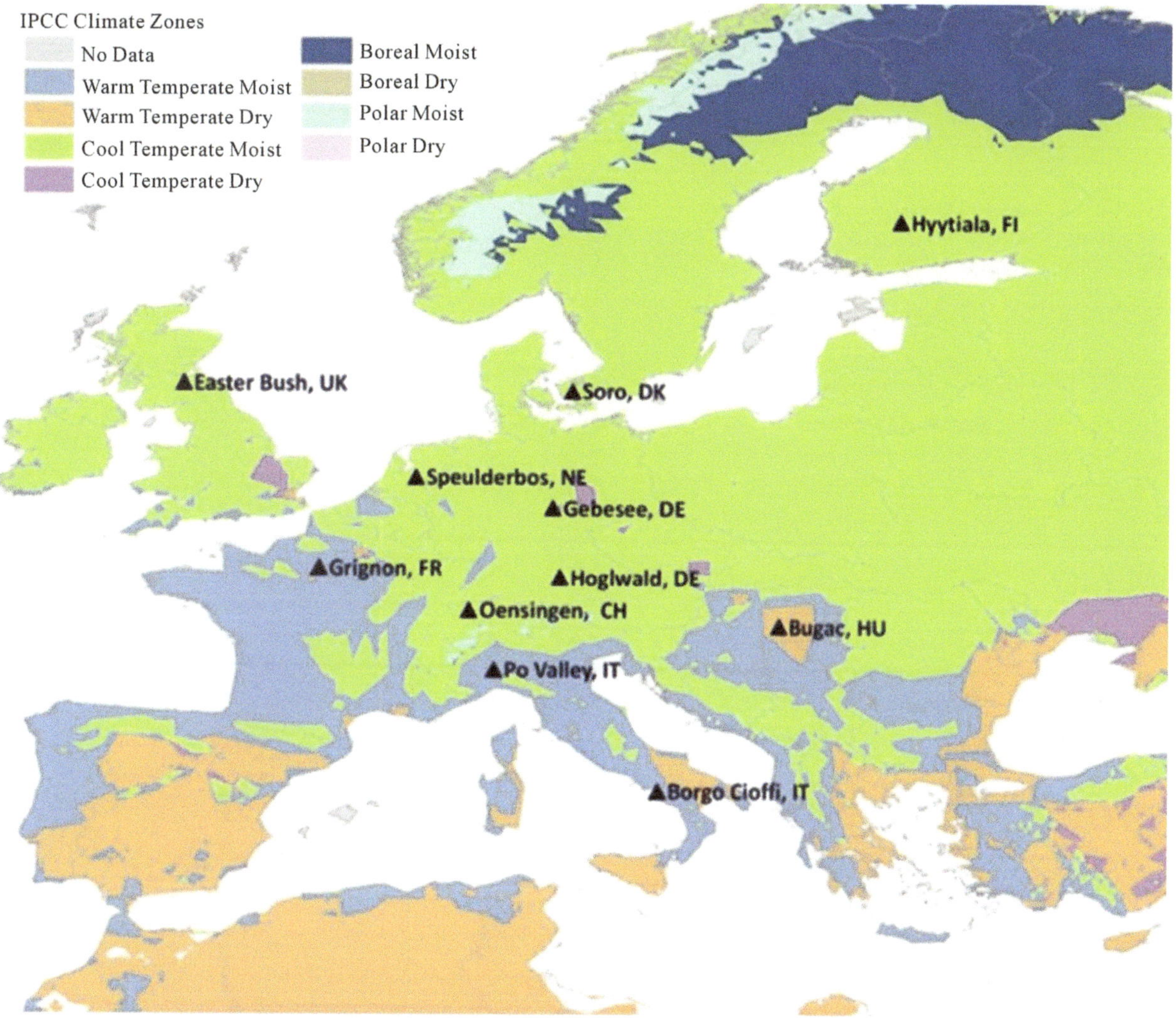

Figure 1. Map with location of sampled sites and IPCC climate zones.

ally, the sites also differ in terms of vegetation, soil properties (e.g. bulk density, texture, pH) and land use, including four forests, three grass lands and four croplands. All croplands and grasslands, except for the grassland in Bugac, are fertilized with different amounts of fertilizer. Grasslands are either grazed or mowed. Forest sites differ in tree species of coniferous or deciduous type. The main site characteristics are summarized in **Table 1**. Additional information can be found in Schaufler *et al.* [36] and Skiba *et al.* [35].

2.2. Soil Sampling and Preparation

To allow for data comparison and integration with other soil and gas measurements at the NEU Level-3 "Super Site" network [35], soil samples were collected in proximity to the six replicate flux measurement chambers which were randomly placed at each site (for details, see [37]). Around each chamber, after removing the litter layer (where present), one soil core was taken at each of the four corners of a 16 m^2 area to a depth of 30 cm with a 5 cm diameter sample probe. Each of the four soil cores was split into four depths (0 - 5 cm, 5 - 10 cm, 10 - 20 cm and 20 - 30 cm) and composited by depth, resulting in one composite sample for each depth for each replicate area (n = 6) per site. Soil samples of all depths were sieved to 2 mm and oven-dried (105˚C) prior to further analyses. One additional soil core was taken at each sampling location for bulk density measurement. Bulk density was measured for each of the four depths, as the ratio of the dry (105˚C) weight of the soil within that depth divided by the volume, measured from the diameter and height of the core, after correcting for any rocks or coarse fragments. No significant core compaction was observed during sampling.

2.3. Bulk Soil Analyses

Oven-dried soil samples were finely ground and analyzed for SOC, TN and isotopic composition (δ^{13}C [‰] and

Table 1. Sampling sites with information on land use, vegetation, location, climate and soil characteristics.

Site	Site code	Vegetation	Site location	Elevation /m a.s.l.	MAT[a]/°C	MAP[b] /mm	Soil texture[c] Sand-silt-clay /%	pH	N-fert[d] /kg ha^{-1} y^{-1}
Cropland sites									
Borgo Cioffi, Italy	IT-BCi	*Zea mays/Medicago sativa/Foeniculum vulgare/Lolium perenne*	40°31'N 14°57'E	15	19.0	500	30-22-48	7.2	545
Castellaro, Italy	IT-Cas	*Zea mays/Oryza sativa*	45°12'N 9°40'E	87	12.7	980	28-44-29	6.9	300
Gebesee, Germany	DE-Geb	*Beta vulgaris L./Solanum tuberosum/Brassica napus*	51°06'N 10°55'E	158	9.7	510	26-61-13	6.2	31
Grignon, France	FR-Gri	*Sinapis alba/Zea mays/Triticum spp./ Hordeum vulgare*	48°51'N 01°58'E	73	11.5	600	30-53-17	6.7	175
Forest sites									
Hyytiälä, Finland	FI-Hyy	*Pinus sylvestris*	61°51'N 24°17'E	181	3.9	710	56-16-18	3.6	-
Höglwald, Germany	DE-Hog	*Picea abies*	48°30'N 11°11'E	530	8.5	890	54-32-12	3.1	-
Sorø, Denmark	DK-Sor	*Fagus sylvatica*	55°29'N 11°38'E	40	8.3	730	48-26-26	3.4	-
Speulder Bos, The Netherlands	NL-Spe	*Pseudotsuga menziesii/ Quercus robur*	52°22'N 05°32'E	27	9.4	460	81-18-1	2.9	-
Grassland sites									
Bugac, Hungary	HU-Bug	*Festuca spp.*	46°41'N 19°36'E	113	10.5	500	93-3-4	6.9	16
Easter Bush, UK	UK-EBu	*Lolium perenne*	55°52'N 03°12'W	271	9.6	850	53-26-20	4.8	171
Oensingen, Switserland	CH-Oen	*Lolium perenne, Trifolium repens*	47°17'N 7°44'E	452	9.0	1100	10-47-43	5.7	230

[a]MAT: mean annual temperature; [b]MAP: mean annual precipitation; [c]Sand: 53-2000 μm; silt: 2-53 μm; clay < 2 μm; [d]N fertilizer inputs from N fertilization (IT-BCi, IT-Cas, DE-Geb, FR-Gri, UK-EBu, CH-Oen) or grazing animals (HU-Bug).

$\delta^{15}N$ [‰]) using an elemental analyzer (EA) (Thermo Finnigan: EA 1112) coupled with an isotope ratio mass spectrometer (IRMS) (Thermo Finnigan: Deltaplus). Carbonates were removed prior to EA and IRMS analysis by exposure to HCl vapor for 6 hours [38].

2.4. Soil Fractionation

All surface samples (0 - 5 cm) were fractionated using a combined size-density fractionation protocol, adopted from Marzaioli *et al.* [39]. This method separates total light fraction (LF) from heavy particulate organic matter (hPOM) by density separation and silt-from clay-associated SOM by size. In brief, 2 mm sieved soil samples underwent a density separation in sodium polytungstate (SPT; ρ = 1.85 g cm^{-3}). This step allows the separation of LF (floats in the SPT) from heavy fraction (settles out in the SPT). The heavy fraction, which is composed of aggregates, sand and POM, was dispersed by glassbeads in de-ionized water, and wet sieved on a 53 μm mesh sieve, to separate the hPOM plus sand fraction (retained on the 53 μm screen) from the silt plus clay fraction. Silt and clay sized fractions were separated by means of wet centrifugation (127 g for 7 min for silt, and 1730 g for 15 min for clay following addition of flocculant, *i.e.* 0.25 M $CaCl_2$-$MgCl_2$). All recovered fractions were analyzed for SOC and TN after carbonate removal [38], as described above.

2.5. Statistical Analyses

To test the assumption of normality, we examined QQ plots of the quantiles of studentized residuals versus standard normal quantiles. Soil organic C and TN stocks and concentrations were root-transformed to achieve normality. Statistical tests were performed using transformed data, but non transformed values were used to report average values in tables and figures. The effects of land use, climate (mean annual temperature (MAT) and precipitation (MAP)), soil properties (pH, texture) and their interactions on SOC and TN stocks were analyzed using the ANCOVA model with the proc mixed module in SAS 9.2 (SAS Institute). The effects of land use on fraction C and N content were analyzed using the ANOVA model with the proc mixed module. The variable "site within land use" was considered a random effect in all analyses to resolve non-independencies and to account for the by-site variation. Separation of means was tested with Tukey's honestly significant difference at a significance level of 0.05. Linear relationships between total SOC and TN content and individual C and N fractions, isolated by the different fractionation methods, were tested using ordinary least squares linear regression.

3. Results

3.1. Bulk Soil Organic Carbon and Total Nitrogen

Soil organic C and TN stocks in the 0 - 30 cm depth ranged between 3.80 and 11.19 kgm^{-2} for SOC and 0.17 and 0.90 kgm^{-2} for TN and differed greatly among sites, even within a land use type (**Table 2**; see also **Tables 3** and **4** for individual depths). The highest SOC and TN stocks were found at the Bugac grassland site and Sorö forest site, respectively, while the lowest SOC and TN stocks were found at the Borgo Cioffi cropland site and Hyytiälä forest site, respectively. Over the entire 30 cm depth, SOC and TN stocks did not significantly differ with land use, climate (MAT, MAP), soil properties (pH, texture) or their interactions.

Significant differences between land use types were observed for SOC and TN concentrations at the different depths (**Figure 2**; **Table 5**). The SOC and TN concentration in the surface 0 - 5 cm depth was significantly lower in cropland sites compared to grassland (SOC: $P = 0.0085$; TN: $P = 0.05$) and forest sites (SOC: $P < 0.0001$). At the deepest depth (20 - 30 cm), TOC and TN concentrations were significantly lower in the forest sites compared to grassland and cropland sites ($P < 0.05$). SOC and TN concentration significantly decreased with depth in the grassland and forest sites ($P < 0.0001$), but not in the cropland sites (**Figure 2**).

3.2. Bulk Soil 13C and 15N

Delta ^{13}C values ranged between −29.9‰ and −20.6‰, whereas $\delta^{15}N$ values ranged between −2.1‰ and 11.0‰ (**Figure 3**). The highest $\delta^{13}C$ and $\delta^{15}N$ values were found at the Borgo Cioffi cropland site, while the lowest $\delta^{13}C$ and $\delta^{15}N$ values were found at the Hyytiälä and Speulderbos forest sites, respectively. At all sites, bulk soil became enriched in ^{13}C up to several parts permil with depth (average increase 2.4‰). Similar results were found for ^{15}N in grassland (as high as 3.6‰ increase with depth) and forest sites (as high as 10.4‰ increase with depth), but for cropland sites, the ^{15}N signature of the soil did not significantly change with depth (**Figure 3**). Land use significantly affected the ^{15}N ($P = 0.0147$) but not the ^{13}C signature of the bulk soil (**Table 5**). The land use effect on ^{15}N was dependent on depth, with forest soils being more depleted in ^{15}N in the 0-5 cm depth (as low as −2.1‰ at Speulderbos) compared to the cropland ($P < 0.0001$) and grassland sites ($P = 0.0079$). At deeper depths, no significant differences among land use types were found in the isotopic composition of the SOC and TN.

3.3. Carbon and Nitrogen Content of Soil Organic Matter Fractions

Soil organic C and TN were differently distributed among the isolated SOM fractions at the different sites (**Figure 4**). In the majority of forest and grassland soils, total POM (LF plus hPOM) accounted for most of the SOC and TN. Over 60% of SOC and TN were found in total POM in the forest soils at Hyytiälä, Hogwald, Sorö and Speulderbos, and even over 90% were found in the grassland soils at Bugac and Easter Bush (**Figure 4**). The grassland site at Oensingen, on the other hand, had the majority of SOC and TN (80%) stored in the mineral fractions (silt and clay). Cropland soils stored most of their SOC and TN in the mineral fractions, with the exception of the cropland soil at Grignon, which stored nearly 70% of SOC and TN in its POM fractions (**Figure 4**).

When fraction C and N were expressed on a per fraction mass basis, the silt and clay fractions were significantly less enriched in SOC and TN in all the cropland sites (Borgo Cioffi, Castellaro, Gebesee and Grignon) compared to the grassland and forest sites (**Figure 5**; **Table 5**). At most sites, more SOC and TN were associated with clay compared to silt minerals (**Figure 5**).

3.4. Relationship between Soil Organic Matter Fractions and Total Carbon and Nitrogen

In order to evaluate the potential of the isolated SOM fractions for predicting SOC and TN changes, we correlated each individual fraction's C and N content against the SOC and TN calculated combining all fractions. A significant linear relationship was found between SOC (and TN) and total POM (LF plus hPOM) C (and N) when all sites were included as well as within a land use type, with the predictive strength (R^2) being much greater for the non-cropped sites ($R^2 > 0.85$) compared to the croplands ($R^2 = 0.46$ for TN; $R^2 = 0.71$ for TC) (**Table 6**). No reasonable relationship was observed when correlating total SOC and TN with any of the other fractions (**Table 6**), suggesting that total POM is the largest con-

Table 2. Total 0 - 30 cm soil organic carbon (SOC) and nitrogen (TN) stocks (kg m^{-2}). Values represent averages of 6 field replicates, with standard deviation in brackets.

Land Use	Site	SOC stock/kg m^{-2}	TN stock/kg m^{-2}
Cropland	IT-BCi	3.80 (0.56)	0.40 (0.08)
	IT-Cas	7.93 (1.13)	0.73 (0.11)
	DE-Geb	8.61 (0.33)	0.78 (0.03)
	FR-Gri	9.29 (0.49)	0.89 (0.07)
Forest	FI-Hyy	4.12 (0.65)	0.17 (0.04)
	DE-Hog	4.33 (1.26)	0.24 (0.10)
	DK-Sor	8.42 (2.05)	0.90 (0.16)
	NL-Spe	9.28 (1.00)	0.30 (0.07)
Grassland	HU-Bu	11.19 (0.83)	0.86 (0.07)
	UK-Ebu	8.77 (1.14)	0.63 (0.13)
	CH-Oen	7.64 (0.28)	0.77 (0.03)

Table 3. Soil organic carbon (SOC) stocks (kg m^{-2}) at 0 - 5 cm, 5 - 10 cm, 10 - 20 cm and 20 - 30 cm depths. Values represent averages of 6 field replicates within a site, with standard deviation in brackets.

Land Use	Site	0 - 5 cm/kg m^{-2}	5 - 10 cm/kg m^{-2}	10 - 20 cm/kg m^{-2}	20 - 30 cm/kg m^{-2}
Cropland	IT-BCi	0.54 (0.05)	0.64 (0.19)	1.40 (0.17)	1.24 (0.01)
	IT-Cas	1.16 (0.37)	1.34 (0.38)	2.17 (0.36)	3.26 (0.70)
	DE-Geb	1.28 (0.09)	1.45 (0.22)	2.90 (0.27)	2.98 (0.22)
	FR-Gri	1.67 (0.17)	1.46 (0.28)	3.43 (0.18)	2.79 (0.18)
Forest	FI-Hyy	0.99 (0.44)	1.00 (0.20)	1.34 (0.58)	0.77 (0.31)
	DE-Hog	1.90 (0.81)	0.80 (0.23)	0.87 (0.14)	0.76 (0.23)
	DK-Sor	2.42 (0.45)	2.19 (0.80)	2.72 (0.93)	1.09 (0.42)
	NL-Spe	3.92 (0.55)	1.70 (0.30)	2.30 (0.38)	1.36 (0.19)
Grassland	HU-Bu	2.85 (0.88)	2.10 (0.28)	4.98 (0.56)	2.14 (0.75)
	UK-Ebu	1.92 (0.39)	1.92 (0.52)	2.72 (0.33)	2.21 (0.63)
	CH-Oen	1.54 (0.25)	1.41 (0.20)	2.35 (0.31)	2.33 (0.09)

Table 4. Total soil nitrogen (TN) stocks (kg m^{-2}) at 0 - 5 cm, 5 - 10 cm, 10 - 20 cm and 20 - 30 cm depths. Values represent averages of 6 field replicates within a site, with standard deviation in brackets.

Land Use	Site	0 - 5 cm/kg m^{-2}	5 - 10 cm/kg m^{-2}	10 - 20 cm/kg m^{-2}	20 - 30 cm/kg m^{-2}
Cropland	IT-BCi	0.059 (0.009)	0.009 (0.067)	0.067 (0.022)	0.022 (0.142)
	IT-Cas	0.117 (0.037)	0.037 (0.130)	0.130 (0.037)	0.037 (0.195)
	DE-Geb	0.117 (0.010)	0.010 (0.126)	0.126 (0.009)	0.009 (0.280)
	FR-Gri	0.154 (0.013)	0.013 (0.132)	0.132 (0.022)	0.022 (0.339)
Forest	FI-Hyy	0.031 (0.012)	0.012 (0.035)	0.035 (0.007)	0.007 (0.053)
	DE-Hog	0.096 (0.065)	0.065 (0.044)	0.044 (0.018)	0.018 (0.052)
	DK-Sor	0.352 (0.176)	0.176 (0.196)	0.196 (0.095)	0.095 (0.177)
	NL-Spe	0.154 (0.040)	0.040 (0.049)	0.049 (0.012)	0.012 (0.063)
Grassland	HU-Bu	0.291 (0.172)	0.172 (0.169)	0.169 (0.038)	0.038 (0.290)
	UK-Ebu	0.153 (0.028)	0.028 (0.134)	0.134 (0.032)	0.032 (0.189)
	CH-Oen	0.147 (0.024)	0.024 (0.148)	0.148 (0.022)	0.022 (0.227)

tributor to surface soil C and N differences across the different sites.

4. Discussion

Recent research efforts supported by the European Union (e.g., CarboEurope and NitroEurope) have initiated large scale assessments of the C and N fluxes from soils across Europe. Our investigation contributes to this effort by providing SOC and TN stocks for 11 NitroEurope flux measurement sites across Europe [34], and allows for a more in-depth evaluation of how soil C and N is stabilized in these systems by evaluating the distribution of C

Table 5. Summary of the results of the analysis of variance on SOC and TN concentrations, SOC and TN distribution among fractions (%), and ^{13}C and ^{15}N.

Source of variation	Degrees of freedom	*P*
SOC concentration		
Land use	2	0.09
Depth	3	<0.0001
Land use x depth	6	<0.0001
TN concentration		
Land use	2	0.13
Depth	3	<0.0001
Land use x depth	6	<0.0001
^{13}C		
Land use	2	0.2107
Depth	3	<0.0001
Land use x depth	6	0.0314
^{15}N		
Land use	2	0.03
Depth	3	<0.0001
Land use x depth	6	<0.0001
TOC concentration in silt and clay fractions		
Land use	2	0.0003
Fraction	1	<0.0001
Land use x fraction	2	<0.0001
TN concentration in silt and clay fractions		
Land use	2	0.003
Fraction	1	<0.0001
Land use x fraction	2	<0.0001

Table 6. Statistical results for linear regressions between total and individual fraction SOC and N in the 0 - 5 cm soil depth.

Fraction-C or -N	R^2	*P*	R^2	*P*
	Versus total fraction SOC		Versus total fraction TN	
Total POM	0.84	<0.0001	0.93	<0.0001
Cropland	0.71	<0.0001	0.46	0.0003
Forest	0.91	<0.0001	0.91	<0.0001
Grassland	0.85	<0.0001	0.92	<0.0001
Total mineral (silt + clay)	0.17	0.0006	0.01	ns
LF	0.44	<0.0001	0.26	0.0005
hPOM	0.47	<0.0001	0.02	ns
Silt	0.16	0.0008	0.003	ns
Clay	0.11	0.007	0.0008	ns

and N among SOM fractions. This level of detail can inform on how SOM will behave when ecosystems undergo changes in management or land use, ultimately helping decision-making on how to best preserve or increase soil C and N. Furthermore, our investigation serves the broader objectives of the NitroEurope (NEU) project, by providing a dataset that can be integrated with other soil and gas flux measurements, enabling full C and N budget assessments at the different measurement sites. Such assessments will generate extensive empirical field data from widely different ecosystems across Europe which can be used in the validation of biogeochemical process models such as DNDC, RothC or DayCent.

4.1. Total SOC and TN Stocks

In the mineral soil surface (0 - 5 cm), SOC and TN stocks were significantly lower in cropland compared to the other land use systems, as has been found in other studies [e.g.,16,40]. This was a direct result of lower soil C and N concentrations and not just a mere effect of differences in bulk density (**Figure 2**). The loss of surface SOC and TN in cropped soils has been linked to lower C inputs to the soil, the removal of crop residues, reduced vegetation cover, enhanced mineralization of SOM, erosion and deteriorated soil aggregation. Whereas SOC and TN concentrations were rather uniform across the 0 - 30 cm profile in the croplands, a clear stratification with depth was observed in the non-cropped systems. Similar differences in profile distribution of SOC among different land uses (*i.e.* more stratified with depth under conservation management than under conventional cropping) were reported by Franzluebbers [41] using an extensive soil survey dataset in Georgia USA, and suggests that SOC stratification should be viewed as an improvement in soil quality.

For assessing whole-ecosystem responses to management, deeper soil depths need to be included to avoid underestimation of total SOC and TN stocks in intensively managed ecosystems [42]. No significant land use effect on SOC and TN stocks could be discerned when considering the entire 0 - 30 cm depth. Against our expectations, we even found higher SOC stocks in a few cropland sites than in some of the forest and grassland soils. This is in contrast to other inventories reporting distinct differences of SOC stocks between cropland and grassland/forest. For example, the European SOC inventories reviewed by Wiesmeier *et al.* [43] that were restricted to the 0 - 30 cm depth, showed on average 40% and 43% lower SOC stocks for cropland compared with grassland and forest, respectively. The large differences across the sites in our study in terms of soil properties, climate, vegetation and management history could have obscured any effects of land use on total profile SOC and TN stocks. Also, our investigation excluded the litter layer. In forests, the O-horizon can account for a large portion of the total SOC stock. For forests in South-East Germany, Wiesmeier *et al.* [43] found 2.5 kg m^{-2} SOC to be stored in the O horizon, or about 35% of what was stored in the A, B and C horizons combined. In addition, C and N contained in litter fragments > 2 mm were not included

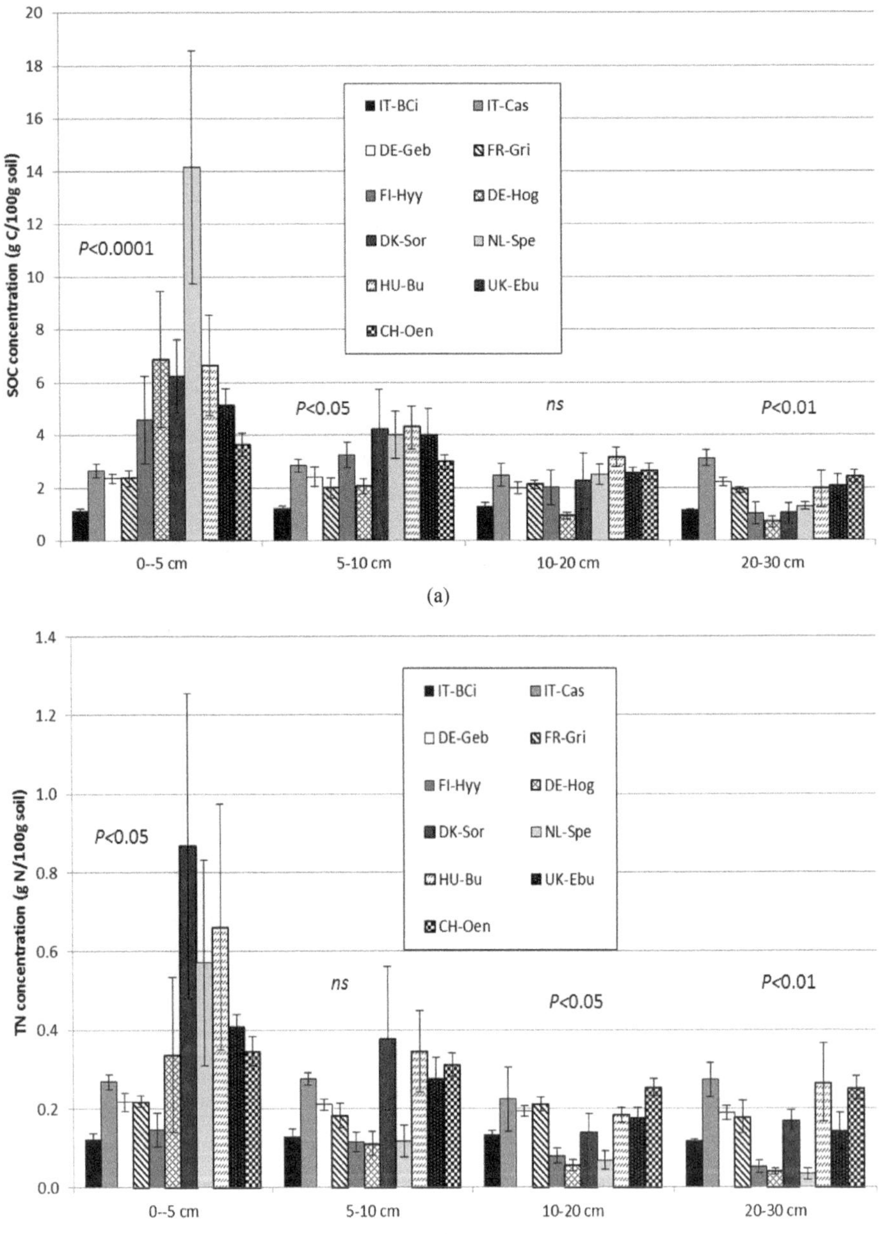

Figure 2. Soil organic carbon (a) and total nitrogen (b) concentrations (g C or N/100 g soil) in each sampling depth. Bars and error bars represent means and standard deviations of six replicate samples within a site. Sites code is as in Table 1. Statistical results are presented for land use effects per depth.

in this study, but could contribute to a significant part of the stocks in grassland and forest ecosystems.

Unlike other continental studies [e.g., 2], no clear relationships were found between SOC and TN stocks and any of the climate or soil parameters available for these sites, likely due to the relatively large number of influencing parameters compared to the number of sites included in this study. The highest SOC stocks were found at the Bugac grassland site, which is characterized by a dry-warm continental climate and low-intensity management in terms of grazing and N inputs. These conditions could promote high C inputs while keeping SOM decomposition rates relatively low. The lowest TN stocks were found at the Hyytiälä coniferous forest site, which is characterized by severe N limitation potentially as a result of the low N deposition compared to other forest sites [36,44].

4.2. 13C and 15N Abundance

The ^{15}N pattern in the forest soils with negative ^{15}N in the surface mineral soils and ^{15}N enrichment at deeper

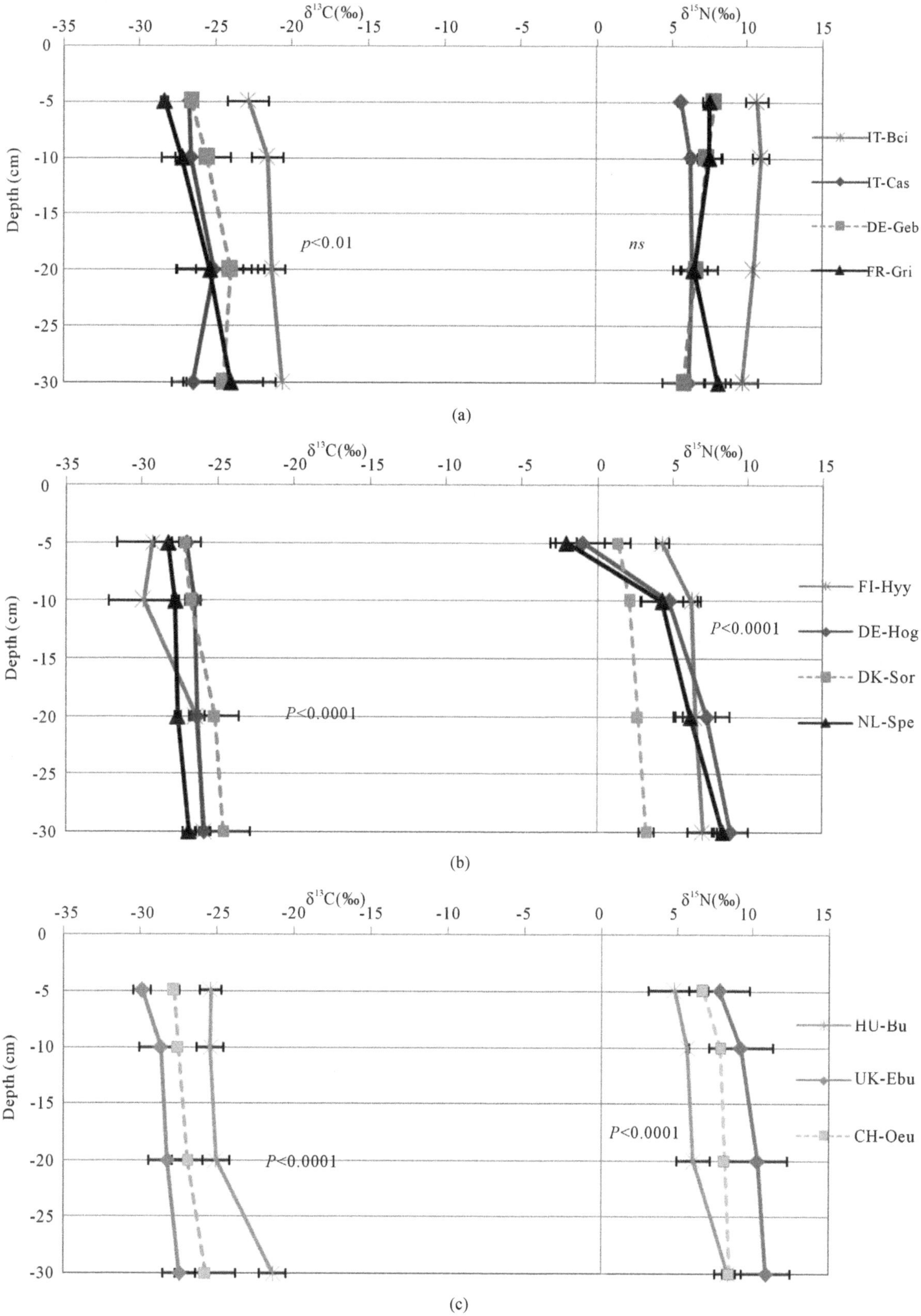

Figure 3. Delta ^{13}C (‰) and delta ^{15}N (‰) values across depths for cropland (a), forest (b) and grassland (c) sites. Symbols and error bars represent means and standard deviations of six replicate samples within a site. Sites code is as in Table 1. Statistical results are presented for depth effects per land use.

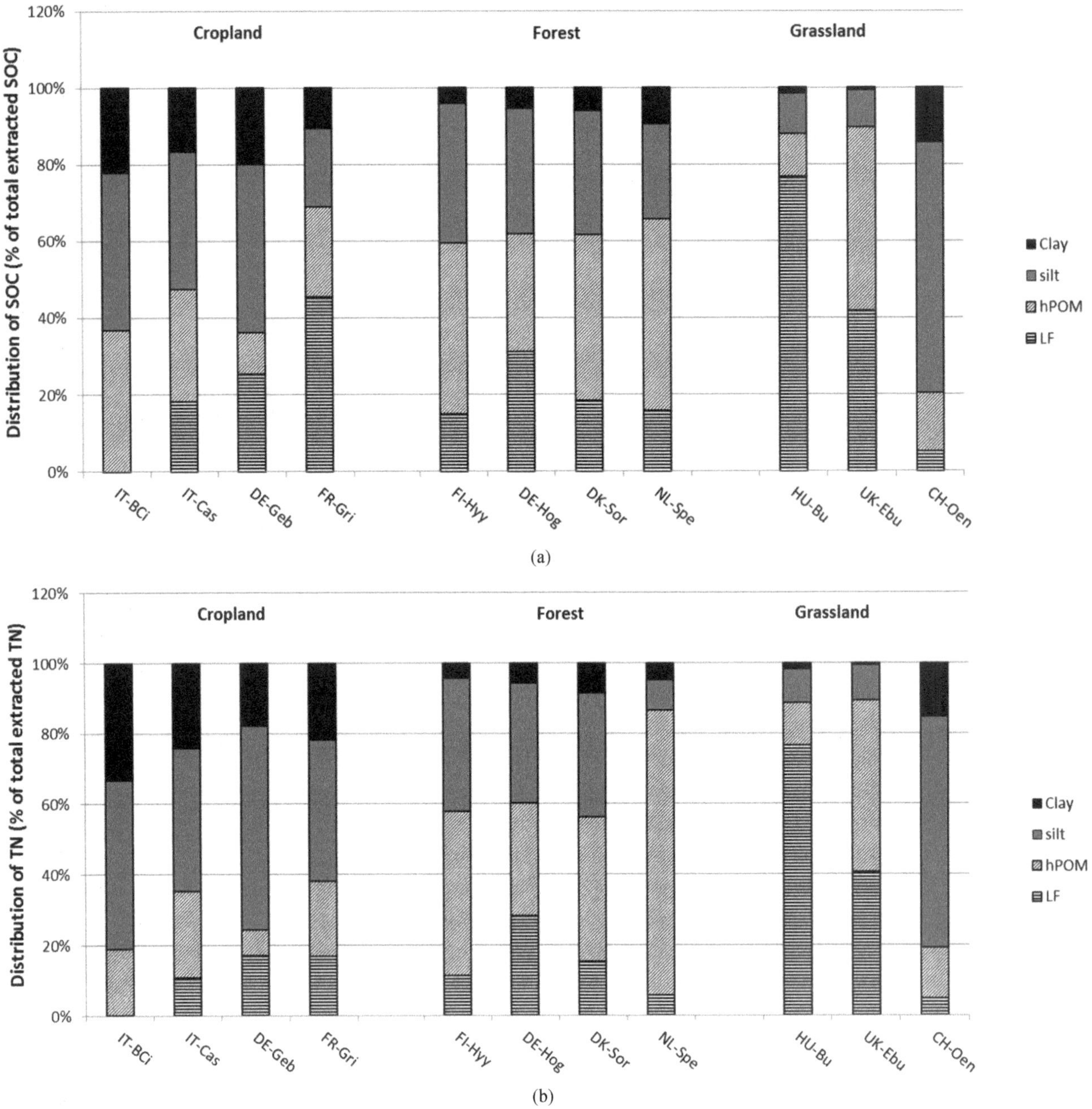

Figure 4. Proportion of soil organic carbon (a) and total nitrogen (b) (%) in isolated soil organic matter fractions (light fraction, LF; heavy particulate organic matter, hPOM; silt; clay) in 0 - 5 cm depth. Bars represent means of six replicates within a site. Sites code is as in Table 1.

depths supports our hypothesis of a closed N cycle as has been found in many other temperate forests [31,32]. The higher ^{15}N values in the surface layers of the grassland and cropland soils suggest greater losses of soil N at these sites through processes which discriminate against ^{15}N in soils. In systems that receive N inputs from fertilizer or animal grazing, soil ^{15}N values may also be indicative of the dominant source of soil N. The more enriched ^{15}N values at Easter Bush compared to the other grassland sites could suggest greater N inputs from animal manure [45] due to the more intense grazing at this site [36]. Similarly, the greater ^{15}N values at Borgo Cioffi compare to all other cropland sites could be a result of the use of animal waste as an organic fertilizer for the winter fodder crop at this site [46] owing to the preferential volatilization of ^{14}N ammonia from manure [45].

^{15}N enrichment with soil depth, as found in all forest and grassland sites, is a common observation which has been related to different mechanisms, including ^{15}N isotope discrimination during microbial N transformations, differential preservation of ^{15}N-enriched SOM components during N decomposition, and more recently to the

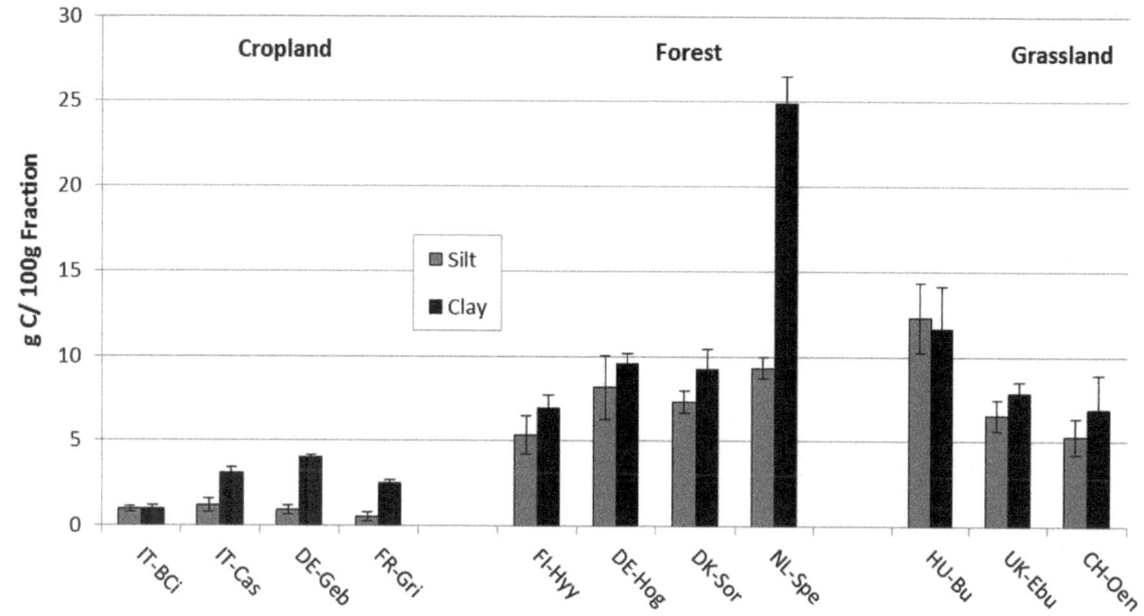

(a)

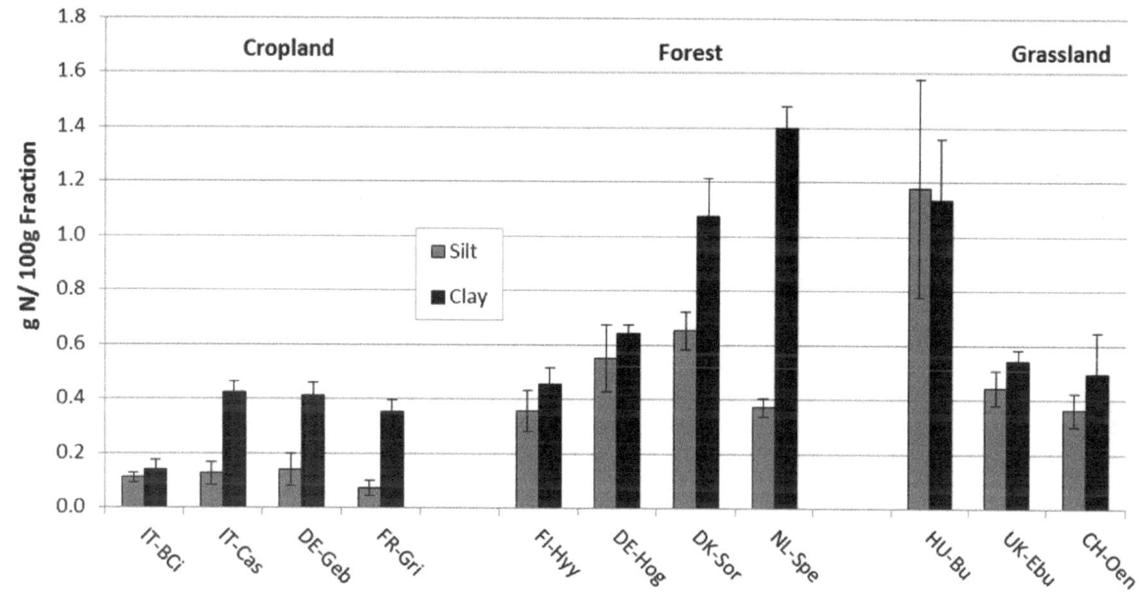

(b)

Figure 5. Concentration of soil organic carbon (a) and total nitrogen (b) in isolated silt and clay fractions (g C or N/100 g fraction). Bars and error bars represent means and standard deviations of six replicate samples within a site. Sites code is as in Table 1.

buildup of microbial ^{15}N-enriched microbial necromass [47]. The lack of ^{15}N variation within the soil profiles of the cropland soils are probably due to mixing of soil layers by plowing. However, this mixing effect was not apparent for ^{13}C, which showed a slight enrichment with depth in all sites, independent of land use. The observed variation in ^{13}C within a depth layer among the cropland sites is likely a result of the different amount of C4 relative to C3 plant C inputs at these sites. The ^{13}C of the grassland sites, all cultivated with C3 grasses, increased from wetter to drier sites, which could be driven by soil moisture differences and associated differences in plant water use efficiency. Plants grown under water stress become more water-use efficient and exhibit higher ^{13}C values than plants grown under adequate moisture conditions [48]. The highest ^{13}C values were found at Bugac, which is characterized by a warm temperate dry climate (**Figure 1**), likely causing higher evapotranspiration and water use efficiency (hence lower ^{13}C discrimination in C3 plants) than at the cool temperate moist sites at Easter

Bush and Oensingen.

4.3. C and N Distribution among SOM Fractions

The soil fractionation data revealed some interesting and clear trends across the different land-use systems, despite the lack of clear trends in the whole-soil, and is a primary reason to fractionate soils [49]. Undisturbed land uses, such as forest and grassland sites (with the exception of Oensingen) stored the majority of SOM in the POM fractions (LF and hPOM), whereas cropland sites (with the exception of Grignon) stored relatively more SOM in the mineral (silt and clay) fractions. Similar results were reported by Franzluebbers & Stuedemann [50] who found lower POM-C as a proportion of TOC in cropland comred to pasture and forest. These observations suggest that less disturbed ecosystems accumulate relatively more C and N in POM fractions, likely as a result of higher C inputs and slower decomposition rates. The distribution of TOC and TN at the grassland site in Oensingen was more similar to that of the croplands with the majority (80%) contained in the mineral fraction. This may be due to the history of intense cropping at this site, with conversion to grass-clover only in 2001, *i.e.* six years prior to soil sampling [51].

Another possible cause for the relatively higher proportion of mineral-associated OM at Oensingen and most cropland sites compared to the other sites could be due to differences in soil texture. Much finer soil textures were found in the cropland sites (between 70% and 74% silt and clay) and Oensingen (90% silt and clay) than at the other sites, hence more mineral surfaces are available in these soils for SOM to accumulate. In fact, a negative relationship between soil texture, as represented by whole-soil silt plus clay content, and the percentage of organic C stored in the POM fraction was observed across all study sites ($P = 0.007$; $r^2 = 0.58$).

Our results are in contrast to other studies, where the mineral-associated OM fraction accounted for most of the total C content independent of land use [e.g., 40]. Relatively high POM proportions have been reported for grassland and forest soils but typically not more than 40% [22,50,52]. This suggests that the European ecosystems included in our study appear to store large amounts of C and N in a rather labile SOM fraction that could potentially be rapidly lost if soils were to be disturbed (e.g., when converted to cropland) [22,50]. The large amounts of POM-C and -N in these systems could also suggest that the mineral (silt and clay) fraction in these systems is C and N saturated [53-55] and cannot store additional C and N inputs. This saturation idea is supported by the higher C and N concentrations in the silt and clay fractions on a per-fraction mass basis in the grassland and forest soils than in the cropland soils (**Figure 5**), suggesting that the non-cropped systems were closer to or at saturation. To further explore this idea, we calculated the theoretical soil C protective capacity of the different sites based on the linear regressions between soil texture and mineral (silt + clay) C content (g kg^{-1} soil) developed by Six *et al.* [54] for different land uses and clay mineralogies. We used the <50 μm regression equations of Six *et al.* [54] for cultivated, grassland and forest to calculate the silt + clay protective capacity (g C kg soil^{-1}) from <53 μm silt + clay contents obtained in the soil textural analyses performed on our soils (**Table 6**). These values were then used to derive an estimate of C saturation deficit of the different soils (**Table 6**) based on Stewart *et al.* [55] as follows:

$$C \text{ saturation deficit} = 1 - \frac{\text{Measured silt} + \text{clay } C \text{ content}}{\text{Protective capacity estimate}}$$

with values closer to 1 indicating a lower degree of silt + clay C saturation, and values closer to 0 indicating a higher degree of silt + clay C saturation.

For all forest and grassland sites, with the exception of Easter Bush, the silt + clay C protective capacity estimates were within the 95% confidence intervals of the measured silt + clay C values (**Table 7**), indicating some degree of silt + clay C saturation in these sites. In contrast, the estimated silt + clay C protective capacity of the cropland soils were greater than the measured silt + clay C values (**Table 7**). Correspondingly, greater C saturation deficits were calculated for these soils (0.43 - 0.72) compared to the forest soils (−1.26 - 0.07) and grassland soils (0.01 - 0.30, with the exception of Easter Bush: 0.72). These results support the idea that most forest and grassland ecosystems were closer to mineral C saturation, so that any additional C inputs will accumulate in the more labile POM fractions. Hence, any increase in SOC would be captured in differences in the non-mineral-associated OM pools, which was confirmed by the strong correlation between total SOC and total POM (LF + hPOM) (**Figure 6**), and by the lack of a similar correlation with the mineral-associated SOM fractions (**Table 6**). A linear relationship between POM-C and TOC was also found by Franzluebbers and Stuedeman [50] for pastures in the Southern Piedmont of the USA, whereas the relationship between non-particulate organic C and TOC flattened off, suggesting mineral C saturation beyond which TOC preferentially accumulates in the POC pool. This would also mean that upon any shift in land-use or any soil disturbance, these ecosystems risk losing significant amounts of SOC from enhanced POM decomposition. If these systems can unlimitedly accumulate C and N in POM pools is not clear and the existence of a saturation level for these pools remains to be tested. Yet, there are some indications that the POM pool can become saturated, in particular through aggregate-protected C saturation [54, 56].

Table 7. Estimates of (1) soil C protective capacity (g C kg^{-1} soil) using relationships between texture (% silt + clay) and mineral (silt + clay) C content (g C kg^{-1} soil) developed by Six *et al.* [54] and (2) C saturation deficit (g C kg^{-1} soil) based on Stewart *et al.* [55], for the 0 - 5 depth of the different soils. Values in between brackets represent 95% confidence interval.

Land Use	Site	Estimated C protective capacity	Measured silt + clay C	Estimated C saturation deficit
		/g C kg^{-1} soil	/g C kg^{-1} soil	
Cropland	IT-BCi	21.18 (±5.84)	6.14 (±0.1.20)	0.71 (0.60; 0.77)
	IT-Cas	21.78 (±5.96)	10.13 (±2.02)	0.54 (0.36; 0.63)
	DE-Geb	19.98 (±5.60)	11.40 (±2.38)	0.43 (0.21; 0.55)
	FR-Gri	21.18 (±5.84)	5.91 (±1.76)	0.72 (0.61; 0.78)
Forest	FI-Hyy	22.00 (±7.93)	20.47 (±4.32)	0.07 (−0.46; 0.32)
	DE-Hog	26.80 (±9.53)	36.92 (±11.39)	−0.38 (−1.14; −0.02)
	DK-Sor	28.72 (±10.17)	26.64 (±11.54)	0.07 (−0.46; 0.32)
	NL-Spe	20.80 (±7.53)	47.11 (±5.84)	−1.26 (−2.55; −0.66)
Grassland	HU-Bu	18.57 (±5.18)	13.00 (±8.50)	0.30 (0.03; 0.45)
	UK-Ebu	31.05 (±7.91)	8.64 (±2.49)	0.72 (0.63; 0.78)
	CH-Oen	45.13 (±10.99)	44.71 (±7.88)	0.01 (−0.31; 0.20)

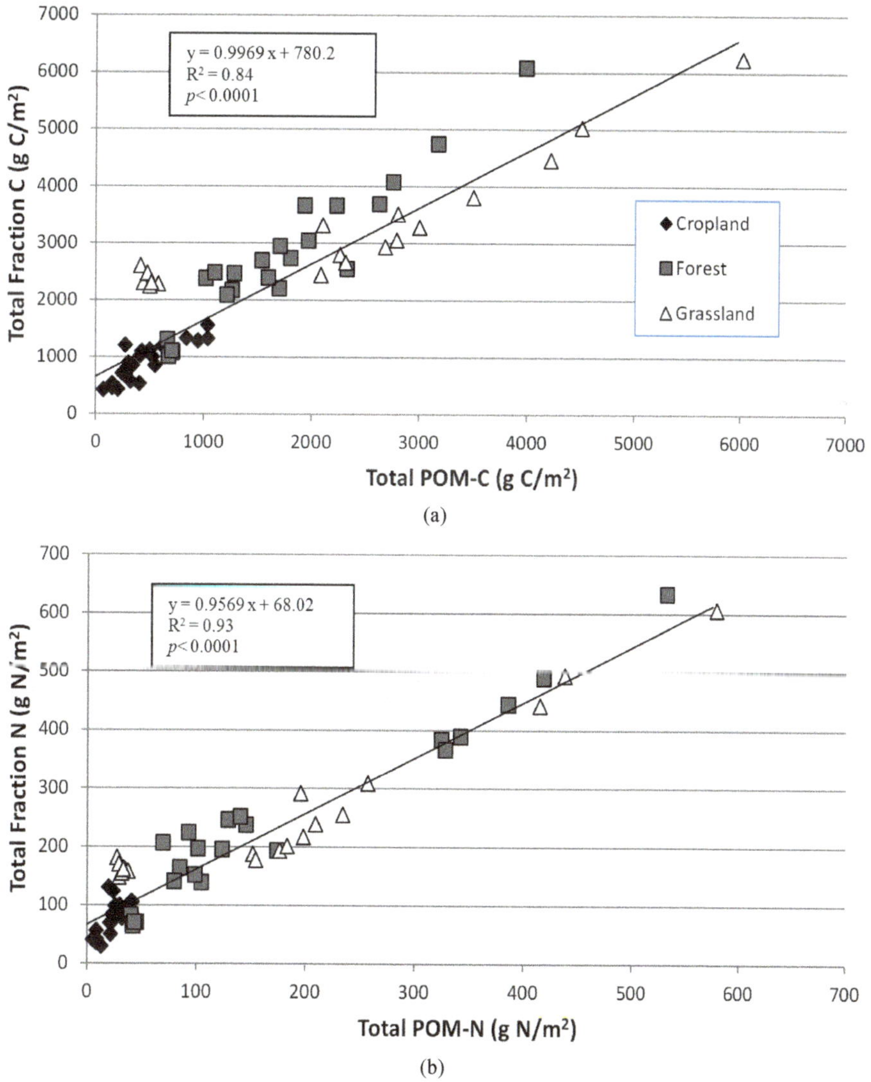

Figure 6. Linear regression of total particulate organic matter and total fraction organic carbon (a) and total nitrogen (b), for the 0 - 5 cm depth.

The lower concentrations of C and N in the silt and clay fractions (**Figure 5**) as well as the greater calculated saturation deficit for the croplands (**Table 7**) indicate that the croplands are further from mineral C saturation compared to the grasslands and forests. These croplands have therefore the potential to sequester additional C and N in mineral fractions if appropriately managed. The continuous physical disturbance in tilled croplands has been shown to stimulate POM decomposition due to enhanced turnover of aggregates, which in turn decreases mineral C and N stabilization compared to minimally disturbed soils [17,19]. We postulate that when these systems are managed in an improved way (less disturbance, e.g. no-tillage, cover crops, etc.), there could be an actual increase in SOC, due to physical protection of POM through aggregate formation and subsequent mineral C association, that could be stabilized in the long term.

We conclude that the applied fractionation scheme was successful in providing insights into how and to what amount C and N are stored and stabilized in the different soils and land uses included in this study. A simple chemical dispersion followed by a mechanical sieving was sufficient to obtain a highly diagnostic fraction for total soil C and N variability among widely varying ecosystems, and supports the idea that the total POM fraction is a good predictor of total SOM in these investigated ecosystems.

5. Conclusion

This study shows that the investigated Nitro Europe network sites store a large fraction of C and N in POM pools that are particularly vulnerable to soil disturbance such as the one caused by land use change. Best management practices that minimize soil disturbance and increase OM inputs are therefore recommended to avoid rapid losses of SOM in these systems. Rough estimates of saturation deficits suggested most forest and grassland sites to be near mineral C and N saturation, explaining the accumulation of soil C and N in the more labile POM pools, and the strong correlation between POM-C and –N and total SOC and TN. The cropland soils, on the other hand, being further from saturation, appeared to have greater potential to sequester C and N in the mineral fractions, which may be promoted under appropriate management. The observed lower C and N concentration in the silt and clay fractions points to the lower C and N stabilization efficiency at these cropland sites, likely due to the continuous physical disturbance and enhanced aggregate turnover.

6. Acknowledgements

Publication of this article was sponsored by the Colorado State University Libraries Open Access Research and Scholarship Fund. We thank Dr. Catherine Stewart for her insightful comments on this work. The study was funded by the European Union, Nitro Europe IP #017841. We are grateful to all the Nitro Europe level 3 site managers for their assistance during sampling.

REFERENCES

[1] N. H. Batjes, "Total Carbon and Nitrogen in the Soils of the World," *European Journal of Soil Science*, Vol. 47, No. 2, 1996, pp. 151-163. http://dx.doi.org/10.1111/j.1365-2389.1996.tb01386.x

[2] R. F. Follett, C. E. Stewart, E. G. Pruessner and J. M. Kimble, "Effects of Climate Change on Soil Carbon and Nitrogen Storage in the US Great Plains," *Journal of Soil and Water Conservation*, Vol. 67, No. 5, 2012, pp. 331-342. http://dx.doi.org/10.2489/jswc.67.5.331

[3] E. A. Davidson, S. E. Trumbore and R. Amundson, "Biogeochemistry—Soil Warming and Organic Carbon Content," *Nature*, Vol. 408, No. 6814, 2000, pp. 789-790. http://dx.doi.org/10.1038/35048672

[4] R. T. Conant, S. M. Ogle, E. A. Paul and K. Paustian, "Measuring and Monitoring Soil Organic Carbon Stocks in Agricultural Lands for Climate Mitigation," *Frontiers in Ecology and the Environment*, Vol. 9, No. 3, 2011, pp. 169-173. http://dx.doi.org/10.1890/090153

[5] S. Sleutel, S. De Neve, G. Hofman, P. Boeckx, D. Beheydt, O. Van Cleemput, I. Mestdagh, P. Lootens, L. Carlier, N. Van Camp, H. Verbeeck, I. Vande Walle, R. Samson, N. Lust and R. Lemeur, "Carbon Stock Changes and Carbon Sequestration Potential of Flemish Cropland Soils," *Global Change Biology*, Vol. 9, No. 8, 2003, pp. 1193-1203. http://dx.doi.org/10.1046/j.1365-2486.2003.00651.x

[6] S. Spencer, S. M. Ogle, F. J. Breidt, J. Goebel and K. Paustian, "Designing a National Soil Carbon Monitoring Network to Support Climate Change Policy: A Case Example for US Agricultural Lands," *Greenhouse Gas Management & Measurement*, Vol. 1, No. 3-4, 2011, pp. 167-178. http://dx.doi.org/10.1080/20430779.2011.637696

[7] W. J. Parton, D. S. Ojima, C. V. Cole and D. S. Schimel, "A General Model for Soil Organic Matter Dynamics: Sensitivity to Litter Chemistry, Texture and Management," *Soil Science Society of America Journal*, Vol. 39, 1994, pp. 147-167.

[8] W. J. Parton, M. Hartman, D. Ojima and D. Schimel, "Daycent and Its Land Surface Submodel: Description and Testing," *Global and Planetary Change*, Vol. 19, No. 1-4, 1998, pp. 35-48. http://dx.doi.org/10.1016/S0921-8181(98)00040-X

[9] K. Coleman, D. S. Jenkinson, G. J. Crocker, P. R. Grace, J. Klir, M. Korschens, P. R. Poulton and D. D. Richter, "Simulating Trends in Soil Organic Carbon in Long-Term Experiments Using RothC-26.3," *Geoderma*, Vol. 81, No. 1-2, 1997, pp. 29-44. http://dx.doi.org/10.1016/S0016-7061(97)00079-7

[10] C. S. Li, "Modeling Trace Gas Emissions from Agricultural Ecosystems," *Nutrient Cycling in Agroecosystems*, Vol. 58, No. 1-3, 2000, pp. 259-276.

http://dx.doi.org/10.1023/A:1009859006242

[11] S. M. Ogle, F. J. Breidt, M. Easter, S. Williams, K. Killian and K. Paustian, "Scale and Uncertainty in Modeled Soil Organic Carbon Stock Changes for US Croplands Using a Process-Based Model," *Global Change Biology*, Vol. 16, No. 2, 2010, pp. 810-822. http://dx.doi.org/10.1111/j.1365-2486.2009.01951.x

[12] E. D. Schulze, S. Luyssaert, P. Ciais, A. Freibauer, I. A. Janssens, J. F. Soussana, P. Smith, J. Grace, I. Levin, B. Thiruchittampalam, M. Heimann, A. J. Dolman, R. Valentini, P. Bousquet, P. Peylin, W. Peters, C. Rodenbeck, G. Etiope, N. Vuichard, M. Wattenbach, G. J. Nabuurs, Z. Poussi, J. Nieschulze, J. H. Gash and T. Carbo Europe, "Importance of Methane and Nitrous Oxide for Europe's Terrestrial Greenhouse-Gas Balance," *Nature Geoscience*, Vol. 2, No. 12, 2009, pp. 842-850. http://dx.doi.org/10.1038/ngeo686

[13] M. W. I. Schmidt, M. S. Torn, S. Abiven, T. Dittmar, G. Guggenberger, I. A. Janssens, M. Kleber, I. Kogel-Knabner, J. Lehmann, D. A. C. Manning, P. Nannipieri, D. P. Rasse, S. Weiner and S. E. Trumbore, "Persistence of Soil Organic Matter as an Ecosystem Property," *Nature*, Vol. 478, No. 7367, 2011, pp. 49-56. http://dx.doi.org/10.1038/nature10386

[14] J. Six and J. D. Jastrow, "Organic Matter Turnover," In: R. Lal, Ed., *Encyclopedia of Soil Science*, Marcel Dekker, New York, 2002, pp. 936-942.

[15] P. Smith, "How Long before a Change in Soil Organic Carbon Can Be Detected?" *Global Change Biology*, Vol. 10, No. 11, 2004, pp. 1878-1883. http://dx.doi.org/10.1111/j.1365-2486.2004.00854.x

[16] I. Del Galdo, J. Six, A. Peressotti and M. F. Cotrufo, "Assessing the Impact of Land-Use Change on Soil C Sequestration in Agricultural Soils by Means of Organic Matter Fractionation and Stable C Isotopes," *Global Change Biology*, Vol. 9, No. 8, 2003, pp. 1204-1213. http://dx.doi.org/10.1046/j.1365-2486.2003.00657.x

[17] K. Denef, J. Six, R. Merckx and K. Paustian, "Carbon Sequestration in Microaggregates of No-Tillage Soils with Different Clay Mineralogy," *Soil Science Society of America Journal*, Vol. 68, No. 6, 2004, pp. 1935-1944. http://dx.doi.org/10.2136/sssaj2004.1935

[18] J. Leifeld and I. Kogel-Knabner, "Soil Organic Matter Fractions as Early Indicators for Carbon Stock Changes under Different Land-Use?" *Geoderma*, Vol. 124, No. 1-2, 2005, pp. 143-155. http://dx.doi.org/10.1016/j.geoderma.2004.04.009

[19] J. Six, E. T. Elliott and K. Paustian, "Soil Macroaggregate Turnover and Microaggregate Formation: A Mechanism for C Sequestration under No-Tillage Agriculture," *Soil Biology & Biochemistry*, Vol. 32, No. 14, 2000, pp. 2099-2103. http://dx.doi.org/10.1016/S0038-0717(00)00179-6

[20] E. G. Gregorich and M. H. Beare, "Physically Uncomplexed Organic Matter," In: M. R. Carter and E. G. Gregorich, Eds., *Soil Sampling and Methods of Analysis*, CRC Press/Taylor & Francis, Boca Raton, 2008, pp. 607-616.

[21] C. E. Stewart, R. F. Follett, J. Wallace and E. G. Pruessner, "Impact of Biosolids and Tillage on Soil Organic Matter Fractions: Implications of Carbon Saturation for Conservation Management in the Virginia Coastal Plain," *Soil Science Society of America Journal*, Vol. 76, No. 4, 2012, pp. 1257-1267. http://dx.doi.org/10.2136/sssaj2011.0165

[22] C. A. Cambardella and E. T. Elliott, "Particulate Soil Organic-Matter Changes across a Grassland Cultivation Sequence," *Soil Science Society of America Journal*, Vol. 56, No. 3, 1992, pp. 777-783. http://dx.doi.org/10.2136/sssaj1992.03615995005600030017x

[23] A. S. Grandy and J. C. Neff, "Molecular C Dynamics Downstream: The Biochemical Decomposition Sequence and Its Impact on Soil Organic Matter Structure and Function," *Science of the Total Environment*, Vol. 404, No. 2-3, 2008, pp. 297-307. http://dx.doi.org/10.1016/j.scitotenv.2007.11.013

[24] M. Helfrich, B. Ludwig, P. Buurman and H. Flessa, "Effect of Land Use on the Composition of Soil Organic Matter in Density and Aggregate Fractions as Revealed by Solid-State C-13 NMR Spectroscopy," *Geoderma*, Vol. 136, No. 1-2, 2006, pp. 331-341. http://dx.doi.org/10.1016/j.geoderma.2006.03.048

[25] P. Sollins, C. Swanston, M. Kleber, T. Filley, M. Kramer, S. Crow, B. A. Caldwell, K. Lajtha and R. Bowden, "Organic C and N Stabilization in a Forest Soil: Evidence from Sequential Density Fractionation," *Soil Biology & Biochemistry*, Vol. 38, No. 11, 2006, pp. 3313-3324. http://dx.doi.org/10.1016/j.soilbio.2006.04.014

[26] J. S. Clemente, A. J. Simpson and M. J. Simpson, "Association of Specific Organic Matter Compounds in Size Fractions of Soils under Different Environmental Controls," *Organic Geochemistry*, Vol. 42, No. 10, 2011, pp. 1169-1180. http://dx.doi.org/10.1016/j.orggeochem.2011.08.010

[27] K. F. Nadelhoffer and B. Fry, "Controls on Natural N-15 and C-13 Abundances in Forest Soil Organic Matter," *Soil Science Society of America Journal*, Vol. 52, No. 6, 1988, pp. 1633-1640. http://dx.doi.org/10.2136/sssaj1988.03615995005200060024x

[28] J. Balesdent, A. Mariotti and B. Guillet, "Natural 13C Abundance as a Tracer for Studies of Soil Organic Matter Dynamics," *Soil Biology & Biochemistry*, Vol. 19, No. 1, 1987, pp. 25-30. http://dx.doi.org/10.1016/0038-0717(87)90120-9

[29] A. Diochon and L. Kellman, "Natural Abundance Measurements of (13)C Indicate Increased Deep Soil Carbon Mineralization after Forest Disturbance," *Geophysical Research Letters*, Vol. 35, No. 14, 2008. http://dx.doi.org/10.1029/2008GL034795

[30] J. R. Ehleringer, N. Buchmann and L. B. Flanagan, "Carbon Isotope Ratios in Belowground Carbon Cycle Processes," *Ecological Applications*, Vol. 10, No. 2, 2000, pp. 412-422. http://dx.doi.org/10.1890/1051-0761(2000)010[0412:CIRIBC]2.0.CO;2

[31] P. Hogberg, L. Hogbom, H. Schinkel, M. Hogberg, C. Johannisson and H. Wallmark, "N-15 Abundance of Sur-

face Soils, Roots and Mycorrhizas in Profiles of European Forest Soils," *Oecologia*, Vol. 108, No. 2, 1996, pp. 207-214.

[32] K. J. Nadelhoffer and B. Fry, "Nitrogen Isotope Studies in Forest Ecosystems," In: K. Lajtha and R. Michener, Eds., *Stable Isotopes in Ecology and Environmental Science*, Blackwell Scientific Publications, Boston, 1994, pp. 23-44.

[33] P. Hogberg, "Tansley Review No 95 - N-15 Natural Abundance in Soil-Plant Systems," *New Phytologist*, Vol. 137, No. 2, 1997, pp. 179-203. http://dx.doi.org/10.1046/j.1469-8137.1997.00808.x

[34] M. A. Sutton, E. Nemitz, J. W. Erisman, C. Beier, K. B. Bahl, P. Cellier, W. de Vries, F. Cotrufo, U. Skiba, C. Di Marco, S. Jones, P. Laville, J.F. Soussana, B. Loubet, M. Twigg, D. Famulari, J. Whitehead, M.W. Gallagher, A. Neftel, C. R. Flechard, B. Herrmann, P. L. Calanca, J. K. Schjoerring, U. Daemmgen, L. Horvath, Y. S. Tang, B. A. Emmett, A. Tietema, J. Penuelas, M. Kesik, N. Bruggemann, K. Pilegaard, T. Vesala, C. L. Campbell, J. E. Olesen, U. Dragosits, M. R. Theobald, P. Levy, D. C. Mobbs, R. Milne, N. Viovy, N. Vuichard, J. U. Smith, P. Smith, P. Bergamaschi, D. Fowler and S. Reis, "Challenges in Quantifying Biosphere-Atmosphere Exchange of Nitrogen Species," *Environmental Pollution*, Vol. 150, No. 1, 2007, pp. 125-139. http://dx.doi.org/10.1016/j.envpol.2007.04.014

[35] U. Skiba, J. Drewer, Y. S. Tang, N. van Dijk, C. Helfter, E. Nemitz, D. Famulari, J. N. Cape, S. K. Jones, M. Twigg, M. Pihlatie, T. Vesala, K. S. Larsen, M. S. Carter, P. Ambus, A. Ibrom, C. Beier, A. Hensen, A. Frumau, J. W. Erisman, N. Bruggemann, R. Gasche, K. Butterbach-Bahl, A. Neftel, C. Spirig, L. Horvath, A. Freibauer, P. Cellier, P. Laville, B. Loubet, E. Magliulo, T. Bertolini, G. Seufert, M. Andersson, G. Manca, T. Laurila, M. Aurela, A. Lohila, S. Zechmeister-Boltenstern, B. Kitzler, G. Schaufler, J. Siemens, R. Kindler, C. Flechard and M. A. Sutton, "Biosphere-Atmosphere Exchange of Reactive Nitrogen and Greenhouse Gases at the NitroEurope Core Flux Measurement Sites: Measurement Strategy and First Data Sets," *Agriculture Ecosystems & Environment*, Vol. 133, No. 3-4, 2009, pp. 139-149. http://dx.doi.org/10.1016/j.agee.2009.05.018

[36] G. Schaufler, B. Kitzler, A. Schindlbacher, U. Skiba, M. A. Sutton and S. Zechmeister-Boltenstern, "Greenhouse Gas Emissions from European Soils under Different Land Use: Effects of Soil Moisture and Temperature," *European Journal of Soil Science*, Vol. 61, No. 5, 2010, pp. 683-696. http://dx.doi.org/10.1111/j.1365-2389.2010.01277.x

[37] K. Pilegaard, T. N. Mikkelsen, C. Beier, N. O. Jensen, P. Ambus and H. Ro-Poulsen, "Field Measurements of Atmosphere-Biosphere Interactions in a Danish Beech Forest," *Boreal Environment Research*, Vol. 8, No. 4, 2003, pp. 315-333.

[38] D. Harris, W. R. Horwath and C. van Kessel, "Acid Fumigation of Soils to Remove Carbonates Prior to Total Organic Carbon or Carbon-13 Isotopic Analysis," *Soil Science Society of America Journal*, Vol. 65, No. 6, 2001, pp. 1853-1856. http://dx.doi.org/10.2136/sssaj2001.1853

[39] F. Marzaioli, C. Lubritto, I. Del Galdo, A. D'Onofrio, M. F. Cotrufo and F. Terrasi, "Comparison of Different Soil Organic Matter Fractionation Methodologies: Evidences from Ultrasensitive (14)C Measurements," *Nuclear Instruments & Methods in Physics Research Section B-Beam Interactions with Materials and Atoms*, Vol. 268, No. 7-8, 2010, pp. 1062-1066. http://dx.doi.org/10.1016/j.nimb.2009.10.098

[40] S. De Gryze, J. Six, K. Paustian, S. J. Morris, E. A. Paul and R. Merckx, "Soil Organic Carbon Pool Changes Following Land-Use Conversions," *Global Change Biology*, Vol. 10, No. 7, 2004, pp. 1120-1132. http://dx.doi.org/10.1111/j.1529-8817.2003.00786.x

[41] A. J. Franzluebbers. "Depth Distribution of Soil Organic Carbon as a Signature of Soil Quality," *Proceedings of the 19th World Congress of Soil Science: Soil Solutions for a Changing World. Symposium* 4.1.1 *Valuing the Soil's Natural Capital*, Brisbane, 2010, pp. 1-4.

[42] R. B. Harrison, P. W. Footen and B. D. Strahm, "Deep Soil Horizons: Contribution and Importance to Soil Carbon Pools and in Assessing Whole-Ecosystem Response to Management and Global Change," *Forest Science*, Vol. 57, No. 1, 2011, pp. 67-76.

[43] M. Wiesmeier, P. Sporlein, U. Geuss, E. Hangen, S. Haug, A. Reischl, B. Schilling, M. von Lutzow and I. Kogel-Knabner, "Soil Organic Carbon Stocks in Southeast Germany (Bavaria) as Affected by Land Use, Soil Type and Sampling Depth," *Global Change Biology*, Vol. 18, No. 7, 2012, pp. 2233-2245. http://dx.doi.org/10.1111/j.1365-2486.2012.02699.x

[44] K. Pilegaard, U. Skiba, P. Ambus, C. Beier, N. Bruggemann, K. Butterbach-Bahl, J. Dick, J. Dorsey, J. Duyzer, M. Gallagher, R. Gasche, L. Horvath, B. Kitzler, A. Leip, M. K. Pihlatie, P. Rosenkranz, G. Seufert, T. Vesala, H. Westrate and S. Zechmeister-Boltenstern, "Factors Controlling Regional Differences in Forest Soil Emission of Nitrogen Oxides (NO and N_2O)," *Biogeosciences*, Vol. 3, No. 4, 2006, pp. 651-661. http://dx.doi.org/10.5194/bg-3-651-2006

[45] A. S. Bateman, S. D. Kelly and T. D. Jickells, "Nitrogen Isotope Relationships between Crops and Fertilizer: Implications for Using Nitrogen Isotope Analysis as an Indicator of Agricultural Regime," *Journal of Agricultural and Food Chemistry*, Vol. 53, No. 14, 2005, pp. 5760-5765. http://dx.doi.org/10.1021/jf050374h

[46] S. Ranucci, T. Bertolini, L. Vitale, P. Di Tommasi, L. Ottaiano, M. Oliva, U. Amato, A. Fierro and V. Magliulo, "The Influence of Management and Environmental Variables on Soil N_2O Emissions in a Crop System in Southern Italy," *Plant and Soil*, Vol. 343, No. 1-2, 2011, pp. 83-96. http://dx.doi.org/10.1007/s11104-010-0674-x

[47] D. Huygens, K. Denef, R. Vandeweyer, R. Godoy, O. Van Cleemput and P. Boeckx, "Do Nitrogen Isotope Patterns Reflect Microbial Colonization of Soil Organic Matter Fractions?" *Biology and Fertility of Soils*, Vol. 44, No. 7, 2008, pp. 955-964. http://dx.doi.org/10.1007/s00374-008-0280-8

[48] K. T. Hubick, G. D. Farquhar and R. Shorter, "Correlation between Water Use Efficiency and Carbon Isotope

Discrimination in Diverse Peanut (Arachis) Germplasm," *Australian Journal of Plant Physiology*, Vol. 13, No. 6, 1986, pp. 803-816. http://dx.doi.org/10.1071/PP9860803

[49] K. Denef, A. F. Plante and J. Six, "Characterization of Soil Organic Matter," In: W. L. Kutsch, M. Bahn and A. Heinemeyer, Eds., *Soil Carbon Dynamics: An Integrated Methodology*, Cambridge University Press, Cambridge, 2009, pp. 91-126.

[50] A. J. Franzluebbers and J. A. Stuedemann, "Particulate and Non-Particulate Fractions of Soil Organic Carbon under Pastures in the Southern Piedmont USA," *Environmental Pollution*, Vol. 116, No. S1, 2002, pp. S53-S62. http://dx.doi.org/10.1016/S0269-7491(01)00247-0

[51] C. Ammann, C. Spirig, J. Leifeld and A. Neftel, "Assessment of the Nitrogen and Carbon Budget of Two Managed Temperate Grassland Fields," *Agriculture Ecosystems & Environment*, Vol. 133, No. 3-4, 2009, pp. 150-162. http://dx.doi.org/10.1016/j.agee.2009.05.006

[52] M. R. Carter and E. G. Gregorich, "Carbon and Nitrogen Storage by Deep-Rooted Tall Fescue (Lolium Arundinaceum) in the Surface and Subsurface Soil of a Fine Sandy Loam in Eastern Canada," *Agriculture Ecosystems & Environment*, Vol. 136, No. 1-2, 2010, pp. 125-132. http://dx.doi.org/10.1016/j.agee.2009.12.005

[53] J. Hassink, "The Capacity of Soils to Preserve Organic C and N by Their Association with Clay and Silt Particles," *Plant and Soil*, Vol. 191, No. 1, 1997, pp. 77-87. http://dx.doi.org/10.1023/A:1004213929699

[54] J. Six, R. T. Conant, E. A. Paul and K. Paustian, "Stabilization Mechanisms of Soil Organic Matter: Implications for C-Saturation of Soils," *Plant and Soil*, Vol. 241, No. 2, 2002, pp. 155-176. http://dx.doi.org/10.1023/A:1016125726789

[55] C. E. Stewart, K. Paustian, R. T. Conant, A. F. Plante and J. Six, "Soil Carbon Saturation: Evaluation and Corroboration by Long-Term Incubations," *Soil Biology & Biochemistry*, Vol. 40, No. 7, 2008, pp. 1741-1750. http://dx.doi.org/10.1016/j.soilbio.2008.02.014

[56] H. Chung, K. J. Ngo, A. F. Plante and J. Six, "Evidence for Carbon Saturation in a Highly Structured and Organic-Matter-Rich Soil," *Soil Science Society of America Journal*, Vol. 74, No. 1, 2010, pp. 130-138. http://dx.doi.org/10.2136/sssaj2009.0097

20

Soil Carbon Changes Influenced by Soil Management and Calculation Method

Maysoon M. Mikha[1*], Joseph G. Benjamin[1], Ardell D. Halvorson[2], David C. Nielsen[1]

[1]US Department of Agricultural, Agricultural Research Service, Central Great Research Station, Akron, USA; [2]US Department of Agricultural, Agricultural Research Service, Northern Plains Soil Plant Nutrient Research, Fort Collins, USA.
Email: *Maysoon.Mikha@ars.usda.gov

ABSTRACT

Throughout the years, many studies have evaluated changes in soil organic carbon (SOC) mass on a fixed-depth (FD) basis without considering changes in soil mass caused by changing bulk density (ρ_b). This study evaluates the temporal changes in SOC caused by two factors: 1) changing SOC concentration; and 2) changing equivalent soil mass (ESM) in comparison with FD. In addition, this study evaluates calculating changes in SOC stock over time using a minimum equivalent soil mass (ESM_{min}) basis from a single sampling event compared with the FD scenario. A tillage [no-tillage (NT) and chisel plow (CP)]-crop rotation (multiple crop and continuous corn), and irrigation (full and delayed)) study was initiated in 2001 on Weld silt loam soil. After seven years, SOC concentration in the 0 - 30 cm depth was 19.7% greater in 2008 compared with 2001. Standardizing the soil mass of 2001 to the ESM of 2008 for each individual treatment showed an average gain in SOC of 5.8 Mg C·ha^{-1} in 2008 compared with 2001. However, the increase in SOC using ESM was twice the SOC gained with the FD calculation, where some treatments lost SOC after seven years of management. Estimating SOC levels using the ESM_{min} and, thereby, eliminating the confounding effect of soil ρ_b indicated that SOC stock was influenced by crop species and their interaction with irrigation, but not by tillage practices. Over all, the ESM calculation appears to be more effective in evaluating SOC stock than the FD calculation.

Keywords: Soil Organic Carbon Stock; Equivalent Soil Mass; Minimum Equivalent Soil Mass; Fixed-Depth; Management Practices

1. Introduction

Numerous studies [1-8] have assessed management practice-induced changes in SOC due to anthropogenic and environmental effects. The majority of previous research [1-6] evaluated SOC mass using SOC concentration and soil bulk density (ρ_b) associated with a specific soil depth. However, management practices that influence SOC concentration may also affect soil ρ_b [5,9-12]. Consequently, researchers have argued that changes in soil ρ_b and its effect on unequal soil mass associated with the fixed depth have a confounding effect on SOC mass estimation when comparing changes in SOC associated with different management practices [13-17]. With an acknowledgment that SOC storage depends on soil mass and soil ρ_b, recent research estimated SOC and other soil nutrients based on their concentration, soil thickness, and soil ρ_b [13-18]. Researchers have suggested alternative methods to calculate SOC and other soil nutrients to account for the changes in soil volume under different management practices [13,14,16].

The equivalent soil mass (ESM) is one of the methods being used to assess SOC and other soil nutrients on a normalized soil mass per unit area basis to account for differences in soil masses caused by soil management [13-16,18]. In the ESM calculation scenario, soil masses associated with different management practices are standardized to a particular soil mass per unit area of a specified layer and the equivalent soil C mass is adjusted to the ESM [13,16,19]. The goal of using the ESM and its associated equivalent C mass calculation is to reduce the SOC error calculated in soil profiles due to changing soil ρ_b under different management practices [13,16]. Lee *et al.* [16] proposed two other scenarios for SOC estimation associated with a single sampling event when the initial SOC or initial (ρ_b) values were not available. The first method was based on using the minimum equivalent soil

*Corresponding author.

mass (ESM_{min}) measured and the second alternative measurement was based on using the maximum equivalent soil mass (ESM_{max}) measured. The ESM_{min} uses the lowest soil mass measured, on an FD basis, to standardize other soil masses associated with different management practices. The ESM_{max} uses the greatest soil mass measured, on an FD basis, among the different management practices as the standard [16].

The concept of soil mass and expressing SOC on an equivalent mass basis has been adopted by researchers for more than a decade, but this approach has not been readily applied to evaluate different management practices [13,16]. There has been limited research on the influence of different management practices on the standardized soil ρ_b, soil mass, and the associated SOC using field measured SOC and soil ρ_b. This type of research is vital due to the fact that, in recent years, there has been a great interest for evaluating SOC stock and improving soil C sequestration as influenced by different management practices. Preventing SOC losses due to soil management practices, especially tillage, and increasing crop residue return to the soil are important parameters in improving soil quality and sustainability. However, the confounding effect of soil ρ_b variability and the accuracy the soil ρ_b measurement are influencing the perceived temporal changes in SOC stock or SOC content in any given sampling period.

This study aims to evaluate different scenarios of SOC calculation where the confounding effect of soil ρ_b variability and its associated soil mass could be eliminated or reduced. We hypothesize that eliminating the differences in soil ρ_b and standardizing the soil masses into an ESM basis may be more effective in predicting the temporal changes in SOC stock or in SOC content at any given time period compared with FD scenario. Therefore, the objectives of this study were to: 1) evaluate the temporal changes in SOC concentration to eliminate the temporal variation in soil ρ_b; 2) evaluate the temporal changes in SOC stock as influenced by the FD and the ESM calculation methods; and 3) evaluate treatment effects on SOC using the ESM_{min} basis of standardization. Overall, this study identifies the temporal changes in SOC stock as influenced by different calculation scenarios after seven years of different management practices and subsequent changes in soil ρ_b and soil mass.

2. Materials and Methods[1]

2.1. Site Description

An irrigation-tillage-crop rotation study was established in 2001 with an individual plot size of 18 m × 9 m at the USDA-ARS Central Great Plains Research Station near Akron, CO [6]. The research station is located within a semiarid climate region with approximate mean annual precipitation of 418 mm. The study site is located at 40°8'N latitude and, 103°9'W longitude with elevation of 1384 m. The soil type is a Weld silt loam (fine, smectitic, mesic Aridic Argiustolls). Treatments were arranged in a split-plot design with three replications. The main plot was an irrigation treatment and the subplot was the tillage and crop system that were randomized within the main irrigation plots. Details of tillage practices, irrigation treatment, previous and current cropping history, and site management were reported in detail by Benjamin *et al.* [6]. Briefly, tillage treatments included NT (directly planting into the previous crop residues; no-till) and CP (fall chisel plow at 35 cm depth with a parabolic-shank deep ripper and spring pass with a mulch treader disrupting the approximately 0 - 5 cm depth). The irrigation treatment consisted of either full or delayed irrigation. The crop system treatments consisted of either continuous corn (CC) or a rotation (Rot) of a variety of crops throughout the study period, red kidney bean (*Phaseolus vulgaris* L.), spring barley (*Hordeum vulgare* L.), sunflower (*Helianthus annuus* L.), corn (*Zea mays* L.), spring pea (*Pisum sativum* L.), winter wheat (*Triticum aestivum* L.).

2.2. Soil Sampling

Soil samples were collected from non-wheel-tracked areas from the 0 - 15 and 15 - 30 cm depths in the spring of 2001 and 2008 before planting. The soil samples were collected from each plot using a 5 cm diam. probe attached to a Giddings hydraulic soil sampler (Giddings Machine Co., Windsor, CO). Soil samples were stored in sterile polypropylene bags placed coolers during field sampling and then stored at 4°C in a walk-in cooler for about a week before processing. Soil samples were prescreened through a 2 mm sieve to remove large pieces of plant material before being air dried and ground to pass through a 2 mm screen using a flail-type soil grinder. Soil samples for determining soil bulk density (ρ_b) were also collected at the same sampling dates in 2001 and 2008. Soil ρ_b samples were collected for each plot using a 7.5 cm diam. by 60 cm long probe containing 7.5 cm diam. by 7.5 cm deep aluminum rings. The probe was attached to a Giddings hydraulic soil sampler for insertion into the soil. The rings were sectioned in the field to ensure an undisturbed soil sample with depth. Data were averaged for the 0 - 15 cm and 15 - 30 cm depths used in this study.

2.3. Soil Total C, Soil Inorganic C, and Soil Organic Carbon

Three subsamples per sampling bag were composited,

[1]Mention of commercial products and organization in this paper is solely to provide specific information. It does not constitute endorsement by USDA-ARS over other products and organization not mentioned. The US Department of Agriculture, Agricultural Research Service, is an equal opportunity/affirmative action employer and all agency services are available without discrimination.

ground to pass a 150 μ screen, and analyzed for different forms of soil carbon. Soil total C (STC) contents from the 0 - 15 and 15 - 30 cm depth were determined by dry combustion using a Carlo Erba C-N analyzer (Haake Buchler Instruments, Inc., Saddle Brook, NJ) at a commercial lab (Ward Laboratories, Kearney, NE). Soil inorganic C (SIC) content was evaluated using a modified pressure-calcimeter method reported by Sherrod *et al.* [20]. Soil organic C (SOC) content was calculated from the differences between STC and SIC. The SOC at 0 - 30 cm depth presented in this study is a sum of SOC associated with 0 - 15 and 15 - 30 cm depths. Data at 0 - 15 and 15 - 30 cm depths for SOC on a fixed depth basis were explained in details by Benjamin *et al.* [6].

2.4. Expression of Soil Organic Carbon (SOC)

The SOC associated with an ESM was calculated using the approach outlined by Ellert and Bettany [13] and Lee *et al.* [16]. The calculations presented in this study represent three scenarios for SOC evaluation. The first scenario estimates the temporal changes only in SOC concentration ($g \cdot C \cdot kg^{-1}$ soil) from 2001 to 2008 (**Table 1**). The second scenario estimates the temporal changes in SOC stock from 2001 to 2008 adjusted to the ESM of 2008 for every treatment combination in the 0 - 30 cm depth. The third scenario evaluates treatment effects on SOC after 7 yrs of differing management practices adjusted to the ESM_{min} as if there was a single sampling in 2008. In addition, changes in SOC in the 0 - 30 cm depth using the second calculation scenario (ESM) will be compared to SOC calculated by the FD method reported by Benjamin *et al.* [6].

2.5. Equivalent Soil Mass (ESM) Scenario

To evaluate the temporal changes in SOC stock between 2001 and 2008 in the 0 - 30 cm depth, soil mass of 2001 was adjusted to the soil mass of 2008 (ESM) as influenced by management practices. In 2001, soil at this study site exhibited a high bulk density, ranging between 1.46 to 1.51 $Mg \cdot m^{-3}$ at the 0 - 30 cm depth. In 2008, soil bulk density at the 0 - 30 cm depth ranged between 1.21 to 1.37 $Mg \cdot m^{-3}$ (**Table 2**). The soil mass in 2008 associated with each treatment was considered the baseline for the ESM calculation.

Table 1. The 2001 and 2008 soil organic C (SOC) concentration ($g \cdot kg^{-1}$) and the changes in SOC from 2001 to 2008 at 0 - 30 cm depth as influenced by irrigation, cropping, and tillage practices.

Irrigation	Tillage	Cropping	SOC 2001	SOC 2008	Δ SOC[a] 2008-2001
			0 - 30 cm		
			$g \cdot kg^{-1}$		
Full	NT[b]	CC[c]	8.47	10.03	1.56
Full	CP	CC	8.29	9.61	1.32
Delayed	NT	CC	7.05	9.42	2.37
Delayed	CP	CC	7.01	8.97	1.96
Full	NT	Rot	7.91	9.00	1.09
Full	CP	Rot	7.75	8.67	0.92
Delayed	NT	Rot	7.60	8.60	1.00
Delayed	CP	Rot	7.50	9.43	1.93
			PR > F		
Irrigation (I)			0.08	0.27	0.22
Cropping (Cr)			0.95	0.004	0.074
Tillage (T)			0.67	0.56	0.94
I × Cr			0.07	0.028	0.64
I ×T			0.86	0.13	0.53
Cr × T			0.98	0.049	0.24
I × Cr × T			0.94	0.07	0.28

[a]Δ SOC = Changes in SOC concentration from 2001 to 2008; [b]NT = No-tillage; CP = Chisel plow; [c]CC = Continuous corn; Rot = Mixed grass and broadleaf crops.

Table 2. The 2001 and 2008 soil organic C (SOC) mass ($Mg \cdot ha^{-1}$), soil bulk density (ρ_b) in $Mg \cdot m^{-3}$, soil mass ($Mg \cdot ha^{-1}$), and changes in SOC from 2001 to 2008 at 0 - 30 cm depth evaluated on a fixed depth (FD) basis and on an equivalent soil mass of the 2008 (ESM_{2008}) as influenced by irrigation, cropping, and tillage practices.

			Fixed depth							ESM_{2008}	
			SOC		ρ_b		Soil mass		Δ SOC[a]	SOC	Δ SOC[b]
Irrigation	Tillage	Cropping	2001	2008	2001	2008	2001	2008	2008-2001	2001	2008-2001
			0 - 30 cm								
			$Mg \cdot ha^{-1}$		$Mg \cdot m^{-3}$		$Mg \cdot ha^{-1}$				
Full	NT[c]	CC[d]	38.3 a[e]	41.4 a	1.505 a	1.370 a	4515 a	4110 a	3.2 a	34.88 a	6.5 a
Full	CP	CC	37.0 a	36.5 a	1.481 ab	1.270 a	4443 ab	3810 a	−0.5 a	31.58 a	4.9 a
Delayed	NT	CC	31.1 a	37.3 a	1.464 b	1.313 a	4392 b	3939 a	6.2 a	27.78 a	9.5 a
Delayed	CP	CC	31.0 a	32.3 a	1.477 ab	1.208 a	4431 ab	3624 a	1.3 a	25.41 a	6.9 a
Full	NT	Rot	35.2 a	34.6 a	1.485 ab	1.295 a	4455 ab	3885 a	−0.6 a	30.72 a	3.9 a
Full	CP	Rot	34.5 a	31.5 a	1.483 b	1.213 a	4449 ab	3639 a	−2.9 a	28.18 a	3.3 a
Delayed	NT	Rot	33.9 a	32.1 a	1.500 a	1.242 a	4500 a	3727 a	−1.8 a	28.32 a	3.8 a
Delayed	CP	Rot	32.7 a	34.6 a	1.463 b	1.226 a	4389 ab	3678 a	1.9 a	27.72 a	7.0 a
			PR > F								
Irrigation (I)			0.08	0.10	0.45	0.17	0.47	0.17	0.61	0.08	0.26
Cropping (Cr)			0.94	0.0007	0.88	0.004	0.87	0.006	0.02	0.34	0.07
Tillage (T)			0.51	0.02	0.17	0.003	0.19	0.004	0.16	0.14	0.85
I × Cr			0.06	0.01	0.16	0.13	0.17	0.15	0.87	0.06	0.70
I × T			0.83	0.07	0.87	0.33	0.96	0.28	0.67	0.60	0.53
Cr × T			0.89	0.01	0.29	0.06	0.30	0.06	0.18	0.63	0.20
I × Cr × T			0.68	0.06	0.02	0.20	0.02	0.19	0.36	0.86	0.29

[a]Δ SOC = Changes in SOC from 2001 to 2008 calculated on a fixed depth basis (FD) as reported by [6]; [b]Δ SOC = Changes in SOC from 2001 calculated on ESM_{2008} basis to 2008 calculated on a fixed depth basis; [c]NT = No-tillage; CP = Chisel plow; [d]CC = Continuous corn; Rot = mixed grass and broadleaf crops; [e]Different letter represents significant ($P < 0.05$) differences among the treatments within the same measurement and year.

Soil masses for each treatment in 2001 and 2008 were calculated on an FD basis using soil depth at 0 - 30 cm and field measured bulk density (**Table 2**) as

$$M_{soil} = d \times \rho_b \times 10^{-4} \quad (1)$$

where M_{soil} represents soil mass measured in ($Mg \cdot ha^{-1}$), d represents soil depth measured in (m), ρ_b represents soil bulk density measured in ($Mg \cdot m^{-3}$) at the FD, and 10^4 is the conversion factor ($m^2 \cdot ha^{-1}$). The SOC mass for the FD of 0 - 30 cm was calculated from field measured SOC concentration as

$$M_C = M_{soil} \times C_{cons} \times 10^{-3} \quad (2)$$

where M_C represents soil C mass ($Mg \cdot C \cdot ha^{-1}$) at the fixed-depth, C_{cons} represents SOC concentration ($kg \cdot Mg^{-1}$), and 10^{-3} is the conversion factor ($Mg \cdot kg^{-1}$).

To account for the difference in soil masses and bulk densities between 2001 and 2008, the soil mass and associated SOC in 2001 were adjusted to the ESM determine in 2008 using a calculation procedure similar to Ellert and Bettany [13] and Lee *et al.* [16]. The equivalent soil mass for each treatment (i) [$ESM_{(i)}$] in 2008 was considered to be the baseline soil mass in 2001 for the same treatment. For each treatment (i) in 2001, the SOC content on an $ESM_{(i)}$ basis was calculated as shown in **Figure 1**. The example calculation in **Figure 1** represents the data from the first treatment in **Table 2** in the 0 - 30 cm depth, Full irrigation with NT and continuous corn (CC) cropping system. In this example, the soil mass in 2001 at 0 - 30 cm depth on a FD basis was reduced by 405 $Mg \cdot ha^{-1}$ due to different management practices compared with the baseline of the same treatment (4110 $Mg \cdot ha^{-1}$) in 2008 (**Table 2**), calculated as

$$M_{sub(i)} = M_{soil(2001)(i)} - ESM_{(i)} \quad (3)$$

where $M_{sub(i)}$ represents the soil mass subtracted from

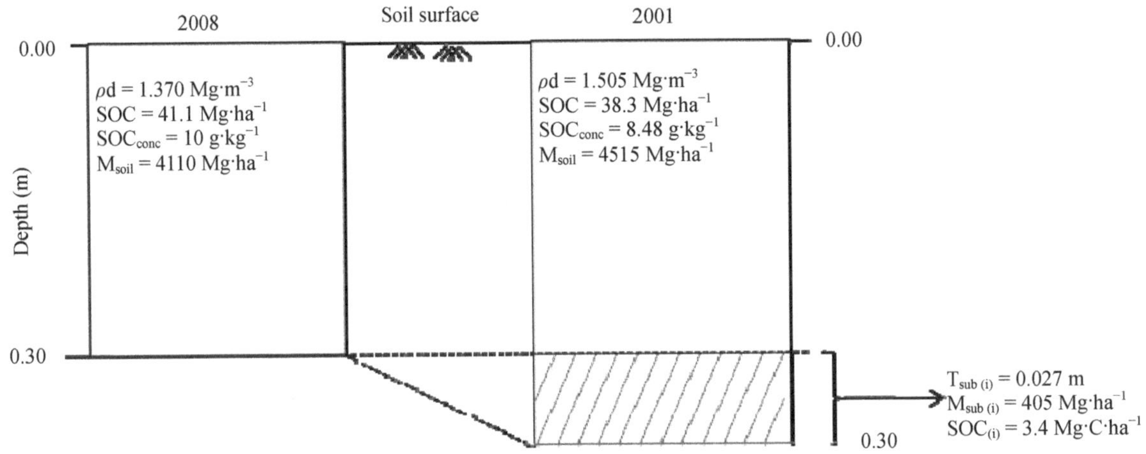

Figure 1. Example calculation for 2001 soil organic C (SOC) on an equivalent soil mass (ESM) basis of the 2008 soil mass at 0 - 30 cm depth. M_{soil} represents soil mass in Mg·ha^{-1} on a fixed-depth basis (FD) for 2001 and 2008, ρ_b represents soil bulk density in Mg·m^{-3}, and SOC value represents soil C concentration (SOC_{conc}) in g·kg^{-1} and SOC mass in Mg·ha^{-1}. The shaded area associated in 2001 represents the soil thickness ($T_{sub(i)}$) measured in m, soil mass ($M_{sub(i)}$) measured in Mg·ha^{-1}, and its corresponding $SOC_{(i)}$ measured in Mg·ha^{-1} that needs to be subtracted from the 2001 value associated with the treatment (i) to standardized to the 2008 soil masses.

2001 (Mg·ha^{-1}) for specific treatment (i), $M_{soil\ (2001)\ (i)}$ represents soil mass on an FD basis (Mg·ha^{-1}) in 2001 associated with a specific treatment (i), $ESM_{(i)}$ represents an equivalent soil mass on an FD basis (Mg·ha^{-1}) in 2008 associated with the same specific treatment (i). The SOC associated with the subtracted layer of 2001 was equivalent to 3.4 Mg·ha^{-1} and was calculated using SOC concentration following Equation (2). Therefore, the SOC in 2001 was reduced to ~34.9 Mg·ha^{-1} to adjust to the $ESM_{(i)}$ for the same treatment (**Table 2**). In addition, as the soil mass of treatment (i) in 2001 was reduced, the soil thickness in 2001 for the same treatment was also reduced. The specific soil thickness (0.02691 m) associated with $M_{sub(i)}$ was also subtracted from 30 cm depth in 2001 (**Figure 1**) calculated as follows:

$$T_{sub(i)} = \frac{M_{sub(i)}}{\rho_{b(2001)(i)}} \times 10^{-4} \quad (4)$$

where $T_{sub(i)}$ represents the soil thickness (m) that needs to be subtracted from 2001 for a specific treatment (i), $M_{sub(i)}$ represents the soil mass subtracted from 2001 (Mg·ha^{-1}) to adjust to the soil mass of 2008 for a specific treatment (i), $\rho_{b\ (2001)\ (i)}$ represents the 2001 bulk density (Mg·m^{-3}) of treatment (i), and 10^{-4} represents the conversion factors (ha·m^{-2}).

2.6. Minimum Equivalent Soil Mass Basis (ESM_{min}) Scenario

To estimate changes in SOC in 2008 among different management practices in the 0 - 30 cm depth, SOC was evaluated on the ESM_{min} as proposed by Lee *et al.* [16]. This type of SOC evaluation can be used when the initial conditions (SOC or bulk density) are not available. In this method, all soil masses associated with different treatments will be adjusted to the minimum observed soil mass and its associated ρ_b measured in 2008 in the 0 - 30 cm depth. The ESM_{min} chosen from the 2008 was 3624 Mg·ha^{-1} that represents the minimum ρ_b (**Table 2**). In this approach, since the minimum soil mass was chosen, a specific amount of soil mass, with its associated SOC, was subtracted from each treatment in 2008 to equal the ESM_{min} (**Table 3**). For each treatment in 2008, the SOC on an ESM_{min} basis was calculated as in Equation (1) and Equation (2), except that $M_{sub(i)}$ was calculated as

$$M_{sub(i)} = M_{soil(i)} - ESM_{min} \quad (5)$$

where $M_{sub(i)}$ represents the soil mass subtracted (Mg ha^{-1}) from each treatment (i) in 2008, $M_{soil(2008)(i)}$ represents soil mass on an FD basis (Mg·ha^{-1}) in 2008 associated with a specific treatment (i), and ESM_{min} represents the chosen minimum equivalent soil mass (Mg·ha^{-1}) in 2008. The specific soil thickness associated with $M_{sub(i)}$ was also subtracted from 30 cm depth in 2008 calculated as follows:

$$T_{sub(i)} = \frac{M_{sub(i)}}{\rho_{b(i)}} \times 10^{-4} \quad (6)$$

where $T_{sub(i)}$ represents the soil thickness (m) that needs to be subtracted from a specific treatment (i) in 2008, $M_{sub(i)}$ represents the soil mass subtracted from 2008 (Mg·ha^{-1}) to adjust to the ESM_{min}, $\rho_{b(i)}$ represents the 2008 bulk density (Mg·m^{-3}) of treatment (i), and 10^{-4} represents the conversion factors (ha·m^{-2}).

2.7. Statistical Analyses

Analyses of variance were calculated using SAS Version 9.2 [21] to determine statistically significant effects of irrigation, tillage, and crop system on SOC content. The data were analyzed as a split plot design, with irrigation being the main effect and the factorial tillage and crop rotation treatments being sub plots within the main plots. The F-test was used to evaluate the treatment factors main effects and interactions. An F-protected t-test was used on a pair-wise comparison to follow up any significant findings. All results were considered significantly different at $p < 0.05$ unless noted otherwise.

3. Results and Discussions

3.1. Soil Organic Carbon on the Concentration Basis

The first scenario, where SOC and changes in SOC are presented on a concentration basis ($g \cdot C \cdot kg^{-1}$ soil) is used to eliminate the temporal variation and the error associated with soil ρ_b measurements [22,23]. This scenario could be used as an option if the temporal changes in soil ρ_b are high or if the soil ρ_b was not measured.

Average SOC in the 0 - 30 cm depth, expressed on a concentration basis, was 19.7% greater in 2008 compared with 2001 (**Table 1**). These data also show that SOC concentration was higher in 2008 than 2001 for every treatment combination, indicating that all management practices improved SOC at this study site. There were no treatment differences in SOC concentrations in 2001, because 2001 represents the initial condition for this study. However, SOC in 2008 was influenced by cropping system ($p = 0.004$), where SOC associated with CC system was 6.5% greater than the Rot system. All treatments gained SOC between 2001 and 2008, and the increase ranged between 0.13 to 0.34 $g \cdot kg^{-1} \cdot yr^{-1}$ (**Table 1**). The increase of SOC with CC plots was, on average 23%, where the increase of the Rot plots was 16%.

3.2. Soil Organic Carbon Stocks on the Equivalent Soil Mass of 2008 Basis (ESM)

Soil ρ_b in the 0 - 30 cm depth significantly decreased with time ($p < 0.0001$) by an average of 14.6% in 2008 compared with 2001 (**Table 2**). Averaged across all the treatment combinations, there was insignificant change in SOC, as evaluated by the FD basis, between 2008 and 2001. The SOC in the CC cropping system increased an average 7.4%, compared with an average of 2.3% with Rot system. Benjamin *et al.* [6] also concluded that SOC, measured on an FD basis, in NT plots gained an average of 2.1 $Mg \cdot ha^{-1}$, whereas the CP plots lost an average of 0.1 $Mg \cdot ha^{-1}$ at the 0- to 30-cm soil depth during the seven-year study period. Many researchers argue the fact that changes in soil ρ_b, and its effect on unequal soil mass associated with the FD, has a confounding effect on the SOC mass estimation [12-14,16,17]. The changes in ρ_b between 2001 and 2008 (**Table 2**) influenced soil mass, SOC, and the temporal change in SOC in the 0 - 30 cm layer, calculated on a fixed-depth basis. The principle behind the ESM calculation scenario is to evaluate the temporal changes in SOC mass by eliminating the changes in soil ρ_b caused by time and different management practices. In this scenario, the soil masses of 2001 were standardized to the soil masses of 2008 for each treatment. The decision to normalize soil mass of 2001 to the mass of 2008 was influenced by two facts: 1) the soil ρ_b and, consequently, the soil masses associated with 2001 were higher than 2008; and 2) no soil samples were measured below 30 cm depth, which provided no additional information about soil ρ_b and SOC below 30 cm for the 2008 sampling period. This is important because to adjust the low soil mass of 2008 to the high soil mass of 2001, a specific soil mass and SOC need to be added to soil mass of 2008 from the layer below 30 cm.

The reduction in soil masses from 2001 to 2008 ranged between 405 $Mg \cdot ha^{-1}$ to 810 $Mg \cdot ha^{-1}$. Likewise, the reduction in SOC from 2001 to 2008 ranged between 3.3 $Mg \cdot ha^{-1}$ to 6.3 $Mg \cdot ha^{-1}$ (**Table 2**). Using ESM as the calculation scenario showed that management practices throughout the seven year period significantly ($p < 0.0001$) increased SOC by an average of 19.7% in 2008 compared with 2001. The NT plots gained an average of 5.9 $Mg \cdot ha^{-1}$ of SOC, whereas the CP plots gained an average of 5.6 $Mg \cdot ha^{-1}$ of SOC between 2008 and 2001. Similarly, the CC treatment gained an average of 21% SOC, and the Rot plots gained an average of 16% SOC between 2008 and 2001. In fact, SOC stock and percent SOC increased over time calculated with this scenario are similar to what we observed previously when SOC was evaluated on a concentration basis.

Using the ESM scenario produced SOC increases in all of the treatments between 2001 and 2008. The SOC gained with this scenario was parallel to what we previously observed with SOC measured on a concentration basis. However, the SOC change between 2001 and 2008 with the ESM calculation scenario averaged 0.47 to 1.36 $Mg\ C\ ha^{-1} \cdot yr^{-1}$, compared with that of Benjamin *et al.* [6], which averaged −0.41 to 0.89 $Mg \cdot C \cdot ha^{-1} \cdot yr^{-1}$, (**Table 2**). Apparently, the soil ρ_b in both time periods had a confounding effect on the changes in how SOC was evaluated. These data agree with our hypothesis that normalizing the soil mass of 2001 to the ESM of 2008 for each treatment reduced the confounding effect of a changing soil ρ_b. Also these data revealed that different soil management practices affected SOC status at this study site.

3.3. Soil Organic Carbon Stocks on the Minimum Equivalent Soil Mass Basis (ESM_{min})

The minimum equivalent soil mass (ESM_{min}) is an approach for estimating SOC levels when the initial data were not previously measured [16]. The ESM_{min} approach is useful when soil ρ_b has decreased with time. In this method, all soil masses from different management practices are standardized to the lowest soil mass and lowest ρ_b across the treatments of interest. The lowest soil mass sampled in 2008, averaged across replications, was 3624 $Mg \cdot ha^{-1}$ (**Table 3**), and, therefore, was considered the ESM_{min}. The ESM_{min} was observed with delayed irrigation with CP tillage and CC rotation. Since the ESM_{min} was the lowest soil mass compared with any other treatment, the soil mass and associated SOC were subtracted from the 0 - 30 cm from all other treatments to adjust for ESM_{min}.

In this scenario, the soil masses were reduced by an average of 0.0 $Mg \cdot ha^{-1}$ to 486 $Mg \cdot ha^{-1}$ compared to soil masses calculated on an FD basis (**Table 3**). The reduction in soil mass, after the ESM_{min} adjustment, was greater with NT practice (300 $Mg \cdot ha^{-1}$) and CC system (246 $Mg \cdot ha^{-1}$) compared with CP practice (73 $Mg \cdot ha^{-1}$) and Rot system (127 $Mg \cdot ha^{-1}$). The greater amount of soil mass that was subtracted from NT practice was a consequence of higher soil ρ_b compared with CP practice (**Table 2**). After 7 yr of NT, soil ρ_b was 6.5% greater ($p = 0.003$) in those plots than in the CP plots, which may be due to yearly soil disturbance associated with CP treatment. Adjusting the soil masses to the ESM_{min} reduced

Table 3. Soil mass ($Mg \cdot ha^{-1}$) subtracted from the original soil mass of 2008 to adjust to the minimum equivalent soil mass (ESM_{min}) of 2008, soil organic C (SOC) mass ($Mg \cdot ha^{-1}$), and changes in SOC between 2008 calculated at ESM_{min} and SOC and on a fixed depth (FD) basis at 0 - 30 cm depth as influenced by irrigation, cropping system, and tillage practices.

			ESM_{min}		ΔSOC_{2008}[b]
Irrigation	Tillage	Cropping	Soil mass subtracted[a]	SOC	ESM_{min}-FD
			≤30 cm		
			$Mg \cdot ha^{-1}$		
Full	NT[c]	CC[d]	486	36.3	−5.1
Full	CP	CC	186	34.8	−1.7
Delayed	NT	CC	315	34.1	−3.2
Delayed	CP	CC	0	32.3	0.0
Full	NT	Rot	261	32.6	−2.0
Full	CP	Rot	15	31.4	−0.1
Delayed	NT	Rot	102	31.2	−1.0
Delayed	CP	Rot	54	34.2	-0.4
			ESM ($Mg \cdot ha^{-1}$)		
Equivalent soil mass			3624[e]		
			PR > F		
Irrigation (I)			0.12	0.29	0.13
Cropping (Cr)			0.009	0.007	0.004
Tillage (T)			0.003	0.59	0.004
I × Cr			0.09	0.039	0.09
I × T			0.42	0.16	0.19
Cr × T			0.048	0.64	0.039
I × Cr × T			0.30	0.09	0.33

[a]Soil mass subtracted from the original soil mass of 2008 at 0 - 30 cm to adjust to the minimum equivalent soil mass (ESM_{min}) basis (**Table 1**); [b]ΔSOC_{2008} = changes in SOC between ESM_{min} (presented in this table) and SOC on a fixed depth basis (**Table 1**) at 0 - 30 cm depth; [c]NT = No-tillage; CP = Chisel plow; [d]CC = continuous corn; Rot = mixed grass and broadleaf crops; [e]Minimum equivalent soil mass (ESM_{min}) represents the lowest soil mass in 2008 (**Table 1**) used to adjust the mass of the other treatments at 0 - 30 cm depth.

SOC an average of 0.0 $Mg \cdot ha^{-1}$ to 5.1 $Mg \cdot ha^{-1}$ (**Table 3**). The reduction in SOC was influenced by tillage ($p = 0.004$), cropping system ($p = 0.004$), and the two way interaction of tillage x cropping system ($p = 0.039$). Eliminating the variability of soil ρ_b by adjusting the soil masses associated with different treatment combinations to an ESM_{min} reduced the influence of tillage practices (**Tables 2** and **3**) on changes in SOC. The differences in SOC between tillage practices disappeared because a significant amount of SOC was subtracted from the NT practices to adjust all treatments to the ESM_{min} standard. SOC evaluated on an ESM_{min} basis was 4.8% lower ($p = 0.005$) than SOC evaluated on an FD basis. However, the effect of cropping system and the two way interaction, irrigation x cropping system, had the same influence on SOC (**Tables 2** and **3**) calculated with ESM_{min} and FD scenarios. Apparently, standardizing soil masses to an ESM_{min} eliminated the influence of tillage practices on SOC evaluation. These data agree with our previous hypothesis that evaluating SOC from a single sampling event on an ESM_{min} was more effective compared with FD scenario basis due to the elimination of the soil ρ_b variation associated with different management practices.

It is important to recognize that there is another scenario of SOC calculation based on a maximum equivalent soil mass (ESM_{max}) proposed by Lee *et al.* [16] for SOC changes from a single sampling event. Briefly, the ESM_{max} is based on using the greatest soil mass, on the fixed-depth basis, among the different management practices to which the other soil masses were normalized [16]. A specific soil mass, soil thickness, and its associated SOC are then added to the treatment that exhibits smaller soil mass than ESM_{max} chosen. The ESM_{max} calculation scenario may give us a different perspective of SOC influenced by different management practices, but this type of calculation was not performed for this data set due to the sampling depth limitations In this study, the lack of available SOC information associated with soil below 30 cm depth made it difficult to adjust soil masses to the ESM_{max}.

4. Conclusion

Evaluation of the temporal changes in SOC stock was influenced by calculation scenarios as a consequence of changes in soil ρ_b. The SOC estimation depends on how one calculates changes in soil mass associated with different sampling periods or management practices. Changes in SOC evaluated on the FD basis could be influenced by soil ρ_b variability. The temporal changes in the percent SOC gained when using the ESM calculation scenario were similar to what we observed when SOC change was calculated on a concentration basis. These data indicate that the ESM was more effective in evaluating SOC stock due to the similarity to the temporal changes in SOC concentration compared with the FD scenario. The ESM_{min} method appears to be an effective scenario for SOC evaluation from a single sampling event. We were unable to compare the ESM_{min} method with other evaluation scenarios due to the sampling depth limitations. Therefore, it is advisable to sample several centimeters below the chosen depth of interest to allow for SOC evaluation with different scenarios. Over all, the ESM scenario, where the temporal changes in soil ρ_b and soil mass were adjusted for each individual treatment, appears to be an effective scenario for evaluating SOC changes under these study conditions. We recommend, however, that the FD scenario also be included in future SOC evaluations so that comparisons with historical studies on changes in SOC with management and time can be made.

REFERENCES

[1] G. A. Peterson, A. D. Halvorson, J. L. Havlin, O. R. Jones, D. J. Lyon and D. L. Tanaka, "Reduced Tillage and Increasing Cropping Intensity in the Great Plains Conserves Soil C," *Soil and Tillage Research*, Vol. 47, No. 3-4, 1998, pp. 207-218. doi:10.1016/S0167-1987(98)00107-X

[2] A. D. Halvorson, G. A. Peterson and C. A. Reule, "Tillage System and Crop Rotation Effects on Dryland Crop Yield and Soil Carbon in the Central Great Plains," *Agronomy Journal*, Vol. 94, No. 6, 2002, pp. 1429-1436. doi:10.2134/agronj2002.1429

[3] M. M. Mikha and C. W. Rice, "Tillage and Manure Effects on Soil and Aggregate-Associated Carbon and Nitrogen," *Soil Science Society of American Journal*, Vol. 68, No. 3, 2004, pp. 809-816. doi:10.2136/sssaj2004.0809

[4] K. A. McVay, J. A. Budde, K. Fabrizzi, M. M. Mikha, C. W. Rice, A. J. Schlegel, D. E. Peterson, D. W. Sweeney and C. Thompson, "Management Effects on Soil Physical Properties in Long-Term Tillage Studies in Kansas," *Soil Science Society of American Journal*, Vol. 70, No. 2, 2006, pp. 434-438. doi:10.2136/sssaj2005.0249

[5] M. M. Mikha, M. F. Vigil, M. A. Liebig, R. A. Bowman, B. McConkey, E. J. Deibert and J. L. Pikul Jr., "Cropping System Influences on Soil Chemical Properties and Soil Quality in the Great Plains," *Renewable Agricultural and Food System*, Vol. 21, No. 1, 2006, pp. 26-35. doi:10.1079/RAFS2005123

[6] J. G. Benjamin, A. D. Halvorson, D. C. Nielsen and M. M. Mikha, "Crop Management Effects on Crop Residue Production and Changes in Soil Organic Carbon in the Central Great Plains," *Agronomy Journal*, Vol. 102, No. 3, 2010, pp. 990-997. doi:10.2134/agronj2009.0483

[7] C. P. Jantalia and A. D. Halvorson, "Nitrogen Fertilizer Effects on Irrigated Conventional Tillage Corn Yields and Soil Carbon and Nitrogen Pools," *Agronomy Journal*, Vol. 103, No. 3, 2011, pp. 871-878. doi:10.2134/agronj2010.0455

[8] A. D. Halvorson and C. P. Jantalia, "Nitrogen Fertilization Effects on Irrigated No-Till Corn Production and Soil

Carbon and Nitrogen," *Agronomy Journal*, Vol. 103, No. 5, 2011, pp. 1423-1431. doi:10.2134/agronj2011.0102

[9] A. D. Halvorson, C. A. Reule and R. F. Follett, "Nitrogen Fertilization Effects on Soil Carbon and Nitrogen in a Dryland Cropping System," *Soil Science Society of American Journal*, Vol. 63, No. 4, 1999, pp. 912-917. doi:10.2136/sssaj1999.634912x

[10] J. A. Amador, Y. Wang, M. C. Savin and J. H. Görres, "Fine-Scale Spatial Variability of Physical and Biological Soil Properties in Kingston, Rhode Island," *Geoderma*, Vol. 98, No. 1-2, 2000, pp. 83-94. doi:10.1016/S0016-7061(00)00053-7

[11] J. G. Benjamin, M. M. Mikha, A. D. Nielsen, M. F. Vigil, F. Calderon and W. B. Henry, "Cropping Intensity Effects on Physical Properties of No-Till Silt Loam," *Soil Science Society of American Journal*, Vol. 71, No. 4, 2007, pp. 1160-1165. doi:10.2136/sssaj2006.0363

[12] M. M. Mikha, M. F. Vigil and J. G. Benjamin, "Long-Term Tillage Impacts on Soil Aggregation and Carbon Dynamics under Wheat-Fallow in the Central Great Plains," *Soil Science Society of American Journal*, Vol. 77, No. 2, 2013, pp. 594-605. doi:10.2136/sssaj2012.0125

[13] B. H. Ellert and J. R. Bettany, "Calculation of Organic Matter and Nutrients Stored in Soil under Contrasting Management Regimes," *Canadian Society of Soil Science*, Vol. 75, No, 4, 1995, pp. 529-538. doi:10.4141/cjss95-075

[14] X.-M. Yang and M. M. Wander, "Tillage Effects on Soil Organic Carbon Distribution and Storage in a Silt Loam Soil in Illinois," *Soil and Tillage Research*, Vol. 52, No. 1-2, 1999, pp. 1-9. doi:10.1016/S0167-1987(99)00051-3

[15] R. M. Gifford and N. L. Roderick, "Soil Carbon Stocks and Bulk Density: Special or Cumulative Mass Coordinates as a Basis for Expression?" *Global Change Biology*, Vol. 9, No. 11, 2003, pp. 1507-1514. doi:10.1046/j.1365-2486.2003.00677.x

[16] J. Lee, J. W. Hopmans, D. E. Rolston, S. G. Baer and J. Six, "Determining Soil Carbon Stock Changes: Simple Bulk Density Fail," *Agriculture, Ecosystems and Environment*, Vol. 134, No. 3-4, 2009, pp. 251-256. doi:10.1016/j.agee.2009.07.006

[17] S. B Wuest, "Correction of Bulk Density and Sampling Method Biases Using Soil Mass per Unit Area," *Soil Science Society of American Journal*, Vol. 73, No. 1, 2009, pp. 312-316. doi:10.2136/sssaj2008.0063

[18] A. Gál, T. J. Vyn, E. Michéli, E. J. Kladivko and W. W. McFee, "Soil Carbon and Nitrogen Accumulation with Long-Term No-Till Versus Mouldboard Plowing Overestimated with Tilled-Zone Sampling Depths," *Soil and Tillage Research*, Vol. 96, No. 1-2, 2007, pp. 42-51. doi:10.1016/j.still.2007.02.007

[19] B. H. Ellert, H. H. Janzen and B. G. McConkey, "Measuring and Comparing Soil Carbon Storage," In: R. Lal, Ed., *Assessment Methods for Soil Carbon, Lewis Publishers*, CRC Press, Boca Raton, 2001, pp. 131-144.

[20] L. A. Sherrod, G. Dunn, G. A. Peterson and R. L. Kolberg, "Inorganic Carbon Analysis by Modified Pressure-Calcimeter Method," *Soil Science Society of American Journal*, Vol. 66, No. 1, 2002, pp. 299-305. doi:10.2136/sssaj2002.0299

[21] R. C. Littell, G. A. Milliken, W. W. Stroup, R. D. Wolfinger and O. Schabenberger, "SAS for Mixed Models," SAS Institute Inc., Cary, 2006.

[22] S. D. Logsdon and C. A. Cambardella, "Temporal Changes in Small Depth-Incremental Soil Bulk Density," *Soil Science Society of American Journal*, Vol. 64, No. 2, 2000, pp. 710-714. doi:10.2136/sssaj2000.642710x

[23] A. Kulmatiski and K. H. Beard, "Reducing Sampling Error in Soil Research," *Soil Biology and Biochemistry*, Vol. 36, No. 2, 2004, pp. 383-385. doi:10.1016/j.soilbio.2003.10.004

Soil Erosion around Foundations of Houses in Four Communities in Ghana

Kofi Agyarko*, **Joseph Adu, Daniel Gyasi, Samuel Kumi, Lydia Mensah**

College of Agriculture Education, University of Education, Winneba, Ghana.
Email: *agyarkokofi@yahoo.com

ABSTRACT

Soil erosion studies around houses in four communities in Ghana were carried out from August to November, 2010. Thirty houses were selected from each of the four communities for the study. Measurements of slope and the depth of exposed foundation of houses were done with the help of a string and a tape measure. Data were also obtained through questionnaire and interview of house owners. The Pearson's correlation coefficient was used to measure some relationships among data. In the selection of site for building a house, higher percentage of the house owners considered no factor. The major cleaning activity around the surrounding of houses was noted to be sweeping with a broom. All the houses in the communities had their foundations exposed, 57% - 93% of the houses had exposed foundations up to a level within 51 - 100 cm which was classified as "Severely Exposed". The extent of building foundation exposed by soil erosion was found to be positively correlated with the age of building and the slope of the land. Most of the respondents controlled soil erosion by blocking water ways with materials such as, sand heaps, stones and sacks filled with soil. Only 10% of the respondents from one of the communities planted grasses around their houses to control soil erosion. Communities need to be educated to plant grasses around their houses to beautify the surroundings and to control soil erosion.

Keywords: Soil Erosion; Foundations; Houses; Exposed

1. Introduction

Soil, a thin layer that covers most of the earth's land surface might appear insignificant relative to the bulk of the earth, yet it is in this thin layer of soil that living kingdoms establish a dynamic relationship [1]. Soil is an important natural resource which is vital for all living organisms as major ecosystems, as well as for food production, development and growth of settlements and population distribution. However, this layer is gradually being destroyed from its natural state through erosion as a result of man's activities in searching for a better living condition. Soil erosion causes tremendous damage to about one-third of world's cropland [2].

Global Assessment of Soil Degradation (GLASOD) study estimated that about 15 per cent of the earth's ice-free land surface is afflicted by all forms of land degradation. Of this, accelerated soil erosion by water is responsible for about 56 per cent and wind erosion is responseble for about 28 percent. This means that the area affected by water erosion is, around 11 million square km, and the area affected by wind erosion is around 5.5 million square km [3].

About 80% of the World's agricultural land suffers moderate to severe erosion and 10% suffers slight to moderate erosion. Soil erosion rates may exceed 100 ton per ha per year in severely overgrazed pastures [4].

Effects of erosion are enormous. At the household level, the onsite effects include damages to buildings, damages to drains and roads, destruction of landscape, and causing of flood. Erosion exposes building foundations which warrant routine repair of houses. In some cases erosion causes complete collapse of buildings and those buildings must be rebuilt and this adds to the expenditure of poor households. Sometimes erosion renders households homeless if houses collapse [5].

Some houses in older communities of Ghana are observed to have their foundations dangerously exposed by soil erosion. The paper attempts to assess the extent of soil erosion of foundation of houses and the steps taken to solve this problem by house owners in four communities in Ghana.

2. Materials and Methods

The study was conducted in four communities in Ghana,

*Corresponding author.

Nkrankwanta in Brong Ahafo Region, Agona and Mampong in Ashanti Region and Asankragua in Western Region from August to November, 2010. Within each of the four communities thirty houses/buildings were randomly selected for the study. Data were obtained through the use of questionnaire and physical measurement of percentage slope and the depth of the base (foundation) of houses exposed.

The questionnaire was made up of both open ended and close ended questions. Age of respondents, age of houses, factors considered in selecting sites for building houses, methods used to clean around houses and erosion control measures taken by house owners were the major elements of the questionnaire. The owners of the selected houses were the respondents.

The extent of foundation of a house exposed by erosion was measured from the ground level to the exposed foundation (**Figure 1**) with a tape measure. Measurement was taken at the part of the foundation where soil erosion seemed to be very severe.

The measurement of the slope of land at the spot of a house was done using the method described by Cowlitz County Building Planning Department [6] (**Figure 2**): Rise (H)/Run (D) × 100 = Slope in percent.

The data were analyzed using the Statistical Package for Social Sciences (SPSS) version 11. The relationships between the extent of foundation exposure of houses and age of houses and slope of land were found using the Pearson's correlation coefficient at 5% significance level.

3. Results and Discussion

As the studies were conducted in the old settlements of the communities many of the houses were found to be more than 30 years of age (**Table 1**). Majority of the owners of the houses studied in the four communities were above 45 years of age (**Table 2**) with most of them being males (67%, 80%, 67% and 57% for Nkrankwanta, Asankragua, Mampong and Agona respectively). All the four communities were Akan speaking societies with matrilineal inheritance system. Among the matrilineal Akan, family property is inherited without subdivision, in the first instance by the oldest surviving brother. When the whole generation of siblings dies out, the estate then goes to the eldest sister's eldest son [7]. It was therefore not surprising that most of the houses were found having males as the house owners.

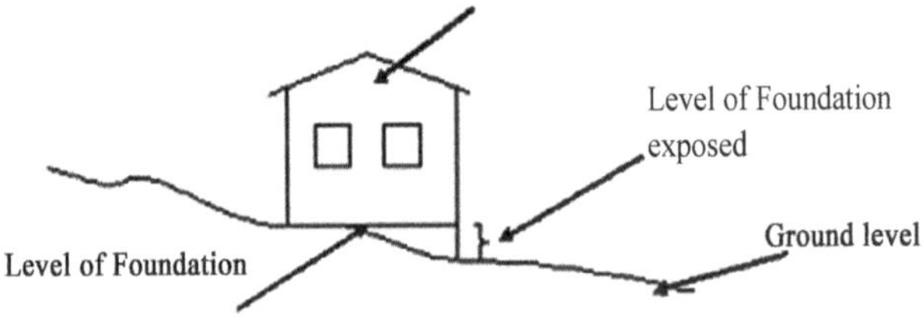

Figure 1. Measurement of the extent of foundation exposed.

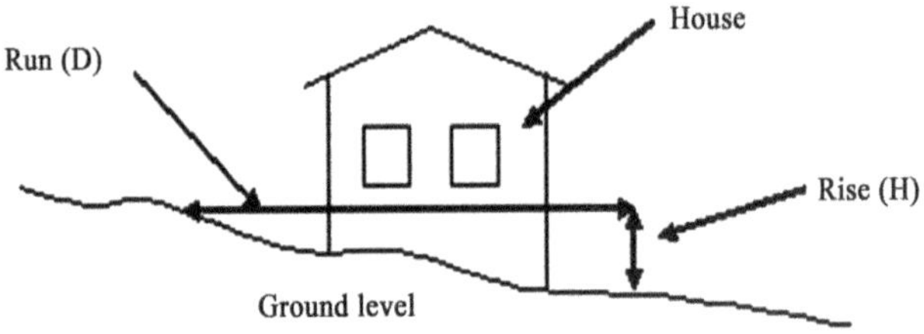

Figure 2. Measurement of slope.

Table 1. Age of houses.

Age (Years)	Community							
	Nkrankwanta		Asankragua		Mampong		Agona	
	Number of Houses	%	Number of Houses	%	Number of Houses	%	Number of Houses	%
Below 10	1	3	0	0	1	3	3	10
11 - 20	3	10	5	17	1	3	3	10
21 - 30	5	17	9	30	8	27	5	17
Above 30	21	70	16	53	20	67	19	63
Total	30	100	30	100	30	100	30	100

Table 2. Age of owners of surveyed houses.

Age of Respondents (Years)	Community							
	Nkrankwanta		Asankragua		Mampong		Agona	
	Frequency	%	Frequency	%	Frequency	%	Frequency	%
Below 25	0	0	0	0	2	7	1	3
26 - 35	6	20	2	7	5	16	5	17
36 - 45	6	20	9	30	3	10	7	23
Above 45	18	60	19	63	20	67	17	57
Total	30	100	30	100	30	100	30	100

In the selection of site for building a house, higher percentage (47%, 40%, 53% and 50% for Nkrankwanta, Asankragua, Mampong and Agona respectively) of the respondents considered no factor. Only few of them considered factors as, soil type and slope (**Table 3**). With the exception of those who purchased land before building a house most of the houses especially in the old settlements were built based on the availability of family land and hence not much emphasis was laid on the factors considered. Consideration of factors prior to building a house is important as these factors may have negative or positive impact on the house and its occupants in the future [8].

The more noted cleaning activity around the houses in the communities was sweeping (60%, 77%, 70% and 83% for Nkrankwanta, Asankragua, Mampong and Agona respectively) with a broom which was sometimes combined with weeding (**Table 4**). These major cleaning activities were seen and are still seen daily in communities in Ghana. Sweepings of these natures add high proportions of sand and stones to the communities' solid waste being carried to refuse dump sites as observed by Agyarko *et al.* [9]. Such removal of soil may act as precursor of soil erosion by water around houses in the communities.

Table 5 shows the level of exposure by soil erosion of foundations of houses in the communities. None of the houses was found to have its foundation exposed by 10 cm or less. All the houses had their foundations exposed to some degree. High percentage (60%, 57%, 93% and 87% for Nkrankwanta, Asankragua, Mampong and Agona respectively) of the houses had their foundations exposed up to a depth between 51 - 100 cm which was considered as "Severely Exposed" on the bases of the classification of the degree of exposed building foundations by the authors (**Table 6**).

Table 3. Factors considered in selecting site for building a house.

Factor	Community							
	Nkrankwanta		Asankragua		Mampong		Agona	
	Frequency	%	Frequency	%	Frequency	%	Frequency	%
Slope	6	20	7	23	11	37	9	30
Soil type	10	33	6	20	3	10	5	17
Drainage	0	0	5	17	0	0	1	3
No factor	14	47	12	40	16	53	15	50
Total	30	100	30	100	30	100	30	100

Table 4. Methods used to clean the surrounding of houses.

Cleaning Method	Community							
	Nkrankwanta		Asankragua		Mampong		Agona	
	Frequency	%	Frequency	%	Frequency	%	Frequency	%
Sweeping	18	60	23	77	21	70	25	83
Weeding	0	0	2	6	0	0	0	0
Sweeping and weeding	12	40	5	17	9	30	5	17
Others	0	0	0	0	0	0	0	0
Total	30	100	30	100	30	100	30	100

Table 5. Depth of foundation of houses exposed.

Depth Exposed (cm)	Community								Classification of exposed depth
	Nkrankwanta		Asankragua		Mampong		Agona		
	Number of Houses	%	Number of Houses	%	Number of Houses	%	Number of Houses	%	
Below 10	0	0.00	0	0.00	0	0.00	0	0.00	Unexposed
10 - 30	2	6.67	2	6.67	0	0.00	0	0.00	Exposed
31 - 50	7	23.33	5	16.66	2	6.67	2	6.67	Highly Exposed
51 - 100	18	60.00	17	56.67	28	93.33	26	86.66	Severely Exposed
Above 100	3	10.00	6	20.00	0	0.00	2	6.67	Extremely Exposed
Total	30	100	30	100	30	100	30	100	

Table 6. Classification of foundation exposure.

Depth Exposed (cm)	Classification
10	Unexposed
10 - 30	Exposed
31 - 50	Highly Exposed
51 - 100	Severely Exposed
100	Extremely Exposed

Most houses in the Mampong and Agona communities were found on the slope between 5% and 9%, most of the houses in the other communities were found on the slope between 1% and 5% (**Table 7**). Though foundation of houses may be exposed by soil erosion with time if proper measures are not put in place, the slope of the land is a major factor to be given a critical consideration. The slope therefore could have attributed to the percentage differences of the number of houses with their foundations considered "Severely Exposed" among the communities—93% and 87% for Mampong and Agona communities respectively and 60% and 57% for Nkrankwanta and Asankragua communities respectively (**Table 5**).

The extent of building foundation exposed by erosion was found to be positively correlated with the age of building and the slope of the land (**Table 8**).The slope had a higher positive correlation figure than the age of the building. Slope is considered to be a critical factor in determining soil erosion and fields with slopes of 10 - 15 percent slopes will likely be very prone to erosion [10]. Though, almost all the houses in the communities were built on land slopes of less than 9%, and may be considered less prone to soil erosion, the length of the slope which is also another critical factor in determining soil erosion could cause even a constant slope of 2 percent to erode severely [10].

Measures taken to control soil erosion around houses in the communities therefore are imperative to prevent collapse of houses which may lead to death of human beings. Apart from the respondents in Mampong where only 27% control soil erosion in their community more than half (53%, 60% and 67% for Nkrankwanta, Asankragua and Agona respectively) of the respondents in the other communities control soil erosion (**Table 9**). **Table 10** shows the various soil erosion control measures embarked by the communities. From the table it is observed that more than half (75%, 61%, 75% and 50% for Nkrankwanta, Asankragua, Mampong and Agona respectively) of the respondents controlled soil erosion by blocking water ways with materials such as, sand heaps, stones and sacks filled with soil. Few of the respondents redirected the flow of running water. Though planting of grasses has been found to be effective in soil erosion control, only 10% of the respondents from Agona community planted grasses around their houses to control soil erosion, the rest of the communities did not embark on this control measure and even weeded any kind of vegetation around their houses.

Table 7. Percentage slope.

Slope (%)	Community							
	Nkrankwanta		Asankragua		Mampong		Agona	
	Number of Houses	%	Number of Houses	%	Number of Houses	%	Number of Houses	%
Below 1	0	0	0	0	0	0	0	0
1 - 3	20	67	18	60	1	3	0	0
3 - 5	10	33	7	23	2	7	1	3
5 - 7	0	0	5	16	10	33	11	37
7 - 9	0	0	0	0	11	37	14	47
Above 9	0	0	0	0	6	20	4	13
Total	30	100	30	100	30	100	30	100

Table 8. Correlation between the extent of foundation exposure of houses and age of houses and slope of land.

	Community							
	Nkrankwanta		Asankragua		Mampong		Agona	
Factor	Age of building	Slope of land	Age of building	Slope of land	Age of building	Slope of land	Age of building	Slope of land
Extent of foundation exposure	0.269	0.703	0.244	0.808	0.479	0.947	0.148	0.555

Table 9. Responses to the control of soil erosion.

Response to erosioncontrol	Community							
	Nkrankwanta		Asankragua		Mampong		Agona	
	Frequency	%	Frequency	%	Frequency	%	Frequency	%
Control erosion	16	53	18	60	8	27	20	67
Do not control erosion	14	47	12	40	22	73	10	33
Total	30	100	30	100	30	100	30	100

Table 10. Erosion control measures.

Erosion control measure	Community							
	Nkrankwanta		Asankragua		Mampong		Agona	
	Frequency	%	Frequency	%	Frequency	%	Frequency	%
Redirection of water ways	4	25	7	39	2	25	8	40
Block water ways	12	75	11	61		75	10	50
Planting grasses	0	0	0	0	0	0	2	10
Total	16	100	18	100	8	100	20	100

4. Conclusions and Recommendations

All the houses in the communities had their foundations exposed, with higher number of them being severely exposed. Efforts were made by the communities to control soil erosion around their houses through blocking of water ways with materials such as, sand heaps, stones and sacks filled with soil, however, planting of grasses in the surrounding of houses was almost absent, surrounding of houses were seen bare without any kind of vegetation.

Communities should be encouraged to plant grasses around their houses to beautify the surroundings and to control soil erosion.

Alternative cleaning methods around the houses should be considered in the communities as the sweeping and weeding methods adopted by the communities carry soil away and induce soil erosion.

REFERENCES

[1] F. R. Troeh and L. M. Thompson, "Soils and Soil Fertility," Blackwell Publishing Ltd., Oxford, 2005.

[2] N. W. Hudson, "Soil Conservation," B.T Brantford Ltd., London, 1989.

[3] D. Favis-Mortlock, "The Extent of Soil Erosion," 2005. http://soilerosion.net/doc/extent_of_erosion.html

[4] D. Pimentel, C. Harvey, P. Resosudarmo, K. Sinclair, D. Kurz, M. McNair, S. Crist, L. Shpritz, L. Fitton, R. Saffouri and R. Blair, "Environmental and Economic Costs of Soil Erosion and Conservation Benefits," *Science, New Series*, Vol. 267, No. 5201, 1995, pp. 1117-1123. doi:10.1126/science.267.5201.1117

[5] J. Gyamfi-Aido and R. G. Congalton, "Application of GIS to the Assessment of Land Degradation and Associated Hazards in the Upper East Region of Ghana," *Proceedings of 2nd International Conference, Extracts and Workshop on GIS*, San Francisco, 26-30 October 1987, pp. 487-497.

[6] Cowlitz County Building Planning Department, "Calculating Slope," 2008. http://www.co.cowlitz.wa.us/buildplan/forms/Calculating%20Slope

[7] Advameg, "Countries and Their Culture—Culture of Ghana—History, People, Clothing, Traditions, Women, Beliefs, Food, Customs, Family," 2011. http://www.everyculture.com/GeIt/Ghana.html#ixzz1bbv67slj

[8] Permalink, "Home Building Remodeling," 2008. http://www.homebuildingremodeling.com/cgi-bin/mt-tb.cgi/6

[9] K. Agyarko, E. O. Akoto, and A. Awuah, "Characterization of Solid Wastes Generated by a Community in Ghana—The Case of Asante Mampong," *Journal of Applied Science and Technology*, Vol. 12, No. 1-2, 2007, pp. 46-51.

[10] Google's Cache, "Factors Affecting Soil Erosion," 2011. http://www.spindigo.com/doctorluke/www/Transcripts/agriculture_eng/2_10_soil_erosion_factors.htm.

22

Morphological and Physico-Chemical Characteristics and Classification of Vertisol Developed on Deltaic Plain

Orhan Dengiz*, Mustafa Sağlam, F. Esra Sarioğlu, Fikret Saygin, Çağla Atasoy

Department of Soil Science and Plant Nutrition, Faculty of Agriculture, Ondokuz Mayıs University, Samsun, Turkey.
Email: *odengiz@omu.edu.tr

ABSTRACT

The name of Vertisol is derived from Latin "vertere" meaning to invert. This case restricts development of soil horizons in profile. These soils have the capacity to swell and shrink, inducing cracks in the upper parts of the soil and distinctive soil structure throughout the soil. The formation of these specific features are caused by a heavy texture, a dominance of swelling clay in the fine fraction and marked changes in moisture content. The swell-shrink behavior is attributed to the wetting and drying of the soil mass. In this study, morphology, physico-chemical characteristics and classification of vertisols that were formed on alluvial delta plains, were investigated. Those soils formed on the Bafra Plain found in the Kızılırmak Delta and located in the central Black Sea region of Turkey. All studied Vertisols are characterised by a dark colour in surface soil, a heavy clayey texture, hardpan formation under top soil (high bulk density a high compaction) and very high COLE values. In addition, they have deep wide-opened desiccation cracks at the surface, slickensides at the middle part of the profiles and a poor differentiation of their horizons. Physico-chemically, the studied soils are slightly basic to very basic, non-saline and poor in organic matter, which is slightly higher in the surface horizon. In addition, cation exchange capacity, sum of exchangeable bases and base saturation of soils are very high. On the basis of morphological and physicochemical analysis, soil profiles were classified as Sodic Haplustert, Typic Calciaquert, Sodic Calciustert according to Soil Taxonomy (Soil Survey Staff, 1975 and 1999) and as Sodic Vertisol and Calcic Vertisol according to FAO/ISRIC (2006) classification systems.

Keywords: Vertisol; Soil Morphology; Soil Classification; Bafra Delta Plain

1. Introduction

The materials that form vertisols can be either allochtonous or autochtonous in origin [1]. The former are geographically more extensive and generally occur in the lower parts of the landscape. Vertisols are known to develop on a wide variety of parent materials such as basalts in Australia [2], calcareous rocks in the West Indies [1], gneisses and sandstones in India [3], deltaic deposits in the United States of America [4], lacustrine deposits in Trinidad [5], glacio-lacustrine in Saskatchewan [6], marine deposits in Guyana [7], and marls [8]. It is reported that in all cases (except for the vertisols developed on lacustrine deposits), the materials were recently deposited and that soil formation was still at its early stages [9]. In the case of vertisols developed from lacustrine deposits in Trinidad, it is believed that extensive weathering and clay-mineral synthesis had occurred prior to deposition [10].

The name of vertisol is derived from Latin "vertere" meaning to invert. This case restricts development of soil horizons in profile [9]. These soils have the capacity to swell and shrink, inducing cracks in the upper parts of the soil and distinctive soil structure throughout the soil [11]. The formation of these specific features are caused by a heavy texture, a dominance of swelling clay in the fine fraction and marked changes in moisture content [12]. The swell-shrink behaviour is attributed to the wetting and drying of the soil mass. Vertisols exhibit cracks a depth of 50 cm down that are at least 1 cm wide and extend upward to the surface or the base of the plough layer or surface crust. These soils exhibit minimal horizon differentiation as a result of pedoturbation. They are also very plastic and sticky when wet [13]. Vertisols are described by Glossary of Soil Science Terms [14] as "mineral soils that have 30% or more clay, deep wide cracks when dry and either gilgai microrelief, intersecting slickenside or wedge-shaped structural aggregates tiled at an angle from the horizon. It was added as an order in US system of soil taxonomy.

Depressions and level to undulating areas, mainly in

*Corresponding author.

tropical, semi arid to sub humid and Mediterranean climates with and alternation of district wet and dry seasons. An estimated 150 million hectares is potential crop land [15]. According to a map called the Turkey Soil Zones Map at the scale of 1:2,000,000 prepared from the results of the Turkey Development Soil Maps Survey at a scale of 1:100,000, vertisols comprise 598,693 hectares or 0.86 percent of the land area of Turkey [16,17].

Vertisols are important agricultural soils in left side of Bafra Delta Plain. Soils are mostly used for rice cultivation. The main aims of this research were to determine the morphological and some physico-chemical characteristics of these soils and to classify the soils according to the USDA Soil Taxonomy and FAO/ISRIC soil classification systems.

2. Material and Methods

2.1. Description of the Study Area

This study was carried out in the left side of Bafra Plains found in the Kızılırmak delta located in the central Black Sea region of Turkey (**Figure 1**). The Bafra Delta Plain is far 30 km from north of the Samsun province. The current climate in the region is semi-humid. The summers are warmer than winters (the average temperature in July is 22.2 and in January is 6.9°C). The mean annual temperature, rainfall and evaporation are 13.6°C, 764.3 mm and 726.7 mm respectively. According to [18], soil temperature regime is mesic and moisture regime is ustic in the study area. The mean annual temperature, rainfall and evaporation are 13.6°C, 764.3 mm and 726.7 mm respectively in Bafra plain that area is mainly flat and slightly

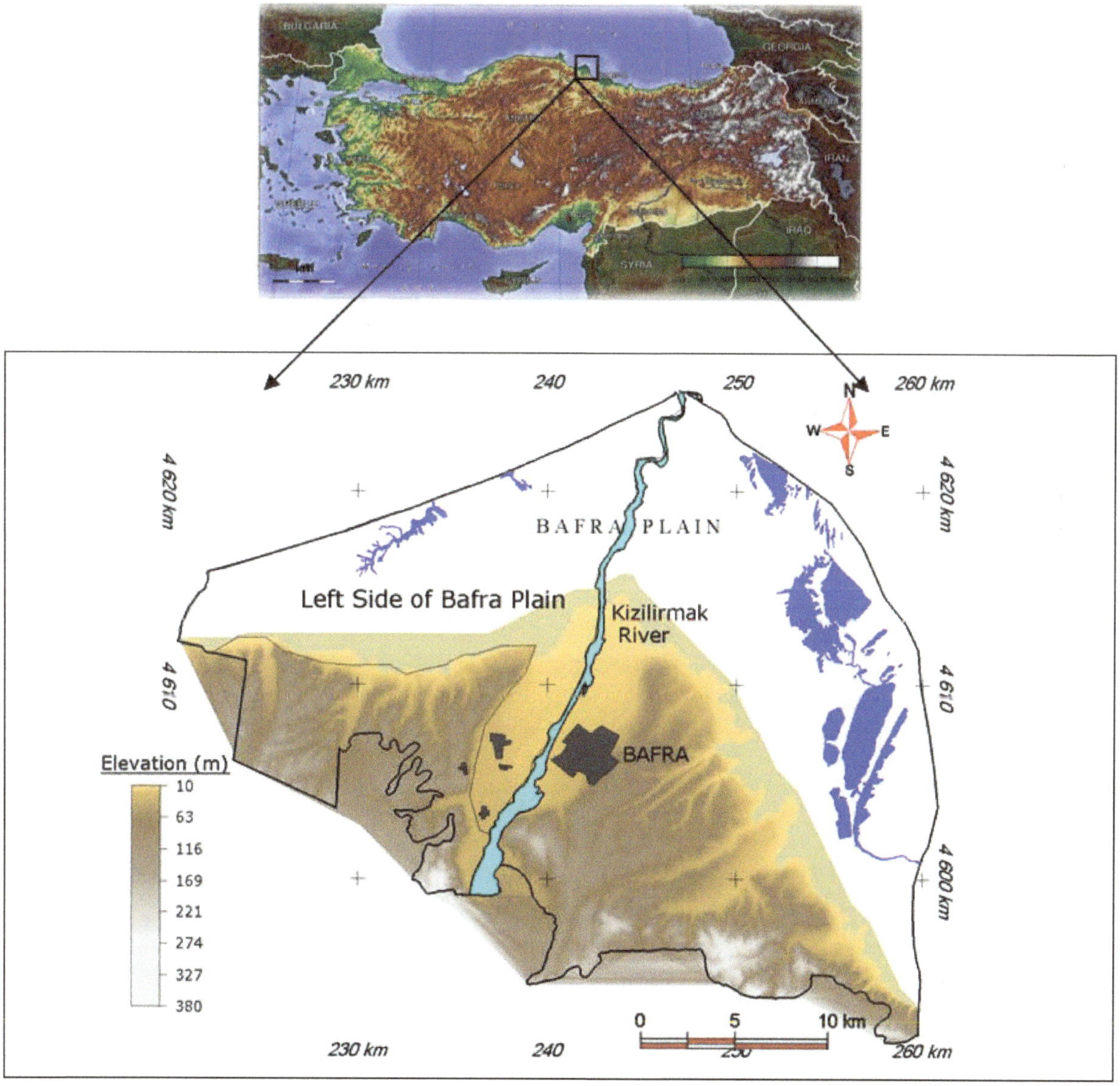

Figure 1. Location map of the study area.

sloped (0% - 2.0%). The study area has been under intensive agricultural activities. Rice, wheat, maize, pepper, watermelon, cucumber and tomato with sprinkler and furrow irrigations in the summer, and cabbage and leek in the winter have been produced in the study area.

2.2. Physical and Chemical Analysis

Four vertisol profiles formed on deltaic deposits were selected for this study in left side of Bafra delta plains. Morphological properties of these four soil profiles in the field were identified and sampled by genetic horizons. Nineteen soil samples were taken to investigate for their physical and chemical properties at the laboratory. Disturbed soil samples were dried under atmospheric condition and passed through a 2 mm sieve to prepare for laboratory analysis. In soil samples, particle size distribution was determined by the hydrometer method [19]. Dry bulk density was determined by the core method [20]. Soil reaction (pH), electrical conductivity (EC) and $CaCO_3$ content were determined by standard procedure [21]. Exchangeable cations and cation exchange capacities (CEC) were measured using a 1 N NH_4OAc (pH 7) extraction method [22]. Organic matter content was determined using the Walkley-Black wet digestion method [23]. Coefficient of linear extensibility (COLE) was determined according to [18]. In addition, these four soil profiles were classified according to [18-24].

3. Results and Discussions

3.1. Soil Morphology and Classification

Two main types of vertisols can be distinguished: Lithomorphic vertisols and topomorphic vertisols [24,25]. Lithomorphic vertisols are formed on various parent rocks whose weathering generates base-rich environments favourable for smectite synthesis, while topomorphic vertisols are formed mainly in favour of low landscape positions which favour the accumulation of bases. In this study, morphological characteristics of selected four soil profiles formed on alluvial delta plains are presented in **Table 1**.

Table 1. Selected morphological characteristics of pedons.

Horizon	Depth (cm)	Boundary	Colors		Structure	Consistence	Special features
			Dry	Moist			
Pedon I (Sodic Haplustert/Sodic Vertisol)							
Ap	0 - 18	as	10YR3/2	10YR3/3	3mgr	sh fi st pt	cracks
Bss1	18 - 45	cw	10YR4/3	10YR4/3	3mabk	h fi st pt	cracks, slickenside
Bss2	45 - 70	dw	2.5YR6/2	2.5YR5/2	3mpr	h fi st pt	cracks, slickenside
Bss3	70 - 116	dw	2.5YR5/3	2.5YR5/3	3cpr	h fi st pt	cracks, slickenside
C	116+	-	10YR5/3	10YR5/4	m	h fi st pt	-
Pedon II (Typic Calciaquert/Calcic Vertisol)							
Ap	0 - 25	aw	2.5YR4/3	2.5YR4/4	3mgr	h fi st pt	cracks
Bss1	25 - 70	cw	5Y5/3	5Y5/2	3mpr	h fi st pt	cracks
Bss2	70 - 116	cs	5Y6/1	5Y6/2	3mpr	h fi st pt	slickenside
Ck	116+	-	2.5YR7/3	2.5YR6/3	sg	so fr ss ps	carbonate mycelium and nodules, water table
Pedon III (Sodic Haplustert/Sodic Vertisol)							
Ap	0 - 15	as	2.5YR4/3	2.5YR4/3	3mgr	sh fi st pt	cracks
Ad	15 - 44	gw	2.5YR5/3	2.5YR4/4	3msbk	h fi st pt	densitic layer
Bssg1	44 - 73	gw	10GY6/1	10GY5/1	3mpr	h fi st pt	slickenside, redoximorphic feature
Bssg2	73 - 108	dw	10GY6/1	10GY5/1	2mpr	h fi st pt	slickenside, redoximorphic feature
Cg	108+	-	2.5YR6/3	2.5YR6/3	m	sh fi ss pt	redoximorphic feature
Pedon IV (Sodic Calciustert/Sodic Vertisol)							
Ap	0 - 18	as	10YR4/2	10YR3/2	3cgr	sh fi st pt	cracks
Ad	18 - 50	aw	10YR 4/3	10YR 3/4	3msbk	sh fi st pt	densitic layer
Bss	50 - 82	cw	2.5YR6/3	2.5YR6/3	3mpr	sh fi st pt	slickenside
Bssk	82 - 125	cw	2.5YR7/3	2.5YR6/3	3mpr	si fi st pt	slickenside, carbonate mycelium and nodules
C	125+	-	2.5YR7/3	2.5YR6/3	sg	so fr ss ps	-

Abbreviations: Boundary: a = abrupt; c = clear; g = gradual; d = diffuse; s = smooth; w = wavy; i = irregular Structure: 1 = weak; 2 = moderate; 3= strong; sg = single grain; m = massive; vf = very fine; f = fine; m =medium; c = coarse; gr = granular; pr = prismatic; abk = angular blocky; sbk = subangular blocky. Consistance: (Dry) lo = loose; so = soft; sh = slightly hard; h = hard; (Moist) lo = loose; vfr = very friable; fr = friable; fi = firm; (Wet) so = nonsticky; ss = slightly sticky; st = sticky; po = nonplastic; ps = slightly plastic; pt = plastic.

[11,26,27] stated that many vertisols have deep and dark topsoil. In addition, [28] reported the dark colour (low chroma) could be related to the strong impregnation of profile by organic matter during pedogenesis or to prolong waterlogging. In this research, colour of the all profiles in hue changed 10YR and 2.5Y, value was 3 - 4 (dry and moisture conditions) in surface soils. Thus, the finding was showed coherent with these researchers. On the other hand, colour changed due to redoximorphic feature (10GY 6/1) in subsurface horizon in Profile III and carbonate accumulation (high value-2.5YR7/3) in Ck and Bssk horizons of PII and PIV. Besides, carbonate nodules and mycelium were also recognised in these pedons within 120 cm depth due to calcification process.

Extensive swelling and shrinking upon wetting and drying is the major characteristics of these soils. This results to pedoturbation or mixing of the soils and minimal horizonation [9]. Shrinkage causes the formation of wide deep cracks and washing of surface materials into these cracks. Vertisols may exhibit open cracks, which are up to 50 cm deep and at least 1 cm wide and extend upward to the surface or the base of the plow layer or surface crust [13]. Moreover, cracks are a unique feature in soils with strong shrink–swell potential and are used as one of the criteria in defining vertisols and vertic intergrades in soil taxonomy [18]. Cracking takes place as a result of seasonal fluctuation. The moisture regime influences the duration and intensity of cracking. The degree and frequency of changes in moisture content of the soil are perhaps the most important parameters that control cracking intensity [6]. Cyclic cracking and swelling contribute to the formation of slickensides. Most of the subsurface horizons (except for C horizons) of profiles were characterised by numerous smooth and shiny surfaces called slickensides that separated different blocks and prismatic; their diameters varied between 5 and 15 cm. While Ap horizons of profiles have strong, medium and coarse granular structures, due to high clay content blocky and prismatic structures were determined in 100 cm of profiles. Consistencies of all B horizon of profiles were very hard when dry and plastic when wet conditions.

Due to the fact that main soil formation process and other features such as slickenside, cracking, carbonate accumulation and redoximorphic symptoms, four soil profiles were classified as Sodic Haplustert, Typic Calciaquert, Sodic Calciustert by taking into consideration of Soil Taxonomy [18]. In addition, these soils were classified as Sodic Vertisol and Calcic Vertisol according to FAO/ISRIC [24].

3.2. Physico-Chemical Characteristics

3.2.1. Chemical Properties

Vertisols are generally found on sedimentary plains as the result of thousands of years eroding the clay content out of the surrounding hills. They can also be found on level land and in depressions. [29] also reported that vertisols are typically developed on alluvial material in flat inland areas. The pedons selected for the present study, it was found significantly vation in pH values among all the soil layers. The pH values varied from 7.84 to 9.39. The increase in pH with depth of pedons is a characteristic feature of all pedons and could imply the proximity of the horizons to the alluvial parent material that is rich in alkaline earth cations (Ca and Mg) and water table fluctuation due to sea effect that includes high sodium ion concentration. In parallel, increasing of sodium concentration with depth caused increasing of ESP values in all pedons. [30] reported that high ESP values in vertisols cause dispersion of fine particles, leading to blockade of pore spaces. This ultimately renders the soils non-porous as reflected by low permeability. Poor drainage conditions lower the rate of organic matter decomposition. The organic matter values varied from 3.2% to 4.8% in surface horizons whereas, these ratios decreased with depth and varied from 0.5% to 2.5%. The higher organic matter at the surface could be linked to the continuous addition of crop residues on the surface of cropped fields in the area. The cation exchange capacity for all the studied soils ranged between 32.6 and 80.7 $cmol \cdot kg^{-1}$ of soils (**Table 2**). It increased regularly with profile depth. Calcium and magnesium were the dominant exchangeable cations, with contents ranging between 25.6 and 73.5 $cmol \cdot kg^{-1}$ of soils. It was followed by sodium (3.0 to 12.2 $cmol \cdot kg^{-1}$), while potassium showed more modest values (0.6 to 1.0 $cmol \cdot kg^{-1}$). The $CaCO_3$ content of the soils increased with depth and values varied from 1.4% to 34.7%. The highest values were observed at the subsurface horizons in Profile II and Profile IV due to calcium carbonate leaching and accumulation process. [31] reported that pedogenic carbonates accumulation in the Texas Gulf Coast Prairie Vertisols form by leaching of detrital carbonate in the fine earth fraction of the sediments to just below the effective meteoric infiltration.

3.2.2. Physical Properties

The soils selected for the present study belong to vertisols and their intergrades. Logically, the sand, silt and clay content should be within the range to qualify them as Vertisols [18]. The contents of sand, silt and clay vary as shown in **Table 3**. The clay content of all studied pedons varies from 52.2% to 85.7% except for C horizons of pedons while, sand content changes between 5.3% - 17.6% in pedogenetic horizons. While the clay contents were the highest in Pedon IV and ranged between 80.1% and 85.7% within 80 cm depth, the highest sand content (20.0%) was observed in Pedon II. The representation of various grain size fractions on [32] Jamagne's textural diagram showed that the all soils layers of pedons had a heavy clayey texture (**Figure 2**).

Table 2. Some chemical properties of pedons.

Horizon	Depth (cm)	pH	EC (dS m^{-1})	$CaCO_3$ (%)	O.M (%)	CEC (cmol·kg^{-1})	ESP (%)	Exchangeable cations (cmol·kg^{-1})		
								Na^+	K^+	$Ca^{++} + Mg^{++}$
Pedon I—Coordinate: North: 4,616,083 m - East: 737,858 m (UTM) (Sodic Haplustert/Sodic Vertisol)										
Ap	0 - 18	7.89	0.88	9.1	3.9	64.2	4.7	3.0	0.9	62.1
Bss1	18 - 45	8.39	0.51	13.4	2.3	60.6	5.9	3.6	0.8	57.6
Bss2	45 - 70	8.60	0.73	12.5	1.8	56.3	8.2	4.6	0.8	52.5
Bss3	70 - 116	9.04	0.45	9.1	1.6	59.8	14.2	8.5	0.9	52.7
C	116+	9.15	0.70	12.4	1.1	53.5	17.4	9.3	0.9	46.0
Pedon II—Coordinate: North: 4,616,541 m - East: 737,542 m (UTM) (Typic Calciaquert/Calcic Vertisol)										
Ap	0 - 25	8.30	0.73	1.4	3.2	65.4	5.0	3.3	0.9	62.9
Bss1	25 - 70	8.52	1.99	6.3	2.5	61.7	12.8	7.9	0.8	55.3
Bss2	70 - 116	8.86	0.91	6.6	1.2	53.7	17.7	9.5	0.8	46.0
Ck	116+	9.39	0.48	34.7	0.5	33.9	13.6	4.6	0.6	30.6
Pedon III—Coordinate: North: 4,617,456 m - East: 736,809 m (UTM) (Sodic Haplustert/Sodic Vertisol)										
Ap	0 - 15	7.84	1.36	11.6	4.6	80.7	7.2	5.8	1.0	72.8
Ad	15 - 44	8.49	1.25	13.7	2.4	73.9	12.3	9.1	0.8	70.4
Bss_1	44 - 73	8.97	1.24	14.1	2.0	54.2	17.9	9.7	0.8	50.8
Bss_2	73 - 108	9.07	0.68	11.0	1.7	58.0	15.0	8.7	0.7	53.2
C	108+	8.94	0.82	21.6	1.3	56.0	16.8	9.4	0.7	50.1
Pedon IV—Coordinate: North: 4,618,055 m - East: 736,735 m (UTM) (Sodic Calciustert/Sodic Vertisol)										
Ap	0 - 18	7.92	1.15	11.7	4.8	78.7	9.0	7.1	1.0	73.5
Ad	18 - 50	8.78	0.36	15.0	2.3	71.0	13.0	9.2	0.8	67.8
Bss	50 - 82	8.89	0.71	10.9	1.9	63.6	18.2	11.6	0.9	58.3
Bssk	82 - 125	9.22	0.54	28.3	1.9	62.8	19.4	12.2	0.8	56.4
C	125+	8.26	2.92	6.7	0.9	32.6	30.1	9.8	0.6	25.6

EC: Electrical conductivity; CEC: Cation exchange capacity; O.M: Organic matter; ESP: Exchangeable sodium percentage.

Table 3. Some physical properties of pedons.

Horizon	Depth (cm)	COLE	Particle size distribution (%)				BD (gr·cm^{-3})
			Clay	Silt	Sand	Class	
Pedon I (Sodic Haplustert/Sodic Vertisol)							
Ap	0 - 18	0.221	68.1	24.2	7.8	C	1.15
Bss1	18 - 45	0.274	66.0	23.4	10.6	C	1.21
Bss2	45 - 70	0.264	65.1	21.8	13.0	C	1.23
Bss3	70 - 116	0.285	79.4	13.2	7.4	C	1.29
C	116+	0.225	72.9	13.7	13.5	C	1.34
Pedon II (Typic Calciaquert/Calcic Vertisol)							
Ap	0 - 25	0.211	68.1	17.4	14.5	C	1.24
Bss1	25 - 70	0.281	72.1	15.7	12.2	C	1.22
Bss2	70 - 116	0.191	61.5	17.5	21.0	C	1.27
Ck	116+	0.025	28.9	35.3	35.8	L	1.35
Pedon III (Sodic Haplustert/Sodic Vertisol)							
Ap	0 - 15	0.280	79.7	13.3	7.0	C	1.23
Ad	15 - 44	0198	80.4	9.1	10.5	C	1.48
Bss1	44 - 73	0.190	55.1	29.4	15.4	C	1.27
Bss2	73 - 108	0.191	52.2	34.9	13.0	C	1.30
C	108+	0.111	40.5	10.7	48.8	C	1.50
Pedon IV (Sodic Calciustert/Sodic Vertisol)							
Ap	0 - 18	0.269	80.5	11.0	8.5	C	1.20
Ad	18 - 50	0.219	85.7	9.0	5.3	C	1.57
Bss	50 - 82	0.231	80.1	2.3	17.6	C	1.31
Bssk	82 - 125	0.174	62.5	25.9	11.6	C	1.35
C	125+	0.021	23.5	53.6	22.9	SiL	1.55

COLE: Coefficient of linear extensibility; BD: Dry bulk density; C: Clay, SiL: Silty loam; L: Loam

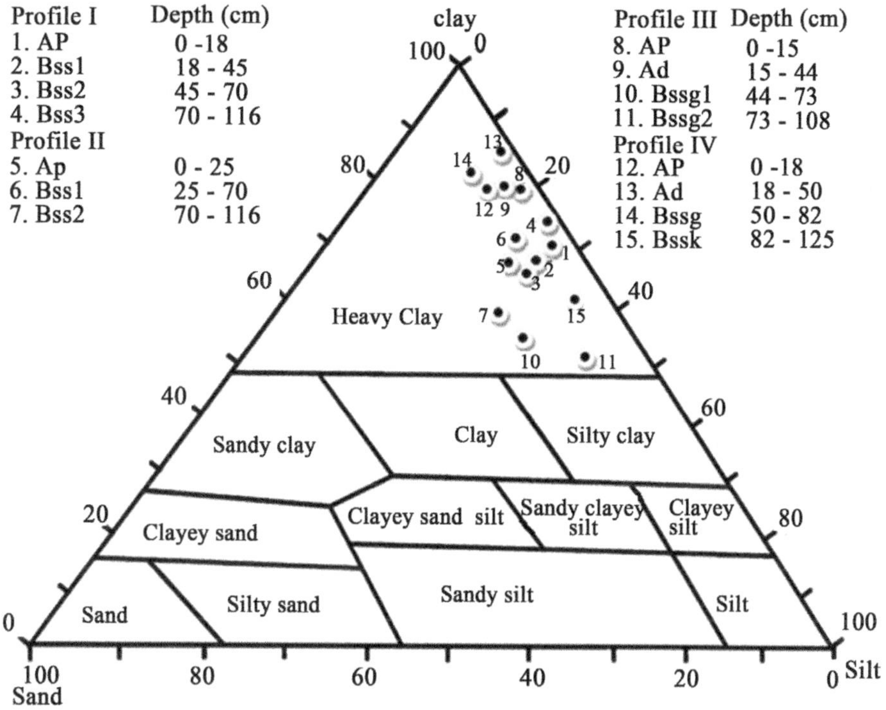

Figure 2. Soil textural classes with reference to Jamagne's triangle (1967).

Bulk density (BD) indicates the weight of all the organic and inorganic materials of a given volume of soil. Higher organic matter lowers BD. High clay content and introduction of farm machinery causes compaction of subsurface layer, which increases the BD. Bulk density also varies with the content of coarser fragments in soils. There is a growing acceptance of the view that BD changes due to change in land use pattern [33]. A study was conducted with soil profiles from arable area used for field crop (rice) and vegetables for 30 - 40 years. The highest BDs were determined in parent materials of all pedons including high sand content. On the other hand, BD of pedogenetic horizons ranges between 1.15 and 1.35 $g \cdot cm^{-3}$ except for Ad horizons of Pedon III and IV that soils have particularly used for rice cultivation. Due to heavy tillage system and field traffic on these soils leading to hard pan formation, BD increased 1.57 $g \cdot cm^{-3}$. Similar results were reported by [34] that over the years, continuous tillage practices resulted bulk density increased from 1.22 $g \cdot cm^{-3}$ to 1.38 $g \cdot cm^{-3}$ in a tropical forest ecosystem of Bangladesh. In addition, due to compression caused by overburden weight, the BD values usually increase with soil depth in all pedons. It has been reported that vertisols may have BD values as high as 2.1 $g \cdot cm^{-3}$ [35]. It has been also shown that at the swelling limit, the gravimetric water content decreases and BD values increase with depth [36].

Coefficient of linear extensibility (COLE) helps to predict the potential of a soil to shrink and swell. The LE of a soil layer is the product of the thickness in centimetres, multiplied by the COLE of the layer in question. The COLE of a soil is the sum of these products for all soil horizons [18]. According to soil taxonomy, a soil should be qualified for *vertic* subgroups if the COLE value is more or equal to 6 between the mineral soil surface and either a depth of 100 cm or a lithic contact, whichever is shallower. In case of vertisols slickensides, cracks and higher COLE values are mutually inclusive. Since higher COLE values indicate the presence of more shrink-swell minerals, namely smectite, a positive correlation between COLE and smectite content can exist [37]. According to COLE classification given in **Table 4**, it was also found that high COLE values should have a positive relation with vertic characteristic properties of shrink-swell soils and cracking and vary from 0.174 to 0.281 in pedogenetic horizons of all pedons in **Table 3**.

Vertisols are important agricultural soils in left side of Bafra Delta Plain. However, vertisols are difficult to work, they are of very hard consistence when dry and very plastic and sticky when wet. Therefore, the workability of the soil is often limited to very short periods of medium (optimal) water status. Vertisols are imperfectly to poorly drained, leaching of soluble weathering products is limited, the contents of available sodium, calcium and magnesium are high and the pH is usually above 7.5. Once they have reached their field capacity, practically no water movement occurs, this is due to the very low hydraulic conductivity of a vertisol. In addition, flooding leading to crop damage can be a major problem in areas with higher rainfall.

Table 4. Coefficient of linear extensibility (COLE) classification.

Class	COLE
Low	<0.03
Moderate	0.03 - 0.06
High	0.06 - 0.09
Very high	>0.09

4. Conclusion

In this study, it was aimed at investigating the morpological and analytical characteristics of vertisols formed on fluvial land in Bafra Deltaic Plain, in order to highlight their particularities as well as to envisage their practical uses and management. From the morphological and physical point of view, the vertisols of the Bafra Deltaic Plain region are characterised by a dark colour, a heavy clayey texture, hardpan formation under top soil (high bulk density a high compaction) and very high COLE value, deep wide-opened desiccation cracks at the surface, slickensides at the middle part of the profiles and a poor differentiation of their horizons. Physico-chemically, the studied soils are slightly basic to very basic, non-saline and poor in organic matter, which is slightly higher in the surface horizon. Cation exchange capacity, sum of exchangeable bases and base saturation are very high. Overall, considering their characteristics in majority similar to those of a majority of world deltaic vertisols, those soils are still unused or are used only for extensive grazing, wood chopping, charcoal burning and the like [24]. In agriculture, optimum yields could be attained if appropriate management techniques are set up for a more efficient exploitation, protection and conservation of these soils. The comparatively good chemical fertility and their occurrence on extensive level plains where reclamation and mechanical cultivation can be envisaged are assets of vertisols. Their physical soil characteristics and, notably, their difficult water management cause problems. The physical properties and the soil moisture regime of vertisols represent serious management constraints. The heavy soil texture and domination of expanding clay minerals result in a narrow soil moisture range between moisture stress and water excess. Tillage is hindered by stickiness when the soil is wet and hardness when it is dry. The susceptibility of vertisols to waterlogging may be the single most important factor that reduces the actual growing period. Especially due to those properties, these vertisol soils have been generally used for rice cultivation in the study area. On the other hand, this case is not suitable other plants. Particularly, due to their very high of COLE values, namely, very high shrink-swell potential plant roots can be damaged from this soil activity. Zero till is commonly advocated as a preferred cropping system to conventional, multicultivation practices. Zero till is particularly attractive on clay soils to minimize compaction and induce natural structure formation [17]. In particular, the soil structure of vertisols has strong potential to attain optimal conditions for plant growth through activation of their inbuilt resiliency via shrink-swell cycles [38,39]. It is accepted that the major purposes of tillage are to reduce bulk density and soil strength and to control pests and diseases [40,41]. However, soil cultivation affects soil quality in various ways. With high clay content, cultivation may lead to the formation of a hard pan below the plough layer that restricts root penetration and downward movement of water [42]. Therefore, many researchers strongly recommend that close attention should be taken into account for the soil cultivation, irrigation system and time depending on the these soil types.

REFERENCES

[1] N. Ahmad and R. L. Jones, "Genesis, Chemical Properties and Mineralogy of Limestone Derived Soils, Barbados, West Indies," *Journal of Tropical Agriculture*, Vol. 46, 1969, pp. 1-15.

[2] J. S. Hosking, "A Comparative Study of the Black Earths of Australia and the Regur of India," *Transactions of the Royal Society of South Australia*, Vol. 59, 1935, pp. 168-200.

[3] D. V. Bal, "Some Aspects of the Black Cotton Soils of the Central Provinces," *Transactions of the 3rd International Congress of Soil Science*, Vol. 3, 1935, pp. 154-158.

[4] G. W. Kunze, H. Oakes and M. E. Bloodworth, "Grumusols of the Coastal Prairie of Texas," *Soil Science Society of America Journal*, Vol. 27, No. 4, 1963, pp. 412-421. doi:10.2136/sssaj1963.03615995002700040019x

[5] C. B. Brown and G. S. Bally, "Land Capability Survey of Trinidad and Tobago," Government Printer, Port of Spain, 1968.

[6] A. R. Mermut, D. F. Acton and C. Tarnocai, "A Review of Recent Research on Swelling Clay Soils in Canada," In: J. M. Kimble, Ed., *Proceeding of the Sixth International Soil Correlation Meeting, Characterization Utilization of Cold Aridisols and Vertisols*, Soil Survey Division USDASCS, Washington DC, 1990, p. 253.

[7] Food and Agriculture Organization of the United Nations, "Report on the Soil Survey Project of British Guiana," FAO, Rome, 1966.

[8] P. Duchafour, "Pedologie 1 Pedogenese et Classification," Masson, Paris, 1977.

[9] N. Ahmad, "Vertisols," In: L. P. Wilding, N. E. Smeck and G. F. Hall, Eds., *Pedogenesis and Soil Taxonomy. II. The Soil Orders. Developments in Soil Science IIB*, Elsevier. Amsterdam, 1983, pp. 91-123.

[10] G. Rodrigues and F. Hardy, "Soil Genesis from a Sedimentary Clay in Trinidad," *Soil Science*, Vol. 64, No. 2, 1947, pp. 127-142. doi:10.1097/00010694-194708000-00005

[11] N. Ahmad and A. Marmut, "Vertisols and Technologies for Their Management," Elsevier, Amsterdam, 1996.

[12] G. D. Hubble, "The Cracking Clay Soils, Definition, Distribution, Nature, Genesis and Use," In: J. W. McGarity, E. H. Hoult and H. B. So, Eds., *The Properties and Utilization of Cracking Clay Soils, Review in Rural Science*, University of New England Armidale, New South Wales, 1984, pp. 3-13.

[13] Soil Survey Staff, "Soil Taxonomy. A Basic System of Soil Classification for Making and Interpreting Soil Surveys," U.S. Department of Agriculture, Washington DC, 1975.

[14] Soil Science Society of America, "Glossary of Soil Science Terms," SSSA, Wisconsin DC, 1997.

[15] P. M. Driessen and R. Dudal, "Geography Formation, Properties and Use of the Major Soils of the World," The Netherlands and Catholic University, Luven, 1989.

[16] Anonymous, "According to FAO-UNESCO and New Soil Taxonomy the GDRS," General Directorate of Rural Services, Ankara, 1999.

[17] G. Özsoy and E. Aksoy, "Characterization, Classification and Agricultural Usage of Vertisols Developed on Neogen Aged Calcareous Marl Parent Materials," *Journal of Biological & Environmental Sciences*, Vol. 1, No. 1, 2007, pp. 5-10.

[18] Soil Survey Staff, "Soil Taxonomy USDA Agricultural Handbook," Natural Resources Conservation Service, Washington DC, 1999.

[19] G. J. Bouyoucos, "A Recalibration of the Hydrometer Method for Making Mechanical Analysis of Soils," *Agronomy Journal*, Vol. 43, 1951, pp. 435-438. doi:10.2134/agronj1951.00021962004300090005x

[20] G. R. Blacke and K. H. Hartge, "Bulk Density," In: A. Klute, Ed., *Methods of Soil Analysis. Part 1. Physical and Mineralogical Methods*, American Society of Agronomy- Soil Science Society of America, Madison, 1986, pp. 363-382.

[21] Soil Survey Staff, "Soil Survey Manual. USDA, Handbook," Natural Resources Conservation Service, Washington DC, 1993.

[22] Soil Survey Staff, "Procedures for Collecting Soil Samples and Methods of Analysis for Soil Survey," U.S. Government Print Office, Washington DC, 1992.

[23] D. W. Nelson, L. E. Sommers, "Total Carbon, Organic Carbon and Organic Matter," In: L. A. Page, R. H. Miller and D R. Keeney, Eds., *Methods of Soil Analysis, Part 2. Chemical and Microbiological Methods*, American Society of Agronomy, Madison, 1982, pp. 539-579.

[24] Food and Agriculture Organization of the United Nations/ International Soil Reference and Information Centre, "World References Base for Soil Resources," World Soil Reports, Rome, 2006.

[25] V. B. Nico, "Soil Formation," 2nd Edition, Kluwer Academic Publisher, Secaucus, 2002.

[26] P. A. Tamfuh, E. D. Woumfo, D. Bitom, D. Njopwouo, "Petrological, Physico-Chemical and Mechanical Characterization of the Topomorphic Vertisols from the Sudano-Sahelian Region of North Cameroon," *The Open Geology Journal*, Vol. 5, 2011, pp. 33-55.

[27] B. P. Yerima, "Weathering and Genesis of Volcanic Ash-Influenced Vertisols and Vertisol-Like Soils of El-Salvador," Master's Thesis, Texas A&M University, College Station, 1983.

[28] H. Ekinci, H. Özcan, Y. Yigini, V. Cavusgil, O. Yüksel and Y. Kavdir, "Profile Developments and Some Properties of Vertisols Formed on Different Physiographic Units," International Soil Congress on Natural Resource Management for Sustainable Development, Erzurum, 2004.

[29] H. Jenny, "The Soil Resource," Springer, New York, 1980.

[30] T. Bhattacharyya, P. Chandran, S. K. Ray, C. Mandal, D. K. Pal, M. V. Venugopalan, S. L. Durge, P. Srivastava, P. N. Dubey, G. K. Kamble, R. P. Sharma, S. P. Wani, T. J. Rego, P. Pathak, V. Ramesh, M. C. Manna and K. L. Sahrawat, "Physical and Chemical Properties of Red and Black Soils of Selected Benchmark Spots for Carbon Sequestration Studies in Semi-Arid Tropics of India," *Global Theme on Agroecosystems Report*, No. 35, 2007, p. 239.

[31] C. A. Stiles, "Pedogenic Relationships in a Vertisol Climosequence Defined by Geochemical Mass Balance of Whole Soil and Chemistry of Iron-Manganese Nodules," Ph.D. Thesis, The University of Tennessee, Knoxville, 2001.

[32] M. Jamagne, "Fundamentals and Techniques of Soil Mapping," *Annals of Agronomy Journal*, Vol. 18, 1967, pp. 271-414.

[33] R. Kızılkaya and O. Dengiz, "Variation of Land Use and Land Cover Effects on Some Soil Physico-Chemical Characteristics and Soil Enzyme Activity," *Zemdirbyste-Agriculture*, Vol. 97, No. 2, 2010, pp. 15-24.

[34] K. R. Islam and R. R. Weil, "Land Use Effects on Soil Quality in a Tropical Forest Ecosystem of Bangladesh", *Agriculture, Ecosystems and Environment*, Vol. 79, No. 1, 2000, pp. 9-16. doi:10.1016/S0167-8809(99)00145-0

[35] R. Dudal, "Dark Clay Soils of Tropical and Subtropical Regions," *Agriculture Development Paper*, No. 83, 1965, p. 161.

[36] D. F. Yule and J. T. Ritchie, "Soil Shrinkage Relationship of Texas Vertisols Small Cores," *Soil Science Society of America Journal*, Vol. 44, No. 6, 1980, pp.1285-1291. doi:10.2136/sssaj1980.03615995004400060031x

[37] S. K. Shirsath, T. Bhattacharyya and D. K. Pal, "Minimum Threshold Value of Smectite for Vertic Properties," *Australian Journal of Soil Research*, Vol. 38, No. 1, 2000, pp. 189-201. doi:10.1071/SR99051

[38] D. McGarry, "The Structure and Grain Size Distribution of Vertisols," In: N. Ahmad and A. Mermut, Eds., *Vertisols and Technologies for Their Management, Developments on Soil Science*, Elsevier, Amsterdam, 1996, pp. 231-259.

[39] U. P. Pillai and D. McGarry, "Structure Repair of a Compacted Vertisol with Wet-Dry Cycles and Crops," *Soil Science Society of America Journal*, Vol. 63, No. 1, 1999, pp. 201-210. doi:10.2136/sssaj1999.03615995006300010029x

[40] C. Culpin, "Farm Machinery," 10th Edition, Grand Publishing Ltd., London, 1981.

[41] R. L. Hill, "Long-Term Conventional Tillage and No-Tillage Effects on Selected Soil Physical Properties," *Soil Science Society of America Journal*, Vol. 54, No. 1, 1990, pp. 161-166. doi:10.2136/sssaj1990.03615995005400010025x

[42] P. K. Singh, and Y. Singh, "Effect of Reduced Tillage on Soil Properties, Root Growth and Grain Yield in Rice-Wheat System," *Indian Journal of Agricultural Research*, Vol. 30, 1996, pp.179-185.

23

Response of the Thermal Conductivity as a Function of Water Content of a Burnt Mediterranean Loam Soil

Carles M. Rubio[1,2], Xavier Úbeda[3], Francesc Ferrer[2]

[1]Department Agri-Food Engineering and Biotechnology, Polytechnic University of Catalonia, Barcelona, Spain; [2]Department Environmental Biophysics and Soils, Lab-Ferrer Soils and Environmental Consulting, Cervera, Spain; [3]Department Physical Geography, University of Barcelona, Barcelona, Spain.
Email: carles.rubio@upc.edu

ABSTRACT

The purpose of this research is to explore the variability on the soil thermal conductivity -λ- after a prescribe fire, and to assess the effects of the ashes on the heat transfer once it's were incorporated into the soil matrix. Sampling plot was located in the Montgrí Massif (NE of Spain). A set of 42 soil samples between surface and 5 cm depth was collected before and after the fire. To characterize the soil chemical and physical variables were analyzed. To determine the variability on the soil λ a dry-out curve per scenario (before and after fire) was determined. SoilRho® method based on ASTM D-5334-08 which was validated by LabFerrer was used. Soil thermal conductivity has shown changes in their values. Indeed, in all moisture scenarios the values of soil λ decreased after soil was burnt. The critical point in the relationship $\theta(\lambda)$ for the soil after fire which always was stronger than soil before to be burnt. Soil with "white" ashes showed a high thermal conductivity. An X-Ray diffractometry analysis allowed to clarify and to verify these results. To sum up, we could say that thermal conductivity presents changes when the scenario changes, *i.e.* before and after to be burnt. On the other hand, the volume of ashes incorporated on the soil increased the differences between no burnt and burnt soil, showing even some improvements on the heat transfer when water content started to govern the process.

Keywords: Thermal Properties; Soil Moisture; Prescribed Fire; Ashes; SoilRHO

1. Introduction

The impact of fire on soils can vary between highly beneficial, when the fire is not too intense and soil heating brief, to irreversible damage, which occurs during deeply penetrating heat pulses and long-term exposures [1,2]. Prescribed fires are used in Catalonia since 1999 as a tool, among others, for managing forested areas with large amounts of fuel in order to prevent high intensity fires. The Montgrí prescribed fire main objective was reduced the shrubland in an ancient and abandoned pine plantation. On the whole of the literature, many researchers have studied the variations on the chemical and physical properties on or above burnt soils, but there are other physical properties as are thermal properties that govern the heat flow transport inside the soil, and affect the aspects mentioned above. Because of the severity of the soil heating during a prescribed fire the impacts can be significant and serious. They include the formation of a hydrophobic layer on the surface of or within the soil, destruction of most of the organic material in the upper few centimeters of soil and the concomitant loss of soil aggregate stability [3], changes in soil pH and soil chemistry, long-term differences in soil moisture amounts, increases in soil bulk density with accompanying decreases in soil porosity, therefore changes on soil structure [2,4,5]. Thus, when biomass on or above a soil surface burns, a heat pulse penetrates the soil. The resulting high soil temperatures can alter soil properties and kill roots and soil microbes [6].

On the other hand, changes in the soil thermal conductivity (λ, a measure a soil's ability to conduct heat) are less obvious, but is not less significant because it is directly to many of the other fire included changes in the soil. Some of these changes would be changes in the structure, because soil thermal conductivity is strongly determined by soil structure [7], and soil composition [8,9], that occur whenever soil organic matter is combusted, therefore soil bulk density changes. Consequently, the purpose of this research is to explore the variability on the soil thermal conductivity -λ- after a prescribed burn for a natural and typical Mediterranean limestone soil to laboratory scale. For achieving the main goal, it was splited into up three operative objectives; 1) to observe the relationship between soil thermal conductivity and water content, 2) to evaluate the influence of several percentages of ashes on the soil thermal conductivity, and 3) to determine which was the impact of the ashes on thermal conductivity when the ashes were incorporated into the soil matrix, taking account several moisture scenarios.

2. Materials and Methods

The study area is located in the north-eastern corner of the Iberian Peninsula in the coastal mountains of Catalonia, within the calcareous Montgrí Massif. The vegetation of this area is typically Mediterranean, composed of Pinus plantation (*Pinus halepensis*) with shrubland of *Quercus coccifera, Cistus albidus, Rosmarinus officinalis and Pistacea lentiscus*. At the time of the fire, the air temperature was 12.5°C with an air relative moisture about 60%. In these conditions a prescribed fire was carried out. A set of 42 soil samples between surface and 5 cm depth was collected before and after the fire (UTM coordinates x: 514555 y: 4659552). The size of the plot is 18 × 4 meters, with a quadrangular structure (see **Figure 1**). Soil samples were taken before, and just after the fire, from 42 points arranged in three transects and three crosses across the central transect.

The fire temperature of the soil surface was measured with a laser thermometer. To characterize the soil chemical and physical variables were analyzed. Particle-size distribution was determined using the wetting sieve method for 2000 to 50 μm and a device by dispersion laser beams (Malvern Mastersizer/E) for particles smaller than 50 μm. Bulk density and porosity were determined from undisturbed sample volume. Calcium carbonate content was determined based on Bernard calcimeter [10], whereas the hygroscopic water content was determined by weight differences after drying the samples at 105°C during 24 hours the pH and conductivity was analysed following extraction with pure water (1:2.5), and measured with a pH-meter and conductimeter [11]. The organic matter content was measured according to the sulfochromic oxidation method [12]. To determine the variability on the soil thermal conductivity -λ- a dry-out curve (understood as the relationship between thermal conductivity and water content) was calculated [13,14]. The type of sample used for the experiment was a composed soil sample. 42 soil samples for each scenario, *i.e.* before and after prescribed burnt were used. The samples were wetted up with different quantities of deionized water, and packed until to

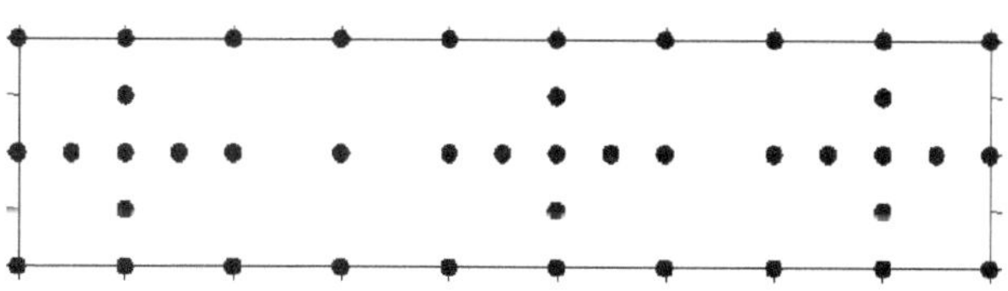

Figure 1. Sampling plot designed. Black points are sampling points.

achieve the target bulk density into a soil column device. Water content was calculated by dried sample in the oven. To determine the thermal conductivity as a function of water content we used the SoilRho® method, which is based on ASTM D-5334-08. This methodology was developed according to the method described by Shiozawa and Campbell [15]. A SH-1 thermal sensor combined with KD2-Pro reader-logger allowed to obtain reliable and accuracy soil thermal conductivity values, thus obtaining a continuous large soil thermal data-set. In all cases, a set of 5 measures per moistured scenario (n = 30) were taken in both composed soil samples. The moisture process gave from air dried up to close to saturation.

The experimental design with ashes was performed on soil column devices, as well. For carrying out the ongoing research, several volumes of two types of ashes (white and black ashes) were used. The colour of ashes imply differences on the temperature reached on the soil surface, therefore, also imply differences in the own composition of ashes. The ashes were mixed with burnt soil on the percentages of 10%, 20% ···, up to 90% in volume of ashes per volume of burnt soil. The differences on soil thermal conductivity for both types of ashes were evaluated only when soil was air dried.

For determining the thermal stability in the most critical case (90% of ashes) a whole of dry-out curves were constructed. In this case, only took into account the higher percentage of two types of ashes (90% white ashes and 90% black ashes). The dry out curves had 8 moistured scenarios with 5 thermal measures per scenario (n = 40). The moisture process gave from air dried up to close to saturation.

3. Results and Discussion

The soil from this plot in the Montgrí massif was classified according to USDA as loam textural class [16]. Mean bulk density was around 1.1 g·cm^{-3}. The chemical and physical properties values before and after prescribed burnt were, respectively: mean total organic carbon content were about 14.7% and 17.2%. The mean electric conductivity increased. On the other hand, the pH of the media did not show any change, and hygroscopic water content was similar, as well. All the values are the average of 42 samples taken before, and 42 samples taken after prescribed burnt. The values are showed in **Table 1**.

Respect to soil thermal conductivity -λ-, it showed changes in its values (**Figure 2**). Indeed, in all cases the values of the dry out curve for soil λ decreased after soil was burnt than native soil. The critical point in the relationship θ (λ) always was stronger when soil samples were burnt than soil before prescribed fire, starting a critical reaction at 8% of water content for samples not burnt, and 6% of water content for burnt samples. Probably, this situation could be explained by the incorporation

Table 1. Physical and chemical characteristics of the studied soil before and after fire. OM = organic matter content; $CaCO_3$ = calcium carbonate content; EC = electrical conductivity and Hw = hygroscopic water content.

Variables	Before Fire	After Fire
Sand (%)	39.3	41.7
Silt (%)	35.1	32.4
Clay (%)	25.6	25.9
E.C. ($\mu s \cdot cm^{-1}$)	330	520
pH	7.0	7.1
O.M. (%)	11.2	10.9
$CaCO_3$ (%)	<3	<3
Hw (%)	1.8	1.9

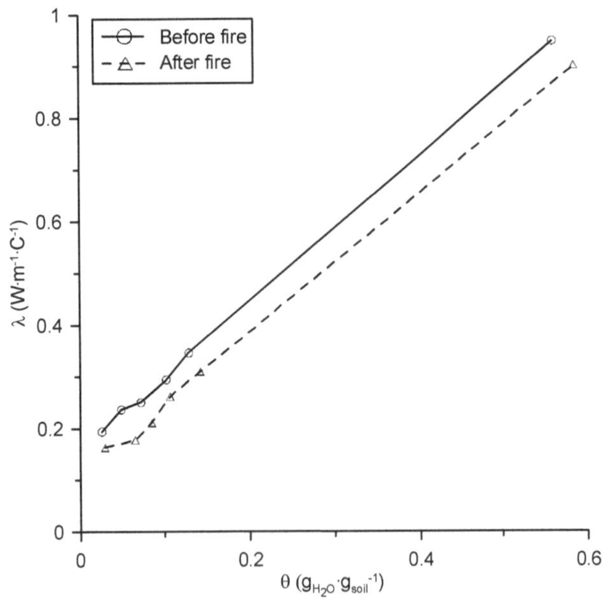

Figure 2. Dry-out curves of the relationship between thermal conductivity and gravimetric water content.

of burnt organic matter on the soil after prescribed fire, such that organic matter behaviour did not transmit well the heat pulse.

Also, the mean temperature values for the soil samples before and after fire during the experiment were about 21.5°C and 18.5°C, respectively. The difference between both values did not affect the soil thermal properties measurements. On the other hand, the coefficient of variation of the thermal conductivity measures indicated that the values were between 0.4% and 2.2% for not burnt soil, and 0.3% and 1.7% for burnt soil. Therefore, the values were considered as acceptable, and very homogeneous.

Finally, a new experiment using the black and white ashes found out over the soil surface after fire was carried out. After the prescribed fire a different quantity of ash patches were found out.

Some of the patches were black ashes where the temperature of the fire was lower, and other patches were white ashes where the temperature of the fire was higher (around 600°C). The new test was used to find differences out between two types of ashes. The soil samples after fire was used to amend and to repack with different quantities of black ashes in volume percentage of soil (0%, 10%, 20%, 30% ··· 100%), and maintaining a similar bulk density. The same test was performed with white ashes.

The relationship between bulk density and porosity using burnt soil and different ashes (black and white) is showed in **Figure 3**. The linear relation between both variables means a well-compacted soil samples, and a decreasing pattern of the macro-porosity when bulk density increase, as well. In any case, two well-defined groups of samples were determined. The soil mixture with black ashes presented less bulk density values than soil with white ashes, which it presented a lower porosity. [3] presented the results about differences in the bulk density for a two soils after fire. Their conclusions were focused in some changes about bulk density, but only within the top section (0.05 m). This fact would evidence the incorporation of a certain quantity of ashes in the few top centimeters of the soil surface, and it would involve a less bulk density than below 0.05 m, such that the compactation effects are limited to upper 0.05 m only [3].

Figure 4, presents different quantities on percentages of two types of ashes mixture with burnt soil, where soil thermal conductivity was measured. The white ashes (dot line) always shown a higher thermal conductivity, meanwhile soil with black ashes was a lower λ. Probably, this fact would be attributed by the large organic carbon content that was not burned during the prescribed fire [17], such that the organic carbon content has a low thermal

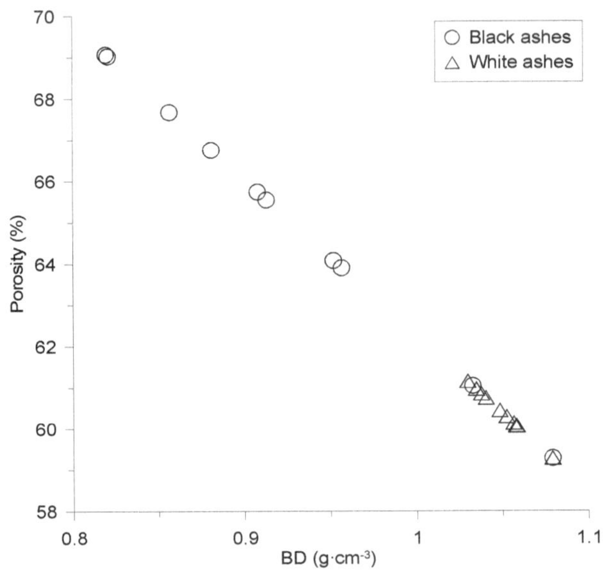

Figure 3. Relationship between bulk density and porosity for a burnt soil mixture with two different types of ashes.

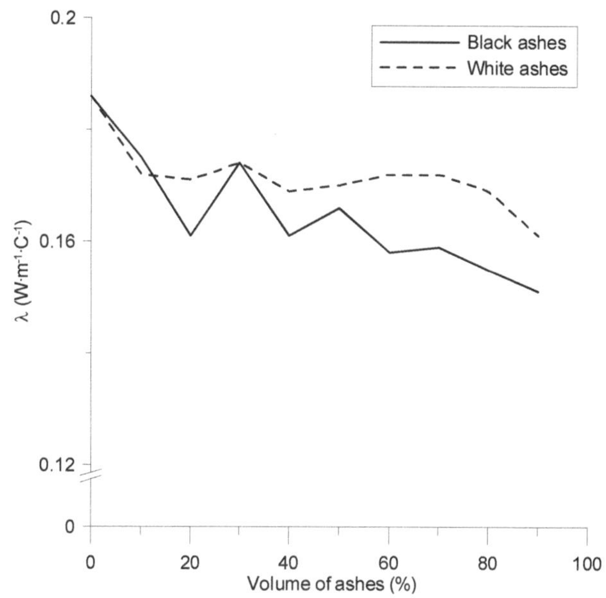

Figure 4. Effects of the different ashes on soil thermal conductivity for a burnt soil.

conductivity, and bigger particle size than ashes [18]. Thus, when organic matter content increase exhibit a decrease of soil λ [19]. However, the fine particle size of the white ashes improved the target bulk density, therefore it contributed to improve the soil thermal conductivity values, except in the two peaks relatives to 30% and 50%, where the quantity of black ashes decreased, it became similar thermal conductivity values than white ashes. On the whole of the experiments, the coefficient of variation indicated that the dispersion was less of 2.4% for whites ashes, and 2.9% for black ashes. Therefore, the overall data set was acceptably homogeneous.

During a burn one may expect changes in soil biota, soil chemistry, and soil minerals [5]. To verify in detail the differences between both soil samples (before and after burnt), an X-ray difractometry was performed. **Figure 5**, shows the results of the analysis. The soil before fire (black line) presented a less quartz than soil after fire, however, exist traces of mica muscovite, which it disappear when soil was burnt.

The values of quartz element increased substantially after burnt, probably the plagioclase were unstructured when temperature rose over 500°C, releasing the quartz and some cations. In this case, the electric conductivity of the soil would be increased, such as indicated the **Table 1**.

Eventually, in **Figure 6**, can observe the relationship between thermal conductivity as a function of water content for a soil sample after burnt with 90% of volume of ashes (white and black ashes).

Clearly, differences exist between white and black ashes when these were incorporated into the soil matrix. In all cases, especially when the soil was closed to saturation the differences were higher. An acceptable explanation could be splited as follows: 1) white ashes presented a fine particle size, hence it increased the water film around the particle, and increased the bulk density as well, reducing the air filled inside the samples. Therefore, a rather high compaction than black ashes involved a better heat transfer; 2) mica muscovite presents dielectric properties, therefore it is a good resistivity material. When mica muscovite disappeared, the thermal conductivity increased.

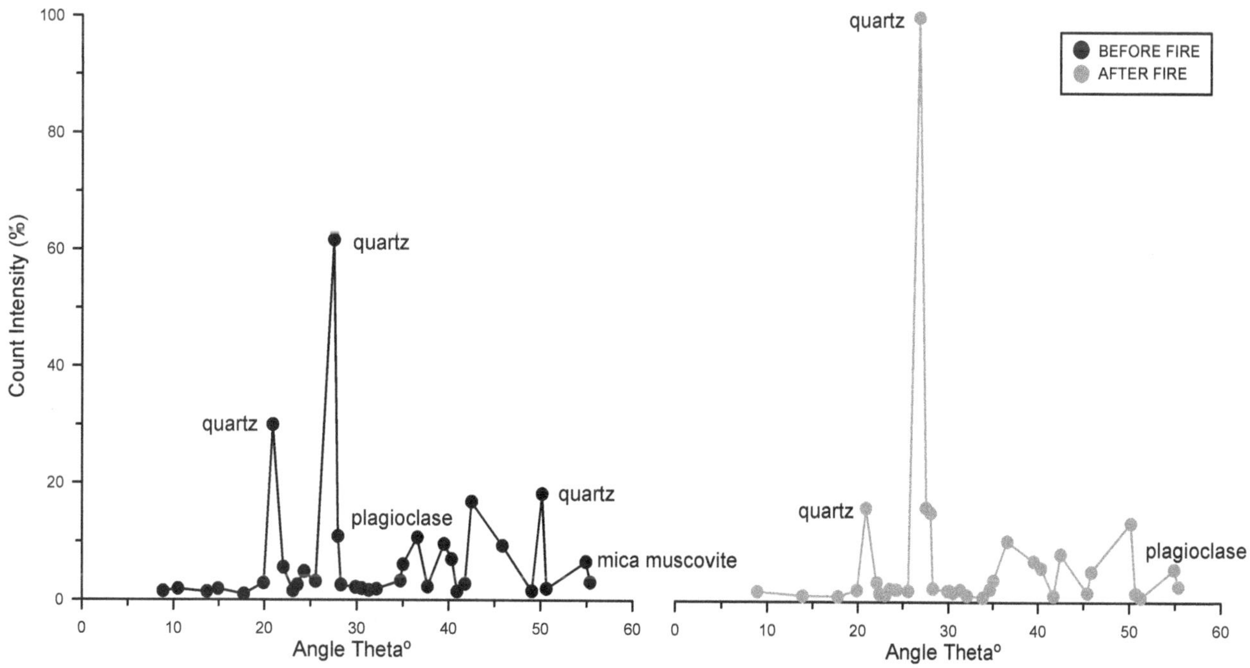

Figure 5. X-Ray difractometry analysis for soil samples before and after prescribed fire.

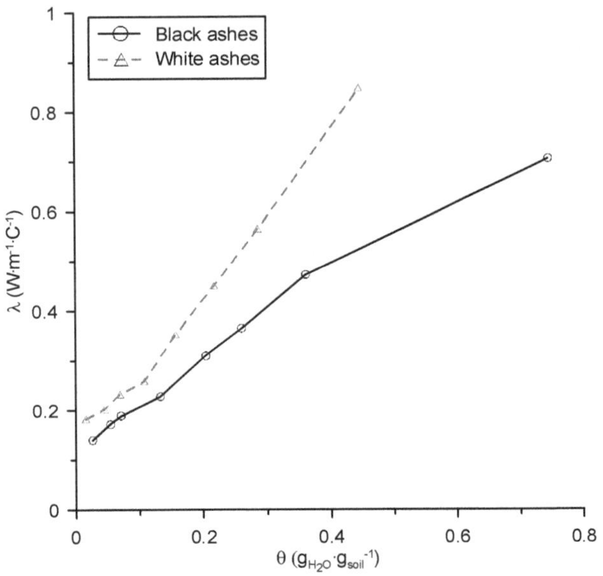

Figure 6. Dry-out curves of the relationship between thermal conductivity and gravimetric water content, when different types of ashes (white and black) were incorporated into the soil matrix.

Finally, note that the particle size of the black ashes were large than white ashes, and also the organic carbon content did not works well when the heat flux transfer was necessary. The measurements of the thermal conductivity of ashes revealed that λ presented a high influence due to water film around the ashes [3,20].

4. Conclusion

As summary, we could say that thermal properties can present changes when the scenario changes, *i.e.* before and after a prescribed fire. Soil after fire always presented a less thermal conductivity, and therefore a less thermal diffusivity, and volumetric specific heat capacity. On the other hand, when the ashes provoked by the fire were incorporated to the soil, the white ashes, which are poorer in organic carbon content, provided a better heat flow transfer. Also, the particle size was relevant in the retention water content. Therefore, when soil is burned its thermal properties change, and a natural or anthropic addition of ashes, especially white ashes, could improve the conductance of the heat flux into the soil, improving the soil bulk density, and water retention content as well.

5. Acknowledgements

The research was founded by Lab-Ferrer Soils and Environmental Consulting Center and University of Barcelona. Likewise, we want to mention the agreements between LabFerrer and the Department of Agri-Food Engineering and Biotechnology of the Universitat Politècnica de Catalunya. We thank the GRAF team for the fieldwork support during the prescribed fire.

REFERENCES

[1] L. F. DeBano, D. G. Neary and P. F. Ffolliott, "Fire's Effects on Ecosystems," JohnWiley & Sons, New York, 1999.

[2] L. F. DeBano, D. G. Neary and P. F. Ffolliott, "Chapter 2: Soil Physical Properties," In: D. G. Neary, K. C. Ryan and L. F. DeBano, Eds., *Wildland Fire in Ecosystems: Effects of Fire on Soil and Water*, United States Department of Agriculture, Ogden, 2005, pp. 29-51.

[3] W. J. Massman, J. M. Frank and N. B. Reisch, "Long-Term Imapcts of Prescribed Burns on Soil Thermal Conductivity and Soil Heating at a Colorado Rocky Mountain Site: A Data/Model Fusion Study," *International Journal of Wildland Fire*, Vol. 17, No. 1, 2008, pp. 131-146. doi:10.1071/WF06118

[4] E. L. Huffman, L. H. MacDonald and J. D. Stednick, "Strength and Persistence of Fire-Induced Soil Hydrophobicity under Ponderosa and Lodgepole Pine, Colorado Front Range," *Hydrological Processes*, Vol. 15, No. 15, 2001, pp. 2877-2892. doi:10.1002/hyp.379

[5] D. G. Neary, K. C. Ryan, L. F. DeBano, J. D. Landsberg and J. K. Brown, "Chapter 1: Introduction," In: D. G. Neary, K. C. Ryan and L. F. DeBano, Eds., *Technical Report RMRS-GTR*-42, United States Department of Agriculture, Ogden, 2005, pp. 1-17.

[6] G. S. Campbell, J. D. Jungbauer, W. R. Bidlake and R. D. Hungerford, "Predicting the Effect of Temperature on Soil Thermal Conductivity," *Soil Science*, Vol. 158, No. 5, 1994, pp. 307-313.

[7] O. T. Farouki, "Thermal Properties of Soils," Trans Tech Publications, Clausthal-Zellerfeld, 1986.

[8] D. A. DeVries, "Thermal Properties of Soils," In: W. R. van Wijk, Ed., *Physics of Plant Environment*, John Wiley & Sons, New York, 1963, pp. 210-235.

[9] G. S. Campbell and J. M. Norman, "An Introduction to Environmental Biophysics," 2nd Edition, Springer-Verlag, NewYork, 1998.

[10] S. I. M.Skinner, R. L. Halstead and J. E. Brydon, "Quantitative Manometric Determination of Calcite and Dolomite in Soils and Limestones," *Canadian Journal of Soil Science*, Vol. 39, No. 2, 1959, pp. 197-204. doi:10.4141/cjss59-025

[11] Ministerio de Agricultura, Pesca y Alimentación, "Métodos Oficiales de Análisis Tomo III, Secretaría General Técnica," M.A.P.A., Madrid, 1986.

[12] A. Walkley and I. A. Black, "An Examination of the Degtjareff Method for Determining Soil Organic Matter and a Proposed Modification of the Chromic Acid Titration Method," *Soil Science*, Vol. 37, No. 1, 1934, pp. 29-38. doi:10.1097/00010694-193401000-00003

[13] C. M. Rubio, R. Josa, J. M. Villar, F. Fonseca and F. Ferrer, "Development of Laboratory Analytical Procedures to Determine Thermal Properties in Soils," *Proceedings of III International Meeting of the European Confederation of Soil Science Societies EUROSOIL*, Vienna, 28 August 2008, p. 308.

[14] C. M. Rubio, R. Josa, D. R. Cobos, C. S. Campbell and F. Ferrer, "Hysteretic Behaviour of Thermal Properties on

Porous Media Advances in Studies on Desertification," University of Murcia, Murcia, 2009.

[15] S. Shiozawa and G. S. Campbell, "Soil Thermal Conductivity," *Remote Sensing Reviews*, Vol. 5, No. 1, 1990, pp. 301-310.

[16] Soil Survey Staff, "Keys to Soil Taxonomy," 8th Edition, US. Government Printing Office, Washington DC, 1998.

[17] N. H. Abu-Hamdeh and R. C. Reeder, "Soil Thermal Conductivity: Effects of Density, Moisture, Salt Concentration and Organic Matter," *Soil Science Society of America Journal*, Vol. 64, No. 4, 2000, pp. 1285-1290. doi:10.2136/sssaj2000.6441285x

[18] T. F. Wall, S. P. Bhattacharya, D. K. Zhang, R. P. Gupta and X. He, "The Properties and Thermal Effects of Ash Deposits in Coal-Fired Furnaces," *Progress in Energy and Combustion Science*, Vol. 19, No. 6, 1993, pp. 487-504.

[19] K. Noborio and K. J. McInnes, "Thermal Conductivity of Salt Affected Soils," *Soil Science Society of America Journal*, Vol. 57, No. 2, 1993, pp. 329-334. doi:10.2136/sssaj1993.03615995005700020007x

[20] L. R. Iverson and T. F. Hutchinson, "Soil Temperature and Moisture Fluctuations during and after Prescribed Fire in Mixed-Oak Forest, USA," *Natural Areas Journal*, Vol. 22, No. 4, 2002, pp. 296-304.

24

A Single-Season Irrigated Rice Soil Presents Higher Iron Toxicity Risk in Tropical Savannah Valley Bottoms

Amadou Keita[1*], Hamma Yacouba[1], Laszlo G. Hayde[2], Bart Schultz[2]

[1]Joint Research Centre for Water and Climate, International Institute for Water and Environmental Engineering (2iE), Ouagadougou, Burkina Faso; [2]Department of Water Science and Engineering, UNESCO-IHE-Institute for Water Education, Delft, The Netherlands.
Email: *amadou.keita@2ie-edu.org, ama_keita@yahoo.fr

ABSTRACT

With the aim of finding the geochemical differences and helping to build alleviating strategies against iron toxicity, two hematite dominant valley bottoms irrigating rice soils were investigated in the Tropical Savannah region of Burkina Faso. The first site was Tiefora, a 15-ha modern double-season irrigated rice system and moderately affected by iron toxicity (10% of the area with a toxicity score of 4). The second site was Moussodougou, a 35-ha traditional single-season irrigated rice valley-bottom, with 50% facing more severe iron toxicity (score 7). Nine soil extracts were taken from three depths—30, 50 and 100 cm—*i.e.* 27 at Tiefora and 27 at Moussodogou. Five techniques were used to measure the data: 1) the ferrous iron concentration was determined using a reflectometer, 2) a pH-meter yielded the pH, 3) clay-proportions were obtained by United States Department of Agriculture (USDA) grain size analysis and densitometry, 4) the organic matter was determined by oven drying (900˚C) and v) the dry bulk density was determined by using undisturbed soil samples. Statistical hypothesis testing of One-way ANOVA and Welch t-test was applied to the data to isolate the similarities and the differences between the two sites. A geochemical analysis followed to find the causes of these differences. The results showed that while oxidation of pyrite leads to a simultaneous increase in Fe^{2+} concentrations and acidity in the soils of coastal floodplains and mangroves, the oxidation of hematite in Tropical savannah valley bottoms decreases Fe^{2+} but also increases acidity during the dry season. As a consequence, it was found that the single-season irrigation scheme of Moussodougou is significantly (p-value 0.4%) more acidic (pH 5.7) than the double-season system of Tiefora (6.4) with also 750 - 1800 mg/l higher ferrous Fe^{2+}. The ferrous iron reached 3000 mg/l in some layers in Moussodougou. This result is a justification to modernize a traditional single-season spate irrigation scheme into a double-season irrigated rice scheme.

Keywords: ANOVA; Burkina Faso; Hematite; Iron Toxicity; Rice; Soil Sampling; Tropical Savannah; Valley Bottoms; Welch T-Test

1. Introduction

Rice is a global cereal facing yield growth challenges in several regions. It is currently consumed by more than 3 billion people and is cultivated on more than 15.5 million ha [1]. Food security challenge is closely linked to rice production since this cereal is essentially self-consumed, being internationally marketed for only 7% [2]. Nevertheless, rice yield is either stagnant or declining in several regions since 1990s [3]. In fact, if the current production trend had to continue, the cereal food availability estimated to 150 kg/person in 2005 will drop to 130 kg/person by 2035 [4].

About 10% of the crop farming areas are found in valley bottoms in the Tropical Savannah of West Africa, where exploitation focuses on rice production but also faces a yield decline [5]. The West African rice production only covers 50% of the needs and thus most of the countries in the region are rice importers [6]. Nigeria, Senegal and Ivory Coast are among the biggest importers [2]. In Burkina Faso, the importation was estimated to more than 210,000 tons in 2010 [7,8]. Iron toxicity is one of the most important constraints to yield growth in Tropical Savannah valley bottoms, threatening up to 60% of the swampy area, with a yield reduction of 10% - 100% [9]. In Burkina Faso, 300 ha ferrous iron intoxicated soils were abandoned in the Vallée du Kou in 1986, most of

*Corresponding author.

which remained uncultivated up to date [10].

Iron toxicity issue in West African Tropical Savannah is currently addressed by various fields of research. In agronomy, Africa Rice devotes an important effort in the development of ferrous iron resistant cultivars [9,11]. Microbiology strives to identify and quantify the bacterial activities involved in the redox processes, occurring in the rice irrigated valley bottoms, in order to propose chemical and/or biotechnological means of reclamation [10,12]. Still, one cannot find in the current research any study comparing a traditional single-season spate irrigated rice soil with a double-season modern irrigation system in terms of acidity and ferrous iron concentrations threats. Such a comparison can lead to a better understanding of the processes and to different alleviation strategies against iron toxicity. Our study was carried out taking this view point and the results are presented in this paper.

2. Materials and Methods

2.1. Sites Location

The two sites present remarkable similar but also contrasting features. For example, both are located in valley bottoms at the heart of the Tropical Savannah zone of Burkina Faso. The coordinates Tiefora are 10°37'33.56"N and 4°33'13.19"W, while Moussodougou is situated at 10°52'18.32"N and 4°55'16.19"W. These zones are Tropical Savannah [13], with 1200 mm/yr rainfall, falling only within four to five months, from May to October. The soils in the environment are essentially made of hematite (Fe_2O_3) [14]. Both sites are exploited by a farmer's organizations, though bigger in Moussodougou (35 ha) than in Tiefora (16 ha). The farm plot sizes range from 0.15 up to 0.50 ha per farmer in both sites. In fact, an important contrast is the number of rice growing seasons: Tiefora with its lined irrigation tertiary canals practices two rice growing seasons using gravity water from a 500,000 m^3 reservoir, while Moussodougou is watered only during the rainy season from the spate of the river passing the middle of the quasi-symmetric valley bottom (**Figure 1**).

2.2. Soil Sampling

The procedure strove to avoid bias while collecting soil extracts from these colluvio-alluvial valley bottoms. Both valleys possess longitudinal and transversal slopes making possible the occurrence of different thicknesses of clay or concentration of ferrous iron for example. Therefore a randomized block sampling was used [15,16]. The valley was divided into three zones longitudinally and three zones transversally. The longitudinal axis was assumed to be the line of the main irrigation canal passing through the middle of the valley in Tiefora, while it coincides with the river bed in Moussodougou. Considering the symmetrical shape of this second valley, the soils extracted from the rightbank of the river are assumed to present similar properties than the corresponding investigated area on the leftbank (**Figure 1**). In Tiefora, the soil extracts where taken from both sides of the asymmetric valley [17]. Twenty seven soil extracts where collected from each of the two valley bottoms.

2.3. Measurements

With a strong concern about data accuracy, one set of measurements was implemented in the field and a second set in the laboratory. For the field side, the pH and ferrous iron concentrations were measured shortly after extraction from the flooded rice soils and their introduction into anaerobic boxes. For this purpose the extracts were diluted 5 times to get the soil solutions [18]. A pH meter was then used to determine the current acidity in the samples, and a reflectometer to quantify Fe^{2+} concentrations [19,20]. For the laboratory side, grain size sorting using standardized American Society for Testing and Materials (ASTM) sieves and sedimentometry yielded the clay proportions (%Clay) for each depth according to US Department of Agriculture (USDA) classification. In addition, the organic matter content (%OM) was assessed by weighting and oven drying at 900°C. Finally, the dry bulk density (Db) at each individual depth was obtained by weighting and oven drying at 105°C.

2.4. Statistical Analysis

Since one of the objectives was to compare Tiefora and Moussodougou on a sound ground, statistical hypotheses testing was required [21]. Prior to the choice and the implementation of the tests, three conditions had to be verified: 1) the equality of the number of samples, 2) the normality of samples underlying populations, and 3) the equality of variances. These checks were performed with the previous five response variables (Fe^{2+}, pH, %Clay, Db and %OM) for both sites and at each of the depths. When the three conditions were fulfilled, a One-Way ANOVA + Fisher grouping could be applied to check if there was any significant difference at the depth under consideration for the two sites. The depth is considered as the independent variable. The null hypothesis was that the five response variables are equivalent when the two sites are compared depth per depth.

In case one of the prerequisite conditions was missing, a non-parametric hypothesis testing model was applied [22].

2.5. Geochemistry

A geochemical analysis endeavored to find the underlying causes of the differences revealed by the statistical hypothesis testing. This analysis considered the chemical

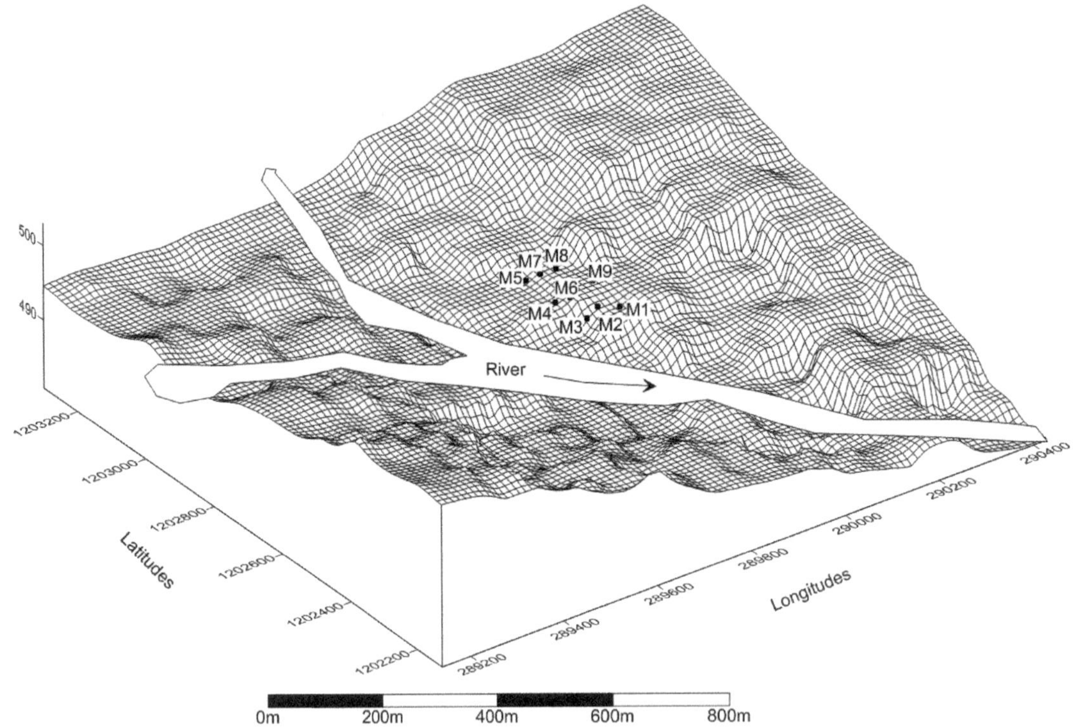

Figure 1. Moussodougou valley with the "M" soil sampling point locations. A randomized block sampling was implemented. Three soil extracts where collected from 30, 50 and 100 cm depth in each random "M" point inside a "block". The blocks represent different longitudinal and transversal sloping conditions. The corresponding area by symmetry on the right bank of the river was assumed to have similar sloping conditions.

reactions related to iron oxido-reduction as well as the bacterial involvement as catalysts [23,24]. Tropical Savannah soils are predominantly hematite (Fe_2O_3) and precipitates of iron(III)-hydroxide ($Fe(OH)_3$). Therefore oxido-reduction processes are alternatively occurring in the wet valley bottom soils [14,25,26]. The key reactions, particularly in hematite dominant environment, were selected and examined in order to predict their impact on the increase and/or decrease of the two main response variables in iron toxicity, *i.e.* Fe^{2+} concentrations and the acidity indicator pH [27]. Consequently, a table was drawn, showing each important reaction involving hematite or iron(III)-hydroxide with its impact on the change of ferrous iron concentration and the acidity. Corresponding to the reactions in this table, crop growing season—wet or dry—in Tiefora and Moussodougou were inserted according to the assumed oxido-reduction process occurring. This analysis led to uncovering the reasons behind the differences between the two sites and to practical proposals to prevent or reduce iron toxicity and acidity.

3. Results and Discussions

3.1. Conditions Check for Hypotheses Testing

The two first conditions required for enabling the application of One-Way ANOVA—equality of sample numbers and normality—were rather well met using Anderson-Darling normality statistic [22]. All the samples of the five response variables have eight or nine elements. If the significance level is set $\alpha = 5.0\%$, all the samples are normally distributed except at 3 depths for 2 variables. These depths are 50 cm/pH and 100 cm/%OM in Tiefora, and 100 cm/%OM in Moussodougou.

The related p-values of Anderson-Darling statistics are slightly smaller than 5% (**Figure 2**). However, this does not constitute a real obstacle to the use of One-Way ANOVA since this test is less sensitive to departure from normality [21].

The equality of variances check constituted a greater obstacle to the application of One-Way ANOVA (**Figure 3**). The computations performed for the five response variables at the three depths in Tiefora and Moussodougou showed that the samples of Fe^{2+} and of %Clay provided Levene p-values smaller than 5% [16,28]. Therefore, the variances are not equal in both Tiefora and Moussodougou for these two variables at the related depths. Consequently, a hypothesis test model less sensitive to the equality of variances was applied, namely the non-parametric Welch 2-sample t-Test [29,30].

3.2. Statistical Results

The application of hypotheses testing for comparing the

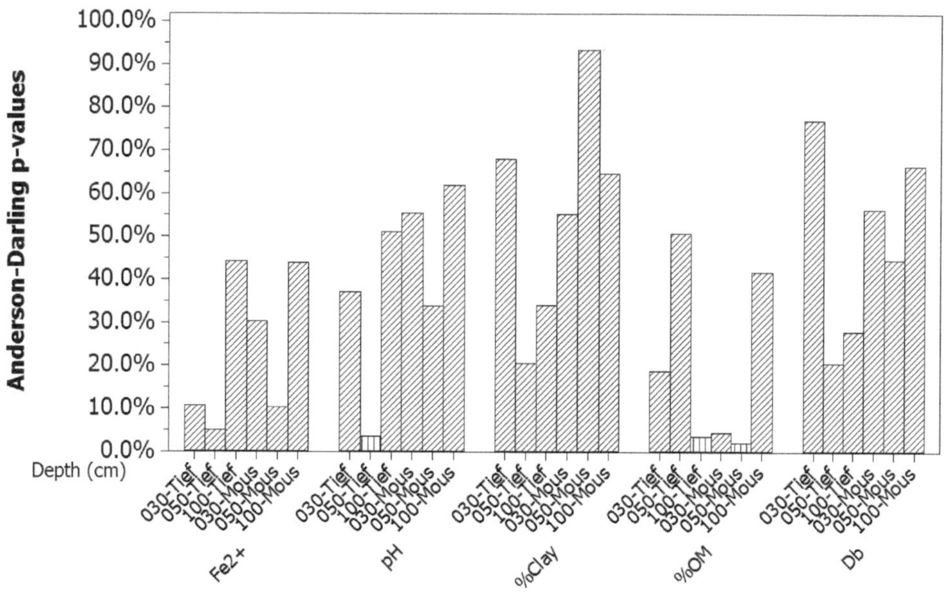

Figure 2. Normality check for the 5 response variables. The p-values of the Anderson-Darling statistics are significant at α = 5% except for the three depths with vertical fill pattern (50 cm/pH and 100 cm/%OM for Tiefora; 100 cm/%OM for Moussodougou). Except these three depth-variables, all the other samples had a normal distribution.

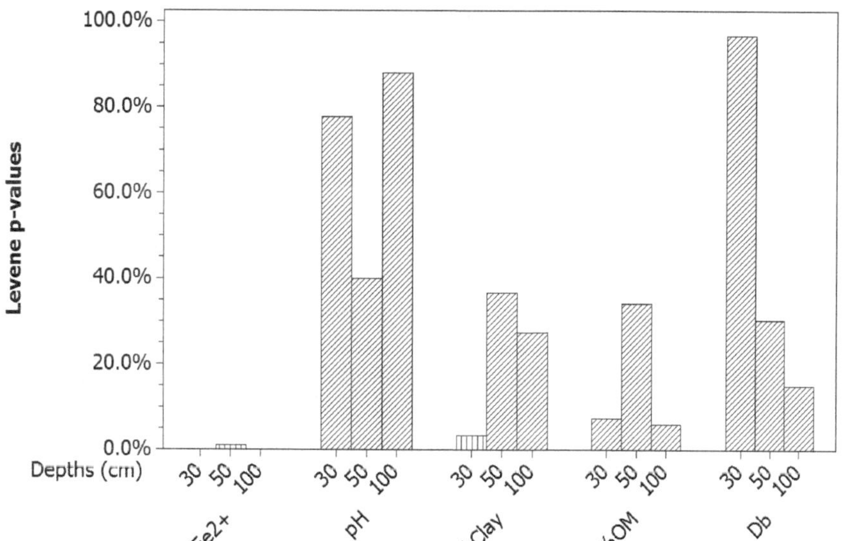

Figure 3. Equality of variances check for the five response variables. The Levene statistic p-values are greater than 5% for three response variables—pH, %OM and Db—but are smaller for Fe^{2+} at all depths and %Clay at depth 30 cm (vertical fill pattern of the bars). For these four depths, the related variables are not equivalent in variance in the two sites of Tiefora and Moussodougou [31].

two sites—Welch 2-samples t-Test, One Way ANOVA + Fisher Grouping—do not lead to significant differences except for Fe^{2+} and the pH. Both sites were similar in the corresponding depths when considering the clay content (%Clay), the organic matter content (%OM) and the dry bulk density (Db). For example, the organic matter content for all the three layers are equivalent in both sites, with an average varying between 6% and 8% and a p-value of 87% (**Table 1** and **Table 2**). Similarly an average 1.3 - 1.5 dry bulk density was found equivalent for both sites, and the clay content %Clay was found not significantly different with the averages varying from 20% - 40%.

As regards with iron, the results show that Moussodougou is more exposed to ferrous iron intoxication than Tiefora. The application of Welch 2-samples t-Test to compare the ferrous iron Fe^{2+} concentrations at the three individual depths in the two sites brings about two important differences (**Figure 4**).

Table 1. One-way ANOVA: OM (%) versus Depth (cm) in Tiefora and Moussodougou.

Source	DF(degre of freedom)	SS(sum of squares)	MS(mean squares)	F of Fisher	p-value
Depth (cm)	5	19.5	3.9	0.37	87.0%[a]
Error	48	506.5	10.6		
Total	53	526.1			

S = 3.2; R^2 = 3.70%; R^2adj = 0.00%. [a] p-value greater than 5%; all the %OM values are similar in Tiefora and Moussodougou.

Table 2. Grouping information using fisher method.

Depth (cm)	Number of observations	Mean(%OM)	Grouping[a]
030-Tief	9	8.0	A
030-Mous	9	7.6	A
050-Tief	9	7.4	A
050-Mous	9	6.1	A
100-Tief	9	7.0	A
100-Mous	9	7.2	A

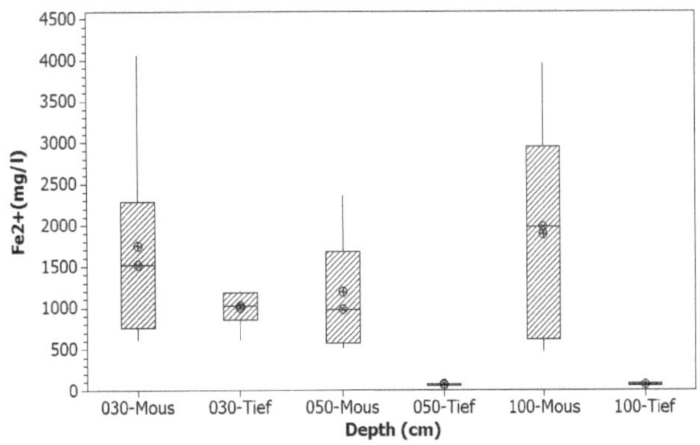

Figure 4. Comparison of ferrous iron concentrations on the sites of Tiefora and Moussodougou. While the Fe^{2+} concentrations were quite uniform in Moussodougou, they dropped in Tiefora below 30 cm depth. Only the top 30 cm had comparable ferrous iron content in both sites.

First, while Fe^{2+} concentrations were not significantly different in the 30 cm top soil in both sites, they were much higher in Moussodougou than in Tiefora at depths 50 and 100 cm. The values were some 1100 mg/l higher in Moussodougou (significant at $\alpha = 5\%$), and even more than 1800 mg/l higher at depth 100 cm (**Table 3**). The second remarkable aspect was that while the ferrous iron gets a quite uniform concentration (no significant p-value) in the whole 100 cm soil profile in Moussodougou, it suddenly drops to less than 100 mg/l underneath 30 cm in Tiefora [17]. Such discrepancies are revelatory of different biochemical reactions that will be addressed in the geochemical analysis section below. At least the high concentrations show why the iron toxicity is a bigger issue—reaching 50% of the farm plots—in Moussodougou than in Tiefora (10% of the farm plots).

The second response variable that made a real difference between the two sites was the acidity indicator, the pH. As the result of the application of One-Way ANOVA + Fisher grouping, it comes again that the top 30 cm draws the attention: Moussodougou with its average 5.7 is significantly more acidic than Tiéfora (p-value = 0.4%). In the layers underneath till 100 cm, it remains more acidic (**Table 4** and **Table 5**). However, compared with for example pyrite soils (pH 2-4) exposed to oxidation, the pH values were rather moderately acid, although the value can decrease to almost 4 in some layers, particularly in Moussodougou (**Figure 5**) [32,33]. Though this is not surprising due to the hematite dominant environment, it leads in combination with a high ferrous iron concentration to a more severe iron toxicity in Moussodougou (score 7) [34].

3.3. Geochemical Analysis

The statistical analysis of the variables making a significant difference between Tiefora (a double-season irrigation scheme) and Moussodougou (a single-season spate irrigation scheme) brings about two important questions: 1) Why is Moussodougou more "ferrous" and more acidic (and thus more exposed to iron toxicity) than Tiefora? 2) How ferrous iron was mainly concentrated in the 30 cm topsoil of Tiefora? To address these questions, some biogeochemical analysis would be helpful.

Though sulfate might be present to some extent, oxido-reduction processes related to iron were the most important in hematite dominant soils, such as found in the valley bottom soils of Tropical Savannah [10,35]. Acid sulfate soils—more often found in coastal flood-plains and mangroves [32,36]—are known to be very acidic when exposed to oxidation. The pH can decrease to less than 3 when exposed to drainage for example. In Tropical Savannah inland valley bottoms, like Tiefora and Moussodougou, the main reactions involve the oxidation and the reduction of hematite (Fe_2O_3) and the precipitation of Iron (III) hydroxide $Fe(OH)_3$. These reactions may be catalyzed (up to million times) by iron reducing or oxidizing bacteria. The predominance of various iron (III) species may also be pH dependent [24,37,38]. The principal reactions involved valleys oxido-reduction are:

Oxidation of pyrite: Simultaneous increase of Fe^{2+} and acidity [39]:

$$FeS_2 + 14Fe^{3+} + 8H_2O \rightarrow 15Fe^{2+} + 2SO_4^{2-} + 16H^+ \quad (1)$$

Precipitation of Fe(III) hydroxide: Simultaneous decrease of Fe^{2+} acidity [40]:

$$15Fe^{2+} + O_2 + H^+ \rightarrow Fe(OH)_3 + 3Fe^{3+} \quad (2)$$

Formation of hematite (oxidation): Decrease of Fe^{2+} and increase of acidity [41,42]:

Table 3. Tiefora versus. Moussodougou-Welch 2-samples t-Test for Fe^{2+} at three depths.

Depths	μ_{Mousso}-μ_{Tiefo}(mg/l)	95% Confidence interval	t-Test of diffrce	Deg. of freedom	p-Value	Iron stratum[b]
30 cm	750	[-100;1600]	2.0	8	7.7%	A-A
50 cm	1110	[601;1625]	5.0	7	0.1%[a]	B-A
100 cm	1820	[855;2780]	4.3	8	0.2%[a]	C-A

[a]Significant at α = 5.0%. The corresponding Fe^{2+} values are significantly different in Tiefora compared to Moussodougou. [b]The first letter stands for Tiefora and the letter for Moussodougou. Fe^{2+} concentrations are comparables at 30 cm on both sites. But at 50 cm and 100 cm, Fe^{2+} concentrations are much higher in Moussodougou than Tiefora.

Table 4. One-way ANOVA: pH versus depth level (cm).

Source:	DF (degre of freedom)	SS (sum of squares)	MS (mean squares)	F of Fisher	P-value
Depth (cm)	5	7.79	1.557	4.08	0.4%[a]
Error	48	18.3	0.381		
Total	53	26.1			

S = 0.617; R^2 = 29.90%; R^2adj = 22.50%; [a]Significant at α = 0.05.

Table 5. Fisher grouping applied to pH in the two sites of Tiefora and Moussodougou.

Depth (cm)	Number of observations.	Mean	Grouping[a]
030-Tief	9	6.4	A
030-Mous	9	5.7	C
050-Tief	9	6.3	AB
050-Mous	9	5.7	C
100-Tief	9	6.6	A
100-Mous	9	5.8	BC

[a]Means that do not share a letter are significantly different.

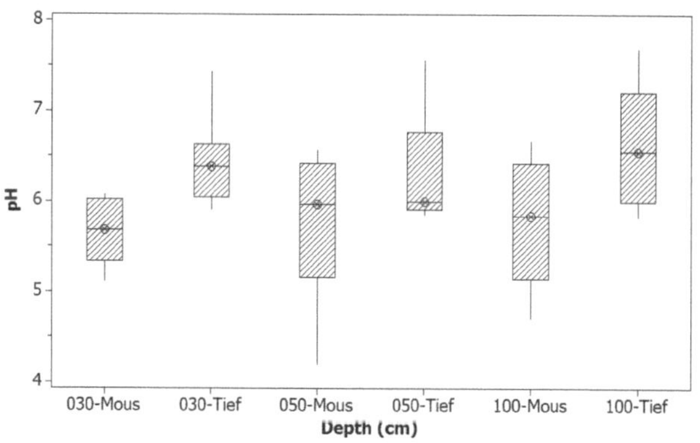

Figure 5. Comparison of pH on the sites of Tiefora and Moussodougou.

$$4Fe^{2+} + O_2 + 4H_2O \rightarrow 2Fe_2O_3 + 3H^+ \quad (3)$$

Reduction of Hematite: Increase of Fe^{2+} and decrease of acidity [43]:

$$2Fe_2O_3 + CH_2O + 8H^+ \rightarrow 4Fe^{2+} + CO_2 + 5H_2O \quad (4)$$

The impacts of the processes—reported in Equations (1) to (4)—on the changes in ferrous iron concentrations and the acidity measured by the pH are of special interest. It is noteworthy that the velocity of these changes can be boosted million times by iron oxidizing (IOB) and iron reducing bacteria (IRB) [23,24]. The oxidation of pyrite (Equation (1)) when exposed to air or leaching leads to a strong simultaneous increase of both Fe^{2+} and acidity. Pyrite is mainly formed in marshy tidal mangroves and coastal floodplains with very low pH [26,39]. In fact, a pH below 3.5 suffices to attest the presence of sulfuric acid resulting from pyrite oxidation [32]. For this reason, it seems reasonable to admit that pyrite is not much in the soils of Tiefora and Moussodougou having a moderate acid pH when exposed to air and which soils were weathered from hematite (**Table 6**). The precipitation of iron III hydroxide (Equation (2)) leads to the simultaneous decrease of acidity and ferrous iron. This phenomenon was observed particularly in Moussodougou [40]. It is essentially observable during spate irrigation in Moussodougou, but also during the two-season irrigation periods in Tiefora (**Table 6**).

From Equations (3) and (4), it can be drawn how the redox processes result in a higher ferrous iron concentration and a more acidic environment in Moussodougou than in Tiefora. The oxidation of hematite will particularly occur in the soil of Moussodougou during the dry season (height months). During that period, ferrous iron concentration decreases, but the acidity increases (Equation (3) and **Table 6**). This observation is confirmed by oxidizing subsurface drainage experiments on rice growing microplots with hematite soils [44]. However, during the wet season with spate irrigation the reduction processes in Moussodougou resume on the hematite (Equation (4)), leading to an increase of the ferrous iron concentration. The degree reached by this increase suggests an important activity iron reducing bacteria (IRB) [38]. This high level of bacterial activity may be linked to river use as water resource for irrigation. The increase of Fe^{2+} concentration in a medium already more acidic due the previous dry season makes the soluble ferrous iron more dominant in the soil of Moussodougou compared

Table 1. Dominant iron redox reactions in valley bottoms.

Reference	Reactions	$[Fe^{2+}]$	Acidity	Site
1	Oxidation Pyrite + IRB (a)	↗↗↗	↗↗↗	
2	Precipitation of Fe(III) hydroxide + IOB (b)	↘	↘	Mousso wet Tiefo wet season
3	Oxidation of Hematite + IOB (b)	↘	↗	Mousso dry season
4	Reduction of Hematite + IRB (a)	↗	↘	Mousso wet season Tiefo wet season

(a) IRB = Iron reducing bacteria; (b) IOB = Iron oxidizing bacteria; ↗ = increase; ↘ = decrease.

with Tiefora. Finally, we suggest that the high Fe^{2+} concentration in the 30 cm top soil of Tiefora (the only one similar to Moussodougou) is due to a higher IRB activity in this layer.

Water management aiming to alleviate iron toxicity in Tropical savannah valley bottoms will have to deal with these three processes: 1) precipitation of $Fe(OH)_3$, 2) oxidation hematite and iii) reduction of hematite. Fortunately, the precipitation of $Fe(OH)_3$, which is one of the most spectacular effects observable at the resurging points in the valley, is actually an ally in this fight. It is the only one simultaneously decreasing both Fe^{2+} and the acidity. It is often observable in open drains in the valleys. Therefore, open channel subsurface drainage will contribute to improve the fertility of the soils. In the same line, oxidation of hematite by bringing oxygenated water in the rice rootzone will decrease ferrous iron concentration, unfortunately along with an increase of acidity in short run. An incorporation of limestone ($CaCO_3$) will certainly reduce acidity in short run.

In very severe cases of iron toxicity, biotechnology resources can be investigated. A few studies have already been executed to identify the type of iron or sulfate reducing bacteria in Tropical Savannah valley bottom irrigated rice fields [10,12,45]. There is a need for more investigations in this climatic zone about the positive impact of iron oxidizing bacteria (Thiobacillus ferrooxidans, Ferrobacillus ferrooxidans ...) in terms of reinstatement of ferric iron Fe^{3+} and the reduction of acidity.

Finally, it appears that practicing two-season rice growing presents less risk to develop iron toxicity in the valley bottoms. Therefore, the numerous traditional valley where farmers are practicing a single-season spate irrigation would have to be developed into at least a double-season modernized rice irrigation scheme.

4. Conclusion

There exists more risk to develop iron toxicity in a single-season compared with a double-season valley bottom irrigated rice in Tropical Savannah hematite rich soils. On the contrary of pyrite oxidation that leads to a simultaneous increase in Fe^{2+} concentration and acidity in costal floodplains and mangroves, the oxidation of hematite decreases Fe^{2+} but moderately increases soil acidity in Tropical Savannah. However, very high ferrous iron concentration may also occur (more than 3000 mg/l). We found a ferrous iron concentration 750 - 1800 mg/l higher in the single-season valley bottom in the 100 cm top soil profile. The pH, with an average of 5.7, was also significantly more acidic in the single season valley bottom, making ferrous iron more available in solution. During the dry season in which no crop was grown and the soil dried, acidity increased in the hematite. Therefore, our results show that there is a well founded interest to modernize traditional single-season spate irrigated rice into modern double-season schemes, equipped with an efficient open/pipe subsurface drainage system.

5. Acknowledgements

We are indebted to several institutions for the support provided to this research project. In this regard, we express our sincere gratitude to the Economic and Monetary Union of West Africa (UEMOA), The Netherlands Organization of International Cooperation in Higher Education (NUFFIC), French Development Agency (AFD), and the International Foundation for Science, Sweden (IFS).

REFERENCES

[1] N. V. Nguyen and A. Ferrero, "Meeting the Challenges of Global Rice Production," Spinger-Verlag, Vol. 4, 2006, pp. 1-9.

[2] E. J. Wailes, "Rice: Global Trade, Protectionist Policies, and the Impact of Trade Liberalization," In: M. Ataman and J. C. Beghin, Eds., *Global Agriculural Trade and Developing Countries*, The Wolrd Bank, Washington DC, 2005, p. 327.

[3] FAO, "Trends on Yield and Productivity of Modern Rice in Irrigated Rice Systems in Asia," The FAO Expert Consultation on Technological Evolution and Impact for Sustainable Rice Production in Asia and the Pacific, Thailand, 1996.

[4] B. Schultz, H. Tardieu and A. Vidal, "Role of Water Management for Global Food Production and Poverty Alleviation," *Irrigation and Drainage*, Vol. 58, No. S1, 2009,

pp. S3-S21.

[5] M. C. Masiyandima, N. van de Giesen, S. Diatta, P. N. Windmeijer, *et al.*, "The Hydrology of Inland Valleys in the Sub-Humid Zone of West Africa: Rainfall-Runoff Processes in the M'be Experimental Watershed," *Hydrological Processes*, Vol. 17, No. 6, 2003, pp. 1213-1225.

[6] F. Lançon and H. D. Benz, "Rice Imports in West Africa: Trade Regimes and Food Policy Formulation," Pro-Poor Development in Low Income Countries, 2007.

[7] Institut National de la Statistique et de la Démographie [INSD-BF], "Indices du commerce extérieur du Burkina Faso [Indices of External Trade of Burkina Faso], Ouagadougou (Burkina Faso)," 2011.

[8] Y. Dembele, H. Yacouba, A. Keïta and H. Sally, "Assessment of Irrigation System Performance in South-Western Burkina Faso," *Irrigation and Drainage*, Vol. 61, No. 3, 2012, pp. 3006-3015.

[9] WARDA, "Iron Toxicity in Rice-based Systems in West Africa, Cotonou (Benin)," 2006.

[10] B. S. Ouattara, "Contribution à L'étude des Bactéries Réductrice du fer et du Sulfate dans les sols de Rizière de la Vallée du Kou (Burkina Faso) [Contribution to the Study of Iron and Sulfate Reducing Bacteria in the Vallee du Kou Valley Bottom Rice Field]," Microbiologie, Celular Biology, Université de Provence Aix-Marseille I, Aix-Marseille, 1992.

[11] E. A. Somado, R. G. Guei and S. O. Keya, "NERICA: The New Rice for Africa—A Compendium," Africa Rice Center, Benin, 2008, 210 p.

[12] D. Dianou, "Etude des Bactéries Sulfato-Réductrices, Méthanotrophes et des Archaeobacteries Méthanogènes dans les sols de Rizière: Influences des Interactions sur la Production du riz et sur L'émission de Méthane Dans L'Atmosphère [Study of the Sulfate Reducing Bacteria, the Methanotroph and the Methanogenic Archaeobacteria in Rice Growing Soils: Influence of the Interactions on Rice Production and the Methane Emission in the Atmosphere]," Applied Biological Sciences, University of Ouagadougou, Ouagadougou, 2005.

[13] M. C. Peel, B. L. Finlayson and T. A. McMahon, "Updated World Map of the Koppen-Geiger Climate Classification," *Hydrology and Earth System Sciences*, Vol. 11, 2007, pp. 1633-1644.

[14] O. C. Spaargaren and J. A. Deckers, "Climate," In: D. Hillel, C. Rosenzweig, D. Powlson *et al.*, Eds., *Encyclopedia of Soils in the Environment*, Academic Press, New York, 2004, pp. 512-519. http://dx.doi.org/10.1016/B0-12-348530-4/00012-6

[15] R. L. Mason, R. F. Gunst and J. L. Hess, "Statistical Design and Analysis of Experiements with Appliations to Engineering and Science," 2nd Edition, Willey-Interscience, 2003.

[16] D. Rumsey, "Statistics II for Dummies," Wiley Pusblishing Inc., 2009.

[17] A. Keïta, B. Schultz, H. Yacouba and L. G. Hayde, "Clay and Ferrous Iron Stratifications in a Tropical Savannah Valley Bottom Soil under Irrigated Rice," *Academia Journal of Agricultural Research*, Vol. 1, No. 11, 2013, pp. 204-210.

[18] M. B. Murray, "Environmental Chemistry of Soils," Oxford University Press, Oxford, 1994.

[19] K. B. Persson, "Soil Phosphate Analysis: A New Technique for Measurement in the Field Using a Test Strip," *Archaeometry*, Vol. 39, No. 2, 1997, pp. 441-443. http://dx.doi.org/10.1111/j.1475-4754.1997.tb00819.x

[20] A. Keïta, "Guide D'utilisation du Réflectomètre RQFlex Plus 10 [The Reflectometer User Guide]," 2013. http://www.youtube.com/watch?v=xH8lISJ6CZU

[21] D. C. Montgomery and G. C. Runger, "Applied Statistics and Probability for Engineers," Wiley and Sons Inc., 2011.

[22] S. Boslaugh and P. A. Watters, "Statistics in a Nutshell," OReilly, 2008.

[23] J. K. Fredrickson and Y. A. Gorby, "Environmental Processes Mediated by Iron-Reducing Bacteria," *Current Opinion in Biotechnology*, Vol. 7, 1996, pp. 287-294. http://dx.doi.org/10.1016/S0958-1669(96)80032-2

[24] D. Emerson, J. V. Weiss and J. P. Megonigal, "Iron-Oxidizing Bacteria Are Associated with Ferric Hydroxide Precipitates (Fe-Plaque) on the Roots of Wetland Plants," *Applied and Environmental Microbiology*, Vol. 65, No. 6, 1999, pp. 2758-2761.

[25] W. R. Fisher and U. Schwertmann, "The Formation of Hematite from Amourphous Iron(III) Hydroxide," *Clays and Clay Minerals*, Vol. 23, 1974, pp. 33-37.

[26] F. R. Moormann and N. V. Breemen, "Rice: Soil, Water, Land, Los Banos Laguna (Philippines)," International Rice Research Institute (IRRI), 1978.

[27] A. R. Conklin Jr., "Introduction to Soil Chemistry Analysis and Instrumentation," Wiley and Sons Inc., 2005. http://dx.doi.org/10.1002/0471728225

[28] J. L. Gastwirth, Y. R. Gel and W. Miao, "The Impact of Levene's Test of Equality of Variances on Statistical Theory and Practice," *Statistical Science*, Vol. 24, No. 3, 2009, pp. 343-360. http://dx.doi.org/10.1214/09-STS301

[29] B. L. Welch, "The Generalization of Student's Problem When Several Different Population Variances Are Involved," *Biometrika*, Vol. 34, No. 1-2, 1947, pp. 28-35. http://dx.doi.org/10.1093/biomet/34.1-2.28

[30] C. H. Brase and C. P. Brase, "Understanding Basic Statistics," Houghton Mifflin Company, 2007.

[31] H. Levene, "Robust Testes for Equality of Variances," In: I. Olkin, Ed., *Contributions to Probability and Statistics*, Stangord University Press, 1960, pp. 278-292.

[32] T. Attanandana and S. Vacharotayan, "Acid Sufate Soils: Their Characteristics, Genesis, Amelioration and Utilization," *Southeast Asian Studies*, Vol. 24, No. 2, 1986, pp. 154-180.

[33] V. A. Jacq and J. C. G. Ottow, "Iron Sulphide Accumulation in the Rhizosphere of Wetland Rice (*Oryza sativa* L.) as Result of Microbial Activities," In: J. Berthelin, Ed., *Diversity of Environmental Biogeochemistry*, La Haye. Elsevier, 1991, pp. 453-468.

[34] IRRI, "Standard Evaluation Systems for Rice," International Rice Research Institute, Philippines, 2002.

[35] H. Bouwer and R. D. Jackson, "Determining Soil Properties," In: J. Shilfgarde van, Ed., *Drainage for Agriculture*, *Agronomy*, 1974, pp. 611-672.

[36] L. J. Pons, N. V. Breemen and P. M. Driessen, "Pysiography of Coastal Sediments and Development of Potential Soil Acidity," In: J. A. Kittrick, D. S. Fanning and L. R. Hossner, Eds., *Acid Sulfate Weathering*, Soil Science Society of America, 1982, pp. 1-18.

[37] F. N. Ponnamperuma, "The Chemistry of Submerged Soils," *Advances in Agronomy*, Vol. 24, 1972, pp. 29-96.

[38] L. Wenjing, W. Hongtao, H. Changyong and W. Reichardt, "Aromatic Compound Degradation by Iron Reducing Bateria Isolated from Irrigated Tropical Paddy Soils," *Journal of Environmental Sciences*, Vol. 20, No. 12, 2008, pp. 1487-1493.

[39] F. X. Suryadi, "Soil and Water Management Strategies for Tidal Lowlands in Indonesia," Delft University of Technology, Deft, 1996.

[40] D. Rickard and G. W. Luther, "Chemistry of Iron Sulfides," *Chemistry Revisions*, Vol. 107, 2007, pp. 514-562.

[41] Y. Kato, K. Suzuki, K. Nakamura, A. H. Hickman, *et al.*, "Hematite Formation by Oxigenated Groundwater More than 2.76 Billion Years Ago," *Earth and Planetary Science Letters*, Vol. 278, No. 1-2, 2008, pp. 40-49.

[42] U. Schwertmann and E. Murad, "Effects of pH on the Formation of Goethite and Hematite from Ferryhidrite," *Clays and Clay Minerals*, Vol. 31, No. 4, 1983, pp. 277-284. http://dx.doi.org/10.1346/CCMN.1983.0310405

[43] N. V. Breemen, "Environmental Aspects of Acid Sulphate Soils," *International Symposium on Acid Sulphate Soils*, Ho Chi Minh City, 1992.

[44] H. Otoidobiga, "Etude de L'Impact du Développement et des Activités des Bactéries Sulfato-Réductrices et des Bactéries Réductrices de Fer sur le Développement Végétatif et les Rendements de riz de Submersion sous Drainage de Sub-Surface [Sudy of the Impact of Sulphate and Iron Reducing Bacteria on the Vegetative Development and the Yields of Flooded Rice under Subsurface Drainage]," Unité de Formation et de Recherche Sciences de la Vie et de la Terre, University of Ouagadougou, Burkina Faso, 2012.

[45] K. R. Butlin, M. E. Adams and M. Thomas, "The Isolation and Cultivation of Sulphate-Reducing Bacteria," *Microbiology*, Vol. 3, No. 1, 1949, pp. 46-59.

25

Evaluation of Factors Influencing the Biomass of Soil Microorganisms and DNA Content

Wolińska Agnieszka, Stępniewska Zofia, Bułaś Aleksandra, Banach Artur

Department of Biochemistry and Environmental Chemistry, Institute of Biotechnology, The John Paul II Catholic University, Lublin, Poland.
Email: awolin@kul.lublin.pl

ABSTRACT

The aim of the study was the statistical evaluation of the impact of water potential (pF), oxygen availability (ODR) and the way of land use on microbial biomass (MB) and soil DNA content. Soil was extracted from the surface (0 - 20 cm) and subsurface (20 - 40 cm) layers of Mollic Gleysol. Soil material was collected in July 2009 from the village Kosiorów (SE part of Poland), from the two distinct neighbouring areas: agriculturally exploited (AE), and fallow land (FL), which served as the control area. Moisture content was determined for a range of pF values (0, 1.0, 1.5, 2.0), which corresponded to availability of water usable by microorganisms and plant roots. Finally, our results revealed significant ($p < 0.001$) positive relationship between DNA and soil MB content, and negative correlations between soil MB and both pF and ODR. Importantly, MB seemed significantly dependent on the different way of land use, and higher MB content was noted in the soil agriculturally exploited ($p < 0.05$) in contrast to the control area.

Keywords: Microbial Biomass; Soil DNA; Water Potential; Agricultural Activity

1. Introduction

The term of microbial biomass is commonly used to describe the total mass of microorganisms present in soil [1]. The importance of the MB in soil functioning is well recognised [2]. Moreover, MB as an integrative measure of the physiologically active part of the soil microflora is recommended by 8 countries of European Union (e.g., Germany, United Kingdom, Austria, Switzerland), as important factor of soil quality, which is included in soil microbial degradation monitoring program [3].

Soil MB is also considered to be a useful criterion for an early indication of environmental stress [1,4] as variability in microbial communities can precede detectable changes in soil properties. An example is the turnover rate of the MB, which is much faster and takes, e.g., 1 - 5 years, than the turnover rate of total soil organic matter [5]. It is partly due to the large pool of relatively inactive and dormant microorganisms, having the potential to reflect the past [6]. Life in the soil environment is constantly influenced by drying and rewetting cycles, as soils are continually exposed to rainfall, wind and snow [7]. A part of the microbial population dies during each drying-and-wetting cycle resulting in the fluctuation of soil microbial composition [8]. Soil water content as a function of the soil water tension is described by pF curve, which provides information about the ability for water retaining by the soil pores at any given water tension, or conversely, how tightly a water is held between soil aggregates [9]. Therefore, also soil aeration status is strongly depended from pF values. It has been also shown experimentally, that ODR factor satisfactorily reflects the supply of oxygen to the plant roots [10,11]. Oxygen availability is among the most important factors affecting soil microbial activeties [12]. There is a specific gas demand for different soil microbes including bacteria, fungi, and other microorganisms [7]. ODR is affected by several factors. It increases with reducing soil water content or increasing suction up to a certain level and then declines with further depletion of water [13]. Soil environment is the major reservoir of microbial genetic diversity and thus should be particularly protected. The breakthrough in soil biology was the discovery of DNA, which is a carrier of biological information and the best characteristics of every organism. Thus, soil DNA analysis is considered to be important and precise tool towards a better recognition of soil microbial functionality and interrelationships among them.

The size of MB was found to be strongly correlated with content of base cations, base saturation, cation exchange capacity, and organic matter quality [14], as well as with soil bulk density, nutrient contents and phosphatase and invertase activities [15]. However, there is a lack of studies on relationships between MB and such important soil factors like pF or ODR, what may be decisive for the course of processes responsible for plant development, and soil fertility. Investigation of the effect of human agricultural activity on MB content is also interesting part of the current study. Thus, the aim of the work was the statistical description of the impact of pF, ODR, and human agricultural activities on MB and DNA content in soil.

2. Material and Method

2.1. Soil and Investigated Area Description

The soil used in the experiment was *Mollic Gleysol* (**Table 1**). Soil was sampled in July 2009 from the village Kosiorów, situated in the Wilków community (SE part of Poland) from depths of 0 - 20 and 20 - 40 cm. To make possible an estimation of the effect of different way of land use on MB, soil samples were collected from two neighbouring plots: one of them was agriculturally exploited with systematic fertilization and pasturage (under human activity), whilst the other one was classified as fallow land and used as a control area (without any human impact).

2.2. Assaying of Soil Retention Curves

The instrument used for determining water retention curves was a steel pressure chamber, inside of which a porous plate saturated with water was located. At the bottom, soil samples, continuously exposed to atmospheric pressure, make the hydraulic contact with the porous plate [16]. The chamber was closed and the desired air pressure *P* was applied, driving away the soil water retained at pressures below *P*, until equilibrium was reached [16]. Soil samples were collected using plastic containers and placed in an airtight chamber (for 10 days), part of a laboratory set LAB o12 (Soil Moisture Equipment Company, USA), before a pressure was applied. The moisture content was determined via the drying process, for the range of water potentials (0, 1.0, 1.5 and 2.0 pF), corresponding to availability of water usable by microorganisms and plant roots.

2.3. ODR Measurement

After determination of proper pF values, ODR was measured by ODR-meter manufactured by the Institute of Agrophysics, Polish Academy of Soil Sciences (Lublin), using Malicki and Bieganowski [17] method. The ODR technique consists of the measurement of the electric current intensity corresponding to the reduction of oxygen on a platinum cathode placed in the soil and negatively polarized with respect to the reference electrode (calomel). As oxygen is consumed at the microelectrode, more oxygen needs to diffuse radially to the electrode in response to the accumulated gradient. This is analogous to oxygen consumption by respiration of root surface or by microbial respiration. Four platinum wire electrodes (0.5 mm × 4 mm) were placed at the depth of 2 cm and polarized to –0.65 V versus saturated calomel electrode for 4 min. The data were recorded in three replicates, for each sample.

2.4. Microbial Biomass

Soil MB was determined by a fumigation-extraction method using $CHCl_3$ as an agent responsible for the cellular death of microorganisms, according to Joregensen [18] procedure. Concentration of head-space CO_2 released by microorganisms which survived incubation with $CHCl_3$ was measured by a gas chromatograph (Varian CP-3800, equipped with a TCD detector).

2.5. DNA Extraction

Soil DNA was extracted from samples at pF 0 (full water capacity conditions) and pF 2.0 (field capacity conditions), using the GeneMatrix soil DNA isolation kit (EURx 1.4, Poland), according to the manufacturer's instructions. This kit was designed specifically for the rapid isolation of pure, humic-free microbial DNA from soil samples, and guaranteed the proper DNA isolation procedure. Concentration of DNA was determined spectrophotometrically at 260 nm (Shimadzu, UV-1800, Japan).

2.6. Data Analyses

Statistical analyses were performed by means of Statistica 8.0 software (STATSOFT, USA). One-way ANOVA test was used to investigate significant ($p < 0.05$) effect of pF, ODR, DNA content on soil MB quantity. Results of the significance differences analyses are presented in **Table 2** only. In **Figures 1-4** the average values with standard deviations are demonstrated.

Table 1. Basic characteristics of the soil.

Place	Depth (cm)	Granulometric composition (%, dia in mm)				pH (H_2O)	TOC (%)
		1 - 0.1	0.1 - 0.02	0.02 - 0.002	<0.002		
Agriculturally exploited (AE)	0 - 20	87	8	3	2	6.7	22.5
	20 - 40	90	7	2	1	6.4	1.4
Fallow land (FL)	0 - 20	91	6	3	0	6.2	20.1
	20 - 40	95	3	2	0	6.5	1.3

Table 2. Statistically significant relationships between MB and analyzed parameters, N = 12.

Object investigated	Depth (cm)	pF	ODR	c DNA
AE	0 - 20	-0.68^{*}	$-0.20^{n.s.}$	0.97^{**}
	20 - 40	-0.53^{*}	$-0.52^{n.s.}$	0.99^{***}
FL	0 - 20	-0.94^{***}	-0.83^{***}	0.86^{*}
	20 - 40	-0.74^{**}	-0.59^{*}	0.89^{*}

*, **, ***—indicate significance at the $p < 0.05$, $p < 0.01$, and $p < 0.001$, respectively; $^{n.s.}$—not significant difference.

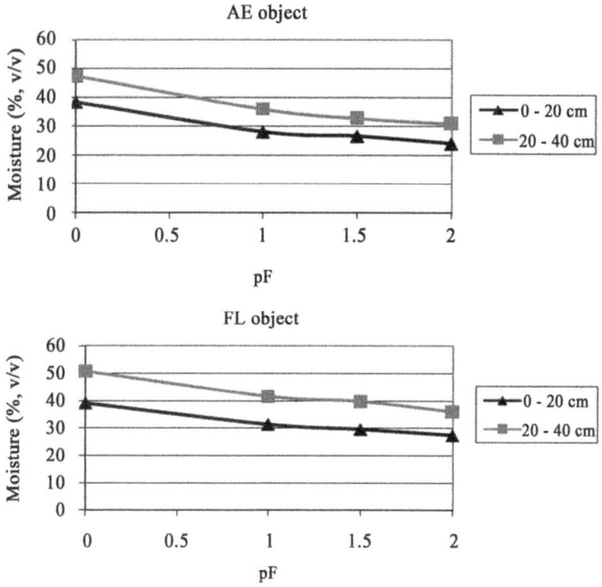

Figure 1. The relationship between soil water content (% v/v) and pF values. The curves are related to two depths (0 - 20; 20 - 40 cm) of the investigated soil areas.

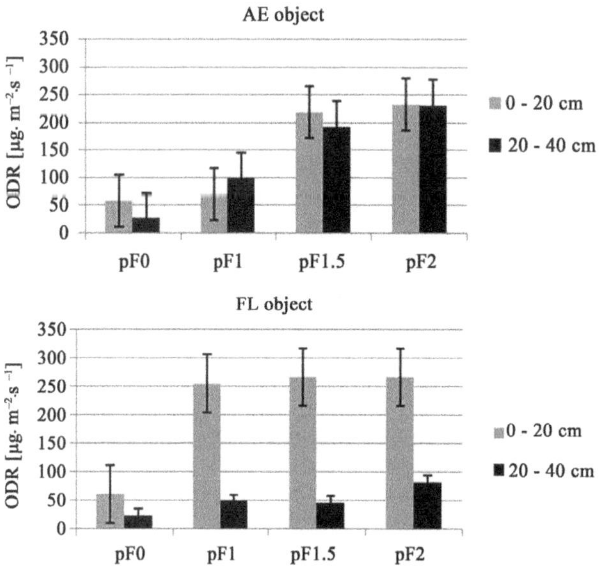

Figure 2. Variability of ODR values as an effect of water potential at two depths of the investigated soil areas. Average values with standard deviations are presented.

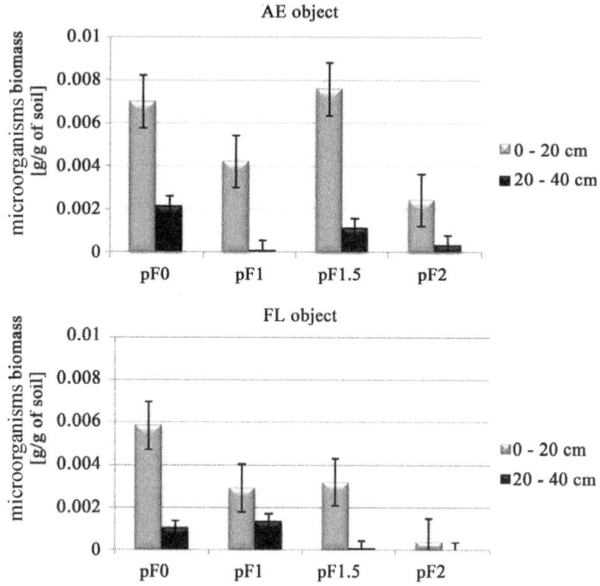

Figure 3. Variability of MB levels as an effect of water potential at two depths of the investigated soil areas. Average values with standard deviations are presented.

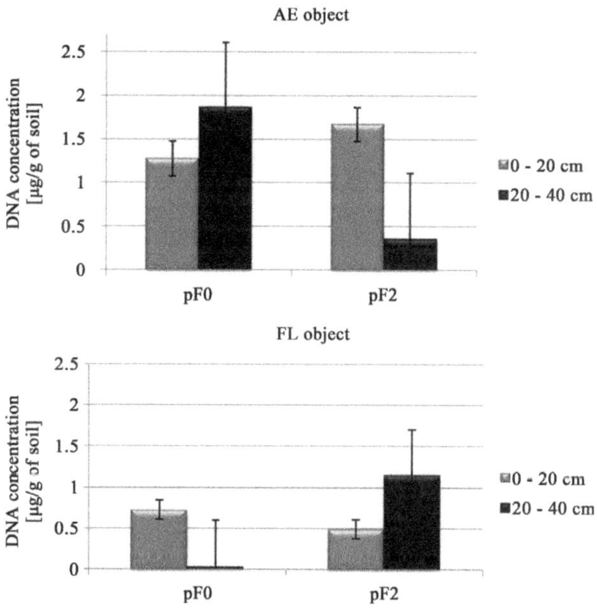

Figure 4. DNA content at two depths of the investigated soil areas as an effect of water potential. Average values with standard deviations are presented.

3. Results and Discuss

3.1. Soil Retention Abilities

An incubation of the soil samples under different controlled moisture conditions altered significantly ODR, MB and DNA concentration. The relationships between soil water content (%, v/v) and pF for the two layers of *Mollic Gleysol*, for both tested areas (AE and FL) are presented in **Figure 1**.

Generally, the investigated soils demonstrated similar abilities to retain water, even though soil from the control area (FL) displayed slightly higher (c.a. 3% more than soil agriculturally exploited) capacity for water keeping. This might be connected with beginning of decay process on AE object, what resulted in looseness of soil structure and higher intensity of mineralization and humification of the soil organic compounds. Moreover, higher ability of holding water in the subsurface layers were noted, and equalled 34% - 47% v/v and 41% - 50% v/v in AE and FL objects, respectively. Similar capability of *Mollic Gleysol* for water maintaining was also indicated [19]. The soil-water interactions are greatly or extremely important to soil fertility and therefore are the subject of interest to agricultural engineers and farmers. Furthermore, information about water holding capacity is important for agronomic and hydrologic characteristic of soils. It expresses, how much water can be stored in the soil for plant use during periods without rain or irrigation [20].

This provides an indication of soil sensitivity to drought and could be used to calculate the probability of occurrence of deep drainage or groundwater recharge process. It was also reported that macropore continuity is very important to the aeration status of the soil, thus the effect of soil compaction on other aeration properties depends on soil hydro-physical status [13,21].

3.2. Oxygen Availability in AE and FL Objects

Based on the performed measurements it was found, that pF constitutes a significant factor ($p < 0.001$) determining ODR levels in the soil environment. Soil desiccation, occurring in the direction from pF 0 to pF 2.0, was the reason of stimulation of ODR (**Figure 2**).

The oxygen availability in relation to the soil water potential indicates, that ODR values at surface layer of AL object fluctuated from 50 to 233 μg $O_2/(m^2{\cdot}s)$ at pF 0 and pF 2.0, respectively. At deeper layer (20 - 40 cm) of agriculturally exploited *Mollic Gleysol* ODR ranged between 26 till to 240 μg $O_2/(m^2{\cdot}s)$, for as follows pF 0 and pF 2.0. Stronger variety of ODR values in the FL object was observed, we noted values between 60 and 266 μg $O_2/(m^2{\cdot}s)$ as in the surface layer, whilst in the subsurface 80% decrease of ODR was found and registered values oscillated from 23 to 81 μg $O_2/(m^2{\cdot}s)$, for pF 0 and pF 2.0, respectively. We assume that way of land use and sys- tematically applied ploughing at AE object may be the reason of higher ODR level as well in surface as in sub- surface layer. FL on the other hand, despite the fact of comparable level of ODR in surface layer, in the sub-sur- face was characterized by rapid reduction of oxygen availability, below 35 μg $O_2/(m^2{\cdot}s)$, which is the minimum level of ODR necessary for proper root growth [11,22]. This might be caused by lack of human agricultural activities, as e.g. ploughing may contribute in soil ventilation improvement. On the contrary, [19] Walczak *et al.* (2001) noted tendency for higher oxygen availability in the deeper layers, rather than in surface of the *Mollic Gleysol*, whereas [23] Stępniewska *et al.* (2003) observed analogous trend in the *Eutric Cambisol*.

It may be caused either by methodical limitations, as the water barriers or water films present on the surface of electrode could be broken off [9], or by the differences in granulometric composition of analyzed soil samples (**Table 1**), as the fact, that large granulation favourable for forming of aeration pores was noted in the subsurface layers.

3.3. Soil Microbial Biomass

Soil MB content was also strongly influenced by pF conditions and the way of land use (**Figure 3**). The significantly higher values (0.0077 and 0.0058 g/g of soil) were stated in the surface layers at full water capacity conditions (pF 0) for AL and FL objects, respectively, as compared to the deeper layers of soil. In subsurface layers the reduction of MB content, c.a. 5 times, was observed.

The highest values of MB estimations in AE object were undoubtedly connected with total organic carbon (TOC) content, what favoured the microorganisms' abundance by supplying sources of energy necessary for activity of soil biota and growing crops. This observation is consistent with other studies [24-26]. Registered almost 85% reduction of MB in the subsurface layers come out of distribution of microorganisms in the soil profile, since microorganisms are mostly confined to the surface soil layer owing to better aeration and greater nutrient availability. Anthropogenic activities and soil management in particular, are mostly responsible for disturbing the chemico-physical and biological equilibrium of soil [27]. A particularly serious problem is the decrease in the organic matter content of agricultural soils, which may endanger soil fertility and enhance erosion. The MB, as a small fraction of soil organic matter, is a source and sink of nutrients and controls soil organic matter mineralization [27,28] Fisk and Fahey (2001), analogically at current study noted higher content of MB (by 20% - 30%) at agricultural areas in response to fertilization, than at fallow ones. Also, other findings pointed out a distinct relationship between soil fertility and soil MB, suggesting that MB measurements provide a valid estimate of soil quality [14,29].

3.4. DNA Content

Similarly to MB distribution of DNA concentrations in surface layers were 2 times higher in AE object (**Figure 4**), than in control soil (not under cultivation) at pF 0. Even more differentiation in subsurface layers were stated, where DNA level reached 12 fold higher values in AE, in relation to FL area. The usage of GeneMatrix soil DNA isolation kit let us to receive 0.4 - 1.8 μg/g and 0.5 - 1.2 μg/g of soil DNA concentrations in AE and FL objects, respectively and revealed that the quality (fragment size and purity) of the extracted DNA was generally very good. However, one should always realize that extraction of DNA from soil samples is never 100% efficient and can vary from a few μg to almost 200 μg DNA per g dry weight of soil. Most of authors, however, reported that the obtained values ranged from 1 to about 50 μg of total DNA per g dry weight of soil depending on the method applied and soil sample studied [30,31].

3.5. Statistical Relationships between Parameters Analyzed

Statistical relationships between MB and investigated parameters (pF, ODR, DNA content, way of land use) described by correlations coefficient (r) are presented in **Table 2**. Significant influence ($p < 0.05$) of tested parameters on MB was found. The positive relationships between MB and DNA content, and negative correlations between MB and soil physical factors like pF and ODR were revealed.

Both our and other studies demonstrated that MB can be highly sensitive to environmental factors. Although, study by [32] Singh and Yadava (2006) confirmed negative correlation between MB and soil moisture, even so prior to our study, rather little attention has been paid to the influence of pF, ODR, DNA content and way of land use on soil MB. Therefore, our work was focused on these relationships and determination of statistical correlations. The positive relationships between MB and DNA content suggest the domination of intercellular DNA form at both AE and FL objects, and MB is considered to be a good factor relating to the total mass of microorganisms present in soil. Our results are supported by the findings of [33] Blagodatskaya *et al.* (2003), who noted that DNA content correlated strongly with the total MB in *Paleosol* soils from Southern Urals ($R^2 = 0.97$), as well as by work of [34] Hartmann *et al.* (2005), who described analogical correlation ($r = 0.91^{***}$). Nevertheless, the interpretation of our results has been challenging because of the lack of enough publications in available literature concerning the determination of r coefficients as a goodness of fit between investigated soil factors and MB. However, determined relationships demand further investigations and selection of the other soil types, for better explanation and confirmation of the obtained correlations.

4. Conclusion

Importantly, the way of land use significantly ($p < 0.05$) influenced on MB and DNA content, and higher content of MB and DNA concentration were noted in the AE ($p < 0.05$), in contrast to the FL area. Significant ($p < 0.05$) positive relationships between soil MB and DNA concentration were revealed, whereas pF and ODR were negatively correlated to MB. Oxygen availability values of 30 - 255 μg $O_2/(m^2 \cdot s)$ and 20 - 80 μg $O_2/(m^2 \cdot s)$ in the AE and FL objects, respectively, suggested higher oxygen availability in the subsurface layer of AE area ($p < 0.05$), what might be connected with human agricultural practices, e.g. traditional regular ploughing, which significantly improves soil aeration status.

5. Acknowledgements

This work was supported by the IUVENTUS PLUS grant (No IP 2010 001070).

REFERENCES

[1] K. Broos, M. L. Macdonald, D. A. Heemsbergen, M. B. Barnes, M. Bell and M. J. McLaughlin, "Limitations of Soil Microbial Biomass Carbon as an Indicator of soil Pollution in the Field," *Soil Biology and Biochemistry*, Vol. 39, No. 10, 2007, pp. 2693-2695. doi:10.1016/j.soilbio.2007.05.014

[2] E. A. Stockdale and P. C. Brookes, "Detection and Quantification of the Soil Microbial Biomass-Impacts on the Management of Agricultural Soils," *The Journal of Agriculture Sciences*, Vol. 144, No. 4, 2006, pp. 285-302. doi:10.1017/S0021859606006228

[3] A. Winding, K. Hund-Rinke and M. Rutgers, "The use of Microorganisms in Ecological Soil Classification and Assessment Concepts," *Ecotoxicology and Environmental Safety*, Vol. 62, No. 2, 2005, pp. 230-248.

[4] M. Schloter, O. Dilly and J. C. Munch, "Indicators for Evaluating Soil Quality," *Agriculture, Ecosystems and Environment*, Vol. 98, No. 1-3, 2003, pp. 255-262. doi:10.1016/S0167-8809(03)00085-9

[5] M. R. Carter, E. G. Gregorich, D. A. Angers, M. H. Beare, G. P. Sparling, D. A. Wardle and R. P. Voroney, "Interpretation of Microbial Biomass Measurements for soil Quality Assessment in Humid Temperate Regions," *Canadian Journal of Soil Science*, Vol. 79, No. 4, 1999, pp. 507-520. doi:10.4141/S99-012

[6] L. Ranjard, F. Poly and S. Nazaret, "Monitoring complex Bacterial Communities Using Culture-Independent Molecular Techniques: Application to Soil Environment," *Research in Microbiology*, Vol. 151, No. 3, 2000, pp. 167-177. doi:10.1016/S0923-2508(00)00136-4

[7] A. Wolińska and Z. Stępniewska, "Microorganisms Abundance and Dehydrogenase Activity as a Consequence of Soil Reoxidation Process," In: M. Miransari, Ed., *Soil Tillage and Microbial Activities*, Research Singpost Publisher, Kerala, 2011, pp. 111-143.

[8] R. Daniel, "The Metagenomics of Soil," *Nature Reviews Microbiology*, Vol. 3, No. 6, 2005, pp. 470-478.

[9] A. Wolińska and R. P. Bennicelli, "Dehydrogenase Activity Response to Soil Reoxidation Process Described as Varied Conditions of Water Potential, Air Porosity and

Oxygen Availability," *Polish Journal of Environmental Studies*, Vol. 19, No. 3, 2010, pp. 651-657.

[10] J. Gliński, W. Stępniewski, Z. Stępniewska, T. Włodarczyk and M. Brzezińska, "Characteristics of Aeration Properties of Selected Soil Profiles from Central Europe," *International Agrophysics*, Vol. 14, No. 1, 2000, pp. 17-31.

[11] W. Stępniewski, Z. Stępniewska, J. Gliński, M. Brzezińska, T. Włodarczyk, G. Przywara, G. Varallay and K. Rajkai, "Dehydrogenase Activity of Some Hungarian Soils as Related to Their Water and Aeration Status," *International Agrophysics*, Vol. 14, No. 3, 2000, pp. 341-354.

[12] G. L. Hutchinson, "Biosphere-Atmosphere Exchange of Gaseous N Oxides," In: R. Lal, J. Kimble, E. Levine and B. A. Steward, Eds., *Soil and Global Change*, CRC Lewis Publisher, Boca Raton, 1995, pp. 219-236.

[13] M. Brzezińska, T. Włodarczyk, W. Stępniewski and G. Przywara, "Soil Aeration Status and Catalase Activity," *Acta Agrophysica*, Vol. 5, No. 3, 2005, pp. 555-565.

[14] J. Zwoliński, "Microbial Biomass versus Soil Fertility in Forest Sites," *Polish Journal of Ecology*, Vol. 52, No. 4, 2004, pp. 5553-5561.

[15] Z. Z. Jin, J. Q. Lei, X. W. Xu, J. L. Fan, S. F. Zhao, H. W. Zhou and F. Gu, "Relationships of Soil Microbial Biomass with Soil Environmental Factors in Tarim Desert Highway Shelter-Forest," in Chinese, *Chinese Journal of Applied Ecology*, Vol. 20, No. 1, 2009, pp. 51-57.

[16] L. F. Pires, O. O. S. Bacchi and K. Reichardt, "Soil Water Retention Curve Determined by Gamma-Ray Beam Attenuation," *Soil and Tillage Research*, Vol. 82, No. 1, 2005, pp. 89-97. doi:10.1016/j.still.2004.06.003

[17] M. A. Malicki and A. Bieganowski, "Chronovoltammetric Determination of Oxygen Flux Density in the Soil," *International Agrophysics*, Vol. 13, No. 3, 1999, pp. 273-281.

[18] R. G. Joergensen, "The Fumigation-Extraction Method to Estimate Soil Microbial Biomass: Calibration of the k_{EN} Value," *Soil Biology and Biochemistry*, Vol. 28, No. 1, 1996, pp. 25-31. doi:10.1016/0038-0717(95)00101-8

[19] R. Walczak, B. Witkowska-Walczak and C. Sławiński, "Water Retention and Conductivity of Polish *Mollic Gleysols*," in Polish, *Acta Agrophysica*, Vol. 53, No. 2, 2001, pp. 211- 223.

[20] J. Brouver and H. Anderson, "Water Holding Capacity of Ironstone Gravel in a Typical Plinthoxeralf in Southeast Australia," *Soil Science Society of America Journal*, Vol. 64, No. 5, 2000, pp. 1603-1608. doi:10.2136/sssaj2000.6451603x

[21] R. Walczak, E. Rovdan and B. Witkowska-Walczak, "Water Retention Characteristics of Peat and Sand Mixtures," *International Agrophysics*, Vol. 16, No. 2, 2002, pp. 161-165.

[22] T. Włodarczyk, W. Stępniewski, M. Brzezińska and G. Przywara, "Impact of Different Aeration Conditions on the Content of Extractable Nutrients in Soil," *International Agrophysics*, Vol. 22, No. 4, 2008, pp. 371-375.

[23] Z. Stępniewska, A. Wolińska and R. P. Bennicelli, "Influence of Soil Water Potential on Microdiffusion of Oxygen in the *Eutric Cambisol*," *Acta Agrophysica*, Vol. 84, No. 1, 2003, pp. 145-152.

[24] A. M. Gajda, "Effect of Different Tillage Systems on Some Microbiological Properties of Soils under Winter Wheat," *International Agrophysics*, Vol. 22, No. 1, 2008, pp. 201-208.

[25] A. M. Gajda, "Microbial Activity and Particulate Organic Matter Content in Soils with Different Tillage System Use," *International Agrophysics*, Vol. 24, No. 2, 2010, pp. 129-137.

[26] N. Jedidi, "Microbial Biomass in a Soil Amended with Different Types of Organic Wastes," *Waste Management Research*, Vol. 22, No. 2, 2004, pp. 293-299. doi:10.1177/0734242X04043930

[27] P. Nannipieri, L. Badalucco, L. Landi and G. Pietramellara, "Measurements in Assessing the Risk of Chemicals to the Soil Ecosystem," In: J. T. Zelikoff, Ed., *Ecotoxicology: Responses, Biomarkers and Risk Assessment*, SOS Publications, New York, 1997, pp. 507-534.

[28] M. C. Fisk and T. J. Fahey, "Microbial Biomass and Nitrogen Cycling Responses to Fertilization and Litter Removal in Young Northern Hardwood Forests," *Biogeochemistry*, Vol. 53, No. 2, 2001, pp. 201-223. doi:10.1023/A:1010693614196

[29] K. Mohammadi, "Soil Microbial Activity and Biomass as Influenced by Tillage and Fertilization in Wheat Production," *American-Eurasian Journal of Agriculture and Environmental Sciences*, Vol. 10, No. 3, 2011, pp. 330-337.

[30] S. Lerat, L. England, J. Klironomos, P. Pauls, C. Swanton and J. T. Trevors, "Real-Time Polymerase Chain Reaction Detection of the Transgenes for Roundup Ready Corn and Soybean in Soil Samples," *Journal of Agricultural and Food Chemistry*, Vol. 53, No. 5, 2005, pp. 1337-1342. doi:10.1021/jf048830+

[31] L. Sagova-Mareckova, L. Cermak, J. Novotna, K. Plhackova, J. Forstova and J. Kopecky, "Innovative Methods for Soil DNA Purification Tested in Soils with Widely Differing Characteristics," *Applied and Environmental Microbiology*, Vol. 74, No. 9, 2008, pp. 2902-2907. doi:10.1128/AEM.02161-07

[32] L. I. Singh and P. S. Yadava, "Spatial Distribution of Microbial Biomass in Relation to Land-Use in Subtropical Systems of North-East India," *Tropical Ecology*, Vol. 47, No. 1, 2006, pp. 63-70.

[33] E. V. Blagodatskaya, O. S. Khokhlova and S. A. Blagodatskii, "Extractable Microbial DNA Pool and Microbial Activity in *Paleosols* of Southern Urals," *Microbiology*, Vol. 72, No. 6, 2003, pp. 750-755. doi:10.1023/B:MICI.0000008380.57233.9e

[34] M. Hartmann, B. Frey, R. Kölliker and F. Widmer, "Semi-Automated Genetic Analyses of Soil Microbial Communities: Comparison of T-RFLP and RISA Based on Descriptive and Discriminative Statistical Approaches," *Journal of Microbiological Methods*, Vol. 61, No. 3, 2005, pp. 349-360. doi:10.1016/j.mimet.2004.12.011

26

Soil Quality Mapping Studies Using Nematodes as Bioindicators

Lakshmy Kamala Sthanu[1*], Jaya Divakaran Sarasamma[2], Radhakrishnan Thanu Iyer[3], Mohandas Chellappan[4]

[1]Centre for Environment and Development, Thozhuvancode, Vattiyoorkavu P.O., Thiruvananthapuram, India; [2]Department of Environmental Sciences, University of Kerala, Kariavattom Campus, Thiruvananthapuram, India; [3]Division of Geo Informatics, Indian Institute of Information Technology and Management, Kerala, Techno Park Campus, Thiruvananthapuram, India; [4]Division of Crop Protection, Central Tuber Crops Research Institute, Sreekaryam, Thiruvananthapuram, India.
Email: [*]hariluxmi@gmail.com, jayvijayds@gmail.com, rkequinox@gmail.com, drcmohandas@gmail.com

ABSTRACT

Soil quality is one of the most important factors in sustaining the global biosphere and developing sustainable agricultural practices. Land use and management practices greatly impact the direction and degree of soil quality changes in time and space. Understanding the effects of land use and management practices on soil quality and its indicators has been identified as one of the most important goals for modern soil science. Soil quality mapping study represents a method for assessing and mapping soil quality changes in time and space in small units. For the present study, changes in the physical, chemical parameters and nematode density of the soils in the rural and urban areas of Thiruvananthapuram district, Kerala, were determined. The soil samples were collected from seven different categories of contaminated soils namely coastal area, sewage disposal area, industrial area, road-side area, agricultural area, market area and gasoline station area, and also from two control stations in rural and urban areas. The soil physico-chemical parameters and nematode density were determined. Geostatistics combined with GIS was applied to analyze the spatial variability of soil physico-chemical characteristics and nematode density. This soil quality mapping study provides a basis for identifying tension zones and serves as a triggering mechanism for implementation of soil contamination mitigating strategies.

Keywords: Soil Quality; Mapping Studies; Nematode Density; Spatial Variability; Contaminated Soils

1. Introduction

Industrialization and urbanization have a strong impact on the environment, and its biotic and abiotic components. Many toxins added to the soils by different anthropogenic activities can build up to concentrations that become serious threats to plant and animal health (Alloway, 1996) [1]. The presence or absence of indicator species or indicator community reflects environmental conditions. Bio monitors provide a direct measurement of a biological effect rather than inferring values using soil extractions (Tarazona *et al.*, 2005) [2]. The study of regional variations and the anthropogenic contamination by metals of soils is very important for environmental planning and monitoring in urban areas. Mapping allows for more efficient approach to remediation and monitoring of soil contaminants (Fortin and Dale, 2005) [3]. Geographical Information System (GIS) is used for the mapping of the contaminants in the study area soils. Based on the GIS analysis, the highest risk sites are identified in which majority of the remediation and monitoring should take place. The major objective of the present study was to monitor the soil quality of selected rural and urban areas in Thiruvananthapuram District, Kerala, using nematodes as bioindicators, and also aimed to provide base line information on the soil health status of rural and urban areas of the capital district of Kerala.

[*]Corresponding author.

2. Materials and Methods

2.1. Study Area

The rural and urban areas in Thiruvananthapuram District, Kerala were selected for the present study (http: www.censusindia.net/2001census results). Thiruvananthapuram

District is located at 8°26'25" N latitude and 76°55'25"E longitude. Thiruvananthapuram, the southernmost District in the State of Kerala is the study area. Its southern most extremity, Kaliaikkavila, is only 54 kms away from Kanyakumari (Cape Comarin), the land's end of India. The district stretches 78 kms along the shores of the Arabian Sea on the west, Kollam district lies on the north with Thirunelveli and Kanyakumari districts of Tamil Nadu on the east and the south respectively. The district can be divided into three geographical regions; highlands, midlands and lowlands. The large forest reserves favorably affect the climate and induce rain. Cold weather is experienced in the mountain ranges whereas lower down, the weather is bracing and in the plains, it is generally hot. Though the mean maximum temperature is only around 900F, it is oppressive in the moisture-laden atmosphere of the plains. Humidity is high and rises to about 90 percent during the south-west monsoon. The average rainfall is around 150 cm per annum. It is significant that the district gets rainfall both from the south-west and the north-east monsoons. According to the final figures of 2011census (Census of India, 2011), the population of the district is 3,307,284. The Taluks in Thiruvananthapuram District are Neyyattinkara, Thiruvananthapuram, Nedumangad and Chirayinkeezhu. The district has three major rivers, several freshwater lakes and more than 300 ponds. The eastern region is forested, northern regions are mostly under rubber cultivation and the remaining areas have mixed dry land crops of coconut, plantain, tapioca, etc. Built up areas and rice fields complete the land use. Based on the land use pattern, Thiruvananthapuram District has been classified into different categories namely, coastal areas, sewage disposal areas, industrial areas, road-side areas, agricultural areas, market areas, gasoline station areas, benign environment, etc. These areas will be receiving different types of wastes based on the source of contamination. The contaminated sites due to different anthropogenic activities were identified in both rural and urban areas of Thiruvananthapuram District after conducting a reconnaissance survey. Sixteen sampling stations in the rural and urban areas of the Thiruvananthapuram District were selected. These include seven different categories of contaminated soils selected in the rural and urban areas namely coastal area, sewage disposal area, industrial area, roadside area, agricultural area, market area and gasoline station area. Control soil samples were collected from comparatively benign environments in both rural and urban areas which were selected as control stations (**Table 1**).

2.2. Soil Collection

Soil samples (0 - 15 cm depth) from the sampling stations were collected during the pre-monsoon, monsoon and post-monsoon seasons of the study period (April 2009 to January 2010). Composite samples were taken by mixing the samples collected from five different sites of each station. Soil sampling and analysis of soil physico-chemical characteristics were carried out according to the standard procedures (Gupta, 1999) [4]. The concentration of the heavy metals viz. lead, chromium, manganese, copper and zinc in the acid digested soil samples were determined (Saxena, 1994) [5] using an Atomic Absorption Spectrophotometer (Model, GBC 932 AA). Karl Pearson's Product Moment Correlation is used for finding significant correlation between nematode densities with all other soil quality parameters in the three different seasons (SPSS v - 16 for WINDOWS).

The Geographical coordinates of the sampling stations were collected using Hand-held GPS (Garmin eTrex Vista HCx). The Coordinate values were entered into a spreadsheet with properly assigning IDs for each location. Another spreadsheet was prepared with physico-chemical characteristics and nematode density. In this spreadsheet, the same IDs were entered against respective stations. Both spreadsheets were added in GIS software (Arc GIS version 10) as tables. The table with Geographic coordinate values were plotted as point layer using "Display XY" method and subsequently exported as a shape file with WGS84 as reference ellipsoid. The physico-chemical characteristics and nematode density table was joined with the attribute table of the point layer using ID as the foreign key using JOIN method in GIS software (Korte, 2001) [6]. The resultant layer comprised the geometry of stations as points and all the results as associated attribute table (**Figure 1**). The soil physico-chemical characteristics were analyzed and mapping of significantly correlated soil physico-chemical characteristics with nematode density was plotted.

3. Results and Discussion

3.1. Mapping of Physico-Chemical Characteristics and Nematode Density in Soil

The results of physical characteristics of the soils in the study stations are given in **Tables 2(a)** and **(b)** respectively and the soil chemical characteristics analyzed are given in **Tables 3(a)** and **(b)** respectively. The results of heavy metal (Pb, Zn, Mn, Cu, Cr) content in soil are given in **Tables 4(a)** and **(b)** respectively. Correlation studies of soil physico-chemical characteristics with nematode density were done.

3.2. Nematode Sampling, Extraction and Density Calculation

One kg soil samples from 0 - 15 cm depth were collected. For soil nematode identification studies, 100 cc (approx. 150 gms) of soil samples were washed and the soil solutions were passed through 60 mesh (250 μ) and 350 mesh

Table 1. Portrait of sampling stations in the study area.

Rural Stations			Urban Stations		
Name	Location		Name	Location	
	Latitude	Longitude		Latitude	Longitude
Vizhinjam (Station 1)	8°33'51"	76°53'25"	Shangumugham (Station 9)	8°27'50"	76°55'55"
Vilappilssala (Station 2)	8°30'18"	76°53'45"	Valiathura (Station 10)	8°33'51"	76°53'25"
Kochuveli (Station 3)	8°30'18"	76°53'45"	Peroorkkada (Station 11)	8°31'35"	76°58'03"
Mangalapuram (Station 4)	8°30'18"	76°53'45"	Thampanoor (Station 12)	8°29'15"	76°57'12"
Palode (Station 5)	8°28'53"	76°57'08"	Sreekaryam (Station 13)	8°30'18"	76°53'45"
Balaramapuram (Station 6)	8°27'06"	76°57'21"	Chalai (Station 14)	8°28'52"	76°57'08"
Pallichal (Station 7)	8°27'06"	76°57'31"	Karamana (Station 15)	8°28'56"	76°57'36"
Anappara (**Control Station** 8)	8°28'53"	76°57'08"	Ambalathara (**Control Station** 16)	8°27'06"	76°57'21"

S1 - S8: Rural Sampling Stations; S9 - S16: Urban Sampling Stations.

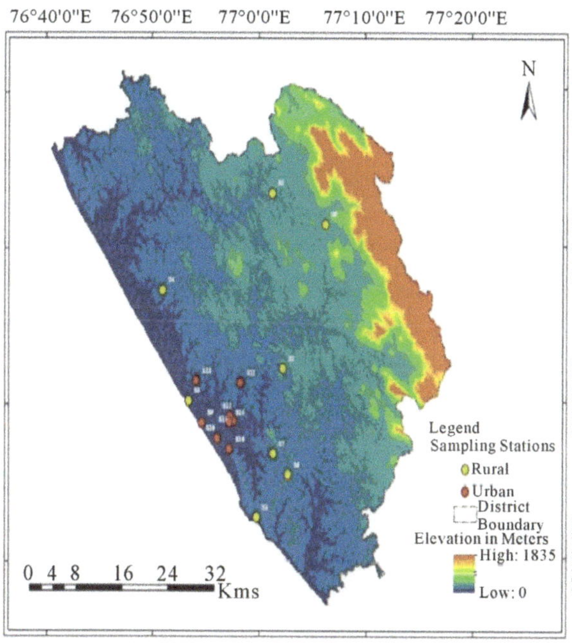

Figure 1. Base map showing the sampling stations in the rural and urban areasof.

(45 μ) sieves (ASTM Endecott Sieves) respectively. The residues from 350 mesh sieve was washed and collected into a beaker and allowed to settle for one hour. Then decanted the supernatant and poured the rest on a wet tissue paper placed on metal gauze in a Petri dish containing water. After two days, collected the nematode containing solution, allowed to settle for decanting and added equal quantity of boiled 4% formalin to kill and fix it. 10 ml of this solution was placed in a nematode Perspex counting dish (Cobb, 1917) [7] and were observed under a Stereomicroscope (Model, Leica LED3000 NVI) and their abundance was calculated (Southwood and Henderson, 2000) [8] during the three seasons as shown in **Figures 2(a)-(c)**. The soil sample solution was made up to 100 ml and stirred well. Pipetted 10 ml of this soil solution from three different depths of this 100 ml soil sample solution kept in a beaker and placed in Doncaster's circular mounting tray (Goodey, 1963) [9]. Nematode density is expressed in Number per 100 ml soil.

Electrical conductivity, water holding capacity, moisture content, organic matter, total nitrogen content, phosphorus content, potassium and copper contents showed positive correlation with nematode density in soil. pH, temperature, pore space, bulk density, chloride, lead, zinc, manganese and chromium contents showed negative correlation with nematode density in soil samples collected from both rural and urban area as shown in **Tables 5(a)** and **(b)** respectively. Of the different physicochemical characteristics studied, water holding capacity, moisture content, organic matter content, pH, zinc and manganese contents only showed significant level of correlation with nematode density. Mapping of significantly correlated soil physico-chemical characteristics with nematode density was plotted. The significance level of correlation coefficient is given in **Table 6**.

3.3. Mapping of Significantly Positively Correlated Physico-Chemical Characteristics and Nematode Density in Soil

Water holding capacity of soils in the selected stations of

Table 2. (a) Physical characteristics of rural and urban soils; (b) Physical characteristics of rural and urban soils.

(a)

St.No.	pH			Conductivity (mho/cm)			Temperature (°C)		
	Pre-monsoon season	Monsoon season	Post-monsoon season	Pre-monsoon season	Monsoon season	Post-monsoon season	Pre-Monsoon season	Monsoon season	Post monsoon season
S1	7**	7.7**	7.1**	0.07**	0.049**	0.032**	32.6**	30**	31.5**
S2	5**	5.7**	5**	1.52	0.88**	0.463**	30	29.7	31
S3	5.7**	6.4**	6.3**	0.26**	0.15**	0.094**	32.5**	30**	31.5**
S4	6.1**	7.1**	6.5**	0.26**	0.082**	0.057**	31.6**	30**	31.48*
S5	5.3*	6*	5.5*	1.09**	0.245**	0.326**	29.9**	29**	31**
S6	4.7**	5**	4.9**	3.14**	1.35**	1.343**	31.1**	30**	31.4**
S7	6.2**	7.3**	6.7**	0.14**	0.072**	0.048**	33**	30**	31.4**
S8	5.4	6.2	5.9	0.44	0.225	0.211	29.5	29.8	31
S9	8.7**	8.3**	7.3**	0.06**	0.042**	0.028**	32.6**	30**	31.5**
S10	5.7	5.7**	5.3**	1.33**	0.528**	0.451**	32**	30**	31.3**
S11	5.7	6.3	6.2	0.27**	0.174**	0.099**	32.2**	30**	31.1
S12	6.1**	7.3**	6.7	0.25**	0.073**	0.049**	31.3**	30**	31.4**
S13	5.3**	6.2	5.7**	0.52**	0.225**	0.214**	32**	29.9	31.2**
S14	4.8**	5.3**	4.9**	1.6**	1.305**	0.865**	31*	30**	31.3**
S15	6.9**	7.5**	7*	0.07**	0.064**	0.04**	33**	30**	31.4**
S16	5.7	6.2	6	0.33	0.197	0.118	31.1	29.8	31

(b)

St.No.	Pore Space (%)			Water holding capacity (%)			Bulk density (g/cc)		
	Pre-monsoon season	Monsoon season	Post-monsoon season	Pre-monsoon season	Monsoon season	Post-monsoon season	Pre-monsoon season	Monsoon season	Post-monsoon season
S1	57.8**	57.8**	57.8**	0.04**	0.05**	0.03**	1.56**	1.58**	1.57**
S2	30.19**	30.2**	30*	1.33**	1.34**	1.32**	0.71**	0.72**	0.71**
S3	38.32**	38.2**	38.1**	0.15	0.17**	0.15**	1.14**	1.14**	1.2
S4	40.12**	40.13**	40.1**	0.15**	0.15**	0.14**	1.23**	1.24**	1.23**
S5	30.28**	30.3**	30.2**	0.79**	0.75**	0.74**	0.5**	0.9**	0.9**
S6	25.43**	25.4**	25.3**	1.84**	1.87**	1.86**	0.2**	0.3**	0.3**
S7	45.25**	45.3**	44.9**	0.13	0.13**	0.11**	1.26**	1.25**	1.25**
S8	32.45	32	32	0.27	0.27	0.25	1.09	1.1	1.11
S9	58.87**	58.86**	58.84**	0.02**	0.03**	0.03**	1.58**	1.6**	1.6**
S10	30.19**	30.2**	30.1**	1.23**	1.25**	1.23**	0.8**	0.8**	0.78**
S11	35.45**	35.5**	35.2**	0.18**	0.2	0.2	1.11**	1.12	1.15**
S12	42.02**	42.1**	41.9**	0.13**	0.14**	0.12**	1.24**	1.25**	1.23**
S13	31.33**	31.34**	31.3**	0.54**	0.54**	0.51**	0.93**	1	0.97**
S14	29.26**	30**	30**	1.84**	1.84**	1.83**	0.3**	0.5**	0.46**
S15	48.3**	48.3**	48.19**	0.11**	0.12**	0.11**	1.28**	1.28**	1.28**
S16	35.09	35.1	34.8	0.25	0.26	0.23	1.11	1.1	1.11

*Significant at $p < 0.05$; **Significant at $p < 0.01$ level.

Table 3. (a) Chemical characteristics of rural and urban soils; (b) Chemical characteristics of rural and urban soils.

(a)

St.No.	Chloride (mg/g)			Moisture content (%)			Organic matter (g %)		
	Pre-monsoon season	Monsoon season	Post-monsoon season	Pre-monsoon season	Monsoon season	Post-monsoon season	Pre-monsoon season	Monsoon season	Post-monsoon season
S1	0.34**	0.338**	0.339**	1.74**	1.81**	1.76**	0.517**	0.591**	0.569**
S2	0.046**	0.046**	0.045**	12.54**	12.6**	11.8**	3.034**	3.839**	3.836**
S3	0.053**	0.052**	0.052**	5.46**	5.55**	5.54**	1.569**	1.772**	1.765**
S4	0.045**	0.045	0.044**	4.78**	4.9**	4.9**	1.397**	1.626**	1.562**
S5	0.046**	0.046**	0.046**	9.42**	9.62**	9.62**	2.569**	3.696**	3.6**
S6	0.046**	0.047**	0.046**	15.17**	15.25**	15.11**	5.655**	5.769**	4.448**
S7	0.046**	0.044**	0.046**	2.38**	2.4**	2.36**	1.155**	1.579**	1.155**
S8	0.046	0.046	0.045	8.55	8.61	8.59	1.69	2.081	2.512
S9	0.41**	0.406**	0.407**	0.19**	0.22**	0.22**	0.345**	0.367**	0.335**
S10	0.047**	0.047**	0.046**	11.85**	11.9**	11.5**	2.672**	2.838**	3.836**
S11	0.046	0.046	0.045	6.79**	6.82**	6.76**	1.569**	1.94**	1.867**
S12	0.044**	0.044**	0.043**	2.7**	2.9**	2.9**	1.328**	1.688**	1.455**
S13	0.046**	0.046**	0.046	8.67**	8.69**	8.6	2.207**	2.491**	2.255**
S14	0.046**	0.047**	0.046**	13.3**	14**	13.6**	3.5**	3.881**	3.87**
S15	0.046**	0.045**	0.046**	2.35**	2.36**	2.29**	0.69**	0.909**	0.905**
S16	0.046	0.046	0.046**	8.06	8.1	8	1.69	2.081	2.41

(b)

St.No.	o			Total Phosphorus (g %)			Total Potassium (g %)		
	Pre-monsoon season	Monsoon season	Post-monsoon season	Pre-monsoon season	Monsoon season	Post-monsoon season	Pre-monsoon season	Monsoon season	Post-monsoon season
S1	24.5**	42**	12.6**	4.42**	3.75**	3.62**	66.52**	64.67**	25.6**
S2	63.1**	105**	182.7**	52.46**	89.82**	75.31**	189.73**	147.41**	106.1**
S3	44.8**	70**	132.3**	6.21**	30.56**	14.41**	125.45**	145.54**	99.33**
S4	43.5**	63**	119.7**	5.31**	27.68**	13.15**	106.25**	143.53**	90.3**
S5	59.6**	91**	157.5**	43.57**	64.11**	69.21**	163.39**	184.6**	100.7**
S6	70**	168**	252**	159.82**	106.43**	76.36**	189.73**	187.32**	155**
S7	30.1**	56**	107.1**	4.45**	7.86**	5.68**	83.48**	128.46**	60.2**
S8	49.2	77	144.9	25.76	44.64	41.65	133.04	168.86	102.88
S9	20.3**	42**	40.8**	3.53**	0.71**	0.78**	34.38	17.08**	25.59**
S10	61**	98**	182.7**	52.46**	66.07**	50.2**	174.55**	398.44**	301**
S11	47**	70	138.6**	10.67**	33.21**	16.41**	127.23	64.51**	102.34**
S12	35.4**	56**	107.1**	5.31**	14.64**	7.78**	85.27	43.53**	84.28**
S13	49.6**	77**	157.5**	40.09**	63.04**	67.95**	144.2**	178.35**	155**
S14	64.4**	154**	195.3**	159.82**	92.14**	75.52**	189.73**	434.82**	415.5**
S15	29.6**	49**	103.83**	4.45**	4.11**	2.73**	66.52	26.56**	52.68**
S16	47.6	70	144.9	12.46	41.96	27.24	130.8	85.38	106.86

*significant at $p < 0.05$; **significant at $p < 0.01$ level.

Table 4. (a) Heavy metal content in rural and urban soils; (b) Heavy metal content in rural and urban soils.

(a)

St.No.	Pb (mg/kg)			Zn (mg/kg)			Mn (mg/kg)		
	Pre-monsoon season	Monsoon season	Post-monson season	Pre-monsoon season	Monsoon season	Post-monsoon season	Pre-monsoon season	Monsoon season	Post-monsoon season
S1	BDL	BDL	BDL	43.2	73.1	40.3	140	100.2	130.3
S2	0.6	0.3	0.44	70.2	352.1	245.4	105.2	310.5	310.6
S3	0.6	0.7	0.56	220.4	100.2	310.1	78.6	69.7	78.3
S4	0.9	0.6	0.76	510	220.9	535.4	401	225.6	512.6
S5	BDL	BDL	BDL	70.4	52.3	92.3	88.6	78.4	89.3
S6	0.6	0.6	0.4	90.1	70	186.5	111.4	142.5	127.9
S7	0.9	0.9	0.86	340.2	470.1	752	540.2	398.6	567
S8	BDL	BDL	BDL	75.4	40.4	72	70.3	63.2	70.2
S9	BDL	BDL	BDL	52.3	86	42	157.4	109.5	510
S10	0.8	0.3	0.4	69	314	252	146.4	152.3	152.1
S11	0.6	0.7	0.61	320	112.3	312.3	65.4	310.5	361.9
S12	0.6	0.6	0.7	360	216.9	502.9	512	213.9	509.4
S13	BDL	BDL	BDL	220	67.5	89.5	88.6	86.5	86.5
S14	0.4	0.5	0.38	98.1	72.7	182.9	115	152.3	138
S15	1	0.8	0.9	360.3	508	760.4	569.8	396.5	538
S16	BDL	BDL	BDL	74.4	52.3	77.1	78.6	68.6	65.8

(b)

St.No.	Cu (mg/kg)			Cr (mg/kg)		
	Pre-monsoon season	Monsoon season	Post-monsoon season	Pre-monsoon season	Monsoon season	Post-monsoon season
S1	19.2	18.7	20.5	0.2	BDL	BDL
S2	90.4	70.6	88.5	0.3	0.2	0.26
S3	79.6	71.9	77.9	0.4	0.2	0.33
S4	78.2	74.7	75.0	0.4	0.4	0.4
S5	79.6	76.2	77.5	0.2	BDL	0.57
S6	80.3	98.4	99.8	0.2	0.3	0.28
S7	95.2	90.5	93.6	0.7	0.6	0.8
S8	90.2	88.1	89.2	BDL	BDL	0.32
S9	20.2	18.9	20.0	0.2	BDL	BDL
S10	92.6	89.8	90.6	0.4	0.3	0.28
S11	86.8	85.4	85.6	0.4	0.2	0.38
S12	82.2	81.2	82.0	0.3	0.3	0.38
S13	87.2	83.7	86.5	0.3	BDL	0.6
S14	95.6	93.3	94.3	0.3	0.3	0.26
S15	98.3	84.1	87.4	0.8	0.6	0.82
S16	88.2	83.3	85.3	0.20	BDL	0.31

BDL-Below Detected Level.

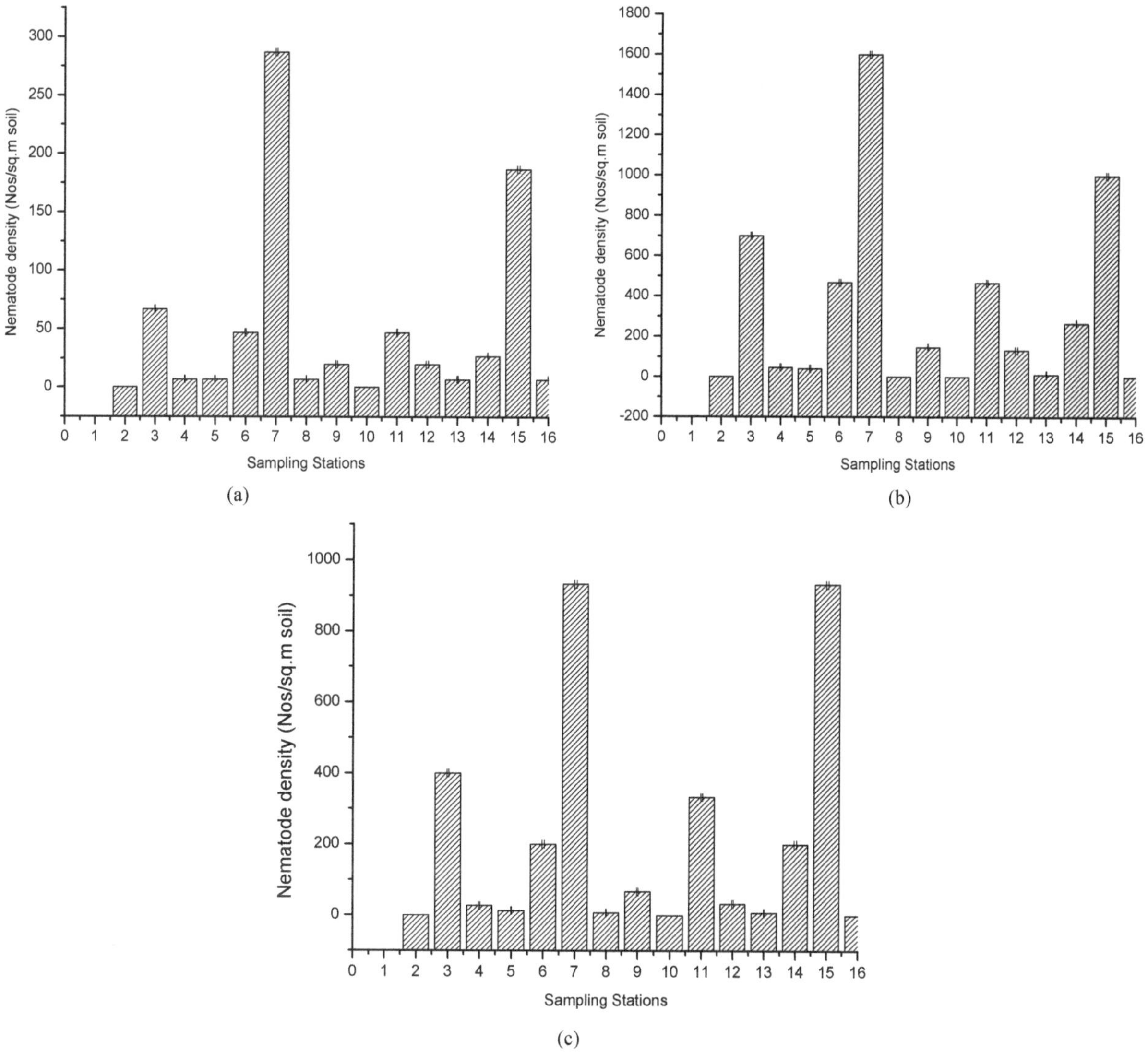

Figure 2. Nematode abundance in the Soils of Study stations during different seasons. (a) Pre-monsoon season; (b) Monsoon season; (c) Post-monsoon season.

the study area showed significant and positive correlation with nematode density. Smiles (1988) [10] in his studies concluded that the rate of soil water movement controls many biological activities such as wilting and germination of plants and hatching of nematode cysts and spores. The spatial variability of water holding capacity and nematode density during the pre-monsoon, monsoon and post monsoon seasons were plotted and the results are shown in **Figure 3**. Water holding capacity showed very high positive correlation with nematode density during the pre monsoon season in all the stations. But in the monsoon season due to the dilution factor, this positive correlation was very high in most of the study stations except in the stations S1, S2, S3, S4 and S16 where this positive correlation was only high. In the post-monsoon season, this positive correlation showed the same pattern as that in the monsoon season except for the stations—S3 and S4 where the positive correlation was reverted back to very high.

Moisture content in the soils of study stations showed a significant and positive correlation with nematode density. Townshend and Webber (1971) [11] showed that the movement and survival of nematodes was maximum in soils with maximum moisture retention. The spatial variability of moisture content and nematode density during the pre-monsoon, monsoon and post-monsoon seasons were plotted and the results are shown in **Figure 4**. Moisture content showed high positive correlation with nematode density during the pre monsoon season in most of the study stations except in the stations, S11, S12, S13, S14 and S15 where this positive correlation was very high. But in the monsoon season due to the dilution factor, this positive correlation was very high all the stations. In the post monsoon season, this positive correla-

Table 5. (a) Correlation of nematode density with physico-chemical characteristics in rural area; (b) Correlation of nematode density with physico-chemical in urban area.

(a)

Parameters	Pre monsoon	Monsoon	Post monsoon
pH	−0.680(**)	−0.841(**)	−0.820(**)
Conductivity	0.974(**)	0.967(**)	0.992(**)
Temperature	−0.223	−0.202	0.000
Pore space	−0.612(**)	−0.692(**)	−0.662(**)
Water holding capacity	0.845(**)	0.972(**)	0.965(**)
Bulk density	−0.817(**)	−0.933(**)	−0.923(**)
Moisture content	0.798(**)	0.892(**)	0.875(**)
Chloride	−0.245	−0.288	−0.277
Organic matter	0.958(**)	0.967(**)	0.804(**)
Total nitrogen	0.719(**)	0.980(**)	0.821(**)
Phosphorus	0.990(**)	0.907(**)	0.766(**)
Potassium	0.680(**)	0.567(**)	0.497(**)
Lead	−0.563(**)	−0.436(*)	−0.774(**)
Zinc	−0.311	−0.183	−0.254
Manganese	−0.275	−0.020	−0.237
Copper	0.333	0.450(**)	0.458(**)
Chromium	−0.464(**)	−0.405(*)	−0.434(**)

(b)

Parameters	Pre monsoon	Monsoon	Post monsoon
pH	−0.307	−0.796(**)	−0.819(**)
Conductivity	0.856(**)	0.989(**)	0.990(**)
Temperature	−0.616(**)	−0.016	−0.059
Pore space	−0.551(**)	−0.641(**)	−0.558(**)
Water holding capacity	0.915(**)	0.978(**)	0.961(**)
Bulk density	−0.893(**)	−0.902(**)	−0.886(**)
Moisture content	0.741(**)	0.875(**)	0.806(**)
Chloride	−0.266	−0.305	−0.256
Organic matter	0.828(**)	0.860(**)	0.756(**)
Total nitrogen	0.711(**)	0.991(**)	0.700(**)
Phosphorus	0.988(**)	0.903(**)	0.814(**)
Potassium	0.693(**)	0.929(**)	0.948(**)
Lead	−0.570(**)	−0.528(**)	−0.767(**)
Zinc	−0.354(*)	−0.254	−0.262
Manganese	−0.313(*)	−0.274	−0.537(**)
Copper	0.342(*)	0.438(**)	0.377(*)
Chromium	−0.138	−0.261	−0.455(**)

Table 6. Significance level of correlation coefficient.

Range	Significance
0.00 - 0.200	Negligible
0.200 - 0.400	Low
0.400 - 0.600	Moderate
0.600 - 0.800	Substantial
0.800 - 1.00	Very high

tion showed the same pattern as that in the monsoon season except for the stations, S7 and S8 where the positive correlation was high.

Organic matter content in the soils of study stations showed a significant and positive correlation with nematode density. This is in agreement with the studies by Gorres *et al.* (1998) [12] which showed that the soils rich in organic matter were reported with the highest nematode density.

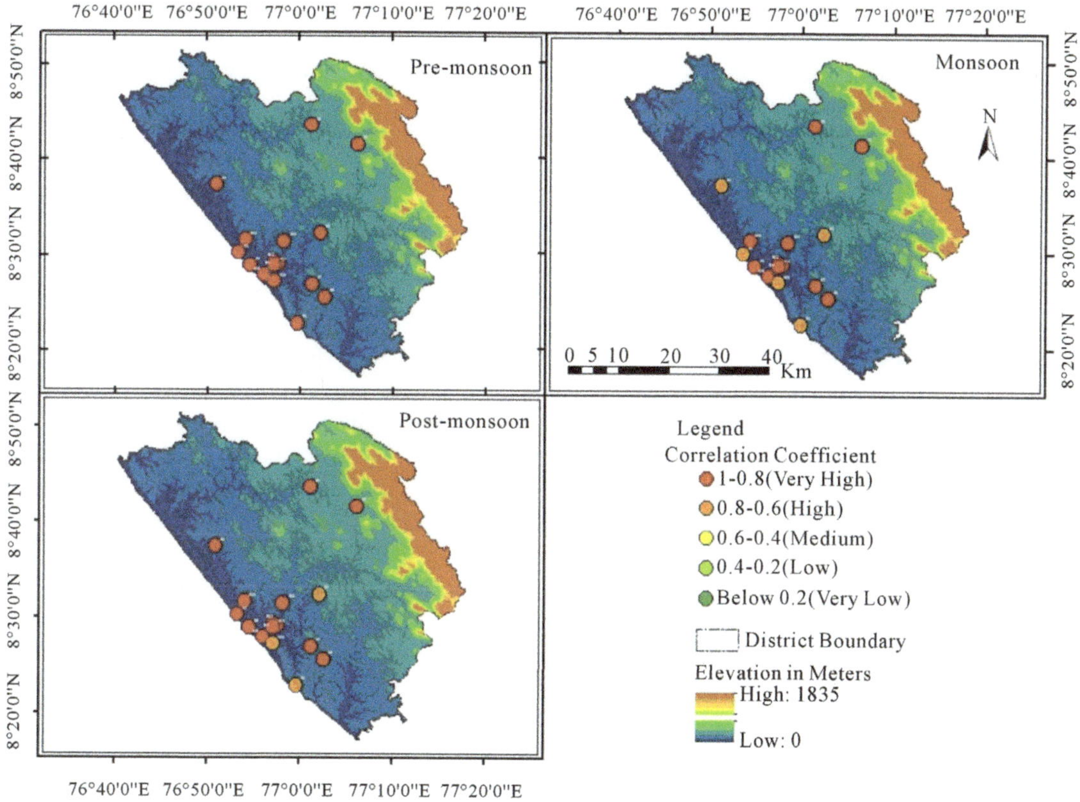

Figure 3. Mapping of water holding capacity and nematode density.

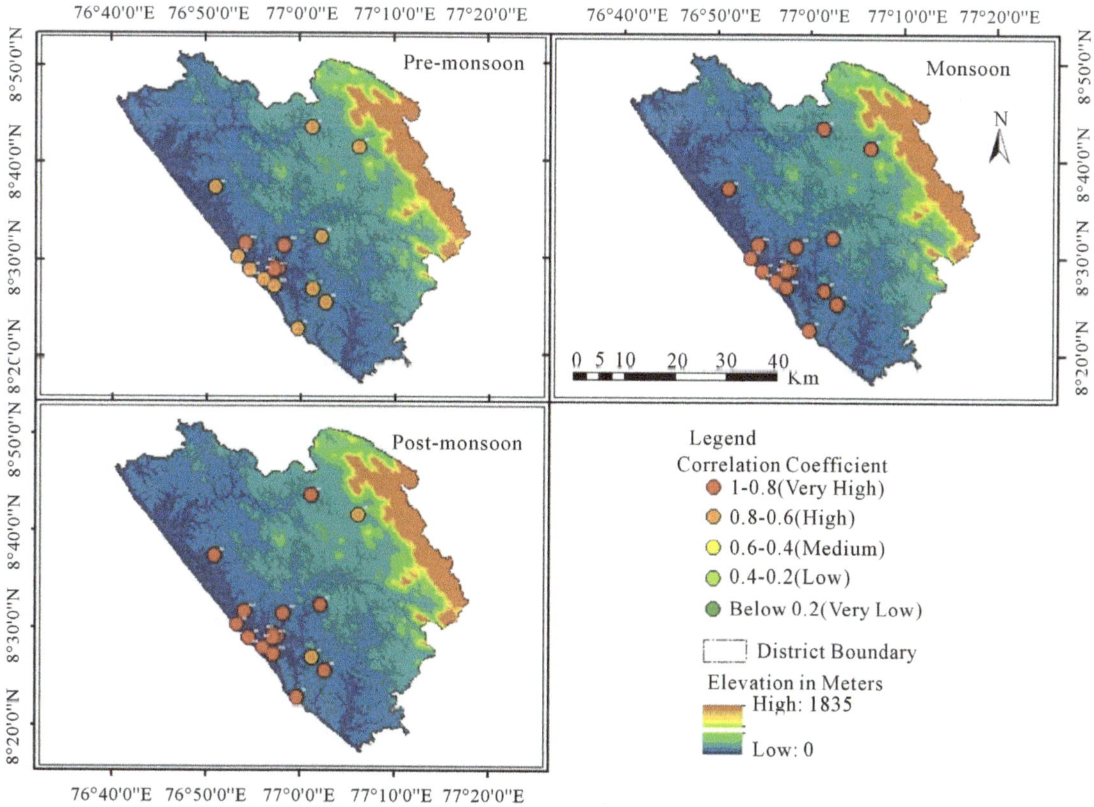

Figure 4. Mapping of moisture content and nematode density.

The spatial variability of organic matter content and nematode density during the pre monsoon, monsoon and post monsoon seasons were plotted and the results are shown in **Figure 5**. Organic matter content showed very high positive correlation with nematode density during the pre monsoon and monsoon seasons in all the stations. In the post monsoon season, this positive correlation showed the same pattern as that in the pre monsoon and monsoon seasons except for the stations, S7 and S9 where the positive correlation was high.

3.4. Mapping of Significantly Negatively Correlated Physico-Chemical Characteristics and Nematode Density in Soil

There is a significant and negative correlation between soil pH and nematode density. This is in agreement with the studies conducted by Murialdo *et al.* (2002) [13] which showed that when the soil pH was high, nematode density was low. The spatial variability of pH content in soil and nematode density during the pre monsoon, monsoon and post monsoon seasons were plotted and the results are shown in **Figure 6**. Soil pH showed very high negative correlation with nematode density during the pre monsoon season in stations, S13, S11, S12, S14 and S15. But in the monsoon season due to the dilution factor, this negative correlation was reduced. In the post monsoon season, the correlation showed the same pattern as that in the monsoon season.

Zinc content in the soils of study stations showed a significant and negative correlation with nematode density. Studies by Shao *et al.* (2008) [14] showed that nematode densities were negatively correlated with zinc concentrations as their population densities decreased with increased zinc concentrations. The spatial variability of zinc content and nematode density during the pre monsoon, monsoon and post monsoon seasons were plotted and the results are shown in **Figure 7**. Zinc content showed high negative correlation with nematode density during the pre monsoon season in the stations, S10, S11, S12 and S13. Due to the dilution factor in the monsoon season, this negative correlation was reduced in all the stations. But this negative correlation was very high in the stations, S14 and S15 in the monsoon season. The effect of raised water holding capacity and moisture content in the monsoon season influenced the soils in the station, S14 resulting in very high negative correlation. In the station, S15 the retention of zinc content in soil will be more due to the dilution factor resulting in the high negative correlation with nematode density. But in the post monsoon season due to the less water content in these soils, the negative correlation will be lesser.

Manganese content in the soils of study stations showed a significant and negative correlation with nematode density. Studies by Norton and Hoffmann (1974) [15] had shown that a negative correlation was observed between manganese content and egg/cyst ratio of nematodes which in turn will be reflected in their population densities. The spatial variability of manganese content in the soils and nematode density during the pre monsoon, monsoon and post monsoon seasons were plotted and the results are shown in **Figure 8**. Manganese content in soils showed the same negative correlation pattern with nematode density in both the pre monsoon and monsoon seasons. But in the post monsoon season, this negative correlation was very high in the station, S14 as these clayey soils were with less water content at this season as compared to that in the monsoon season and will augment the manganese content in soils. This is in agreement with the studies conducted by Rajmohan and Elango (2005) [16] which concluded that soils with high clay content will retain more manganese especially in the post monsoon season. In stations, S4, S6 and S9, the negative correlation was high. In station, S6 with high clay content in soils, the manganese retention will be more in this season. As S6 is a rural market area and S14 is an urban market area, the manganese contamination will be more in the urban market area due to different anthropogenic activities compared to the corresponding rural area. So station, S6 showed only high level of correlation with nematode density in the post monsoon season. In station, S4 which is a road-side area, the manganese deposition will be more in the post monsoon season. Previous studies by Singh and Singh (2006) [17] had shown that the manganese deposition in soils will be more in the post monsoon season in the road-side soils in India. Station, S9 which is an urban coastal area, this negative correlation will be more due to the hyper saline nature of these soils. This is in agreement with the studies by Alongi *et al.* (2005) [18] which showed that hyper saline soils will retain more manganese content in soils during the post monsoon season.

4. Conclusion

The soil physico-chemical characteristics which showed significant (positive and negative) correlations with nematode density were spatially plotted using the Arc GIS 10 to show their spatial variability. Water holding capacity, moisture content and organic matter content showed a significant and positive correlation with nematode density. PH, zinc content and manganese content showed a significant and negative correlation with nematode density. These soil mapping plots showed variations in the monsoon season compared to the pre monsoon and post monsoon seasons due to the dilution factor. In the study area, the organic matter content of urban soils with industrial activities and gasoline service stations was found to be significantly less compared to that of the rural area

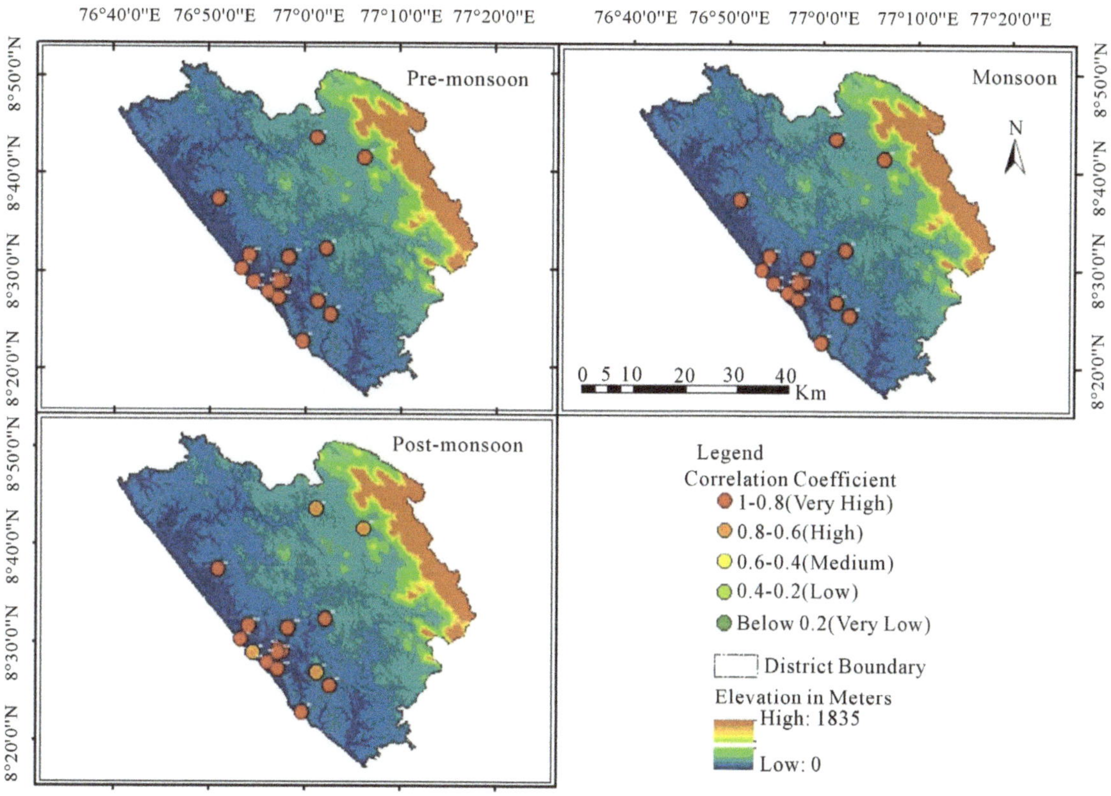

Figure 5. Mapping of organic matter content and nematode density.

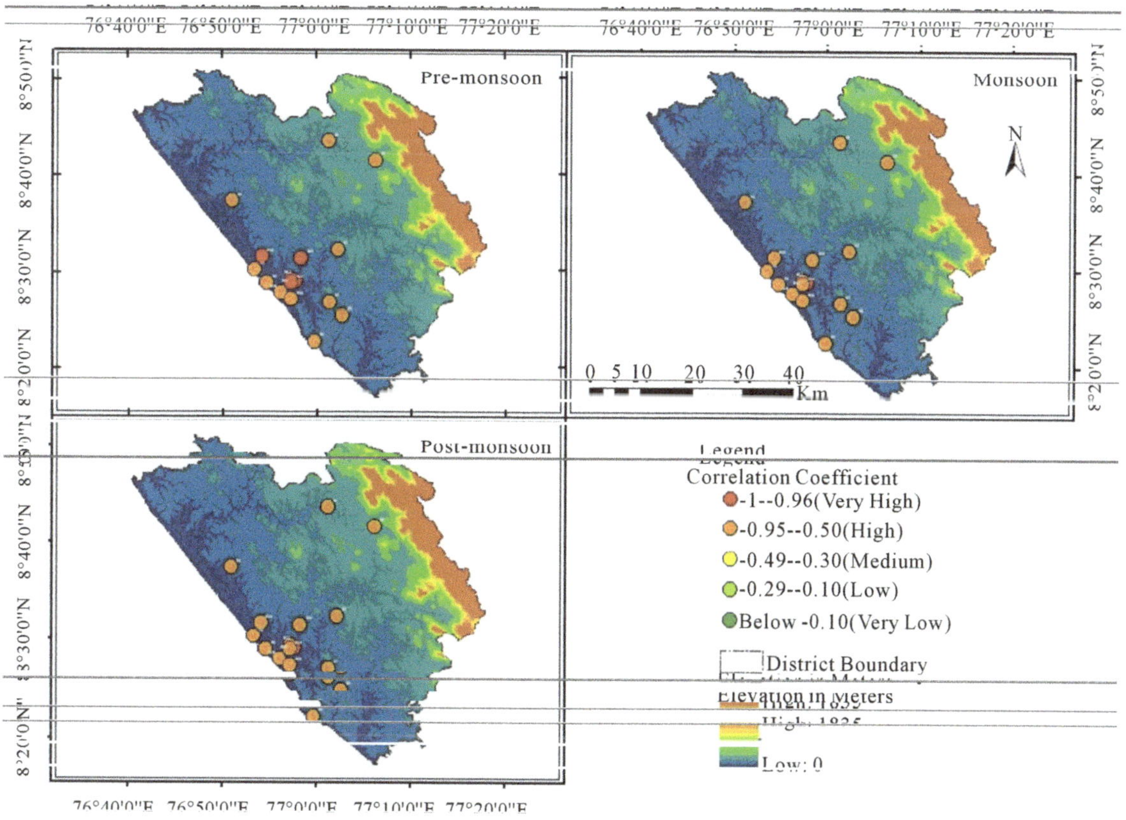

Figure 6. Mapping of pH and nematode density.

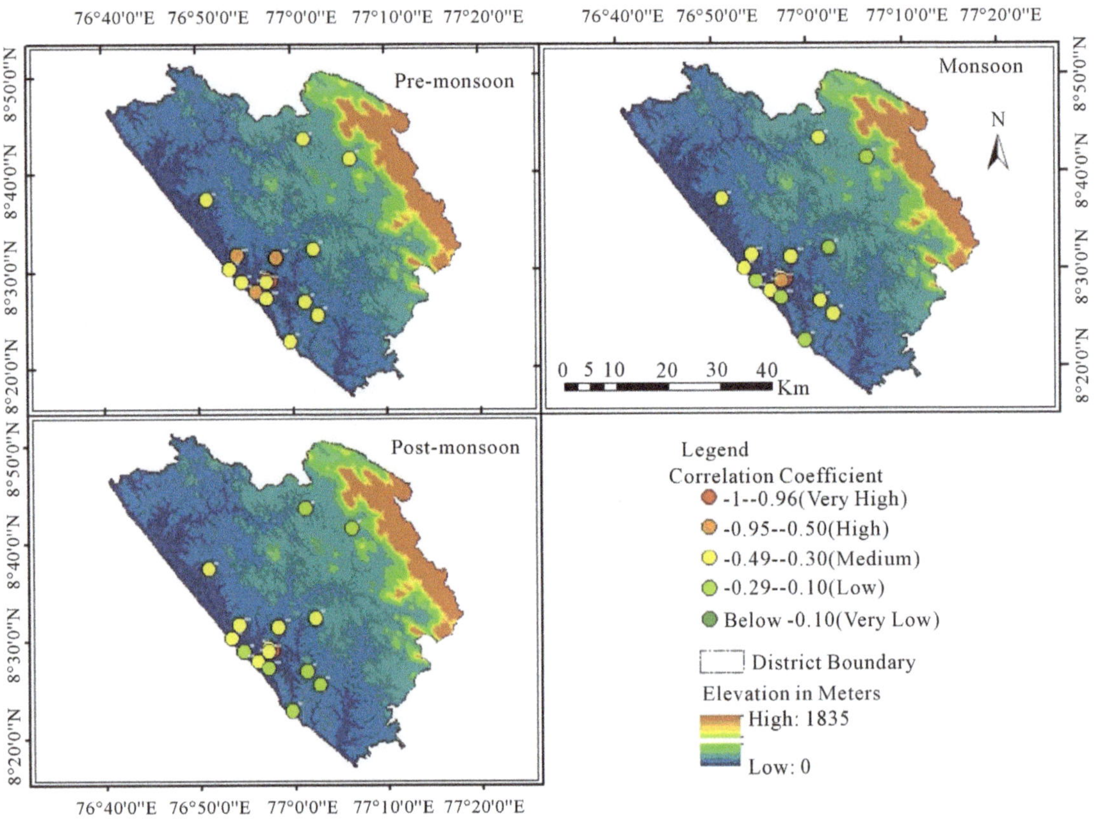

Figure 7. Mapping of zinc content and nematode density.

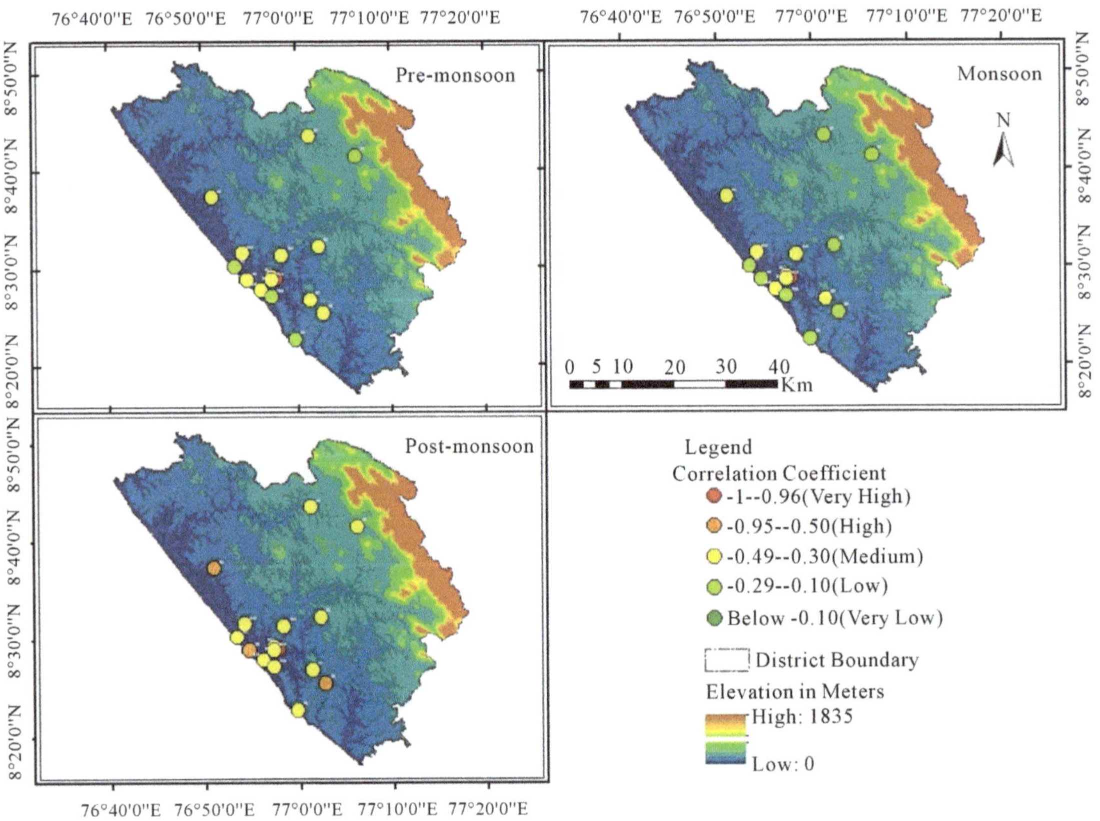

Figure 8. Mapping of manganese content and nematode density.

soils with the same land use patterns. Also the urban area soils were prone to more heavy metal contamination than the rural area soils especially in the gasoline station and roadside areas of Thiruvananthapuram District. This will reduce the soil microbial activity and soil health which in turn will be reflected in the crop production. This soil quality mapping provides a basis for identifying such tension zones and serves as a triggering mechanism for implementation of mitigating strategies.

5. Acknowledgements

The authors sincerely thank Prof. M.S. Rajasree, Director, Indian Institute of Information Technology, TechnoPark, Thiruvananthapuram for providing facilities to conduct the spatial studies. Also extend sincere thanks to Mr. Vishnu, Project Fellow, Indian Institute of Information Technology, Thiruvananthapuram for the help rendered in doing the spatial analysis.

REFERENCES

[1] B. J. Alloway, "Pollution-Causes, Effects and Control," 3rd Edition, The Royal Society of Chemistry Publishers, Cambridge, 1996.

[2] J. V. Tarazona, M. D. Fernández and M. M. Vega, "Regulation of Contaminated Soils in Spain e a New Legal Instrument," *Journal of Soils Sediments*, Vol. 5, No. 3, 2005, pp. 121-124.

[3] M. J. Fortin and M. Dale, "Spatial Analysis—A Guide for Ecologists," The Edinburgh Building, Published by Cambridge University, Cambridge, 2005.

[4] P. K. Gupta, "Soil, Plant, Water and Fertiliser Analysis," Agro Botanica Publishers and Distributors, New Delhi, 1999, p. 438.

[5] N. N. Saxena, "Environmental Analysis of Water, Soil and Air," Agro Botanical Publishers, 1994.

[6] G. B. Korte, "The GIS Book," 5th Edition, SAN 694-0269, On Word Press Publishers, 2001.

[7] N. A. Cobb, "Notes on Nematology," *Controlled Science Nematology*, Vol. 5, No. 5, 1917, pp. 117-128.

[8] T. R. E. Southwood and P. A. Henderson Ecological Methods," Published by Blackwell Science Ltd., London, 2000, pp. 218-240.

[9] J. B. Goodey, "Laboratory Methods for Work with Plant and Soil Nematodes, Technical Bulletin No.2, Ministry of Agriculture, Fisheries and Food," Nematology Department, Rothamsted Experimental Station, Harpenden, Herts, London, 1963, pp. 21-22.

[10] D. E. Smiles, "Aspects of the Physical Environment of Soil Organisms," *Biology and Fertility of Soils*, Vol. 6, No. 3, 1988, pp. 204-215.

[11] J. L. Townshend and L. R. Webber, "Movement of Pratylenchuspenetrans and the Moisture Characteristics of Three Ontario Soils," *Nematologica*, Vol. 17, No. 1, 1971, pp. 47-57. http://dx.doi.org/10.1163/187529271X00404

[12] J. F. Gorres, M. J. Dichiaro, J. B. Lyons and J. A. Amador, "Spatial and Temporal Patterns of Soil Biological Activity in a Forest and an Old Field," *Soil Biological Biochemistry*, Vol. 30, No. 2, 1998, pp. 219-230.

[13] S. E. Murialdo, M. G. Fuertes, J. F. Gonza'lez and E. Chavez, "Nematodes as Indicators of Wetland Pollution," *Journal of Environmental Biology*, Vol. 23, No. 4, 2002, pp. 423-428.

[14] Y. Shao, W. Zhang, J. Shen, L. Zhou, H. Xia, W. Shu, H. Ferris and S. Fu, "Nematodes as Indicators of Soil Recovery in Tailings of a Lead/Zinc Mine," *Soil Biological Biochemistry*, Vol. 40, No. 2, 2008, pp. 2040-2046.

[15] D. C. Norton and J. K. Hoffmann, "Distribution of Selected Plant Parasitic Nematodes Relative to Vegetation and Edaphic Factors," *Journal of Nematology*, Vol. 6, No. 1, 1974, pp. 81-86.

[16] N. Rajmohan and L. Elango, "Distribution of Iron, Manganese, Zinc and Atrazine in Groundwater in Parts of Palar and Cheyyar River Basins, South India," *Journal of Environmental and Monitoring Assessment*, Vol. 107, No. 1-3, 2005, pp. 115-131. http://dx.doi.org/10.1007/s10661-005-5307-0

[17] A. K. Singh and M. Singh, "Lead Decline in the Indian Environment Resulting from the Petrol-Lead Phase-Out Programme," *Science Total Environment*, Vol. 368, No. 2-3, 2006, pp. 686-694. http://dx.doi.org/10.1016/j.scitotenv.2006.04.013

[18] D. M. Alongi, A. L. Ramanathan, L. Kannan, F. Tirendi, L. A. Trott and M. Bala Krishna Prasad, "Influence of Human-Induced Disturbance on Benthic Microbial Metabolism in the Pichavaram Mangroves, Vellar-Coleroon Estuarine Complex, India," *Marine Biology*, Vol. 147, No. 4, 2005, pp. 1033-1044. http://dx.doi.org/10.1007/s00227-005-1634-5

Does Soil Disturbance Affect Soil Phosphorus Fractions?

Yonathan D. Redel[1,2], Rudolf Schulz[1], Torsten Müller[1]

[1]University of Hohenheim, Institute of Crop Science, Fertilisation and Soil Matter Dynamics (340i), Stuttgart, Germany; [2]Scientific and Technological Bioresource Nucleus (Bioren-UFRO), Universidad de La Frontera, Temuco, Chile.
Email: yonathan.redel@ufrontera.cl

ABSTRACT

Increased turnover of organic matter as a result of soil disturbance (e.g. by soil tillage) is described in principle, but the direct influence of soil disturbance on soil P turnover especially for organic farming systems has not been sufficiently proven. The objective of the study was to evaluate the short term effect of soil disturbance on different soil P fractions in a soil shaking experiment. Four soils were incubated for 10 days in the dark with three different disturbance treatments: 1) no disturbance, 2) overhead shaking for 2 h at the beginning of the experiment and 3) continuous overhead shaking at 5 r. p. m. The four investigated soils were: 1) a silty loam soil with long term bio-compost application and 2) the corresponding soil without bio-compost application, 3) a long-term organically managed clay loam soil and 4) a clay loam soil with long time application of pig manure, all not and from Baden-Württemberg, Germany. We determined $NaHCO_3$-, NaOH- and H_2SO_4-extractable inorganic and organic P fractions (P_i and P_o, resp.) in a sequential extraction. Furthermore, the potentially plant available P as Calcium-acetate-lactate-extractable P (CAL-P) and P extractable by electro-ultra-filtration (EUF-P), and *aqua regia* extractable total P (P_T) were determined. Furthermore, we determined microbial biomass carbon (MBC), nitrogen (MBN) and phosphorus (MBP), and acid phosphatase activity in soil. The organically managed soil had the highest P_T contents (1300 $mg \cdot kg^{-1}$). The soil with pig manure application had the smallest potentially labile P fractions ($NaHCO_3$-P_i and -P_o and NaOH-P_i). The ecologically managed soil had the biggest organic P fractions (114 $mg \cdot kg^{-1}$ $NaHCO_3$-P_o and 463 $mg \cdot kg^{-1}$ NaOH-P_o), but, this soil was the lowest in CAL-P (5 $mg \cdot kg^{-1}$). Short term soil disturbance had effects on labile organic P fractions of two of the four analyzed soils, but inorganic P was rather unaffected. In the compost amended COMP(+) soil, there was an incorporation of P from the less available NaOH-P fractions into the more available $NaHCO_3$-P_o fraction. However, if taking all investigated soils and treatments into account, the effects of soil disturbance were limited and not consistent.

Keywords: Phosphorus Fractions; Soil Disturbance; Microbial Phosphorus; Organic Phosphorus; Microbial Nitrogen; Acid Phosphatase Activity

1. Introduction

Investigating nutrient mobilization and immobilization as a result of soil tillage, it has to be distinguished between long term and short term effects [1]. Long term effects include changes in organic matter and nutrient distribution as well as organic matter and nutrient turnover within the soil profile after conversion from highly intensive to less intensive tillage systems, e.g. from ploughing to no-till, and *vice versa*. Many investigations are published on this subject including the effect of nutrient re-stratification [2-8]. In contrast to these investigations, this study focused on short term effects, *i.e.* the direct influence of soil disturbance on soil P turnover.

Increased turnover of organic matter as an immediate result of soil disturbance is described in principle [9,10]. Aggregate crushing usually has positive effects on C and N mineralization as summarized in the review presented by [11]. N mobilization by soil disturbance and tillage is well documented in literature. An influence of soil tillage on N availability was postulated by [12] after comparison of measured and model simulated values. An influence of soil tillage intensity on N mineralization was reported by other authors (e.g. [13,14]). A considerable influence of different intensities of simple soil mixing on the adenylate energy charge during incubation for eight days at 25°C was reported by [9]. The energy level experienced by soils during a typical cultivation operation by repeated dropping of defined weight from a defined height on a soil core was imitated by [15]. The authors found a clear positive effect on soil CO_2 evolution during the first hours following this treatment which was negatively correlated with the total soil C content. This loss of C by an increment of microbial activity as measured by soil respiration can be due to the fact that tillage/disturbance can open up pores, thus exposing previously protected soil organic matter to the attack by microorganisms [16].

Most investigations focusing on the effect of soil disturbance/tillage on soil P refer to long term field experiments with combined effects of fertilization and tillage. Soil disturbance enhances P allocation in inorganic fractions due to mineralization [17] whereas no tillage promotes an increase of enzymatic activities including acid phosphatase [18] and a nutrient stratification [7,8], showing more P fertilization placement effects rather than tillage effects. Other researches indicate that an adequate tillage can prevent topsoil P accumulation and P losses by erosion, but too heavy tillage enhances leaching of P in particulate P forms [19,20]. In contrast to C and N, however, no detailed knowledge exists about P mobilization by soil disturbance and tillage in the short term, and in particular considering organically and long-term compost managed soils.

In organic farming, compost and manure are important sources as P fertilizers. Compost/manure added to soil can enhance the availability of phosphorus through the direct release of P from the compost or indirectly through the release of humic acids to the soil. Humic substances can promote release of P by complexing Fe and Al derived from Fe- and Al-phosphates [21], and by the formation of Ca-humates when reacting with Ca-phosphates [22], both resulting in a release of the corresponding phosphate anions. Furthermore, organic farming promotes accumulation of organic matter, and this also leads to improved physical properties such as a lower bulk density with lower dry strength and increased friability [23]. Moreover, to organic farming currently higher levels of biological activity were associated [24]. However, a positive effect of soil tillage on soil nutrient availability was postulated especially for organic farming systems but was not sufficiently proven [25].

Chemical sequential extraction procedures are widely used to subdivide extractable soil P into inorganic (P_i) and organic (P_o) fractions differing in extractability [26, 27]. In this procedure, the inorganic pool extracted with 0.5 M sodium bicarbonate is attributed mainly to being labile P_i in the solid soil phase attached to crystalline surfaces [27]; moderately labile P_i extracted with 0.1M NaOH comprises chemisorbed P_i of lower plant-availability associated with amorphous and crystalline Fe and Al hydroxides [26]; and no labile primary Ca-bound P is extractable through 0.5 M HCl or H_2SO_4 [28]. Regarding organic fractions, easily mineralisable $NaHCO_3$-P_o and moderate labile NaOH-P_o comprise P linked to organic matter [28,29]. The non-extractable recalcitrant P_i and P_o comprise residual P of very low availability [30]. Microbial P is extracted within or besides sequential fractionation procedure after soil fumigation by $CHCl_3$ fractionation and $NaHCO_3$ extraction [31].

The objective of this study was to evaluate the short term effect of soil disturbance on different soil P fractions in a soil shaking experiment. It was hypothesized that soil disturbance (here soil shaking) mobilizes soil P, eventually leading to a redistribution of soil P towards more labile fractions.

2. Materials and Methods

2.1. Soils and Compost

Four soils were sampled at 0 - 30 cm depth from three different sites in Baden-Württemberg, Southern Germany at November 2009. Five sub-samples of each 20 cm diameter core were taken across the sampling sites (20 × 50 m).

1) COMP(−): Silty loam soil (Loess derived Luvisol) taken from a long term (since 1997) compost fertilization experiment at the experimental farm "Heidfeldhof", belonging to the University of Hohenheim, located in the south-west of Stuttgart, Germany (48°43'00"N; 9°11'40"E), control variant without compost application.

2) COMP(+): The same soil as COMP(−) but with high annual application rates of mature bio-compost corresponding to 400 kg total N ha^{-1} a^{-1} or approx. 40 t DM ha^{-1} a^{-1}. The compost derived mainly from kitchen waste and garden plant residues digested in a central compost plant. The compost contained around 4500 mg P kg^{-1} DM, with an average of 1566 $mg \cdot kg^{-1}$ $NaHCO_3$-P, 342 $mg \cdot kg^{-1}$ NaOH-P and 2472 $mg \cdot kg^{-1}$ H_2SO_4-P, and 84% P_i of the total P (average of years 2000 and 2008).

3) ECO: Clay loam soil (Regosol) from a site under long term bio-dynamic farming (ECO) according to the regulations of the German "Demeter" association; taken 2 km south of Rosenfeld, Germany (48°16'32"N; 8°43'00"E).

4) MAN: Clay loam soil (Gleyic Vertisol) with long term liquid pig manure application from a site at the experimental farm of the Agricultural Technology Centre (LTZ) in Karlsruhe-Durlach, Germany (49°14'25"N; 9°59'43"E).

All soils were sieved (<2 mm Ø), adjusted to approx. 50% of max water holding capacity and stored at 4°C until the experiment started. Three treatments were carried out in five replicates during 10-days at 15°C in the dark: 1) control soil with no disturbance during incubation (T-0), 2) soil disturbance by shaking during the first two hours of incubation on an overhead multi-axle rotating shaker at 5 r. p. m. (T-1), 3) soil disturbance by continuous shaking of the soil during 10 days on an overhead multi-axle rotating shaker at 5 r. p. m. (T-2).

The shaking treatments energy can be estimated from the size of the bottles (250 mL), and the number of rotations for each treatment. Considering a bottle size of approx. 10 cm height, 600 rotations in the 2 h and 72,000 in the 10 d treatment, the energy applied was according the formula K = *m* * *g* * *h* (K: kinetic energy, *m*: weight; *g*: gravity acceleration = 9.8 $m \cdot s^{-2}$ and *h*: height) of 294 $J \cdot kg^{-1}$ and 35,280 $J \cdot kg^{-1}$, respectively. The 2 h treatment can be compared to the energy of 150 $J \cdot kg^{-1}$ of typical soil cultivation [15,32], but the 10 d treatment energy input was several times higher than for average soil cultivation.

For all treatments, 50 g moist soil was incubated in 250 mL polyethylene bottles. After ten days, soils in all bottles were homogenized and further processed as described below. From the years 2000 and 2008 of the long term compost experiment, compost samples were available and subject to P fractionation in the same way as the soil samples (see below).

2.2. Soil and Compost Analyses

Soil pH was determined in 0.01 M $CaCl_2$ in a soil- solution relation of 1 g: 2.5 mL. Organic C (C_{org}) and total N where measured with an C-N-auto-analyzer ("Vario MAX CN" type, "Elementar" Company). Microbial biomass C and N (MBC and MBN, respectively) were estimated by fumigation–extraction [33]. Briefly, one 25 g portion soil was fumigated for 24 h at 25°C with ethanol-free $CHCl_3$. Following fumigant removal, the soil was extracted with 100 mL 0.5 M K_2SO_4 by 30 min horizontal shaking at 200 r. m. p. and filtered yielding a clear filtrate. Another non-fumigated 25 g portion of soil was extracted similarly at the time fumigation commenced. The extracts were frozen until analyses for total C and N concentrations on a TOC/TIC analyzer (Multi N/C 2100, Analytik Jena, Germany). Microbial biomass C (MBC) was calculated as follows: MBC = EC/k_{EC}; where EC = (organic C extracted from fumigated soils)-(organic C extracted from non-fumigated soils) and k_{EC} = 0.45 [34, 35]. Microbial biomass N (MBN) was calculated the same way but using k_{EN} = 0.54 [36].

Phosphorus in soils and compost was fractionated according to a modified "Hedley" fractionation method [26, 37]. Briefly, 0.5 g soil were extracted with 0.5 M $NaHCO_3$ followed by extraction with 0.1 M NaOH. Finally, the remaining soil was extracted with 0.5 M H_2SO_4. The "inorganic" molybdate reactive phosphorous (P_i) in all extracts was measured by the method of [38] at 700 nm and pH 5.0 spectrophotometrically (Hitachi U-3300), whereas "organic" phosphorous (P_o) in all extracts was calculated as the difference between total P and P_i. Thus, this fractionation procedure yielded six different analytical fractions: $NaHCO_3$-P_i, $NaHCO_3$-P_o, NaOH-P_i, NaOH-P_o, H_2SO_4-P_i and H_2SO_4-P_o. Total P (P_T) in soil samples were extracted with aqua regia (1:3 HNO_3 and HCl) extraction) and determined with ICP-OES ("Vista Pro" type, "Varian" company) [39].

Microbial biomass P (MBP) was estimated by Chloroform Fumigation Extraction [31]. Fumigation was done as described for microbial C. Following fumigant removal, the soil was extracted with 0.5 M $NaHCO_3$ by 30 min. horizontal shaking at 200 r. p. m. and filtered. Non-fumigated soil was treated similarly, except that the desiccator contained no $CHCl_3$, and was not evacuated. Inorganic P was analyzed in aliquots of the extracts by the ammonium molybdate-ascorbic acid method described by [38]. A spike of KH_2PO_4 equivalent to 25 μg P g^{-1} soil was used to correct for P_i fixation for each soil during the $NaHCO_3$ extraction. Biomass P was calculated MBP = EP/k_{EP}/recovery, where EP is the difference between $NaHCO_3$-P_i extracted from fumigated and non-fumigated soil, and recovery is the recovering factor obtained for correct P_i fixation mentioned above [31,40] and k_{EP} is 0.4 [41].

Plant available P was characterized by Calcium Acetate Lactate (CAL) extraction [42]. 5 g soil was extracted by shaking for 90 min with 100 ml of a 0.05 M Ca acetate-0.05 M Ca lactate-0.3 M CH_3COOH pH 4.1 solution followed by filtration. P in the extract was determined colorimetrically by using ammonium molybdate reduced with ascorbic acid and Zn-chlorhydrate and followed by spectrophotometrical measurement of the absorbance at 480 nm (Hitachi U-3300).

Acid phosphatase (orthophosphoric-monoester phosphomonoesterase) activity (P-ase) was determined by the *p*-nitrophenyl phosphate method with modifications, as described by [43,44]. Briefly, soil samples (1 g) were incubated with 1 mL 50 mM p-nitrophenol phosphate and 4 mL 0.1 M tris buffer pH 5.5 for 1 h at 20°C. At the end of the incubation period, 1 mL 0.5 M $CaCl_2$ was added, and the solution was quickly filtered. Then, the filtrate was treated with 4 mL 0.5 M NaOH. Samples were homogenized and centrifuged at 2500 g for 10 min. *p*-nitrophenol released was determined spectrophotometrically by measuring the absorbance of the supernatant at 400 nm (Hitachi U-3300).

Soil P as determined by electro ultrafiltration (EUF) was extracted according [45,46]. The first P fraction (P-1) was collected the first 30 min. at 20°C with an electric current of max. 200 V and 15 mA; this fraction is assumed to contain available P. The second P fraction (P-2) was collected during minutes 30 and 35 at 80°C with an electric current of max. 400 V and 150 mA; this fraction is assumed to contain labile P [45].

2.3. Statistics

Results of the soil analyses are presented as mean values of five repetitions. The data were subjected to analyses of variance using the ANOVA procedures of the SAS/STAT, version 6 [47]. Data were arcsine transformed to meet requirements of normality if needed and further compared by the Duncan's multiple range test. Statistical significance was determined at $P \leq 0.05$. Pearson correlations and principle component analysis (PCA) were performed using SAS/STAT, version 6 [47].

3. Results

Several years of compost addition (COMP(+)) increased concentrations of available CAL-P and also P determined by electro-ultrafiltration (P1(EUF) and P2(EUF)) compared to COMP(−) (**Table 1**). However, the increase of P_t by 130 mg (kg soil)$^{-1}$ did not affect most of the sequential P-Fractions. Only H_2SO_4-P_i was increased by 180 mg kg soil^{-1}. H_2SO_4-P_o was extremely small and only detectable in the COMP(−) soil (not indicated in **Table 1**). Furthermore, compost addition enhanced microbial biomass (MBC and MBN) almost two-fold (**Table 1**).

The other two soils, ECO and MAN, were not included in the statistical analysis as they have different parental materials and textures and soils differed consid-

Table 1. Soil quality indicators and phosphorus fractions in the original soils.

Soil	unit	COMP (−)	COMP (+)	ECO	MAN
pH ($CaCl_2$)		6.46 **b**	7.22 **a**	6.22	7.50
C_{org}	mg·g^{-1}	9.8	17.7	36.2	24.3
N_T	mg·g^{-1}	1.26	2.12	3.76	2.52
C_{org}/N_T		7.7	8.3	9.6	9.6
P_T	mg·kg^{-1}	789 **b**	922 **a**	1300	1060
CAL-P	mg·kg^{-1}	67 **b**	144 **a**	5	23
P1 (EUF)	mg·kg^{-1}	32.4 **b**	48.6 **a**	11.1	9.8
P2 (EUF)	mg·kg^{-1}	9.3 **b**	15.1 **a**	4.1	3.5
P fractions					
$NaHCO_3$-P_i	mg·kg^{-1}	185	185	49	40
$NaHCO_3$-P_o	mg·kg^{-1}	51	68	115	91
NaOH-P_i	mg·kg^{-1}	197	186	143	50
NaOH-P_o	mg·kg^{-1}	112	79	464	176
H_2SO_4-P_i	mg·kg^{-1}	212 **b**	392 **a**	316	370
Microbial biomass					
MBC	mg·kg^{-1}	148 **b**	241 **a**	589	463
MBN	mg·kg^{-1}	20 **b**	40 **a**	63	52
MBP	mg·kg^{-1}	19	11	5	3
MBC/C_{org}	%	1.5	1.3	1.6	1.9
MBN/N_T	%	11.7	12.4	5.5	5.3
MBP/P_T	%	2.1	1.9	1.7	2.1
Molar C:N:P		20:2:1	57:8:1	304:28:1	399:38:1
Acid P-ase	mg PNF kg^{-1}·h^{-1}	105 **b**	441 **a**	946	376

COMP(−): Silty loam soil without compost application; COMP(+): Silty loam soil with long term compost application; ECO: Clay loam soil under long term bio-dynamic farming; MAN: Clay loam soil with long term liquid pig manure application; C_{org}: organic carbon; N: total N, P_T: total P extractable with aqua regia. CAL-P: extractable with calcium acetate lactate solution; P1(EUF): P extractable by electro-ultra-filtration, 1st. fraction; P2(EUF): P extractable by electro-ultra-filtration, 2nd. fraction; $NaHCO_3$-P_i: inorganic P extractable with $NaHCO_3$; $NaHCO_3$-P_o: organic P extractable with $NaHCO_3$; NaOH-P_i: inorganic P extractable with NaOH; NaOH-P_o: organic P extractable with NaOH; H_2SO_4-P_i: inorganic P extractable with H_2SO_4; MBC: microbial biomass C; MBN: microbial biomass N; MBP: microbial biomass P; acid P-ase: acid phosphatase activity; PNF: *p*-nitrophenol). Different letters in the same P fraction show significant differences determined by *t* test ($P < 0.05$, MAN and ECO soil were not included).

erably in the determined P fractions. In general, ECO and MAN contain more P_o and less P_i in both $NaHCO_3$- and NaOH-extractable fractions compared to COMP(−) and COMP(+) soils. The ECO soil had the highest concentration of P_T and of the P_o fractions compared with the other soils. Furthermore, ECO and MAN have in general higher microbial biomasses and P-ase activities than COMP(−) and COMP(+) soils (**Table 1**). MBC/C_{org}, MBN/N_T and MBP/P_T ratios were fairly constant among all soils, except MBN/N_T that was almost twice as high in ECO and MAN compared to COMP(−) and COMP(+) soils (**Table 1**).

Significant effects of the soil shaking treatments on the investigated variables were limited and not consistent (**Table 2**). In the COMP(+) soil, NaOH-P_o decreased while $NaHCO_3$-P_o, H_2SO_4-P_i (2 h shaking only) and MBN (10 d shaking only) increased. In the ECO soil, H_2SO_4-P_i and CAL-P increased after 10 d shaking. Only in the ECO soil, acid P-ase activity increased after 10 d shaking. In the MAN soil, $NaHCO_3$-P_i decreased (10 d shaking only) while $NaHCO_3$-P_o increased.

Furthermore, both shaking treatments led to a decrease of soil microbial biomass (significant for MBC and MBN). The COMP(−) soil was unaffected by soil shaking, except an increase of MBP after 10 d shaking.

In general, $NaHCO_3$-P_i is correlated with CAL-P and MBP (**Table 3**). $NaHCO_3$-P_o is positively correlated with NaOH-P_o and P-ase and MBC and MBN, but negatively with CAL-P and MBP. The H_2SO_4-P_i fraction did not correlate with any other parameter. P-ase correlated positively with microbial elements, negatively with available P_i but positively with P_o forms.

PCA analysis (**Figure 1**) shows that organic C, microbial C (MBC) and N (MBN), P-ase, total C and N, and P_o (NaOH and $NaHCO_3$) were strongly related to PC1, whereas available and labile P_i fractions were negatively related to PC1 and positively to PC2. No labile P_i and pH were dispersed on the plot.

As indicated in **Figure 2**, A correlation could be detected between CAL-P (as an indicator of plant available P) and the $NaHCO_3$-P_i fraction (considered as labile P_i). However, data were strongly clustered due to the four

Table 2. Phosphorus fractions and microbial biomass variations after the disturbance experiment.

Soil	-------COMP(−)------			------COMP(+)------			--------ECO--------			--------MAN-------		
Treatment	no shaking	2 h	10 d	no shaking	2 h	10 d	no shaking	2 h	10 d	no shaking	2 h	10 d
unit	mg·kg^{-1}	*% change*		mg·kg^{-1}	*% change*		mg·kg^{-1}	*% change*		mg·kg^{-1}	*% change*	
Total P	840	1	2	1005	−5	−6	1415	4	1	1076	8	1
CAL-P	62	−3	−1	119	0	0	2	70	304*	4	0	36
P fractions												
$NaHCO_3$-P_i	156	−5	−2	208	2	6	35	15	0	23	9	−31*
$NaHCO_3$-P_o	8	58	−1	7	149*	235*	49	2	−22	26	38*	45*
NaOH-P_i	136	−8	0	167	9	−1	163	−5	−1	66	4	5
NaOH-P_o	99	−1	18	158	−74*	−69*	241	0	−6	91	−5	8
H_2SO_4-P_i	232	−10	6	284	20*	−5	222	31	35*	265	12	−12
Microbial biomass												
MBC	148	23	−24	171	57	58	442	43	17	480	−17*	−23*
MBN	23	−26	−19	40	−15	−23*	52	20	26	54	−48*	−37*
MBP	37	3	44*	46	2	3	29	−16	7	23	−3	−15
Molar C:N:P	11:1:1	11:1:1	7:1:1	10:1:1	15:2:1	15:1:1	40:4:1	68:6:1	43:5:1	55:5:1	47:3:1	49:4:1
MBP/P_T (%)	4.4	2	41	4.6	7	10	2.0	−13	6	2.1	−10	−16
Acid P-ase$^{(1)}$	226	3	17	238	−2	−13	596	7	41*	307	−13	−2

COMP(−): Silty loam soil without compost application; COMP(+): Silty loam soil with long term compost application; ECO: Clay loam soil under long term bio-dynamic farming; MAN: Clay loam soil with long term liquid pig manure application; C_{org}: organic carbon; N: total N, P_T: total P extractable with aqua regia. CAL-P: extractable with calcium acetate lactate solution; P1(EUF): P extractable by electro-ultra-filtration, 1st. fraction; P2(EUF): P extractable by electro-ultra-filtration, 2nd. fraction; $NaHCO_3$-P_i: inorganic P extractable with $NaHCO_3$; $NaHCO_3$-P_o: organic P extractable with $NaHCO_3$; NaOH-P_i: inorganic P extractable with NaOH; NaOH-P_o: organic P extractable with NaOH; H_2SO_4-P_i: inorganic P extractable with H_2SO_4;MBC: microbial biomass C; MBN: microbial biomass N; MBP: microbial biomass P; acid P-ase: acid phosphatase activity; PNF: *p*-nitrophenol). Different letters in the same P fraction show significant differences determined by *t* test ($P < 0.05$, MAN and ECO soil were not included).

Table 3. Pearson correlation table for soil phosphorus fractions and microbial biomass after incubation.

		1	2	3	4	5	6	7	8	9	10
1	$NaHCO_3$-P_i										
2	$NaHCO_3$-P_o	**−0.71**									
3	NaOH-P_i	**0.57**	−0.12								
4	NaOH-P_o	**−0.48**	**0.56**	**0.28**							
5	H_2SO_4-P_i	0.11	0.04	0.25	−0.12						
6	Total P	**−0.67**	**0.81**	0.10	**0.75**	0.26					
7	CAL-P	**0.97**	**−0.65**	**0.64**	**−0.44**	0.18	**−0.59**				
8	MBC	**−0.72**	**0.86**	−0.03	**0.64**	0.17	**0.88**	**−0.60**			
9	MBN	**−0.51**	**0.62**	0.22	**0.68**	0.28	**0.85**	−0.30	**0.80**		
10	MBP	**0.77**	**−0.57**	**0.53**	−0.31	0.14	**−0.54**	**0.71**	**−0.63**	−0.36	
11	Acid P-ase(1)	**−0.57**	**0.62**	0.26	**0.84**	0.10	**0.84**	**−0.54**	**0.72**	**0.76**	−0.36

Abbreviations: see **Table 1**. Bold numbers denotes significant correlations at $p < 0.05$, $n = 60$.

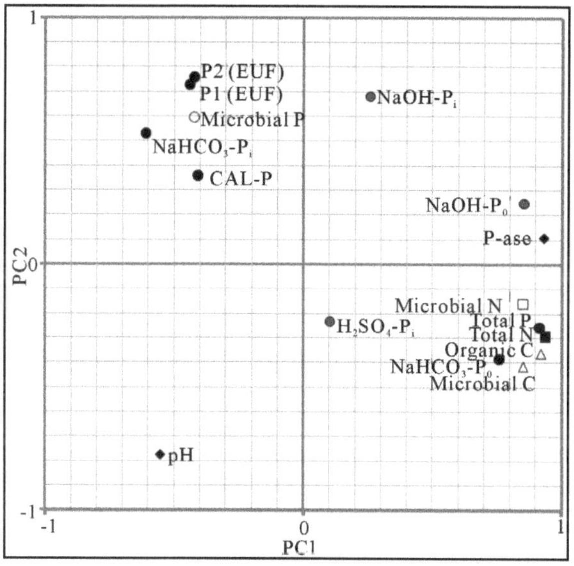

Figure 1. PCA analysis of incubated soils. Only the two principal components were considered. Abbreviations: see Table 1.

soils used in this investigation.

4. Discussion

The different soils had contrasting soil properties concerning P fractions. The high compost amendment resulted in a higher H_2SO_4-P_i and, as a consequence, in a higher P_T content. However, the other fractions were almost unaffected, except available CAL-P. It is reported that P derived from organic fertilizer can be found as P_i after few hours [48], or as labile and recalcitrant P_o after 56 ds in soil [49] and after 25 years [50] or 34 years [51]. However, how compost P is transformed after application may depends on the nature of the residues, soils, crop P uptake, management, climate and others factors. Under high yields, soil P_o can be depleted [52] as well as available P [53]. High P_i in compost could be responsible that large parts of P could be found in the P_i but not in the P_o fraction. Nevertheless, P-ase was four times higher in the COMP(+) soil compared to COMP(−) soil. Thus, microbial activity might be responsible for high available P (CAL-P) in this soil.

CAL-P was relatively high in COMP(+) and COMP(−) soils considering the P classification of official extension services (optimal and elevated, respectively, [50]). CAL measurements resulted in 5 to more than 10 fold higher P concentrations for COMP soils compared to ECO and MAN. Same differences between the soils were visible for other fractions considered as easily available (EUF and $NaHCO_3$-P_i), although, not to the same extent. Although CAL-P and $NaHCO_3$-P_i were closely correlated, ($r = 0.90$, **Table 3**) for all soils, this relationship disappears if focusing on a single soil only. A relationship between the P-fractions considered as labile (EUF, $NaHCO_3$-P_i, CAL-P and MBP) is also indicated by a common cluster in the PCA (**Figure 1**).

CAL-P measurements separate COMP (+) and COMP (−) more clearly than the $NaHCO_3$-P_i-Fraction (**Figure 2**). This may be due the higher pH of COMP (+). In soils high in pH, the acidic CAL solution (pH 4.1) may extract certain amounts of less plant available Ca-Phosphate.

Disturbance affected the labile $NaHCO_3$-P_o fractions mostly in the COMP(+) and MAN soils. As reported earlier, MBP is positively related to $NaHCO_3$-P_o fractions [41], but this was not the case in our study. In contrast, MBP could rather be related to $NaHCO_3$-P_i (**Table 3**), and this was also confirmed by the PCA analysis (**Figure**

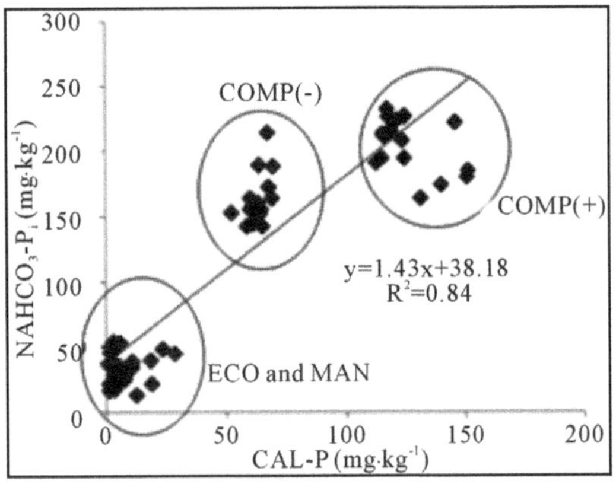

Figure 2. Regression between CAL-P and NaHCO3-Pi. Abbreviations: see Table 1.

1). This relation was also maintained if respective PCA's were done for each soil separately (not shown). Nevertheless, $NaHCO_3$-P_0 could be closely related to P-ase indicating a participation in the microbial turnover (**Figure 1**).

Both COMP(+) and MAN soils have the highest pH, and the highest H_2SO_4-P_i, fraction related to Ca bound Phosphates [28]. This can be a consequence of the slightly alkaline pH of both, compost residues and pig manure in COMP an MAN soils, respectively [54], and due to MAN soil natural properties. An apparent translocation from NaOH-P_0 to $NaHCO_3$-P_0 is visible for the COMP(+) soil indicating an increase of availability of P_0. However the absolute decrease in NaOH-P_0 cannot fully be explained by the increase in $NaHCO_3$-P_0 or changes in any other fraction, indicating that further fractions are existing as a (preliminary) sink of originally organic P. Reference [55] proposed soil microbial residues as an intermediate sink for N which might also be true for P. Higher P mobilization in the COMP(+) compared to the other soils can be due to a higher pH, lower resistance to mechanical disruption or more labile organic matter [56], lower aggregate size and higher friability [57], as a consequence of compost addition.

In MAN soil, on the other hand, labile $NaHCO_3$-P_i was apparently converted to $NaHCO_3$-P_0. Microorganisms could have be grown and died between the time elapsed between incubation and determination, stimulated by soil rewetting by incubation, thus absorbing P_i and releasing P as organic components recovered by $NaHCO_3$-P_0 [55, 58]. As microbes died and cell P was reconverted to P_i no change in MBP could be observed, but decrements of MBC and MBN as these elements cycles faster, as they are parts from no structural components of the cell. This conversion of labile P_i to P_0 has not been observed for the other soils. This can be due to the nature and structure of the soil or due to the effects of the applied pig manure.

Regarding ECO soil, there was an increase of available CAL-P with 10 d as similarly observed with soil tillage [59]. But this enhancement was not reflected by changes in the P fractionation, not even in the most available $NaHCO_3$-P_i fraction. The enhancement of P availability in the ECO soil may be attributed to an enhanced enzymatic activity after 10 d shaking. The COMP(−) soil was the less affected by disturbance. Only MBP was enhanced after 10 d shaking. This P uptake by microbes was not reflected in any change of the other P fractions. None of the other soils showed similar significant MBP changes.

In their meta study, [60] reported that a mean molar C:N:P ratio of 60:7:1 for the soil microbial biomass can be found and that differences from this general ratio are less pronounce than similarities over a wide range of ecological properties. While the original COMP(+) soil is very similar to this ratio, the C:N:P ratio of the original COMP(−) soil indicates a much narrower relationship between P and the two other elements. In the original MAN and ECO soils, however, the ratios are much wider than reported by [60]. After incubation, however, the ratios decreased in all soils indicating a relative enrichment of soil microbial biomass in all treatments.

Our values for MBP and MBP/P_T in COMP(+) and COMP(−) were similar to those reported by [61] for German soils. MBP and MBC were negatively correlated (**Table 3**), and MBP can be related to the other P_i forms, but MBC rather to MBN and P_0, as shown by PCA analysis. This can be explained by different distributions of C and P throughout the cell structures, thus, differing in cycling and microbial acquisition, or by the development of different microbial communities [62,63]. It is reported that environmental changes affect microbial communities more than microbial activity, and that microbial P is more susceptible to changes than N and C, resulting in a faster microbial P recycling [62,64], but in our study microbial P was almost unaffected in spite of changes for microbial C and N in ECO soil and a redistribution of P fractions in COMP(+) soil. Almost all changes were in microbial C and N as well as in P_0 fractions, thus confirmed by a closer relationship between these variables as revealed by PCA analysis.

5. Conclusion

Short term soil disturbance had effects on labile organic P fractions of two of the four analyzed soils, but inorganic P was rather unaffected. These changes can be associated to changes in the microbial biomass C and N as shown by PCA analysis, but microbial P was rather unaffected. In the compost amended COMP(+) soil, there was an incorporation of P from the less available NaOH-P_0 fractions into the more available $NaHCO_3$-P_0 fraction. However, if taking all investigated soils and treatments into account, the effects of soil disturbance are limited and not consistent. Soils with high organic fertilization do not necessarily have higher organic P fractions and higher enzymatic activity than those without organic fertilization.

6. Acknowledgements

This work was supported by the Becas Chile Bicentenario Grant from CONICYT (Comisión Nacional de Investigación Científica y Tecnológica de Chile). We thank Mr. M. Kränzler, manager of the organic farm "Schönberger Hof", and Dr. M. Mokry, Agricultural Technology Centre (LTZ) in Karlsruhe-Durlach, for providing two of the investigated soils. The EUF analyses were performed by Dr. D. Horn, "EUF working group for the advancement of soil fertility and soil health" in Ochsenfurt, Germany. The ICP-OES P analyses were carried out in the State Institute of Agricultural Chemistry of the University of Hohenheim.

REFERENCES

[1] C. Pekrun, H.-P. Kaul and W. Claupein, "Soil Tillage for Sustainable Nutrient Management," In: A. Titi, Ed., *Soil Tillage in Agroecosystems, Advances in Agroecology Series*, Vol. 9, CRC Press, Boca Raton, 2002,. http://dx.doi.org/10.1201/9781420040609.ch4

[2] B. Pezzarossa, M. Barbafieri, A. Benetti, G. Petruzzelli, M. Mazzoncini, E. Bonari and M. Pagliai, "Effects of Conventional and Alternative Management Systems on Soil Phosphorus Content, Soil Structure, and Corn Yield," *Communications in Soil Science and Plant Analysis*, Vol. 261, No. 17-18, 1995, pp. 2869-2885. http://dx.doi.org/10.1080/00103629509369494

[3] Z. R. Zheng, R. Simard, J. Lafond and L. E. Parent, "Pathways of Soil Phosphorus Transformations after 8 Years of Cultivation under Contrasting Cropping Practices," *Soil Science Society of American Journal*, Vol. 66, No. 3, 2002, pp. 999-1007. http://dx.doi.org/10.2136/sssaj2002.0999

[4] S. Daroub, F. Pierce and B. Ellis, "Phosphorus Fractions and Fate of Phosphorus-33 in Soils under Plowing and No-Tillage," *Soil Science Society of American Journal*, Vol. 64, No. 1, 2000, pp. 170-176. http://dx.doi.org/10.2136/sssaj2000.641170x

[5] Z. Zheng, J. A. MacLeod and J. Lafond, "Phosphorus Status of a Humic Cryaquept Profile in a Frigid Continental Climate as Influenced by Cropping Practices," *Biology and Fertility of Soils*, Vol. 39, No. 6, 2004, pp. 467-473. http://dx.doi.org/10.1007/s00374-004-0735-5

[6] L. M. Zibilske and J. M. Bradford, "Tillage Effects on Phosphorus Mineralization and Microbial Activity," *Soil Science*, Vol. 168, No. 10, 2003, pp. 677-685. http://dx.doi.org/10.1097/01.ss.0000095141.68539.c7

[7] Y. Redel, R. Rubio, J. Rouanet and F. Borie, "Phosphorus Bioavailability Affected by Tillage and Crop Rotation on a Chilean Volcanic Derived Ultisol," *Geoderma*, Vol. 139, No. 3-4, 2007, pp. 388-396. http://dx.doi.org/10.1016/j.geoderma.2007.02.018

[8] Y. Redel, M. Escudey, M. Alvear, J. Conrad and F. Borie, "Effects of Tillage and Crop Rotation on Chemical Phosphorus Forms and Some Related Biological Activities in a Chilean Ultisol," *Soil Use and Management*, Vol. 27, No. 2, 2011, pp. 221-228. http://dx.doi.org/10.1111/j.1475-2743.2011.00334.x

[9] R. G. Joergensen and M. Raubuch, "Adenylate Energy Charge and ATP-to-Microbial Biomass C Ratio in Soils Differing in the Intensity of Disturbance," *Soil Biology and Biochemistry*, Vol. 35, No. 9, 2003, pp. 1161-1164. http://dx.doi.org/10.1016/S0038-0717(03)00150-0

[10] N. Morris, P. Miller, J. Orson and R. J. Froud-Williams, "The Adoption of Non-Inversion Tillage Systems in the United Kingdom and the Agronomic Impact on Soil, Crops and the Environment—A Review," *Soil Tillage Research*, Vol. 108, No. 1-2, 2010, pp. 1-15. http://dx.doi.org/10.1016/j.still.2010.03.004

[11] J. Balesdent, C. Chenu and M. Balabane, "Relationship of Soil Organic Matter Dynamics to Physical Protection and Tillage," *Soil Tillage Research*, Vol. 53, No. 3-4, 2000, pp. 215-230. http://dx.doi.org/10.1016/S0167-1987(99)00107-5

[12] T. Müller, K. Thorup-Kristensen, J. Magid, L. S. Jensen and S. Hansen, "Catch Crops Affect Nitrogen Dynamic in Organic Farming Systems without Livestock Husbandry—Simulations with the DAISY Model," *Ecological Modelling*, Vol. 191, No. 3-4, 2006 pp. 538-544. http://dx.doi.org/10.1016/j.ecolmodel.2005.05.026

[13] S. O. Ojeniyi, "Nutrient Availability and Maize Yield under Reduced Tillage Practices," *Soil Tillage Research*, Vol. 26, No. 1, 1993, pp. 89-92. http://dx.doi.org/10.1016/0167-1987(93)90089-8

[14] J. Raupp, C. Letalik and F. Haunz, "Nitrate Release under a Silage Maize Crop after Turning in a Clover-Grass Crop in the Spring," *Journal of Agronomy and Crop Sciences*, Vol. 166, No. 3, 2001, pp. 181-190. http://dx.doi.org/10.1111/j.1439-037X.1991.tb00902.x

[15] C. W. Watts, S. Eich and A. R. Dexter, "Effects of Mechanical Energy Inputs on Soil Respiration at the Aggregate and Field Scale," *Soil Tillage Research*, Vol. 53, No. 3-4, 2000, pp. 231-243.

[16] C. W. Watts, P. D. Hallett and A. R. Dexter, "Effects of Stresses and Strains on Soil Respiration," In: J. Berthelin, P. M. Huang, J. M. Bollag and F. Andreux, Eds., *The Effect of Mineral-Organic-Microorganism Interactions on Soil and Freshwater Environments*, Plenum Press, New York, 1999, pp. 305-316. http://dx.doi.org/10.1007/978-1-4615-4683-2_33

[17] A. Wright, "Phosphorus Sequestration in Soil Aggregates after Long-Term Tillage and Cropping," *Soil Tillage Research*, Vol. 103, No. 2, 2009, pp. 406-411. http://dx.doi.org/10.1016/j.still.2008.12.008

[18] H. Omidi, Z. Tahmasebib, H. Torabic and M. Miransaric, "Soil Enzymatic Activities and Available P and Zn as Affected by Tillage Practices, Canola (*Brassica napus L.*) Cultivars and Planting Dates," *European Journal of Soil Biology*, Vol. 44, No. 4, 2008, pp. 443-450. http://dx.doi.org/10.1016/j.ejsobi.2008.05.002

[19] W. Kingery, C. Wood and J. Williams, "Tillage and Amend-

ment Effects on Soil Carbon and Nitrogen Mineralization and Phosphorus Release," *Soil Tillage Research*, Vol. 37, No. 4, 1996, pp. 239-250. http://dx.doi.org/10.1016/0167-1987(96)01009-4

[20] B. Beavers, Z. Liu, M. Cox, W. Kingery, G. Brink, P. Gerard and K. McGregor, "Phosphorus Dynamics in Two Poultry-Litter Amended Soils of Mississippi under Three Management Systems," *Pedosphere*, Vol. 20, No. 2, 2010, pp. 217-228. http://dx.doi.org/10.1016/S1002-0160(10)60009-4

[21] O. K. Borggaard, B. Raben-Lange, A. L. Gimsing and B. W. Strobel, "Influence of Humic Substances on Phosphate Adsorption by Aluminium and Iron Oxides," *Geoderma*, Vol. 127, No. 3-4, 2005, pp. 270-279. http://dx.doi.org/10.1016/j.geoderma.2004.12.011

[22] C. Singh and A. Amberger, "Humic Substances in Straw Compost with Rock Phosphate," *Biological Wastes*, Vol. 31, No. 3, 1990, pp. 165-174. http://dx.doi.org/10.1016/0269-7483(90)90156-M

[23] C. W. Watts and A. R. Dexter, "The Influence of Organic Matter in Reducing the Destabilization of Soil by Simulated Tillage," *Soil Tillage Research*, Vol. 42, No. 4, 1997, pp. 253-275. http://dx.doi.org/10.1016/S0167-1987(97)00009-3

[24] B. Hansen, H. Alrøe and E. Kristensen, "Review: Approaches to Assess the Environmental Impact of Organic Farming with Particular Regard to Denmark," *Agriculture, Ecosystems and Environment*, Vol. 83, No. 1-2, 2001, pp. 11-26. http://dx.doi.org/10.1016/S0167-8809(00)00257-7

[25] F. Wald, C. Pekrun and W. Claupein, "Einfluss der Bodenbearbeitung nach Mehrjährigem Leguminosen-Grasgemengeanbau auf die N-Mineralisierung unter den Bedingungen des Organischen Landbaus," In: H. J. Reents, Ed., *Beiträge zur* 6. Wissenschaftstagung zum Ökologischen Landbau "Von Leit-linien zu Leit-bildern", Köster, Berlin, 2001, pp. 425-428.

[26] H. Hedley, J. Steward and B. Chauhuan, "Changes in Organic and Inorganic Soil Phosphorus Fractions Induced by Cultivation Practices and by Laboratory Incubations," *Soil Science Society of American Journal*, Vol. 46, No. 5, 1982, pp. 970-976. http://dx.doi.org/10.2136/sssaj1982.03615995004600050017x

[27] A. Cross and W. Schlesinger, "A Literature Review and Evaluation of the Hedley Fractionation: Applications to the Biogeochemical Cycle of Soil Phosphorus in Natural Ecosystems," *Geoderma*, Vol. 64, No. 3-4, 1995, pp. 197-214. http://dx.doi.org/10.1016/0016-7061(94)00023-4

[28] A. Johnson, J. Frizano and D. Vann, "Biogeochemical Implications of Labile Phosphorus in Forest Soils Determined by the Hedley Fractionation Procedure," *Oecologia*, Vol. 135, 2003 pp. 487-499.

[29] B. Turner, B. Cade-Menun, L. Condron and S. Newman, "Extraction of Soil Organic Phosphorus," *Talanta*, Vol. 66, No. 2, 2005, pp. 294-306. http://dx.doi.org/10.1016/j.talanta.2004.11.012

[30] M. Herlihy and D. McGrath, "Phosphorus Fractions and Adsorption Characteristics in Grassland Soils of Varied Soil Phosphorus Status," *Nutrient Cycling in Agroecosystems*, Vol. 77, No. 1, 2007, pp. 15-27. http://dx.doi.org/10.1007/s10705-006-6687-4

[31] P. Brookes, D. Powlson and D. Jenkinson, "Phosphorus in the Soil Microbial Biomass," *Soil Biology and Biochemistry*, Vol. 16, No. 2, 1984, pp. 169-175. http://dx.doi.org/10.1016/0038-0717(84)90108-1

[32] D. E. Patterson, W. C. Chamen and C. D. Richardson, "Long-Term Tillage Experiments with Tillage Systems to Improve the Economy of Cultivations for Cereals," *Journal of Agricultural Engineering Research*, Vol. 25, No. 1, 1980, p. L-36. http://dx.doi.org/10.1016/0021-8634(80)90044-X

[33] E. D. Vance, P. C. Brookes and D. S. Jenkinson, "An Extraction Method for Measuring Soil Microbial Biomass C," *Soil Biology and Biochemistry*, Vol. 19, No. 6, 1987, pp. 703-707. http://dx.doi.org/10.1016/0038-0717(87)90052-6

[34] J. Wu, R.G. Joergensen, B. Pommerening, R. Chaussod and P. Brookes, "Measurement of Soil Microbial Biomass C by Fumigation-Extraction-An Automated Procedure," *Soil Biology and Biochemistry*, Vol. 22, No. 8, 1990, pp. 167-169. http://dx.doi.org/10.1016/0038-0717(90)90046-3

[35] R. G Joergensen, "The Fumigation-Extraction Method to Estimate Soil Microbial Biomass: Calibration of the k_{EC} Value," *Soil Biology and Biochemistry*, Vol. 28, No. 1, 1996, pp. 25-31. http://dx.doi.org/10.1016/0038-0717(95)00102-6

[36] R. G. Joergensen and T. Mueller, "The Fumigation-Extraction Method to Estimate Soil Microbial Biomass: Calibration of the k_{EN} Value," *Soil Biology and Biochemistry*, Vol. 28, No. 1, 1996, pp. 33-37. http://dx.doi.org/10.1016/0038-0717(95)00101-8

[37] H. Tiessen and J. O. Moir, "Characterization of Available P by Sequential Extraction," In: M. R. Carter, Ed., *Soil Sampling and Methods of Analysis*, Lewis Publishers, Boca Raton, 1993, pp. 75-86.

[38] J. Murphy and J. P. Riley, "A Modified Single Solution Method for the Determination of Phosphate in Natural Waters," *Analytica Chimica Acta*, Vol. 27, No. 1, 1962, pp. 31-36. http://dx.doi.org/10.1016/S0003-2670(00)88444-5

[39] VDLUFA, "Methodenbuch Band VII Umweltanalytik," 1. Auflage, 1. Teillieferung. VDLUFA-Verlag, Darmstadt, 1996.

[40] S. Heinze, M. Oltmanns, R. G. Joergensen and J. Raupp, "Changes in Microbial Biomass Indices after 10 Years of Farmyard Manure and Vegetal Fertilizer Application to a Sandy Soil under Organic Management," *Plant and Soil*, Vol. 343, No. 1-2, 2011, pp. 221-234. http://dx.doi.org/10.1007/s11104-010-0712-8

[41] K. S. Khan and R. G. Joergensen, "Changes in Microbial Biomass and P Fractions in Biogenic Household Waste Compost Amended with Inorganic P Fertilizers," *Bioresources Technology*, Vol. 100, No. 1, 2009, pp. 303-309. http://dx.doi.org/10.1016/j.biortech.2008.06.002

[42] VDLUFA, "Methodenbuch Band I. Die Untersuchung der Böden," 4. Auflage, 1. und 2. Teillieferung. VDLUFA-Verlag, Darmstadt, 1997.

[43] R. Rubio, E. Moraga and F. Borie, "Acid Phosphatase Activity and Vesicular-Arbuscular Infection Associated with Roots of Four Wheat Cultivars," *Journal of Plant Nutrition*, Vol. 13, No. 5, 1990, pp. 585-598. http://dx.doi.org/10.1080/01904169009364102

[44] Y. Redel, R. Rubio, R. Godoy and F. Borie, "Phosphorus Fractions and Phosphatase Activity in an Andisol under Different Forest Ecosystems," *Geoderma*, Vol. 145, No. 3-4, 2008, pp. 216-221. http://dx.doi.org/10.1016/j.geoderma.2008.03.007

[45] K. Németh, "Recent Advances in EUF Research (1980-1983)," *Plant and Soil*, Vol. 83, No. 1, 1985, pp. 1-19. http://dx.doi.org/10.1007/BF02182711

[46] D. Steffens, T. Leppin, N. Luschin-Ebengreuth, Z. Yang and S. Schubert, "Organic Soil Phosphorus Considerably Contributes to Plant Nutrition But Is Neglected by Routine Soil-Testing Methods," *Journal of Plant Nutrition and Soil Science*, Vol. 173, No. 5, 2010, pp. 765-771. http://dx.doi.org/10.1002/jpln.201000079

[47] SAS/STAT, "SAS/STAT Users Guide: Statistics," Version 6, 4th Edition, SAS Institute, Cary, 1990.

[48] M. Malik, K. Khan, P. Marschner and S. Ali, "Organic Amendments Differ in Their Effect on Microbial Biomass and Activity and on P Pools in Alkaline Soils," *Biology and Fertility of Soils*, Vol. 49, No. 4, 2013, pp. 415-425. http://dx.doi.org/10.1007/s00374-012-0738-6

[49] M. Malik, P. Marschner, K. Khan and S. Ali, "Addition of Organic and Inorganic P Sources to Soil—Effects on P Pools and Microorganisms," *Soil Biology and Biochemistry*, Vol. 49, No. 1, 2012, pp. 106-113. http://dx.doi.org/10.1016/j.soilbio.2012.02.013

[50] A. Y. Lopez-Contreras, I. Hernandez-Valencia and D. Lopez-Hernandez, "Fractionation of Soil Phosphorus in Organic Amended Farms Located on Savanna Sandy Soils of Venezuelan Amazonian," *Biology and Fertility of Soils*, Vol. 43, No. 6, 2007, pp. 771-777. http://dx.doi.org/10.1007/s00374-006-0162-x

[51] M. Park, A. Singvilay, W. Shin, E. Kima, J. Chung and T. Saa, "Effects of Long-Term Compost and Fertilizer Application on Soil Phosphorus Status under Paddy Cropping System," *Communications in Soil Science and Plant Analysis*, Vol. 35, No.11-12, 2004, pp. 1635-1644. http://dx.doi.org/10.1081/CSS-120038559

[52] X. H. Li, X. Z. Han, H. B. Li, C. Song, J. Yan and Y. Liang, "Soil Chemical and Biological Properties Affected by 21-Year Application of Composted Manure with Chemical Fertilizers in a Chinese Mollisol," *Canadian Journal of Soil Science*, Vol. 92, No. 3, 2012, pp. 419-428. http://dx.doi.org/10.4141/cjss2010-046

[53] C. Welsh, M. Tenuta, D. N. Flaten, R. Thiessen-Martens and M. H. Entz, "High Yielding Organic Crop Management Decreases Plant-Available But Not Recalcitrant Soil Phosphorus," *Agronomy Journal*, Vol. 101, No. 5, 2009, pp. 1027-1035. http://dx.doi.org/10.2134/agronj2009.0043

[54] F. X. Zhu, W. P. Wang, C. L. Hong, M. G. Feng, Z. Y. Xue, X. Y. Chen, Y. L. Yao and M. Yu, "Rapid Production of Maggots as Feed Supplement and Organic Fertilizer by the Two-Stage Composting of Pig Manure," *Bioresource Technology*, Vol. 116, No. 1, 2012, pp. 485-491. http://dx.doi.org/10.1016/j.biortech.2012.04.008

[55] T. Mueller, J. Magid, L. S. Jensen, H. Svendsen and N. E. Nielsen, "Soil C and N Turnover after Incorporation of Chopped Maize, Barley Straw and Blue Grass in the Field: Evaluation of the DAISY Soil-Organic-Matter Submodel," *Ecological Modelling*, Vol. 111, No. 1, 1998, pp. 1-15. http://dx.doi.org/10.1016/S0304-3800(98)00094-5

[56] W. Cookson, D. Murphy and M. Roper, "Characterizing the Relationships between Soil Organic Matter Components and Microbial Function and Composition along a Tillage Disturbance Gradient," *Soil Biology and Biochemistry*, Vol. 40, No. 3, 2008, pp. 763-777. http://dx.doi.org/10.1016/j.soilbio.2007.10.011

[57] J. Miller, B. W Beasley, C. F Drury and B. J. Zebarth, "Long-Term Effect of Fresh and Composted Cattle Manure on the Size and Nutrient Composition of Dry-Sieved Soil Aggregates," *Canadian Journal of Soil Science*, Vol. 92, No. 6, 2012, pp. 865-882. http://dx.doi.org/10.4141/cjss2011-107

[58] B. Turner, J. Driessen, P. Haygarth and I. Mckelvie, "Potential Contribution of Lysed Bacterial Cells to Phosphorus Solubilization in Two Rewetted Australian Pasture Soils," *Soil Biology and Biochemistry*, Vol. 35, No. 1, 2003, pp. 187-189. http://dx.doi.org/10.1016/S0038-0717(02)00244-4

[59] B. Nguyen and P. Marschner, "Effect of Drying and Rewetting on Phosphorus Transformations in Red Brown Soils with Different Soil Organic Matter Content," *Soil Biology and Biochemistry*, Vol. 37, No. 8, 2005, pp. 1573-1576. http://dx.doi.org/10.1016/j.soilbio.2005.01.015

[60] C. Cleveland and D. Liptzin, "C:N:P Stoichiometry in Soil: Is There a 'Redfield Ratio' for the Microbial Biomass?" *Biogeochemistry*, Vol. 85, No. 3, 2007, pp. 235-252. http://dx.doi.org/10.1007/s10533-007-9132-0

[61] K. S. Khan and R. G. Joergensen, "Relationships between P Fractions and the Microbial Biomass in Soils under Different Land Use Management," *Geoderma*, Vol. 173-174, No. 1, 2012, pp. 274-281. http://dx.doi.org/10.1016/j.geoderma.2011.12.022

[62] D. Achat, C. Morel, M. Bakker, L. Augusto, S. Pellerin, A. Gallet-Budynek and M. Gonzalez, "Assessing Turnover of Microbial Biomass Phosphorus: Combination of an Isotopic Dilution Method with a Mass Balance Model," *Soil Biology and Biochemistry*, Vol. 42, No. 12, 2010, pp. 2231-2240.http://dx.doi.org/10.1016/j.soilbio.2010.08.023

[63] E. Bünemann B. Prusisz and K. Ehlers, "Chapter 2. Characterization of Phosphorus Forms in Soil Microorganisms," In: E. K. Bünemann, A. Oberson and E. Frossard, Eds., *Phosphorus in Action*, Springer-Verlag, Berlin, 2011, pp. 37-57.

[64] J. DeForest and L. Scott "Available Organic Soil Phosphorus Has an Important Influence on Microbial Community Composition," *Soil Science Society of American Journal*, Vol. 74, No. 6, 2010, pp. 2059-2066. http://dx.doi.org/10.2136/sssaj2009.0426

28

Soil Organic C:N vs. Water-Extractable Organic C:N

Richard L. Haney[1*], Alan. J. Franzluebbers[2], Virginia. L. Jin[3], Mari-Vaughn. Johnson[4], Elizabeth. B. Haney[5], Mike. J. White[1], Robert. D. Harmel[1]

[1]USDA-Agricultural Research Service (ARS), Grassland, Soil & Water Research Laboratory, Temple, USA; [2]USDA-ARS, J. Phil Campbell, Sr. Natural Resource Conservation Center, Watkinsville, USA; [3]USDA-ARS, Agroecosystem Management Research Unit, Lincoln, USA; [4]USDA-NRCS, Resources Inventory and Assessment Division, Temple, USA; [5]Texas AgriLife Research, Blackland Research & Extension Center, Temple, USA.
Email: [*]Rick.Haney@ars.usda.gov

ABSTRACT

Traditionally, soil-testing laboratories have used a variety of methods to determine soil organic matter, yet they lack a practical method to predict potential N mineralization/immobilization from soil organic matter. Soils with high microbial activity may experience N immobilization (or reduced net N mineralization), and this issue remains unresolved in how to predict these conditions of net mineralization or net immobilization. Prediction may become possible with the use of a more sensitive method to determine soil C:N ratios stemming from the water-extractable C and N pools that can be readily adapted by both commercial and university soil testing labs. Soil microbial activity is highly related to soil organic C and N, as well as to water-extractable organic C (WEOC) and water-extractable organic N (WEON). The relationship between soil respiration and WEOC and WEON is stronger than between respiration and soil organic C (SOC) and total organic N (TON). We explored the relationship between soil organic C:N and water-extractable organic C:N, as well as their relationship to soil microbial activity as measured by the flush of CO_2 following rewetting of dried soil. In 50 different soils, the relationship between soil microbial activity and water-extractable organic C:N was much stronger than for soil organic C:N. We concluded that the water-extractable organic C:N was a more sensitive measurement of the soil substrate which drives soil microbial activity. We also suggest that a water-extractable organic C:N level > 20 be used as a practical threshold to separate those soils that may have immobilized N with high microbial activity.

Keywords: Soil Microbial Activity; C:N Ratios; Soil Organic C; N Mineralization; N Immobilization; Soil Testing

1. Introduction

Soil microorganisms are the centerpiece of biogeochemical cycling of nutrients in soil. Soil fertility is directly related to, and defined by, the heterotrophic activity of soil microbes as a whole [1]. While shifts in soil microbial community composition and structure are indicators of altered environmental conditions or management [2-4], the link between microbial composition and soil function is highly variable and thus limited as a general parameter for predicting soil activity rates. Thus, the integrated response of the soil microbial community is needed to elucidate and predict soil nutrient availability for the management of one of our most important resources, soil.

The metabolically-active component of soil can be measured in its simplest form as emission of CO_2, which corresponds to nutrient availability, moisture, and temperature, and can be rapidly quantified [5]. Emission of CO_2 can reveal the broader impacts of management, crop biodiversity, and climatic changes, but also has predictive capabilities for specifically assessing soil-nutrient release. Measurement of soil microbial activity, in conjunction with other soil physical and chemical properties and processes, can be a valuable tool for developing a complete profile for soil fertility and may be used to increase the efficiency of fertilizer recommendations.

Microbial mineralization/immobilization of soil N can be broadly estimated using soil organic C:N [6]. Based on years of research on conversion rates of decomposable organic matter by soil fungi and bacteria, soil organic C:N of 20 is generally considered to be a threshold point where either net N mineralization or net N immobilization occurs [7,8]. In reality, both N mineralization and immobilization occur simultaneously in soil, making it difficult to predict the amount of available soil N from net N mineralization alone. Correlation between the soil C:N ratio and N immobilization and mineralization is unclear, partly due to early work that was limited to measurements of net N rates. Variations of the C:N ra-

[*]Corresponding author.

tios of different pools of organic matter and variations of the C and N assimilation efficiency of the microbial biomass may also confound the usefulness of the soil C:N ratio to predict gross N transformation rates [8].

Current literature suggests that soil respiration rates can be used to predict soil N mineralization/immobilization rates and provide an estimate of soil N mineralization potential [9,10]. While soil CO_2-C respiration rates have been shown to be correlated to N mineralization [8,11], respiration alone is not an indicator of N immobilization and may not accurately predict net N mineralization/immobilization. As a result, a modified approach couples soil organic C:N with the soil respiration rate for modeling net N mineralization rates in soils [12,13]. High soil microbial activity does not always lead to high N mineralization due to immobilization that can occur; however, determining the C:N ratio from a much smaller more active pool of C and N to soil microbial activity could increase the accuracy of predicting the net mineralization/immobilization. A more sensitive and effective approach may be to assess the much smaller fractions of water-extractable organic C and N, which are highly related to soil microbial activity [14]. Results of a review paper on N cycling focus the importance of both substrate quantity (as C and N concentration) and quality (as C:N ratio) for N cycling rates [9].

In this study, we wanted to explore the link between the release of CO_2 following rewetting of dried soil and soil organic C:N vs. water-extractable organic C:N. We hypothesized: 1) soil microbial activity may be more strongly correlated with water-extractable organic C and N concentrations than soil organic C and N concentrations and 2) C:N ratio calculated from the water-extractable organic fraction may be an additional tool in conjunction with the flush of CO_2 following rewetting of dried soil to better predict plant available N and N immobilization.

2. Materials and Methods

Soil was collected from agricultural fields in Idaho, Georgia, Maine, Mississippi Oklahoma, Texas, and Wyoming. Crop management varied: (till/no-till, continuously cropped/crop rotation) along with soil type. Soils had clay content from 10 to 55% (data not shown), pH from 5.56 to 8.02 (**Figure 1**), and soil organic C from 3.63 to 41.31 g·C·kg^{-1} soil (**Figure 2**).

All samples were dried overnight at 50°C and ground to pass a 2-mm sieve. Soil organic C (SOC) and total N (TN) were determined on 2 g subsamples using dry combustion (Elementar; Hanau, Germany). Water-extractable organic C (WEOC) and water-extractable N (WEN) were determined from 4 g of dry soil with 40 mL of deionized water and shaking for 10 minutes on a mechanical shaker (Eberbach, Ann Arbor, MI). Samples were then centri-

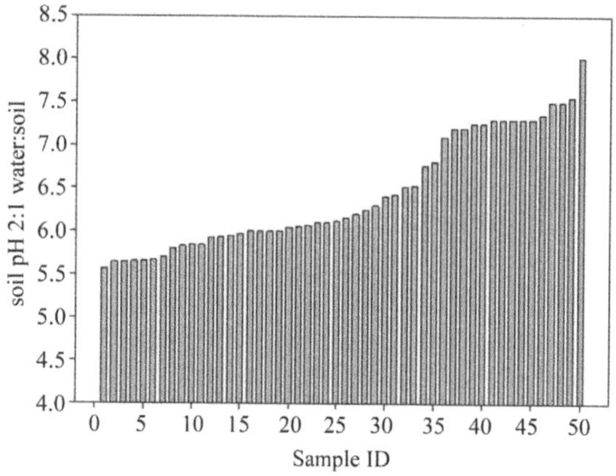

Figure 1. Soil pH range.

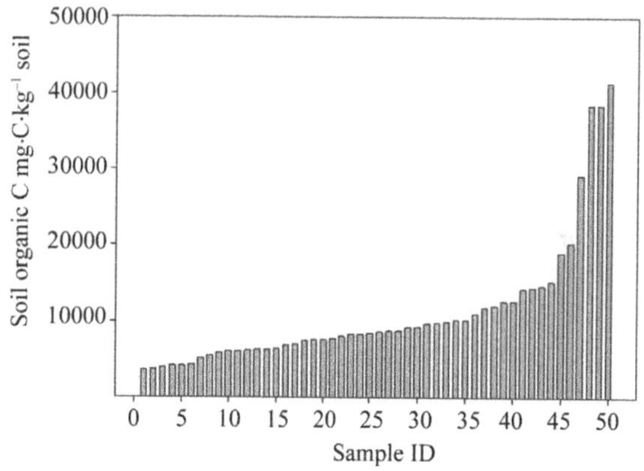

Figure 2. Range in soil organic C.

fuged for 5 minutes at 3500 rpm, filtered through Whatman 2 V paper, and analyzed for WEOC and WEN (Apollo 9000, Teledyne-Tekmar; Mason, Ohio). Inorganic NH_4-N, and NO_3-N concentrations were also determined (Flow Solution IV, OI Analytical; College Station, TX). Soil organic N (SON) was calculated by subtracting inorganic N content (NH_4-N and NO_3-N) from total N. Water-extractable organic N (WEON) was calculated by subtracting inorganic N content (NH_4-N and NO_3-N) from WEN.

The release of CO_2 from 24 hour incubation after rewetting dried soil is directly related to the fertility of a given soil; the method is designed to mimic the natural soil drying and rewetting cycle and is designed to be readily adopted by soil testing labs.The flush of CO_2 in 24 hours (1-day) following rewetting of dried soil was determined from 40 g subsamples in 50 ml polypropylene disposable beakers (Fisherbrand Cat. No. 01-291-10) with four to five 6.35-mm holes drilled in the bottom. A Whatman GF/D 4.25-cm glass microfiber filter (Cat No. 1823-042) was placed in the bottom of each beaker to prevent soil loss. The beaker and Solvita® gel paddle were placed in a gas-tight 250-mL glass jar filled with 25 mL of water and sealed with a screw-top lid; the jar had a convex bottom to allow for drainage. Capillary action was used to rewet soil according to its water holding capability [15]. Soils were incubated at 25°C, and respired CO_2 was trapped during 24 h. The quantity of 1-day CO_2-C released was determined using a digital-color reader (DCR) (www.solvita.com). The Solvita system for esti- mating the flush of 1-day CO_2 following rewetting of dried soil has been shown to be highly correlated with the commonly used titration method and 1-day CO_2-C IRGA method [5]. SigmaStat imbedded in SigmaPlot ver. 12.1 was used for linear regression analysis.

3. Results and Discussion

Soil organic C and SON were highly related ($r^2 = 0.93$), as were WEOC and WEON ($r^2 = 0.84$). Interestingly, ratios of SOC to SON and WEOC to WEON were nearly similar (10.8 and 10.9, respectively), suggesting that WEOC and WEON are a subset of the much larger SOC and SON pools (**Figures 3** and **4**). Some soil test labs currently use SOC data determined from various methods (loss on ignition, titration, and combustion) to estimate potential N release, while few if any use SOC:SON to estimate potential N mineralization. Since both substrate availability and SOC:SON have a strong influence on N mineralization rates [9], it is conceivable that WEOC: WEON could be a better method for determining the state of potential N mineralization/immobilization as an alternative to SOC:SON. However, our data indicate that SOC:SON and WEOC:WEON were poorly related (**Figure 5**). This weak relationship may be attributed to the fact that SOC and SON were roughly 40 times larger than WEOC and WEON fractions (**Figures 3** and **4**). The water-extractable organic fraction, though significantly smaller than total SOC and SON, is a critical component used by soil microorganisms that drive the nutrient cycling system [16-19].

We found that soil microbial activity measured as the flush of 1-day CO_2 following rewetting of dried soil was significantly correlated to SOC, SON, WEOC, and WEON. Water-extractable organic C and WEON, however, had a stronger relationship with the flush of 1-day-CO_2 than SOC and SON (**Table 1**).

This finding is consistent with documented findings [16-19] and the authors' theory that the water-extractable C and N pool is the more readily available energy pool for soil microbes as compared to the total SOC and TON pools. In this data set, organic N accounted for 97.4% of the total soil N, while organic N accounts for 53% of the water extractable total N. While SOC:SON may be a useful indicator of soil quality/fertility, it may not be sensitive enough to reliably quantify the quality of organic matter that soil microbes are actively mineralizing due to the sheer size of the C and N pools compared to the water extractable C and N pools. Therefore, WEOC:WEON could be considered a more accurate property to predict variations in mineralization/immobilization potential of a given soil as compared with SOC: SON.

It may be possible to combine the flush of CO_2 following rewetting of dried soil with WEOC:WEON and develop a potential N mineralization/ immobilization

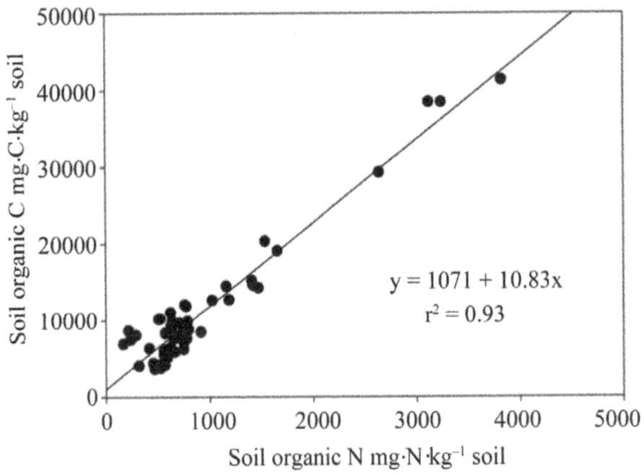

Figure 3. Soil organic N vs. soil organic C.

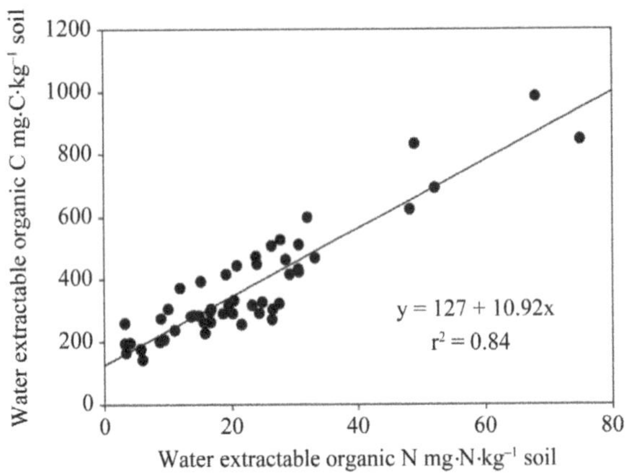

Figure 4. Water extractable organic N vs. water extractable organic C.

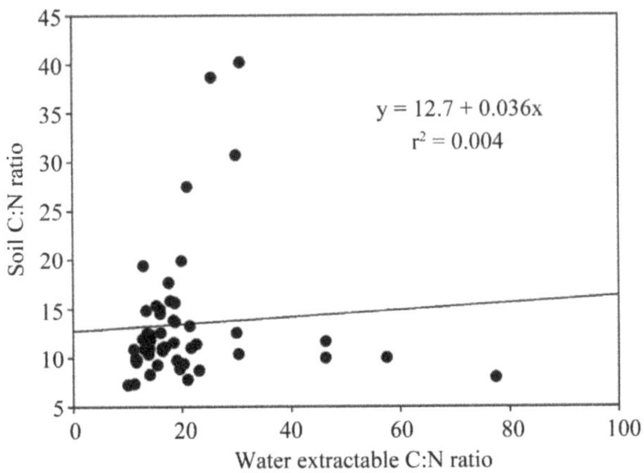

Figure 5. Soil C:N ratio vs. water extractable soil C:N ratio.

Table 1. Correlation table showing correlation coefficients (r) comparing 1-day CO_2-C ($mg\cdot C\cdot kg^{-1}$ soil) and total and water-extractable organic C and N concentrations ($mg\cdot C\cdot kg^{-1}$ soil; $mg\cdot N\cdot kg^{-1}$ soil). All correlations were significant ($P \leq 0.0001$; $n = 50$).

	TOC	TON	WEOC	WEON
1-Day CO_2-C	0.704	0.682	0.874	0.869
TOC	.	0.965	0.755	0.737
TON	.	.	0.725	0.736
WEOC	.	.	.	0.919

indicator (**Figure 6**). Soil organic C:N > 20 reflects reduced N availability due to greater immobilization of N [20,21], although the breakpoint can be as high as 30, depending on the [22]. Using a threshold SOC to SON value of 20, above which no net N mineralization would occur, our results indicate N immobilization in 4 of 50 (8%) soil samples, whereas when using WEOC to WEON, 16 of 50 (30%) soil samples immobilized N (**Figure 6**). This information suggests that the WEO C:N ratio may be more indicative of the quality (water extractable organic C) of the organic C as opposed to quantity (soil organic C) of substrate available for soil microbial activity, which is an important point since easily decomposable substrate should translate into N mineralization release. The importance of characterizing the quality vs. the quantity of organic C is a furtherance of the sensitivity that is revealed by a nutrient cycling system that is driven by C. We define the active soil C pool (WEOC) as one that is easily mineralizable and therefore a direct driver of microbial activity (1-day CO_2) responsible for providing N and P mineralization in soil. Utilizing these three pools (WEOC, WEON, 1-day CO_2) in a soil testing environment could increase our awareness of soil microbial activity and possibly help us account for an often overlooked source of N that can be easily subtracted from fertilizer recommendations. This will have a two-fold effect: 1) Save producers input costs in terms of reduced fertilizer input and 2) reduce the N and P loading in soil which is subject to loss from erosion, leaching, denitrification, and leaching that will ultimately affect the quality of our drinking water and water bodies, both freshwater and saltwater.

A more sensitive C:N indicator combined with a rapid method for soil microbial activity would help soil test labs offer reliable estimates of the mineralization/immobilization state of soil. Soil samples are usually taken prior to planting, which is a time when this information would be most needed for fertilizer decisions by producers. Spring sampling would be at a time of the year when the soil microbial community would have stabilized over winter following decomposition of the previous crop.

Currently there are no standardized universal soil test

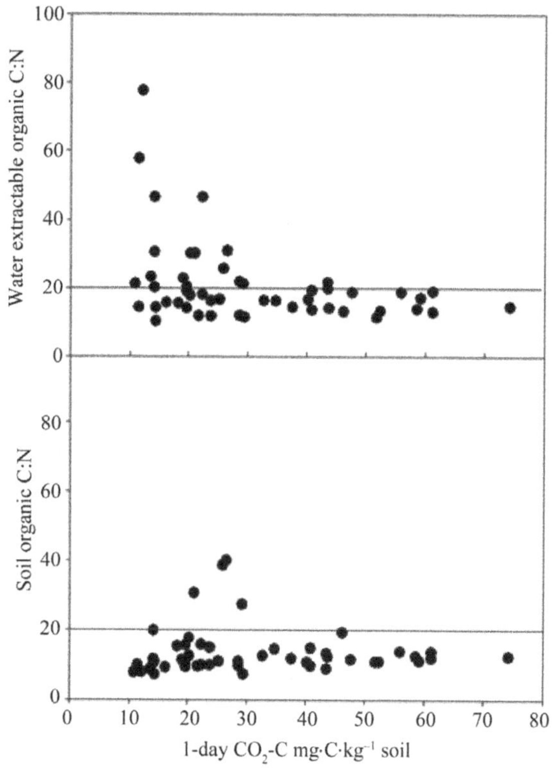

Figure 6. 1-day CO_2-C vs. soil organic C:N and water extractable organic C:N.

methods for quantifying the active portion of soil. Consequently the majority of soil test labs ignore the active soil C pools. In terms of N, soil labs often test for total and inorganic N, while only a few test for ammonium. On the other hand, some labs do not soil test for N at all. Because the cost for fertilizer usually accounts for roughly 30% - 40% of the total budget of a modern farming system, it is a shame that all available N resources are not measured and accounted for. Developing soil test methods to account for soil C and N pools that contribute to plant-available nutrients is a necessary step towards improving resource efficiency and reducing input cost. In addition to using WEOC:WEON instead of SOC:SON, in future research we intend to investigate the possibility of using a sliding scale for WEOC: WEON that can be developed and compared to yield in unfertilized and unamended plots. The sliding scale would begin with a linear response and represent increased N mineralization potential as WEOC:WEON declines. This sliding scale could be quickly evaluated by geographic area using WEOC:N and CO_2-C respiration data and test plots for a variety of crops and forages. For example, higher C:N ratios would predict less N mineralization as compared to low C:N ratios. Setting a break-point of 20:1 above which no potential N mineralization occurs could reduce the possibility of under-fertilizing based on recommendations from soil microbial activity alone. This would represent a safety factor to guard against high soil microbial activity (CO_2 respiration) with little N released when N is immobilized due to high WEOC:WEON.

In preliminary investigations, we found that soils with low WEOC:N ratios released more N than soils with high WEOC:N ratios. We experienced this in fields here in Texas that had legume cover crops vs. no cover crops for two years. The C:N ratio from a water extraction averaged 10:1 in cover crop soils vs. 16:1 in non-cover crop soils. Both of these fields were in a replicated corn-wheat rotation. Wheat yields increased 10 bu and corn yields increased 20 bu in the cover crop vs. no cover-crop fields, which corresponds to the WEO C:N ratios from the two treatments (data not shown). The sliding scale WEOC:N ratio could be developed by geographic areas based on local climatic conditions which would require crop or forage yield, a total C:N analyzer for water extraction, Solvita DCR and paddle kit, and inorganic N analyzer (both NH_4-N and NO_3-N). Overall this future research could provide insight into the application of the previously described methods and their application to both conventional and organic farming systems.

4. Conclusion

Our data suggests the C:N ratios determined from soil water extractions are likely to be more sensitive than total soil C:N to microbial activity and therefore can be a better measurement of the impact of management inputs. Determining the quality as opposed to the quantity of organic C could enhance our ability to rapidly measure impacts on soil microbial activity which is an active reflection of soil health. This information can be applied to track the relative impact of management decisions on soil fertility ultimately improving the efficiency of management decisions. Combining WEO C:N ratios with microbial activity could be used as a rapid soil biology indicator with healthy soil having a WEO C:N ratio below 20: 1 and microbial activity between 30 - 50 ppm C. Tracking soil biology improvement or degradation is currently difficult due to the lack of a proper tool and rapid, accurate analytical methods. Introducing management schemes to improve the C:N ratio and increase microbial activity should result in increased soil fertility/soil biology and highly productive and sustainable systems. Using soil testing labs to measure soil biology and tracking management inputs based on soil biology/soil fertility improvements would be a step towards true sustainability that can be easily acquired through adoption of both the water extraction method for determining organic C and N and the Solvita soil respiration method.

REFERENCES

[1] K. Alef, T. H. Beck, L. Zelles and D. Kleiner, "A Comparison of Methodsto Estimate Microbial Biomass and N-Mineralization in Agricultural and Grassland Soils," *Soil Biology and Biochemistry*, Vol. 20, No. 4, 1988, pp. 561-565. doi:10.1016/0038-0717(88)90073-9

[2] K. P. Dierksen, G. W. Whittaker, G. M. Banowetz, M. D. Azevedo, A. C. Kennedy, J. J. Stein and S. M. Griffith, "High Resolution Characterization of Soil Biological Communities by Nucleic Acid and Fatty Acid Analyses," *Soil Biology and Biochemistry*, Vol. 34, No. 12, 2002, pp. 1853-1860. doi:10.1016/S0038-0717(02)00198-0

[3] W. R. Cookson, D. A. Abaye, P. Marschner, D. V. Murphy, E. A. Stockdale and K. W. T. Goulding, "The Contribution of Soil Organic Matter Fractions to Carbon and Nitrogen Mineralization and Microbial Community Size and Structure," *Soil Biology and Biochemistry*, Vol. 37, No. 9, 2005, pp. 1726-1737. doi:10.1016/j.soilbio.2005.02.007

[4] B. Drigo, G. A. Kowalchuck and J. A. van Veen, "Climate Change Goes Underground: Effects of Elevated Atmospheric CO_2 on Microbial Community Structure and Activities in the Rhizosphere," *Soil Biology and Biochemistry*, Vol. 44, No. 5, 2008, pp. 667-679. doi:10.1007/s00374-008-0277-3

[5] R. L. Haney, W. F. Brinton and E. Evans, "Soil CO_2 Respiration: Comparison of Chemical Titration, CO_2 IRGA Analysisand the Solvita Gel System," *Renewable Agriculture and Food Systems*, Vol. 23, No. 2, 2008, pp. 171-176. doi:10.1017/S174217050800224X

[6] E. A. Paul and N. G. Juma, "Mineralisation and Immobilization of Soil Nitrogen by Microorganisms," In: F. E. Clark and T. Rosswall, Eds., *Ecological Bulletin*, Vol. 33, Stockholm, 1981, pp. 197-204.

[7] R. L. Tate, "Soil Microbiology," Wiley, New York, 1995.

[8] G. Bengtston, P. Bengtson and K. F. Mansson, "Gross Nitrogen Mineralization-, Immobilization-, and Nitrification Rates as a Function of Soil C:N Ratio and Microbial Activity," *Soil Biology and Biochemistry*, Vol. 35, No. 1, 2003, pp. 143-154. doi:10.1016/S0038-0717(02)00248-1

[9] M. S. Booth, J. S. Stark and E. Rastetter, "Controls on Nitrogen Cycling in Terrestrial Ecosystems: A Synthetic Analysis of Literature Data," *Ecological Monographs*, Vol. 75, No. 2, 2005, pp. 139-157. doi:10.1890/04-0988

[10] J. Luxhøi, S. Bruun, B. Stenberg, T. A. Breland and L. S. Jensen, "Prediction of Gross and Net N Mineralization-Immobilization-Turnover from Respiration," *Soil Science Society of America Journal*, Vol. 70, No. 4, 2006, pp. 1121-1128. doi:10.2136/sssaj2005.0133

[11] J. E. Barrett and I. C. Burke, "Potential N Immobilization in grassland Soils across a Soil Organic Matter Gradient," *Soil Biology and Biochemistry*, Vol. 32, No. 11-12, 2000, pp. 1707-1716. doi:10.1016/S0038-0717(00)00089-4

[12] J. A. van Veen, J. N. Ladd and M. J. Frissel, "Modelling C and N Turnover through the Microbial Biomass in Soil," *Plant and Soil*, Vol. 76, No. 1-3, 1984, pp. 257-274. doi:10.1007/BF02205585

[13] B. H. Janssen, "Nitrogen Mineralization in Relation to C:N Ratio and Decomposability of Organic Materials," *Plant and Soil*, Vol. 181, No. 1, 1996, pp. 39-45.

doi:10.1007/BF00011290

[14] R. G. Woodmansee and D. A. Duncan, "Nitrogen and Phosphorus Dynamics and Budgets in Annual Grasslands," *Ecology*, Vol. 61, No. 4, 1980, pp. 893-904. doi:10.2307/1936759

[15] R. L. Haney and E. B. Haney, "Simple and Rapid Laboratory Method for Rewetting Dry Soil for Incubations," *Communications in Soil Science and Plant Analysis*, Vol. 41, No. 12, 2010, pp. 1493-1501. doi:10.1080/00103624.2010.482171

[16] J. R. Burford and J. M. Bremner, "Relationships between the Denitrification Capacities of Soils and Total, Water-Soluble and Readily Decomposable Soil Organic Matter," *Soil Biology and Biochemistry*, Vol. 7, No. 6, 1975, pp. 389-394. doi:10.1016/0038-0717(75)90055-3

[17] E. A. Davidson, L. F. Galloway and M. K. Strand, "Assessing Available Carbon: Comparison of Techniques across Selected Forest Soils," *Communications in Soil Science and Plant Analysis*, Vol. 18, No. 1, 1987, pp. 45-64. doi:10.1080/00103628709367802

[18] R. G. Qualls and B. L. Haines, "Biodegradability of Dissolved Organic Matter in Forest Throughfall, Soil Solution and Stream Water," *Soil Science Society of America Journal—Abstract*, Vol. 56, No. 2, 1992, pp. 578-586. doi:10.2136/sssaj1992.03615995005600020038x

[19] T. DeLuca and D. R. Keeney, "Soluble Organics and Extractable Nitrogen in Paired Prairie and Cultivated Soils of Central Iowa," *Soil Science*, Vol. 155, No. 3, 1993, pp. 219-228. doi:10.1097/00010694-199303000-00008

[20] Alexander and Martin, "Introduction to Soil Microbiology," 2nd Edition, Wiley, New York, 1977.

[21] D. M. Sylvia, J. J. Fuhrmann, P. G. Hartel and D. A. Zuberer, "Principles and Applications of Soil Microbiology," Prentice Hall, Upper Saddle River, 1988.

[22] L. E. Jackson, J. P. Schimel and M. K. Firestone, "Short-Term Partitioning of Ammonium and Nitrate between Plants and Microorganisms in an Annual Grassland," *Soil Biology and Biochemistry*, Vol. 21, No. 3, 1989, pp. 409-416. doi:10.1016/0038-0717(89)90152-1

Analysis of the Bacterial Communities in Lime Concretion Black Soil upon the Incorporation of Crop Residues

Shao-Qiang Tao, Qiang Xia, Lin Zhu, Jing-Jing Chen, Ya-Nan Wang, Bing Qin

College of Resources and Environmental Sciences, Anhui Agricultural University, Hefei, China.
Email: zhulin_2002.163@163.com

ABSTRACT

To analyze the bacterial communities in lime concretion black soil upon the incorporation of crop residues for two years in wheat-maize system, total DNA was directly extracted and PCR-amplified with the F357GC and R518 primers targeting the 16S rRNA genes of V3 region. The amplified fragments were analyzed by perpendicular DGGE. Analyzing of species richness index S and Shannon diversity index H revealed that there was a high diversity of soil bacterial community compositions among all treatments after incorporation of crop residues and fertilizing under field conditions. Eleven DGGE bands recovered were re-amplified, sequenced. Phylogenetic analysis of the representative DGGE fingerprints identified four groups of the prokaryotic communities in the soil by returning wheat residues and fertilizing under field conditions. The bacterial communities belonged to *gamma proteobacterium*, *Cupriavidus sp*, *halophilic eubacterium*, *Acidobacterium sp*, *Sorangium sp*, *delta proteobacterium*, *Streptococcus sp and Streptococcus agalactiae* were main bacterial communities. Principal Component Analysis (PCA) showed that there were the differences in DNA profiles among the six treatments. It showed that wheat residue returning, maize residue returning and fertilizing all can improve bacterial diversity in varying degrees. As far as improvement of bacterial diversity was concerned, wheat residue returning was higher than fertilizing, and fertilizing higher than maize residue returning.

Keywords: Crop Residues; Bacterial Community; Lime Concretion Black Soil; Denaturing Gradient Gel Electrophoresis (DGGE); 16S rDNA; Wheat-Maize System

1. Introduction

Plant residues are an important source of nutrients in both natural and agricultural ecosystems, where synchronous plant growth and residue decomposition are essential for soil fertility. Fresh plant material (e.g. litter) represents a readily available substrate for both soil fauna and soil microorganisms. The main mineralization activity is performed by soil microbial communities, and the composition of the organic residues controls the decomposition rate and the related release of nutrients [1].

It has been reported that only a small fraction of microorganisms in nature are culturable. Therefore, to study the total microbial communities involved in organic matter decomposition, culture-independent and culture-dependent are needed.

Molecular techniques offer new opportunities for analysis of the structure of a microbial community [2]. In particular, sequence variation in rRNA genes has been exploited for inferring phylogenetic relation-ships among microorganisms [3] and may be used to estimate the genetic diversity of complex microbial communities in natural ecosystems [4-7]. All prokaryotes have 16S rRNA genes whose average length is about 1500 bp. There are both conserved and variable regions (the V1-V9 regions), and sufficient information has been compiled with which to conduct for reliable phylogenetic analyses [8,9]. Denaturing gradient gel electrophoresis (DGGE) allows one to directly determine the richness and evenness of the dominant microbial species using 16S rRNA gene amplicons and, thus, to profile the corresponding microbial populations in both a qualitative way and a semiquantitative way [10-13]. The diversity can be estimated from the number of 16SrRNA gene sequence similarity groups (*i.e.* the number of DNA bands on the DGGE gel [4] and the intensity of DGGE bands. Each band is assumed to represent an operational taxonomic unit, which is called a species for simplicity [14].

It is important for the typical wheat-maize cropping system of two crops a year in the Huaibei Plain (HP) to the food security of Anhui, China. The crop residues produced by the systems used to burn and bring about environment pollution, which seriously affect the sustainable development of agriculture. A field experiment was conducted to study the diversity of soil bacteria communities in the Wheat-maize system under incorporation of crop residues.

2. Materials and Method

2.1. Experimental Site and Soil

A field experiment was conducted in rice-wheat cropping systems for two years (2008-2010) in Mengcheng, Anhui, China. The climate is semi-tropical monsoon with hot and rainy summers and cold winters. The area receives 850 mm annual rainfall, about 80% of which occurs from June to September. The mean annual temperature is between 14.3°C to 16.7°C.

The alluvial soil of experimental site was a typical lime concretion black soil, well drained with pH 6.5 and silty clay loam in texture. The soil had soil organic matter 1.25%, and available N, P and K 80.2, 15.4 and 100.3 $mg \cdot kg^{-1}$, respectively. Soils were collected from the top layer (0 - 20 cm) of the field at mid filling stage of maize. The soil was homogenized by sieving (2-mm mesh), and stored at 4°C before use.

2.2. Treatments and Crop Management

Treatments were arranged according to a completely randomized design with three replicates (**Table 1**).

Take the quantity of corn straw returning by grinding and burying. The amount of maize residue was about 12,000 $kg \cdot ha^{-1}$. Wheat straw was returned to the field by grinding and bestrowing. The amount of wheat residue was about 7500 $kg \cdot ha^{-1}$. Wheat (cultivars Yan Nong 19) was transplanted in the plots at row spacing of 20 cm. Planting density reach to 2.7×10^6 ha^{-1} individual plant, Maize (cultivars Zheng Dan 958) at row spacing of 60 cm and planting density reach to 6.75×10^4 ha^{-1}. During wheat growth period, Nitrogen as urea was applied at 240 $kg \cdot N \cdot ha^{-1}$, base fertilizer ratio to additional fertilizer was 5.5:4.5, combined with landpreparation 600 $kg \cdot ha^{-1}$ compound fertilizer and 240 $kg \cdot N \cdot ha^{-1}$ were applied in related plot. Other 45% Nitrogen as additional fertilizer was applied in shooting period of wheat. During maize growth period, Nitrogen as urea was applied at 300 $kg \cdot N \cdot ha^{-1}$, 450 $kg \cdot ha^{-1}$ compound fertilizer ($N:P_2O_5:K_2O$ = 15:15:15) as base fertilizer was applied, 112.5 and 120 $kg \cdot N \cdot ha^{-1}$ were applied respectively at six and twelve leaves expanded.

Table 1. The field trial design of wheat and maize residue returning.

Number	Treatment
CK	No wheat and maize residue returning, No fertilizing
CK-F	No wheat and maize residue returning, Fertilizing
M-F	Maize residue returning and no wheat residue returning, Fertilizing
WM-NF	Wheat and maize residue returning, No fertilizing
WM-F	Wheat and maize residue returning, Fertilizing
W-F	Wheat residue returning and no maize residue returning, Fertilizing

2.3. DNA Extraction of the Soil

Total community DNA extraction of the soil was carried out by the method of SDS-based DNA extraction as described by Zhou *et al.* (1996) with some modification [15]. Extraction buffer for removing humic acids (5 ml of 100 $mmol \cdot L^{-1}$ Tris-HCl, 100 mmol/L $Na_4P_2O_7$, 100 $mmol \cdot L^{-1}$ sodium EDTA, 100 $mmol \cdot L^{-1}$ NaCl, 4% degrease milk powder, pH 10.0) was added with 0.5 g (wet weight) of soil, and mixed by vortexing for 3 min. The mixture was incubated at 65°C for 10 min in a water-bath, with intermittent vortexing. The mixture was centrifuged at 3000 g for 5 min. The supernatant was discarded and pellet was washed with the removing humic acids extraction buffer once more up to the color of the supernatant almost same as the buffer. The pellet was mixed with 0.54 ml DNA extraction buffer (100 $mmol \cdot L^{-1}$ Tris-HCl, 100 $mmol \cdot L^{-1}$ Na_3PO_4, 100 $mmol \cdot L^{-1}$ sodium EDTA, 1.5 $mol \cdot L^{-1}$ NaCl, 1% CTAB, pH 8.0) and added with 20 μl proteinase K (10 $mg \cdot mL^{-1}$). The mixture was shaked at 225 $rpm \cdot min^{-1}$ for 0.5 h at 37°C in table concentrator. Then, 60 μl 20% SDS was added to the mixture, incubated at 65°C for 2 h in a water-bath, with intermittent gently mixing. The sample was centrifuged at 6000 g for 10 min at room temperature. The supernatant was collected and the soil pellet re-extracted with further extraction buffer (90 μl), incubated at 65°C for 10 min and centrifuged as above. The supernatant fluid was transferred to centrifuge tubes (2 ml) and added with a equal-volume of phenol:chloroform:Isoamyl alcohol (25:24:1, v/v/v), mixed, freezing in ice-bath for 10 min. Mixtures were centrifuge at 16,000 g for 10 min at 4°C. The aqueous phase was extracted with hydroxybenzene:chloroform:Isoamyl alcohol two times. After then, the upper aqueous phase was added equal volume of hydroxybenzene:chloroform:Isoamyl alcohol (25:24:1, v/v/v), centrifuging at 16,000 g for 10 min at 4°C. The partially purified nucleic acid pellet resuspended in 20 mL of TE (10 $mmol \cdot L^{-1}$ Tris-HCl, 1 $mmol \cdot L^{-1}$ sodium EDTA, pH 8.0). Potassium acetate (7·5 $mol \cdot L^{-1}$) was added to a final concentration of 0·5 $mol \cdot L^{-1}$. The sample was transferred to ice for 5 min, then centrifuged (16,000 g, 30 min) at 4°C to precipitate proteins and polysaccharides, and DNA was precipitated by adding 0·6 volume isopropanol. After 1 h at room temperature, DNA was precipitated by centrifugation (16,000 g for 30 min) and resuspended in TE (20 μL).

2.4. PCR

The total DNA preparation of each sample was amplified

by PCR with a 2720 Thermal Cycler (Applied Bio-systems, USA). The variable V3 region of 16S rRNA gene sequences from nucleotide 8 to nucleotide 1512 (*Escherichia coli* numbering) was amplified by PCR by using eubacterial primers 27 F (5'-AGAGTTTGATCC-TGGCTCAG-3') and 1492 R (5'-GGTTACCTTGTT-ACGACTT-3') at first, and then nested PCR was performed by using primers F357(5'-GC-CCTACGGG-AGGCAGCAG-3') together with a GC clamp (5' CgCCCgCCgC-gCgCggCgggCggggCgggggCACggggg-g-3') and R518 (5'-ATT ACC GCG GCT GCT GG-3') and the hot-start touchdown protocol described by Muyzer *et al* [2]. DNA extracted from the soil was amplified with a PCR mixture (25 μl) containing 2.5 μl of Mg-containing 10 × buffer, 2 μl of a 2.5 $mmol \cdot L^{-1}$ de-oxynucleoside triphosphate mixture, 5 pmol of forward and reverse primers respectively, 1 μl of the DNA solution (1:10 dilution of DNA extraction), and 0.75 μl of Expand High Fidelity DNA polymerase ((TaKaRa). The polymerase was added after a hot-start procedure (5 min at 94°C, followed by 5 min at 80°C). PCR was performed with a 2720 Thermal Cycler by using the following protocol: 1 min at 94°C (denaturation), 45 s at 65°C (annealing,), and 3 min at 72°C (elongation) with a 0.5°C touchdown every cycle during annealing for 20 cycles, followed by 10 cycles with an annealing temperature of 55°C and a final cycle consisting of 10 min at 72°C.

The absorbances of the DNA at wavelengths of 230, 260, 280 and 320 nm were determined with a spectrophotometer (NanoDrop® ND-1000, Thermo, USA) and 1% agarose gels were run to visualize the integrity of the DNA.

2.5. DGGE

DGGE was performed using Dcode™ Universal Mutation Detection System (Bio-Rad, Hercules, CA, USA). For each sample, 400 ng of PCR product was loaded onto a poly-acrylamide gel (8%, wt/vol) with a 45% to 65% denaturant gradient, A denaturing gradient consisting of 100% denaturant is defined as 7 M urea with 40 % (v/v) Electrophoresis was carried out at 100 v for 16 h (with an initial 30 min at 40 v) at 60°C in 0.5 × TAE buffer (4.84 g of Tris base per liter, 11.42 ml of acetic acid per liter, 20 ml of 0.5 M EDTA per liter; pH 8.0). Gels were stained in 1:10,000 (v/v) SYBR® Green I for 30 min and visualized under UV irradiance. Gel images were captured with a Gel Doc-It™ documentation system. DGGE bands migrating to the same position in different lanes in the gel were considered to have the same sequence. The scanned gels containing DNA band profiles were analyzed to determine the intensity of each band using Quantity One 4.6 software (Bio-Rad, Hercules, CA, USA). Selected bands were cut from DGGE gels and transferred to 1.5 ml micro tubes and purified by using Agarose Gel DNA purification Kit (TaKaRa), and were sub cloned into the pMD18-T vector, then transformed into *E. coli* strain DH5α and sequenced.

2.6. Data Analyses

DNA sequence analyses were edited and reformatted using the ChromasPro software. A similarity search for the nucleotide sequences of 16S rDNA genes of the test isolates was carried out online at http://www.ncbi.nlm.nih.gov using the BLAST search program for the nucleotide database maintained in GenBank. Finally, the test sequences were compared with those of selected related species and the phylogenetic relationship of these sequences was determined by MEGA4 software. Bacterial diversity in the samples was estimated in two different ways: as S, species richness (the number of DGGE fragments detected disregarding their relative intensities), and H, the Shannon index of bacterial diversity. The Shannon diversity index was calculated as $(H) = -\Sigma Pi\ln Pi = -\Sigma(Ni/N)\ln(Ni/N)$ based on the relative band intensities, as formulated by Eichner *et al.* [16]. Pi was defined as ni/N, where Ni is the area of a peak in intensity and N the sum of all peak areas in the lane profiles.

To estimate the diversity of bacterial communities in lime concretion black soil after the incorporation of crop residues in wheat-maize system, the data obtained from the DGGE patterns, based on band intensity and position, were analyzed by principal component analysis. Principal component analysis was performed using SPSS (version 16.0).

3. Results

3.1. DNA Extraction from Soil Samples and Specific PCR Amplification of 16S rRNA Genes

DNA extracted from each soil sample of 6 treatments has much high integrality. The DNA contents of the crude extracts were measured. The A_{260}/A_{230} absorbance ratio indicates polysaccharide or polyphenolic contamination, and the A_{260}/A_{280} ratio indicates protein contamination [17]. The A_{260}/A_{280} ratios of DNA extraction from soil samples were all about 1.4, but the A_{260}/A_{230} ratios were all about 0.7. It suggested that humic acid had yet not been wiped out fully from the DNA extraction The target gene for the bacterial communities could successfully be amplified in all extracts Using primers 27 F and 1492 R, a special DNA fragment amplified by first round PCR was gained in each soil samples The length of the special DNA fragment was about 1533 bp. At second round PCR Using primers F357 and R518, a special DNA fragment (234 bp) was obtained in each soil samples by diluting the product of the first round PCR 1:100 as template.

Equal quantities of amplified DNA (234 bp) were loaded into the slots of the different lanes on the gel for comparison of the diversity of soil bacteria in different treatment by DGGE.

3.2. DGGE

The banding profile of 16S rDNA DGGE analysis revealed that the populations of bacteria varied among all soils of different treatments (**Figure 1**).

The different DNA fragments were the same length but had different base sequences (operational taxonomic units or species). Between 28 and 38 bands were distinguished in the soils of all treatments (**Table 2**). The lowest number of bands was in the control treatment CK with no wheat and maize residue returning, no fertilizing, and the highest number was obtained from the treatment W-F with wheat residue returning and fertilizing, but no maize residue returning. Dominant bands were present particularly WM-NF, WM-NF and W-F, in contrast to CK, CK-F, M-F, where the bands were more similar and had higher intensity. Disregarding the relative intensities of detected fragments, S (species richness) among WM-NF, WM-NF and W-F were higher than CK, CK-F and M-F. It indicated that band number depended on wheat residue returning, in addition, fertilizing can also add the band number in some degree. Considering to the soil samples we got at maize growth period, it was wheat residues but maize residues that had more effects on the bacterial diversity and quantity. Shannon index H of CK treatment was the lowest as 2.69, where no crop residue returning and no fertilizing, and the one of CK-F, WM-NF, WM-NF and W-F treatment was higher than CK treatment. The index H of CK-F treatment was the highest to 3.06 (**Figure 2**).

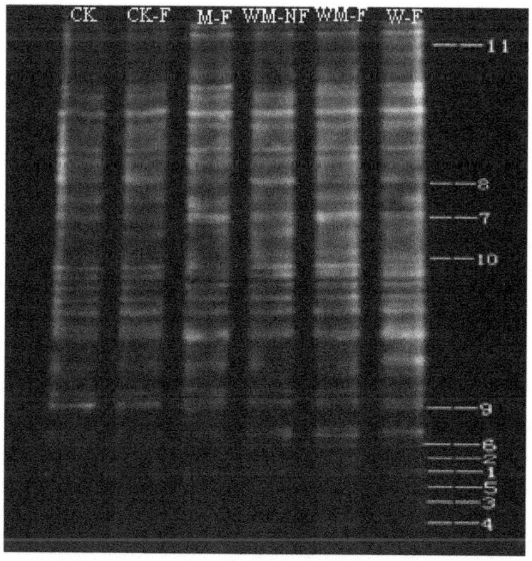

Figure 1. DGGE patterns of 16S rRNA gene sequences of soil samples in lime concretion black soil after the incorporation of crop residues in wheat-maize system.

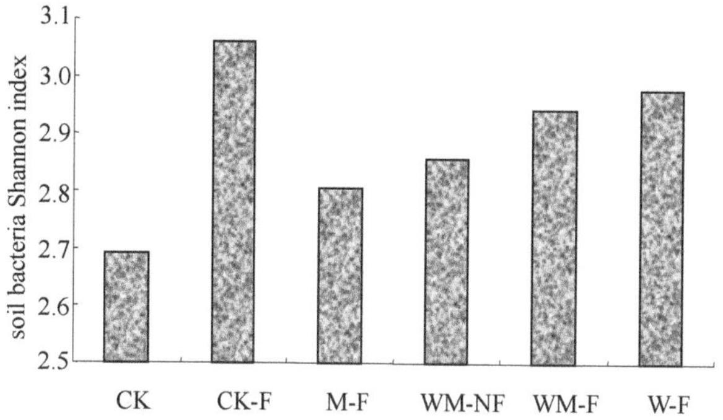

Figure 2. Soil bacteria communities of different species diversity index.

Table 2. Number of DGGE bands in DGGE profiles of bacteria.

Treatment	CK	CK-F	M-F	WM-NF	WM-NF	W-F
Bands	28	29	31	35	36	38

Since we were just considering the technique to describe the genetic diversity associated with the effect of wheat residues inoculation to soil and fertilizing under field conditions, we decided to use the most representative banding profile to construct phylogenetic relationship. In lane W-F of **Figure 1**, eleven selected bands were cut from DGGE gels and sequenced. The phylogenetic relationship of these sequences was showed as **Figure 3**. Using the program BLAST, sequences with most similarity to reference strains were found in the GenBank database. The result showed that the bacterial communities belonged to *Cupriavidus* sp, *gamma proteobacterium*, *Streptococcus* sp, *Sorangium* sp, *Acidobacterium* sp, delta proteobacterium during the incorporation of crop residues in wheat-maize system, thereinto, *Streptococcus* sp and *Acidobacterium* sp were main bacterial communities.

Principal Component Analysis (PCA) showed that there were the (qualitative) differences in DNA profiles among the six treatments (**Figure 4**). The crop residue returning and fertilizing showed divergent development of bacterial communities which was associated with different diversity indices. PC1 contributed much more than PC2 to bacterial community. PC1 and PC2 constitute 63.15% of total variation. There were much similarity between CK-F and W-F treatment. Though there was no incorporation of crop residue in CK-F treatment, it had been applied by fertilizer. This indicates that bacterial communities developed in the process of crop litter returning or fertilizing.

4. Discussion

Denaturing gradient gel electrophoresis (DGGE) is a gel electrophoresis method used to separate DNA fragments

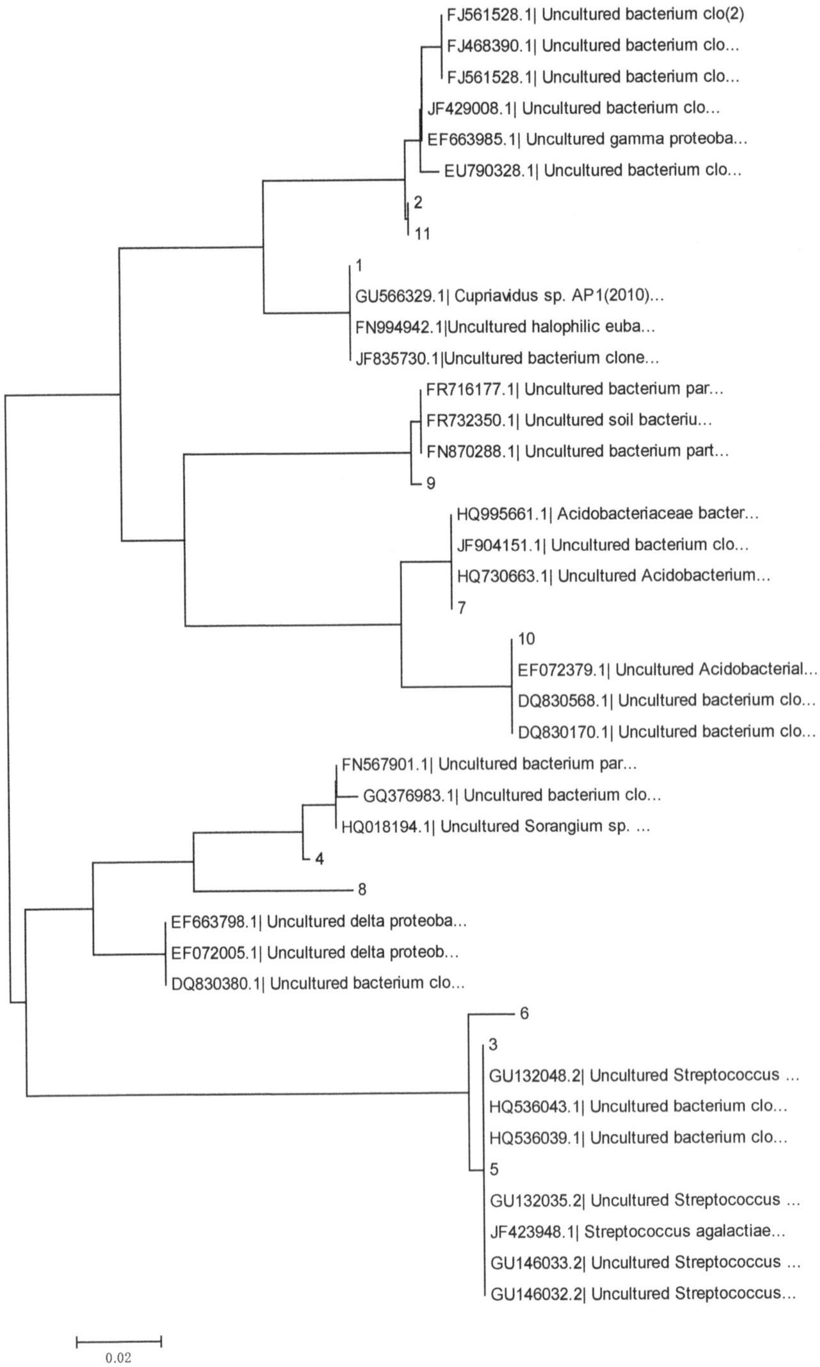

Figure 3. Dendrogram of phylogenetic relationship based on the denaturing gradient gel electrophoresis (DGGE) schematic banding profiles.

of the same length, but containing different base-pair sequences; it is used to determine the presence and abundance of different microbial species in a mixed population [18]. It was put forward first by Fischer and Lerman to detect single base substitutions within long DNA sequences [19]. Muzyer *et al.* apply this approach to analyzing the genetic diversity of complex microbial populations, which is based on the separation of polymerase chain reaction-amplified fragments of genes coding for 16S rRNA ribosomes, all the same length, by DGGE [6]. DGGE exploits base-pair sequence difference to separate amplified DNA on a high-resolution polyacrylamide gel along a denaturing gradient. Sequences differing in base composition denature at different locations on this gel. The number of bands on the gel is therefore indicative of the gene diversity in the original sample, such as a DNA

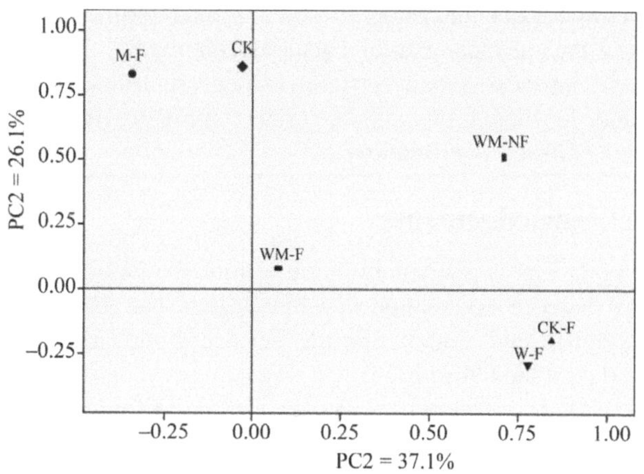

Figure 4. Qualitative PCA of 16S rRNA gene denaturing gradient gel electrophoresis (DGGE) profiles for bacterial communities.

extract from soil.

Crop residues have complex structure and are combinations of many different ingredients such as hemicellulose, cellulose, lignin and water-soluble polysaccharide. The degradation of these substances in the soil demand associative action of multi-microorganisms. Kimura's study showed that there existed complicated microflora in the process of stalk decomposition [20]. Only an estimated 0.1% - 1% of the naturally occurring bacteria have been isolated and characterized so far [21]. Selective enrichment cultures fail to mimic the conditions that particular microorganisms require for proliferation in their natural habitat. Furthermore, many microorganisms are bound to sediment particles and are thus not detected by conventional microscopy [2].

Molecular biological techniques offer new approach for the analysis of the structure and species composition of microbial communities. In particular, sequence variation rRNA has been exploited for inferring phylogenetic relationships among microorganisms [3]. These techniques have also been applied to determining the genetic diversity of microbial communities and to identifying several uncultured microorganisms [22,23]. They constitute the cloning of ribosomal copy DNA or polymerase chain reaction (PCR amplified ribosomal DNA (rDNA) followed by sequence analysis of the resulting clones. By using 16S rDNA in this community DNA for analysis, similarly it is possible to reveal the soil microbial community after the incorporation of crop residues in wheat-maize system.

Extraction of DNA from soils always results in coextraction of humic substances which interfere with DNA detection and measurement. This contamination can inhibit Taq DNA polymerase in PCR interfere with restriction enzyme digestion [24,25], and reduce transformation efficiency [26] and DNA hybridization specificity [27]. Since humic substances are difficult to remove, DNA purification is a critical step following direct extraction to obtain DNA of sufficient purity. Zhou *et al.* indicated that both CTAB and PVPP can effectively remove humic materials [15]. In our study, we had also made use of the de-humic effect of sodium pyrophos-phate and degrease milk powder for removing humic acids, besides adding CTAB and PVPP. Finally, the bacterial variable V3 region of 16S rDNA was amplified by diluted template and nested PCR This simple and practicable method can eliminate inhibition of humic substances to Taq DNA polymerase on the whole.

The DNA band pattern obtained by DGGE is an attractive way to study complex communities in environmental samples because most of the bacterial genotypes (species) gave one band [11]. The intensities of the DNA bands reflected the relative levels of the bacterial strains. Bacterial DNA profiles obtained by DGGE can be used as a semiquantitative measure of bacterial diversity, and we therefore used this method to study bacterial diversity after incorporation of crop residues and fertilizing under field conditions in agricultural soils.

Analyzing of DGGE profiles reveal that either crop residue returning to soil or fertilizing can increase species richness index, S and the Shannon index of bacterial diversity, H. Moreover, Shannon index H was consistent with species richness index S as a whole among all treatments (exception for CK-F treatment). The disappearance of dominating bands and the subsequent development of a more uniform band pattern of CK-F treatment can be interpreted as follows: r strategists (opportunists) that prevailed in soil of fertilizing were replaced by a variety of K strategists (persisters).This resulted in a special high H value of CK-F treatment.

The DNA band patterns obtained from amplified 16S rDNA V3 region gene sequences and DGGE indicated that the structure and diversity of bacterial communities changed significantly by crop residue returning or fertilizing. The numbers of distinct DGGE bands for all treatments were between 28 and 38. The lowest number was in the control treatment CK.

The number of DGGE bands was taken as an indication of species in each sample. The relative intensity of each DGGE band and the sum of all the surfaces for all bands in a sample were used to estimate species abundance [28,29]. The Shannon index of diversity (H) was used to determine the complexity of the DGGE bands from soils of different treatments. A higher diversity was observed in the soil of CK-F treatment than others. By using the Shannon index of diversity in combination with the cluster analysis of the DGGE banding patterns based on the similarity coefficient, we were able to monitor a whole range of community responses from all the treatment soils. Genetic diversity of soil samples was measured as bands on denaturing gradient gel electrophoresis (DGGE) of amplified 16S rDNA sequences from the soil community DNAs. When analysed by Shannon index (H), the highest genetic biodiversity (H = 3.061) was found in CK-F treatment with no wheat and maize residue returning but fertilizing; the poorest biodiversity (H = 2.692) in control treatment (CK) with no wheat and maize residue returning, no fertilizing. The Shannon-Weaver diversity index (H) was calculated from the DGGE profiles for each treatment. High H indicates a high diversity in the microbial community. In the present study, considerable qualitative differences in the community structure were found in the six treatments. With CK-F and W-F (wheat residue returning and no maize residue returning, fertilizing) treatments, the bacterial diversity was higher than with the other treatments, indicating that the fertilizing and wheat residue returning possibly enhanced the microbial community.

Simultaneously, the bacterial diversity of CK treatment was lower than the one of other treatments according to Shannon index of diversity. We can infer that the bacterial diversity be closely related to the crop residue returning and fertilizing which add nutriment for bacteria. The number and intensity of bands on denaturing gradient gel electrophoresis among all treatments were different. It showed that the predominant bacteria in each treatment were different.

Wheat residue returning, maize residue returning and fertilizing all can improve bacterial diversity in varying degrees. Contribution of PC1 was about 37.1%, and the one of PC2 was about 26.1% (**Figure 4**). The result of principal component analysis (PCA) showed that bacterial community of CK and M-F treatment was different from the one of the others. The difference of them lied with that there was no wheat residue returning in CKand M-F treatment. On the other hand, CK-F treatment had the highest index H to 3.06 (**Figure 2**). It suggested that fertilizing have much effect on bacterial community under no crop residue returning. In conclusion, as far as improvement of bacterial diversity is concerned, wheat residue returning was higher than fertilizing, and fertilizing higher than maize residue returning.

The results of this study clearly demonstrate the impacts of crop residue returning and fertilizing on the bacterial community composition of soils. The extensive cloning and sequencing of PCR-DGGE bands proved to be a powerful tool in assessing community structure differences in soils. Our future objectives in studying the impacts of management on microbial populations are aimed at attempting to understand the relationships which may exist between microbial community structure and function.

In conclusion, microbial diversity increased during the course of crop residue returning to soil or fertilizing. Not only crop residue returning to soil but also fertilizing affected the development of bacterial community. The effect of interaction between crop residue returning to soil and fertilizing on soil bacterial community need making a further investigation.

5. Acknowledgements

This work was supported by grants From the National Natural Science Foundation of China (Grant No. 2007-BAD89B10) and Anhui Nature Science Foundation, China (Grant No.090411026).

REFERENCES

[1] C. L. Neely, M. H. Beare, W. L. Hargrove and D. C. Coleman, "Relationships between Fungal and Bacterial Substrate-Induced Respiration, Biomass and Plant Residue Decomposition," *Soil Biology and Biochemistry*, Vol. 23, No. 3, 1991, pp. 947-954. doi:10.1016/0038-0717(91)90175-J

[2] G. Muyzer, E. C. De Waal and A. G. Uitterlinden, "Profiling of Complex Microbial Populations by Denaturing Gradient Gel Electrophoresis Analysis of Polymerase Chain Reaction-Amplified Gene Coding for 16S rRNA," *Applied and Environmental Microbiology*, Vol. 59, No. 3, 1993, pp. 695-700.

[3] C. R. Woese, "Bacterial Evolution," *Microbiological Reviews*, Vol. 51, No. 2, 1987, pp. 221-271.

[4] S. Kim-Jong, M. Sakai, A. Hosoda, T. Matsuguchi and J. S. Kim, "Application of DGGE Analysis to the Study of Bacterial Community Structure in Plant Roots and in Nonrhizosphere Soil," *Soil Science and Plant Nutrition*, Vol. 45, No. 2, 1999, pp. 493-497. doi:10.1080/00380768.1999.10409364

[5] A. E. McCaig, L. Glover and J. I. Posser, "Molecular Analysis of Bacterial Community Structure and Diversity in Unimproved and Improved Upland Grassland Pastures," *Applied and Environmental Microbiology*, Vol. 65, No. 4, 1999, pp. 1721-1730.

[6] G. Muyzer, "DGGE/TGGE: A Method for Identifying Genes from Natural Ecosystems," *Current Opinion Microbioogy*, Vol. 2, No. 3, 1999, pp. 317-322. doi:10.1016/S1369-5274(99)80055-1

[7] U. Nübel, F. Garcia-Pichel, M. Kühl and G. Muyzer, "Quantifying Microbial Diversity: Morphotypes, 16S rRNA Genes, and Carotinoids of Oxygenic Phototrophs in Microbial Mats," *Applied and Environmental Microbiology*, Vol. 65, No. 2, 1999, pp. 422-430.

[8] R. I. Amann, W. Ludwig and K. H. Schileifer, "Phylogenetic Identification and *in Situ* Detection of Individual Microbial Cells without Cultivation," *Microbiology and Molecular Biology Reviews*, Vol. 59, No. 1, pp. 143-169.

[9] J. M. Neefs and R. De Wachter, "A Proposal for the Secondary Structure of a Variable Area of Eukaryotic Small Ribosomal Subunit RNA Involving the Existence of a Pseudoknot," *Nucleic Acids Research*, Vol. 18, No. 19, 1990, pp. 5695-5704. doi:10.1093/nar/18.19.5695

[10] J. Bloem and A. M. Breure, "Bioindicators/Biomonitors-Principles, Assessment, Concepts," In: B. A. Markert, A. M. Breure and H. G. Zechmeister, Eds., *Microbial Indicators*, Elsevier, Amsterdam, 2003, pp. 259-282.

[11] O. Dilly, J. Bloem, A. Vos and J. C. Munch, "Bacterial Diversity in Agricultural Soils during Litter Decomposition," *Applied and Environmental Microbiology*, Vol. 70, No. 1, 2004, pp. 468-474. doi:10.1128/AEM.70.1.468-474.2004

[12] B. S. Griffiths, K. Ritz, R. Wheatly, H. L. Kuan, B. Boag, S. Christensen, F. Ekelund, S. J. Sørensen, S. Muller and J. Bloem, "An Examination of the Biodiversity-Ecosystem Function Relationship in Arable Soil Microbial Communities," *Soil Biology and Biochemistry*, Vol. 33, No. 12-13, 2001, pp. 1713-1722.

[13] K. Westergaard, A. K. Müller, S. Christensen, J. Bloem and S. J. Sørensen, "Effects of Tylosin as a Disturbance on the Soil Microbial Community," *Soil Biology and Biochemistry*, Vol. 33, No. 15, 2001, pp. 2061-2071. doi:10.1016/S0038-0717(01)00134-1

[14] J. B. Hughes, J. J. Hellmann, T. H. Ricketts and B. J. M. Bohannan, "Counting the Uncountable: Statistical Approaches to Estimating Microbial Diversity," *Applied and Environmental Microbiology*, Vol. 67, No. 10, 2001, pp. 4399-4406. doi:10.1128/AEM.67.10.4399-4406.2001

[15] J. Zhou, M. A. Bruns and J. M. Tiedje, "DNA Recovery from Soils of Diverse Composition," *Applied and Environmental Microbiology*, Vol. 62, No. 2, 1996, pp. 316-322.

[16] C. A. Eichner, R. W. Erb, K. N. Timmis and I. Wagner-DÖbler, "Thermal Gradient Gel Electrophoresis Analysis of Bioprotection from Pollutant Shocks in the Activated Sludge Microbial Community," *Applied and Environmental Microbiology*, Vol. 65, No. 1, 1999, pp. 102-109.

[17] J. Logemann, J. Schell and L. Willmitzer, "Improved Method for the Isolation of RNA from Plant Tissues," *Analytical Biochemistry*, Vol. 163, No. 1, 1987, pp. 16-20. doi:10.1016/0003-2697(87)90086-8

[18] L. D. Stetzenbach and M. V. Yates, "Dictionary of Environmental Microbiology," Academic Press, London, 2003.

[19] S. G. Fischer and L. S. Lerman, "DNA Fragments Differing by Single Basepair Substitutions Are Separated in Denaturing Gradient Gels: Correspondence with Melting Theory," *Proceeding of the National Academy Sciences of the United States of America*, Vol. 80, No. 6, 1983, pp. 1579-1583. doi:10.1073/pnas.80.6.1579

[20] M. Kimura and C. C. Tun, "Microscopic Observation of the Decomposition Process of Leaf Sheath of Rice Straw and Colonizing Microorganisms during the Cultivation Period of Paddy Rice," *Soil Science and Plant Nutrition*, Vol. 45, No. 2, 1999, pp. 427-437. doi:10.1080/00380768.1999.10409357

[21] D. M. Ward, M. M. Bateson, R. Weller and A. L. Ruff-Roberts, "Ribosomal RNA Analysis of Microorganisms as They Occur in Nature," *Advances in Microbial Ecology*, Vol. 12, 1992, pp. 219-286. doi:10.1007/978-1-4684-7609-5_5

[22] S. J. Giovannoni, T. B. Britschgi, C. L. Moyer and K. G. Field, "Genetic Diversity in Sargasso Sea Bacterioplankton," *Nature*, Vol. 345, No. 3, 1990, pp. 60-63. doi:10.1038/345060a0

[23] D. M. Ward, R. Weller and M. M. Bateson, "16S rRNA Sequences Reveal Numerous Uncultured Microorganisms in a Natural Community," *Nature*, Vol. 345, No. 3, 1990, pp. 63-65. doi:10.1038/345063a0

[24] K. Smalla, N. Cresswell, L. C. Mendonca-Hagler, A. Wolters and J. D. Van Elsas, "Rapid DNA Extraction Protocol from Soil for Polymerase Chain Reaction-Mediated Amplification," *Journal Applied Microbiology*, Vol. 74, No. 1, 1993, pp. 78-85.

[25] Y. L. Tsai and B. H. Olson, "Rapid Method for Separation of Bacterial DNA from Humic Substances in Sediments for Polymerase Chain Reaction," *Applied and Environmental Microbiology*, Vol. 58, No. 7, 1992, pp. 2292-2295.

[26] C. C. Tebbe and W. Vahjen, "Interference of Humic Acids and DNA Extracted Directly from Soil in Detection and Transformation of Recombinant DNA from Bacteria and Yeast," *Applied and Environmental Microbiology*, Vol. 59, No. 8, 1993, pp. 2657-2665.

[27] R. J. Steffan and R. M. Atlas, "DNA Amplification to Enhance Detection of Genetically Engineered Bacteria in Environmental Samples," *Applied and Environmental Microbiology*, Vol. 54, No. 9, 1988, pp. 2185-2191.

[28] N. Fromin, J. Hamelin, S. Tarnawski, D. Roesti, K. Jourdain-Miserez, N. Forestier, S. Teyssier-Cuvelle, F. Gillet, M. Aragno and P. Rossi, "Statistical Analysis of Denaturing Gel Electrophoresis (DGE) Fingerprinting Patterns," *Environmental Microbiology*, Vol. 4, No. 11, 2002, pp. 634-643. doi:10.1046/j.1462-2920.2002.00358.x

[29] H. Sekiguchi, M. Watanabe, T. Nakahara, B. Xu and H. Uchiyama, "Succession of Bacterial Community Structure along the Changjiang River Determined by Denaturing Gradient Gel Electrophoresis and Clone Library Analysis," *Applied and Environmental Microbiology*, Vol. 68, No. 10, 2002, pp. 5142-5150. doi:10.1128/AEM.68.10.5142

30

Microarray Technology and Its Applicability in Soil Science—A Short Review

Stella Asuming-Brempong

Department of Soil Science, School of Agriculture, University of Ghana, Legon, Ghana.
Email: sbrempong@yahoo.com

ABSTRACT

The GeoChip is a glass slide containing oligonucleotide probes targeting genes that confer specific function to microorganism. The GeoChip has been used to dissect the microbial community functional structure of environmental samples. The PhyloChip is a glass slide containing oligonucleotide probes of the 16S rRNA genes and it offers tremendous potential to monitor microbial population. Below ground microbial community can be linked to the above ground plant community by the use of these Chips in a high throughput manner. This review seeks to determine the various roles of the GeoChip and the PhyloChip in soil microbial ecology studies. During biostimulation of uranium in groundwater, microbial community dynamics was linked to functional processes and in global warming studies, microbial response to functional gene structure has been possible by the use of the GeoChip. The PhyloChip, on the other hand, provides more comprehensive survey of the microbial diversity, composition and structure and are less susceptible to the influence of dominance in microbial community. Some of the concerns regarding the use of compost in agricultural soils *i.e.* the spread of human, animal and plant pathogens were reduced when the PhyloChip was used to monitor composting.

Keywords: GeoChip; PhyloChip; Biodiversity; Microarray

1. Introduction

The soil environment is highly complex with the diversity of soil microorganisms being extremely high. Through DNA re-association kinetics, it is known that a gram of soil contains more than 4000 different genomes. Diverse systems such as soil, maybe more resilient to perturbation because removal of a portion of microbial components or the microbial component being compromised in some way, others that prevail will be able to compensate. However, more diverse systems may be less efficient since a greater proportion of available energy is used in generally countering competitive interactions between the various microbial components. Microorganisms play unique roles in ecosystem functions such as biogeochemical cycling of carbon, nitrogen, sulfur, phosphorus and various metals. Microorganisms also regulate nitrous oxide emissions in soil. However, the precise roles of many of the microorganisms in these cycles are unknown [1]. Owing to their extremely high diversity and their as yet uncultivated status, microbial detection, characterization and quantification in natural systems are difficult especially in a large scale and in a parallel and high throughput manner. Also, owing to the high versatility, rapid adaptation of microbial populations, high heterogeneity and microscale diversity in soils, high-throughput methods have to be applied in order to understand population shifts at a finer level and to be better able to link microbial diversity with functioning of ecosystems.

Microarrays represent a powerful tool for the parallel, high-throughput identification of many microorganisms in different environmental samples. A microarray is made up of thousands of spots on a slide with each spot containing multiple copies of unique nucleic acid sequences that correspond to a single gene. Microarray technology facilitates the detection of genetic sequences or expressed gene from particular samples in a high throughput format. In the case of expressed genes, microarrays are popular due to their unique ability to query the mRNA expression levels of thousands of genes (potentially all of the genes of an organism) simultaneously with relatively high specificity, providing a snapshot in time of the overall gene expression of the system under study. Compared to conventional membrane-based hybridization methods, microarrays offer the additional advantages of rapid detection, low cost, automation and low background level [2].

Living organisms contain thousands of genes that control cellular activities in their cells. Studying one component of the cell at a time will not give a complete picture of the cellular activities, it is only through a comprehensive integration of the entire molecular machinery controlling the cell that a thorough understanding can be gained. Similarly, a comprehensive integration of the microbial activities in environmental samples such as soil can be captured by the use of microarray technology and thus obtaining a holistic view of the microbial community. The traditional soil microbiological approaches, analyze material derived from microbial growth such as liquid cultures or colonies obtained by plating. However, such methods have often met with strong limitations, the reason being that only a small fraction of the microbiota (<0.01%) in soil can be accessed on the basis of cultivation, thus a complete picture of the microbial community is not obtained and about 99% of the microbial population in soil still remains unknown.

Also, some soil microbiologists have focused on evidence of processes and activities such as respiration and enzymatic transformation of adding substrates to soil. Measurements of soil processes give insight into microbial mediated transformations in soils. Such microbial mediated transformations do not inform us of the mechanisms, microbial functional composition and diversity that underlie the process level differences. Thus, relating microbial diversity and function to ecological processes remains a critical issue in the study of soil microbial ecology [1,3]. This review seeks to determine the various roles of the functional gene array (the GeoChip) and the PhyloChip in soil microbial ecology studies.

2. GeoChip and Its Relevance

The GeoChip is made of a glass slide containing thousands of bound oligonucleotide probes (of about 40 - 70 mers long) targeting genes that confer specific function to the microorganism. Thus, one can monitor the levels of thousands of functional genes simultaneously thereby gaining a window into the soil microbial community function of an environmental sample. The GeoChips 2 and 3 have so far been developed for soil microbial ecological studies. Geochip 2.0 contains over 24,000 probes covering more than 10,000 genes distributed among more than 150 functional groups involved in nitrogen, carbon, sulphur cycling, phosphorus utilization, metal resistance, metal reduction and organic contaminant degradation [4]. GeoChip 2.0 is useful for studying biogeochemical processes and functional activities of microbial communities which is important to human health, agriculture, energy, global climate change, ecosystem management and environmental clean-up and restoration. Because the arrays contain probes from genes with known biological function, they are useful in linking microbial diversity to ecosystem processes and functions. This array allows for a detailed analysis of the biogeochemical gene profiles of soil microbe and is ideal for understanding how these profiles change in response to environmental perturbations and experimentally imposed conditions [5,6]. To increase the confidence of detection, multiple probes for each sequence or each group of sequences were designed for the GeoChip 2.0. The positive controls are made of 16S rRNA gene probes (192 probes) and negative controls with 10 probes from human genes (960 spots) and blanks [4]. Later experiments showed that the GeoChip 2.0 was highly specific to their corresponding targets at 45°C to 50°C and with 50% formamide during hybridization.

The GeoChip 3.0 is a more comprehensive microarray than GeoChip 2.0 which currently is available for microbial community studies. The developed GeoChip 3.0 can be used as a generic high throughput tool to address various biological questions in different systems such as bioreactors, soils, groundwater, marine, sediments and animal guts. The GeoChip 3.0 has about 28,000 probes covering 57,000 gene variants from 292 functional gene families involved in carbon, nitrogen, phosphorus and sulphur cycles, energy metabolism, antibiotic resistance, metal resistance and organic contaminant degradation. It has several other distinct features and one of such is the *gyrB* gene for phylogenetic analysis [7]. The *gyrB* gene, encodes DNA gyrase β-subunit gene that has been used to differentiate closely related species/strains. Phylogenetic tree based on *gyrB* results in a magnitude higher resolution than a tree based on 16S rRNA gene [8,9]. GeoChip 3.0 contains eight degenerate probes for the 16S rRNA genes and 672 unique probes designed from hypothetical genes of seven sequenced genomes of hyperthermophiles for negative controls. In addition, a 50 mer common oligonucleotide reference standard (CORS) is mixed with all these probes, including gene probes and controls and co-spotted on GeoChip 3.0 as a common reference standard for data normalization and comparison [10].

In addition, the GeoChip probes are selected from coding sequences of functional genes, GeoChip can be used not only for measuring the abundance, but also for the expression of functional genes in a microbial community if high quality of mRNAs can be recovered from environmental samples. Thus, probing mRNA with the developed GeoChip will provide valuable insight into functions of the genes/populations in critical geochemical and ecological processes. Such information will be useful in establishing mechanistic linkages between diversity of microbial genes/populations and ecosystem functions.

3. The Usefulness of the GeoChip in Soil Microbial Ecological Studies

Studies have demonstrated that GeoChip is an ideal tool for dissecting the microbial community functional structure in both natural and contaminated environments [11-13]. Using the GeoChip 2.0, Liang *et al.* [14] found high abundance of genes involved in organic contaminant degradation in an oil-contaminated site indicating the biodegradation potential of the indigenous microorganisms for oil contaminant degradation. Also, the existence of key genes at that contaminated site (such as genes encoding alkane monooxygenase and benzene dioxygenase) involved in hydrocarbon degradation across the oil-contaminated site implied that stimulating indigenous microorganisms could be a valid option for remediating oil-contaminated sites. However, the degradation process might be influenced by low nutrients [15,16]. Nitrogen could be limited because of the decrease in nitrogen cycling genes with oil concentration which may indicate decrease in nitrogen-cycling activity. Adjustment of the carbon/nitrogen ratio by adding nitrogen maybe important for in situ bioremediation of oil contaminated fields.

He *et al.* [4] monitored microbial community dynamics in groundwater undergoing *in situ* biostimulation for uranium reduction by using the GeoChip. Their results showed that the GeoChip is able to reveal microbial community differences and that it could track bioremediation processes for linking microbial populations to functional processes. During the uranium reduction period, both FeRB (iron reduction bacteria) and SRB (sulphate reduction bacteria) populations reached their highest levels followed by a gradual decrease over 500 days. Consequently, the uranium in groundwater and sediments reduced and thus uranium concentrations in groundwater decreased. Because Geobacter-type FeRB and some SRB can use U(VI) as electron acceptor by obtaining energy for growth. The uranium concentrations in the ground water were significantly correlated with the total abundance of c-type cytochrome genes from Geobacter-type FeRB and Desulfovibrio-type and with the total abundance of dsrAB (dissimilatory sulfite reductase). Experimental results from GeoChip analysis suggested that Geobacter-type FeRB and SRB played significant roles in uranium reduction suggesting that uranium remediation using indigenous microorganisms could be a valid option in heavily uranium contaminated sites.

Other types of microarrays have been developed for application in bioremediation studies. An example of such an array was developed by Rhee *et al.* [17], that comprised of 1662 unique and group-specific 50 mer probes targeting most of the genes and pathways known to be involved in biodegradation and metal resistance. Its applicability was demonstrated in naphthalene-amended enrichment cultures as well as in soil microcosm experiments, the soil containing polychlorinated biphenyl.

A three year experimental field warming (+0.5˚C to 2˚C) to determine microbial response to global warming using the GeoChip microarray analyses showed significant warming effects on functional communities, specifically in the N-cycling microorganisms [18]. The number of functional genes detected on the GeoChip was significantly lower in the plots subjected to higher temperature as compared with the controls. For a range of gene families (*amo A*, cellulase, chitinase, laccase, *nif H*, *nir K*, *nir S*, *nos Z*, *pmo A* and urease) the number of variants detected on the GeoChip was generally lower in the plots subjected to higher temperature as compared with the control plots.

Understanding the factors influencing methanotrophs diversity and activity is of high importance in order to adapt environmental remediation strategies for optimal methane oxidation. Methanotrophs play an essential role in mitigating the greenhouse effect by metabolizing most of the methane produced for example in landfill sites. The gene encoding the particulate methane monooxygenase (*pmo A*) the key enzyme in methane oxidation was chosen for the development of a microarray to identify methanotrophs [19], bacteria that are capable of utilizing methane as their sole source of carbon and energy. The improved *pmo A* microarray contained 68 (18 - 28 mer) probes targeting all known methanotrophs including uncultivated members as well as the related ammonium monooxygenase (*amo A*) genes of ammonium oxidizing bacteria. The *pmo A* microarray identified *Methylocystis* spp. was dominating and was an efficient methane oxidizer.

The use of the functional gene array provided insight into the forces driving important processes of terrestrial Antartic nutrient cycling [20]. In the Antarctica, denitrification genes were linked to higher soil temperatures and N_2 fixation genes were linked to plots mainly vegetated by lichens. The relative detection of cellulose degradation genes was correlated with temperature and microbial carbon fixation genes were more present in plots principally lacking vegetation. Yergeau *et al.* [20] also showed a significant correlation between cellulase activity and the number of cellulase gene variants determined by the functional gene array. In a similar study, Reeve *et al.* [21] observed a significant correlation between cellulase gene signal intensity and cellulase activity in the soil ($p < 0.01$), correlation between dehydrogenase gene signal intensity and dehydrogenase activity, urease gene signal intensity and urease activity and so forth demonstrating that functional gene array can to some extent complement soil process measurements.

4. Some Challenges to Addressing the Use of GeoChip

1) High quality community DNA is required to minimize experimental variations for improving microarray-bored quantitative accuracy. Impure community DNA

with humic acid can affect amplification reactions. Hence to increase sensitivity to amplification reactions, pre amplification by rolling circle can be included in the methodology. This allows for the amplification of low microbial biomass communities before microarray hybridization and thereby increasing the signal levels from such environmental samples [22]. The reaction involves the use of spermidine and single-strand binding protein added to the reaction mix to facilitate amplification. The reactions are then incubated and the enzymatic reaction stopped and the amplification product was used for labeling.

2) The target sequences in public database increase exponentially and hence the GeoChip needs to be continuously updated. That could mean the quantity of data generated by microarray studies of environmental samples will be enormous but rapidly processing, comparing, interpreting hybridization data still remain difficult endeavors.

3) A large component of the soil microbial population may be inactive. Soil DNA hybridizations cannot differentiate between active and inactive microbial cells and potential contribution to signal intensity of the inactive *i.e.* spores or dead biomass or damaged copies of genes cannot be determined. For this reason, caution should be used when interpreting DNA functional gene array [23]. To overcome this criticism, researchers are beginning to use RNA for environmental microarray analysis [17,24]. Analysis of mRNA would allow more direct connection to be drawn. Recent research on environmental samples using both mRNA and genomic DNA microarrays has shown that the dominant species identified by mRNA arrays are also the most abundant in terms of genomic DNA [25]. This suggests that connection drawn between genomic DNA and biogeochemical cycles is reasonable. Yergeau *et al.* [20] had microarray-based results that were confirmed for a number of gene families using specific real-time PCR, enzymatic assays and process rate measurements suggesting a quantitative relationship between microarray signals and environmental gene densities. The significant correlations between the enzymatic activities measured in soil and the microarray data provided some indication that the detected genes are also expressed in the soil system examined.

5. The PhyloChip

The PhyloChip is made of slide on which are attached thousands of oligonucleotide probes (of about 50 mers long) of the 16S rRNA genes. The PhyloChip microarray allows the molecular biologist to monitor the levels of 16S rRNA genes (thousands of them) simultaneously thereby giving an 'inner picture' of microbial community in an environmental sample such as soil. PhyloChip (G2) consists of 506,944 probe features, and of these features, 297,851 are oligonucleotide perfect match or mismatch (MM) probes of 16S rRNA genes [26,27]. Depending on the type of probe set used, the PhyloChip can allow the parallel detection of up to several thousand microbial strains, species, genera or higher taxonomic groups in a single experiment [19,28,29]. The parallel detection of numerous 16S rRNA genes makes the PhyloChip useful for environmental studies of phylogenetically diverse microbial groups.

In a variety of environments, such as contaminated sites [26,28] air [27] water [30] soil [31-33], the PhyloChip has been used to detect microorganisms. In addition, the PhyloChip can detect much more bacterial taxa as compared with the 16S rRNA gene-based clone library approach [28,34] suggesting that the PhyloChip provides more comprehensive surveys of microbial diversity, composition and structure. Furthermore, such microarray-based approaches are less susceptible to the influence of dominance in microbial communities, whereby sequences of more abundant members mask the presence of other numerically significant taxa and rare species [34]. PhyloChip has been considered a powerful tool to comprehensively and rapidly analyze microbial communities.

6. The Use of the PhyloChip in Soil Microbial Community Studies

Among the concerns regarding the composting process and the use of compost in agriculture and horticulture are the survival and spread of animal, human and plant pathogens. Thus any composting process must be capable of eliminating any health risk that may be present in the end product. The microarray technology offers tremendous potential to monitor the detection of pathogens and of beneficial microbial populations during composting and this helps in the management of the compost before being sold to the public. A microarray was designed targeting the species of microorganisms usually encountered in compost [35] and it offered potential for process monitoring, and the detection of pathogens as well as of beneficial microbes [35]. This microarray contained probes targeting actinomycetes and other organisms in the composting process and 35 probes specific to other pathogens. The use of this microarray reduced the concerns regarding the use of composts on agricultural soils and the spread of human, animal and plant pathogens.

He *et al.* [36] used the PhyloChip to determine the impact of elevated CO_2 on the diversity and function of soil microbial communities. Richness of soil microbial communities at the Phylum, Class, Orders, Families and Subfamilies levels *i.e.* at the different taxanomic levels was detected. Thus, the taxonomic structure of microbial communities was linked with soil and plant properties through Mantel and such tests to know the extent, the soil and plant properties helped to shape the taxanomic structure. Shifts in the richness, composition and structure of soil microbial communities under elevated carbon dioxide were observed [36]. As noted by Cheneby *et al.* [37] and Hallin *et al.* [38], shifts in diversity will not necessarily alter the ability of soil microbes to perform biogeochemical functions.

The PhyloChip allows for the simultaneous detection of thousands of bacterial and archeal taxa and has been shown to reveal a broader range of diversity than modesty sized 16S rRNA gene libraries for soil, water and aerosol samples [33]. PhyloChip analyses also offer the opportunity to link microbial community composition to analyses of enzyme activity, density of functional gene families and the distribution of nutrient cycle-related functional gene sequences. It is possible to use both the GeoChip and the PhyloChip in an experiment [39]. To determine whether phylogenetic community structure, based on PhyloChip analysis was related to the distribution of microbial genes involved in nutrient cycling, the PhyloChip data was compared to the GeoChip data. Results showed that communities with more similar taxa composition were also more closely related in their functional genes supporting the notion that the functional genes detected in soils are strongly linked to community composition as determined by 16S rRNA gene-based. Such analysis provides evidence for a strong link between composition and functional gene distribution in Antarctic soils.

Like other high-throughput technologies, however, PhyloChip has its limitations. For example, PhyloChip only detects known sequences already present in a database at the time of probe design, so the G2 PhyloChip may not fully cover the species richness of soil microbial communities. Another limitation might be to improve the sensitivity and selectivity of the analysis. To discover unkown 16S rRNA genes, future investigations may use high-quality, full-length sequencing as a complementary approach to further understand the taxonomic and phylogenetic diversity, composition, structure and function of the soil microbial communities.

Integral to most methods of microbial community analysis is PCR amplification of small-subunit rRNA genes, undertaken primarily to obtain a sufficient mass of genetic material for analysis. This manipulation has well-known inherent biases and potentially unknown effect. The biggest bias is associated with multi-template PCR, in which the relative abundances of 16S rRNA gene signatures are distorted during PCR amplification [40]. The choice of primer pairs as well as the number of amplification cycles strongly influence the ratios of amplicons in the final pool when mixed templates are amplified by PCR [41]. Uneven amplification of mixed templates precludes both accurate estimation of evenness in communities and estimates of fold change in response to perturbation or experimental manipulation. Other problems include formation of chimeric amplicons and deletion and point mutations and amplification of contaminating DNA.

7. Methodology in the Use of the PhyloChip and the GeoChip

PhyloChip analysis includes three major steps: 1) Amplification of the target genomic DNA using 16S rRNA primers; 2) Adding an amount of the amplified DNA (50 - 500 ng PCR products) and hybridizing to the PhyloChip [26,27]; 3) Hybridization data being processed prior to statistical analysis. PCR amplification for microarray hybridization is carried out using a bacterial specific 16S rRNA primer e.g. 27 F1 and 1492 R and an archeal-specific 16S rRNA primers. Many independent PCRs are performed in a thermocycler with different annealing temperatures (eg. 48˚C, 51.9˚C, 54.4˚C and 58˚C). The samples are pooled per treatment then concentrated to a smaller volume. The pooled PCR product of each sample is spiked with known concentration of amplicons derived from yeast and bacterial metabolic genes serving as internal controls during the process of normalization. This mixture is fragmented to 50 - 200 bp with DNase 1 and One-Phor-All buffer following the manufacturer's protocol. The mixture is normally labeled with biotin or Cy5 or Cy3. Next, the labeled DNA is denatured at high temperature (for instance at 99˚C) for 5 min and hybridized to custom made Gene Chips. PhyloChip washing and staining are performed according to the manufacturer's prescription. Each PhyloChip is scanned and recorded as a pixel image and the initial data acquisition and intensity determination performed using standard Affymetrix (or type of platform software used). Background subtraction, data normalization and probe pair scoring are done.

To use GeoChip, soil DNA is extracted after mechanical lysis in a CTAB buffer using a phenol-chloroform purification protocol [42]. Other similar methods of soil DNA extraction such as the one by Zhou *et al.* [43] have been used. The genomic soil DNA can be labeled with cystidine-5 (Cy-5) dye or the Cy-3 dye. Hybridization of the labeled soil genomic DNA to a custom made GeoChip can be carried out at a hybridization station for instance TECAN US, Durham, NC, USA. The first wash is carried out followed by the prehybridization, hybridization and post hybridization washes. Scanning and imaging are then done.

One must take into consideration both the biological and the technical replications in performing experiments using the GeoChip and the PhyloChip. Environmental samples such as soil, the source of biological variation include macro environmental differences such as those caused by growth room/greenhouse effects (light, heat, humidity, location etc.) watering/fertilizing programs, soil conditions, pathogen/herbivore pressures, etc. Sample pooling and replication are the primary methods used to account for biological variation. Biologcal replication is necessary: 1) to estimate the biological variation within an experiment for downstream statistical analysis; 2) to extend the generality of the conclusions beyond the tested samples to the untested population as a whole. Technical variations include differences in labeling efficiencies, amplification reactions and the methodologies involved in hybridization.

8. Conclusion

Soil has been considered as a black box all this while. Especially on earlier the twenty century, it was difficult to establish the link between microbial community structure and function and even to link them to the above ground plant community. With the advent of microarray for microbial ecology studies, such linkages can be established. The time is drawing closer when the soil will no longer be considered as a blackbox.

9. Acknowledgements

I am very grateful to African Women in Agricultural Research and Development (AWARD) for awarding me fellowship to do my postdoctoral studies at the Dow AgroSciences, Indianapolis, Indiana, USA. I am also very grateful to Dr. Delkin Orlando Gonzalez, of the Microarray Laboratory, Dow AgroSciences for reviewing this manuscript thoroughly for me.

REFERENCES

[1] A. H. Fitter, G. A. Gilligan, K. Hollingworth, A. Kleczkowski, R. M. Twyman and J. W. Pitchford, "Biodiversity and Ecosystem Function and Soil," *Functional Ecology*, Vol. 19, No. 3, 2005, pp. 369-377. doi:10.1111/j.0269-8463.2005.00969.x

[2] D. Shalon, S. J. Smith and P. O. Brown, "A DNA Microarray System for Analyzing Complex DNA Samples Using Two-Color Fluorescent Probe Hybridization," *Genome Research*, Vol. 6, No. 7, 1996, pp. 639-645. doi:10.1101/gr.6.7.639

[3] J. J. Kelly, "Molecular Techniques for the Analysis of Soil Microbial Processes: Functional Gene Analysis and the Utility of DNA Microarrays," *Soil Science*, Vol. 168, No. 9, 2003, pp. 369-377. doi:10.1097/01.ss.0000090800.06903.2e

[4] Z. He, T. J. Gentry, C. W. Scheldt, L. Wu, J. Liebich, S. C. Cheng, Z. Huang, W. Wu, B. Gu, P. Jardie, C. Criddle and J. Zhou, "GeoChip, a Comprehensive Microarray for Investigating Biogeochemical, Ecological and Vironemtal Processes," *ISME Journal*, Vol. 1, No. 1, 2007, pp. 67-77. doi:10.1038/ismej.2007.2

[5] L. Wu, X. Liu, C. W. Schadt and J. Zhou, "Microarray Based Analysis of Subnanogram Quantitites of Microbial Community DNA by Using Whole-Community Genome Amplification," *Applied and Environmental Microbiology*, Vol. 72, No. 7, 2006, pp. 4931-4941. doi:10.1128/AEM.02738-05

[6] L. Wu, D. K. Thompson, G. Li, R. A. Hurt, J. M. Tiedje and J. Zhou, "Development and Evaluation of Functional Gene Arrays for Detection of Selected Genes in the Environment," *Applied and Environmental Microbiology*, Vol. 67, No. 12, 2001, pp. 5780-5790. doi:10.1128/AEM.67.12.5780-5790.2001

[7] Z. He, Y. Deng, D. J. Van Nostrand, Q. Tu, M. Xuy, C. L. Hemme, X. Li, L. Wu, T. J. Gentry, Y. Yin, J. Liebich, T. C. Hazen and J. Zhou, "GeoChip 3.0 as a High Throughput Tool for Analyzing Microbial Community Composition, Structure and Functional Activity," *The ISME Journal*, Vol. 4, No. 9, 2010, pp. 1167-1179. doi:10.1128/AEM.67.12.5780-5790.2001

[8] S. Yamamoto and S. Harayama, "Phylogenetic Analysis of *Pseudomonas Putida* Strains Deduced from the Nucleotide Sequences of *gyrB*, *rpoD*, and 16S rRNA," *International Journal of Systematic Bacteriology*, Vol. 48, No. 3, 1998, pp. 813-819. doi:10.1099/00207713-48-3-813

[9] L. Wang, F. Lee, C. Tai and H. Kasai, "Comparison of *gyrB* Gene Sequences, 16S rRNA Gene Sequences and DNA-DNA Hybridization in the *Bacillus Subtilis* Group," *International Journal of Systematic and Evolutionary Microbiology*, Vol. 57, No. 8, 2007, pp. 1846-1850. doi:10.1099/ijs.0.64685-0

[10] Y. Liang, Z. He, L. Wu, Y. Deng, G. Li and J. Zhou, "Development of a Common Oligo Reference Standard (CORS) for Microarray Data Normalization and Comparison across Different Microbial Communities," *Applied and Environmental Microbiology*, Vol. 76, No. 4, 2010, pp. 1088-1094. doi:10.1128/AEM.02749-09

[11] M. B. Leigh, V. H. Pellizari, O. Uhlik, R. Sutka, J. Rodrigues, N. E. Ostrom, J. Zhou and J. M. Tiedje, "Biphenyl-Utilizing Bacteria and Their Functional Genes in a Pine Root Zone Contaminated with Polychlorinated Biphenyls (PCBs)," *The ISME Journal*, Vol. 1, No. 2, 2007, pp. 134-148. doi:10.1038/ismej.2007.26

[12] J. Zhou, S. Kang, C. W. Schadt and C. T. Garten Jr., "Spatial Scaling of Functional Gene Diversity across Various Microbial Taxa," *Proceedings of National Academy of Science USA*, Vol. 105, No. 22, 2008, pp. 7768-7773. doi:10.1073/pnas.0709016105

[13] J. D. Van Nostrand, W. M. Wu, L. Wu, Y. Deng, J. Carley, S. Carroll, Z. He, B. Gu, J. Luo, C. S. Criddle, D. B. Watson, P. M. Jardine, T. L. Marsh, J. M. Tiedje, C. T. Hazen and J. Zhou, "GeoChip-Based Analysis of Functional Microbial Communities during the Reoxidation of a Bioreduced Uranium-Contaminated Aquifer," *Environmental Microbiology*, Vol. 11, No. 10, 2009, pp. 2611-2626. doi:10.1111/j.1462-2920.2009.01986.x

[14] Y. T. Liang, J. D. Van Nostrand, Y. Deng, Z. L. He, L. Y. Wu, Z. Xu, G. H. Li and J. Z. Zhou, "Functional Gene

Diversity of Soil Microbial Communities from Five Oil-Contaminated Fields in China," *ISME Journal*, Vol. 5, No. 3, 2011, pp. 403-413. doi:10.1038/ismej.2010.142

[15] R. Atlas, "Microbial-Degradation of Petroleum Hydrocarbons: An Environmental Perspective," *Microbiological Review*, Vol. 45, No. 1, 1981, pp. 180-209.

[16] Y. Oh, D. Sim and S. Kim, "Effects of Nutrients on Crude Oil Biodegradation in the Upper Intertidal Zone," *Marine Pollutant Bulletin*, Vol. 42, No. 12, 2001, pp. 1367-1372. doi:10.1016/S0025-326X(01)00166-7

[17] S. K. Rhee, X. Liu, L. Wu, S. C. Chang, X. Wan and J. Zhou, "Detection of Genes Involved in the Biodegradation and Biotransformationin Microbial Community by Using 50 Mer Oligonucleotide Microarrays," *Applied and Environmental Microbiology*, Vol. 70, No. 7, 2004, pp. 4303-4317. doi:10.1128/AEM.70.7.4303-4317.2004

[18] E. Yergeau, S. Bokhorst, S. Kang, J. Zhou, C. W. Greer, R. Aerb and G. A. Kowalchuk, "Shifts in Soil Microorganisms in Response to Warming Are Consistent across a Range of Antarctic Environments," *ISME Journal*, Vol. 6, No. 3, 2012, pp. 693-702. doi:10.1038/ismej.2011.124

[19] L. Bodrossy, N. Stralis-Pavese, J. C. Murrell, S. Rasjwqaki, A. Weilharter and A. Sessitsch, "Development and Variation of a Diagnostic Microbial Microarray for Methanotrophs," *Environmental Microbiology*, Vol. 5, No. 7, 2003, pp. 566-582. doi:10.1046/j.1462-2920.2003.00450.x

[20] E. Yergeau, S. Kang, Z. He, J. Zhou and G. A. Kowalchuk, "Functional Microarray Analysis of Nitrogen and Carbon Cycling Genes across Antarctic Latitudinal Transect," *ISME Journal*, Vol. 1, No. 2, 2007, pp. 163-179. doi:10.1038/ismej.2007.24

[21] J. R. Reeve, C. W. Schadt, L. Carpenter-Boggs, S. Kang, J. Zhou and J. P. Reganold, "Effects of Soil Type and Farm Management on Soil Ecological Functional Gene and Microbial Activities," *ISME Journal*, Vol. 4, No. 9, 2010, pp. 1099-1107. doi:10.1038/ismej.2010.42

[22] L. Wu, X. Liu, C. W. Schadt and J. Zhou, "Microarray-Based Analysis of Subnanogram Quantities of Microbial Community DNAs by Using Whole-Community Genome Amplification," *Applied and Environmental Microbiology*, Vol. 72, No. 7, 2006, pp. 4931-4941. doi:10.1128/AEM.02738-05

[23] T. J. Genty, G. S. Wickham, C. W. Schadt and J. Zhou, "Microarray Applications in Microbial Ecology Research," *Microbial Ecology*, Vol. 52, No. 2, 2006, pp. 159-175. doi:10.1007/s00248-006-9072-6

[24] H. Gao, Z. K. Yang, T. J. Gentry, L. Wu, C. W. Schadt and J. Zhou, "Microarray-Based Analysis of Microbial Community RNAs by Whole-Community RNA Amplification," *Applied and Environmental Microbiology*, Vol. 73, No. 2, 2007, pp. 563-571. doi:10.1128/AEM.01771-06

[25] S. E. Bulow, C. A. Francis, G. A. Jackson and B. B. Ward, "Sediment Denitrifier Community Composition and *nirS* Gene Expression Investigated with Functional Gene Microarrays," *Environmental Microbiology*, Vol. 10, No. 11, 2008, pp. 3057-3069. doi:10.1111/j.1462-2920.2008.01765.x

[26] E. L. Brodie, T. Z. DeSanti, D. C. Joyner, S. M. Baek and J. T. Larsen, "Application of a High Density Oligonucleotide Microarray Approach to Study Bacterial Population Dynamics during Uranium Reduction and Reoxidation," *Applied and Environmental Microbiology*, Vol. 72, No. 9, 2006, pp. 6288-6298. doi:10.1128/AEM.00246-06

[27] E. L. Brodie, T. Z. DeSantis, J. P. Parker, I. X. Zubiettza and Y. M. Piceno, "Uuban Aerosols Harbor Diverse and Dynamic Bacterial Population," *Proceedings of National Academy of Science USA*, Vol. 104, No. 1, 2007, pp. 299-304. doi:10.1073/pnas.0608255104

[28] A. Loy, A. Lehner, N. Lee, J. Adamczyk, H. Meier, J. Ernest, K. H. Schleifer and M. Wagner, "Oligonucleotide Microarray for 16S rRNA Gene-Based Detection of All Recognized Lineages of Sulfate-Reducing Prokaryotes in the Environment," *Applied and Environmental Microbiology*, Vol. 68, No. 10, 2002, pp. 5064-5081. doi:10.1128/AEM.68.10.5064-5081.2002

[29] J. Peplies, O. F. Glockner and R. Amann, "Optimization Strategies for DNA Microarray-Based Detection of Bacteria with 16S rRNA Targeting Oligonucleotide Probes," *Applied and Environmental Microbiology*, Vol. 69, No. 3, 2003, pp. 1397-1407. doi:10.1128/AEM.69.3.1397-1407.2003

[30] M. Hery, H. Sanguin, F. S. Perez, T. M. Vogel, E. Paul and S. Alfenore, "Monitoring of Bacterial Communities during Low Temperature Thermal Treatment of Activated Sludge Combining DNA Phylochip and Respirometry Techniques," *Water Research*, Vol. 44, No. 20, 2010, pp. 6133-6143. doi:10.1016/j.watres.2010.07.003

[31] K. Cruz-Martinez, K. B. Suttle, E. L. Brodie, M. E. Power, G. L. Andersen and J. F. Banfield, "Despite Strong Seasonal Responses, Soil Microbial Consortia are More Resilient to Long-Term Changes in Rainfall than Overlying Grassland," *ISME Journal*, Vol. 3, No. 6, 2009, pp. 738-744. doi:10.1038/ismej.2009.16

[32] K. M. DeAngelis, E. L. Brodie, T. Z. DeSantis, G. L. Andersen, S. E. Lindow and M. K. Firestone, "Selective Progressive Response of Soil Microbial Community to Wild Oat Roots," *ISME Journal*, Vol. 3, No. 2, 2009, pp. 168-178. doi:10.1038/ismej.2008.103

[33] E. Yergeau, S. A. Schoondermark-Stolk , E. L. Brodie, S. Dejean, T. Z. DeSantis, O. Gonclaves, Y. M. Piceno, G. L. Andersen and G. A. Kowalchuk, "Environmental Microarray Analyses of Antarctic Soil Microbial Communities," *ISME Journal*, Vol. 3, No. 3, 2009, pp. 340-351. doi:10.1038/ismej.2008.111

[34] T. DeSantis, E. Brodie, J. Moberg, I. Zubieta, Y. Piceno and G. Andersen, "High-Density Universal 16Sr RNA Microarray Analysis Reveals Broader Diversity than Typical Clone Library When Sampling the Environment," *Microbial Ecology*, Vol. 53, No. 3, 2007, pp. 371-383. doi:10.1007/s00248-006-9134-9

[35] I. Franke-Whittle, S. Klammer and H. Insam, "Design and Application of an Oligonucleotide Microarray for the Investigation of Compost Microbial Communities," *Journal of Microbiological Methods*, Vol. 62, No. 1, 2005, pp. 37-56. doi:10.1016/j.mimet.2005.01.008

[36] H. He, Y. Piceno, Y. Deng, M. Xu, Z. Lu, T, DeSantis, G. Andersen, S. E. Hobbie, P. B. Reich and J. Zhou, "The Phylogenetic Composition and Structure of Soil Microbial Communities Shifts in Response to Elevated Carbon Dioxide," *ISME Journal*, Vol. 6, No. 2, 2012, pp. 259-272. doi:10.1038/ismej.2011.99

[37] D. Cheneby, A. Brauman, B. Rabary and L. Philippot, "Differential Responses of Nitrate Reducer Community Size, Structure and Activity to Tillage Systems," *Applied and Environmental Microbiology*, Vol. 75, No. 10, 2009, pp. 3180-3186. doi:10.1128/AEM.02338-08

[38] S. Hallin, C. M. Jones, M. Schloter and L. Philippot, "Relationship between N-Cycling Communities and Ecosystem Functioning in a 50-Year-Old Fertilization Experiment," *ISME Journal*, Vol. 3, No. 5, 2009, pp. 597-605. doi:10.1038/ismej.2008.128

[39] E. Yergeau, S. A. Schoondermark-Stolk, E. L. Brodie, S. Déjean, T. Z. DeSantis , O. Gonçalves, Y. M. Piceno, G. L. Andersen and G. A. Kowalchuk, "Environmental Microarray Analyses of Antarctic Soil Microbial Communities," *ISME Journal*, Vol. 3, No. 3, 2009, pp. 340-351. doi:10.1038/ismej.2008.111

[40] A. Engelbrektson, V. Kunin, K. C. Wrighton, N. Zvenigorodsky, F. Chen, H. Ochman and P. Hugenholtz, "Experimental Factors Affecting PCR-Based Estimates of Microbial Species Richness and Evenness," *ISME Journal*, Vol. 4, No. 5, 2010, pp. 642-647. doi:10.1038/ismej.2009.153

[41] M. T. Suzuki and S. J. Giovannoni, "Bias Caused by Template Annealing in the Amplification of Mixtures of 16S rRNA Genes by PCR," *Applied and Environmental Microbiology*, Vol. 62, No. 2, 1996, pp. 625-630.

[42] E. Yergeau, S. Kang, Z. He, J. Zhou and G. A. Kowalchuk, "Functional Microarray Analysis of Nitrogen and Carbon Cycling Genes across and Antarctic Latitudinal Transect," *ISME Journal*, Vol. 1, No. 2, 2007, pp. 163-179. doi:10.1038/ismej.2007.24

[43] J. Zhou, M. A. Bruns and J. M. Tiedje, "DNA Recovery from Soils of Diverse Composition," *Applied and Environmental Microbiology*, Vol. 62, No. 2, 1996, pp. 316-322.

31

Tillage, Crop Rotation, and Cultural Practice Effects on Dryland Soil Carbon Fractions

Upendra M. Sainju[1*], Andrew W. Lenssen[2], Thecan Caesar-TonThat[1], Jalal D. Jabro[1], Robert T. Lartey[1], Robert G. Evans[1], Brett L. Allen[1]

[1]US Department of Agriculture, Agricultural Research Service, Northern Plains Agricultural Research Laboratory, Sidney, USA; [2]Agronomy Department, Iowa State University, Ames, USA.
Email: *upendra.sainju@ars.usda.gov

ABSTRACT

Information is needed on novel management practices to increase dryland C sequestration and soil quality in the northern Great Plains, USA. We evaluated the effects of tillage, crop rotation, and cultural practice on dryland crop biomass (stems and leaves) yield, surface residue, and soil C fractions at the 0 - 20 cm depth from 2004 to 2008 in a Williams loam in eastern Montana, USA. Treatments were two tillage systems (no-tillage [NT] and conventional tillage [CT]), two crop rotations (continuous spring wheat [*Triticum aestivum* L.] [CW] and spring wheat-barley [*Hordeum vulgaris* L.] hay-corn [*Zea mays* L.]-pea [*Pisum sativum* L.] [W-B-C-P]), and two cultural practices (regular [conventional seed rates and plant spacing, conventional planting date, broadcast N fertilization, and reduced stubble height] and ecological [variable seed rates and plant spacing, delayed planting, banded N fertilization, and increased stubble height]). Carbon fractions were soil organic C (SOC), particulate organic C (POC), microbial biomass C (MBC), and potential C mineralization (PCM). Crop biomass was 24% to 39% greater in W-B-C-P than in CW in 2004 and 2005. Surface residue C was 36% greater in NT than in CT in the regular practice. At 5 - 20 cm, SOC was 14% greater in NT with W-B-C-P and the regular practice than in CT with CW and the ecological practice. In 2007, POC and PCM at 0 - 20 cm were 23% to 54% greater in NT with CW or the regular practice than in CT with CW or the ecological practice. Similarly, MBC at 10 - 20 cm was 70% greater with the regular than with the ecological practice in NT with CW. Surface residue, PCM, and MBC declined from autumn 2007 to spring 2008. No-tillage with the regular cultural practice increased surface residue and soil C storage and microbial biomass and activity compared to conventional tillage with the ecological practice. Mineralization reduced surface residue and soil labile C fractions from autumn to spring.

Keywords: Carbon Sequestration; Dryland Soil; Labile Carbon; Management Practices; Residue Carbon

1. Introduction

Sequestering atmospheric CO_2 through plant photosynthesis in the soil as organic matter is getting increased attention, because reduction in CO_2 concentration can reduce global warming [1,2]. In croplands, C sequestration occurs when non-harvested crop residues, such as stems, leaves, and roots, are returned to the soil, followed by their placement at the soil surface using NT that reduces the mineralization of residue and SOC [3-5]. Carbon sequestration rates, however, depend on the balance between the amounts of plant residue C input and the rate of C mineralized in unmanured soil [6,7]. Other benefits of increasing C storage include enhancement of soil structure and soil water-nutrient-crop productivity relationships [8].

Incentives by governments and private companies for the compensation of C credit in agricultural soils have led to increased need for C sequestration and storage using novel management practices. Some of these practices are NT management, diversified crop rotations, and improved cultural practices. Studies have shown that NT with continuous cropping can increase dryland soil C storage compared to CT with crop-fallow [3-5]. In the northern Great Plains, NT with continuous cropping system can increase C sequestration by as much as 233 kg C $ha^{-1}\cdot yr^{-1}$ compared to a loss of 141 $kg\cdot C\cdot ha^{-1}\cdot yr^{-1}$ in CT with crop-fallow [9]. Sherrod *et al.* [4] found that increased cropping intensity in NT increased dryland SOC in the central Great Plains after 12 yr. Conversion from CT to NT can sequester C at 570 ± 140 $kg\cdot C\cdot ha^{-1}\cdot yr^{-1}$, reaching equilibrium in 15 to 20 yr and enhanced crop rotation can sequester at 200 ± 120 $kg\cdot C\cdot ha^{-1}\cdot yr^{-1}$, reaching equilibrium in 40 to 60 yr [10]. The NT management can also increase annual crop yields due to efficient wa-

*Corresponding author.

ter use as a result of increased soil water conservation from residue accumulation at the soil surface compared to CT [11,12].

Continuous cropping of diversified crops in the rotation can efficiently utilize water and nutrients and sustain yields compared to monocropping in water-limited dryland farming systems [13,14]. One such crop is pea, which uses less soil water than spring wheat and barley, thereby resulting in more water available for succeeding crops and increasing their yields [11,14]. Pea, being a legume, also fixes N from the atmosphere and has higher N concentration than spring wheat or barley [12,15]. Because of greater N supply, N fertilization rates to crops following pea can be reduced [12,14]. Other benefits of crop diversification include effective control of weeds, diseases, and pests [13,14], reductions in the risk of crop failure, farm inputs, and duration of fallow, and improvement in economic and environmental sustainability [16,17].

Increased C storage can also improve biological soil quality as indicated by enhanced microbial biomass and activity [5,18,19]. These can be estimated by measuring labile C fractions, such as MBC and PCM [20,21]. The MBC refers to C storage in the body of microorganisms and also known as microbial biomass. Similarly, PCM refers to soil C mineralization and is related to microbial activity. Because of a large pool size and inherent spatial variability, SOC changes slowly with management practices [18]. As a result, measurement of SOC alone does not adequately reflect changes in soil quality and nutrient status [18,19]. Active C fractions, such as MBC and PCM that change rapidly with time due to management practices, can better reflect changes in soil quality and productivity that alter nutrient dynamics due to immobilization-mineralization [20,21]. Similarly, POC that contains coarse organic matter,has been considered as an intermediate fraction of SOC between active and slow fractions that changes rapidly due to management practices [22]. The POC also provides substrates for microorganisms and influences soil aggregation [23].

In the northern Great Plains, traditional farming systems, such as CT with wheat-fallow, for the last 50 to 100 yr have decreased SOC levels by 30% to 50% [7,24]. Both tillage and fallowing reduce SOC by increasing soil organic matter mineralization [25,26] and reducing the amount of plant residue returned to the soil [10,27]. As a result, traditional farming systems have become unsustainable and uneconomical [16]. Controlling weeds in NT cropping systems is also a major concern in dryland farming systems. Some of the cultural practices used for weed control include high crop seeding rates, banded fertilization, delayed planting and harvest, and increased retention of crop residue [28,29]. Although information is available about the use of NT with continuous cropping in increasing soil C storage compared to CT with crop-fallow, little is known about the combined effects of tillage, diversified crop rotation, and cultural practice on dryland surface residue and soil C fractions.

Because of the short growing season due to cold weather and lack of enough time for sampling before crop planting in the spring, soil samples are normally taken after harvest in the autumn for testing nutrient levels and recommending fertilizer rates for succeeding crops in the next year in the northern Great Plains [16,30]. As a result, fertilizers are applied in the autumn instead of spring in the succeeding year. This results in substantial loss of nutrients, especially N and P, from fertilizers due to leaching, surface runoff, volatilization, and denitrification. To reduce this loss, it is recommended that soil tests be done and fertilizers applied before planting in the spring [30]. Since C mineralization can continue during the winter, it could be possible that C levels vary in surface residue and soil samples collected in the autumn and spring. We hypothesized that NT with diversified crop rotation and the ecological cultural practice would increase surface residue and soil C fractions compared to CT with monocropping and the regular cultural practice, and that surface residue and labile C fractions would decline more than SOC from the autumn to the succeeding spring due to mineralization and lack of additional inputs, regardless of management practices. Our objectives were to: 1) determine the amount of dryland crop biomass (stems and leaves) returned to the soil as influenced by tillage, crop rotation, and cultural practice from 2004 to 2007 in eastern Montana, 2) quantify their effects on surface residue and soil C fractions (SOC, POC, PCM, and MBC) at 0 - 20 cm, and 3) examine changes in surface residue and soil C fractions as influenced by management practices from autumn 2007 to spring 2008.

2. Materials and Methods

2.1. Experimental Site and Treatments

From 2004 to 2008, the experiment was carried out on a dryland farm site, 8 km northwest of Sidney (47°46′N, 104°16′W; elevation 690 m), Montana, USA. The climate is semiarid, with mean monthly air temperature ranging from −8°C in January to 23°C in July and August, and mean annual precipitation (68-yr average) of 357 mm, 80% of which occurs during the crop growing season (April-November). The soil is a Williams loam (fine-loamy, mixed, superactive, frigid, Typic Argiustolls) with 350 $g \cdot kg^{-1}$ sand, 325 $g \cdot kg^{-1}$ silt, 325 $g \cdot kg^{-1}$ clay, and 6.1 pH at the 0 - 20 cm depth. The SOC contents at 0 - 5 and 5 - 20 cm prior to the initiation of the experiment in April 2004 were 8.8 and 21.7 $Mg \cdot C \cdot ha^{-1}$, respectively. Previous cropping system for the last 10 yr was CT with spring wheat-fallow.

Treatments included two tillage practices (NT and CT),

two crop rotations (continuous spring wheat [CW] and spring wheat-barley hay-corn-pea [W-B-C-P]), and two cultural practices (regular and ecological). The cultural practices used for crops in the rotation are described in **Table 1**. The NT plots were left undisturbed, except for applying fertilizer and planting crops in rows. The CT plots were tilled one to two times a year with a field cultivator to a depth of 7 - 8 cm for seedbed preparation and weed control. In the four-year crop rotation in W-B-C-P, each phase (spring wheat, hay barley, corn, and pea) of the crop rotation was present in every year. Treatments were arranged in a split-plot design in randomized complete block with three replications. Tillage was considered as the main-plot and the factorial combination of crop rotation and cultural practice as the split-plot treatment. The split plot size was 12.2 m × 12.2 m.

2.2. Crop Management

At planting in April, 2004 to 2007, P fertilizer as monoammonium phosphate (11% N, 23% P) at 56 $kg·P·ha^{-1}$ and K fertilizer as muriate of potash (60% K) at 48 $kg·K·ha^{-1}$ were banded to a depth of 5 cm to all crops. Nitrogen fertilizer as urea (46% N) (either broadcast or banded [**Table 1**]) and monoammonium phosphate (banded) was applied at 101 $kg·N·ha^{-1}$ for spring wheat, 67 $kg·N·ha^{-1}$ for barley hay, and 78 $kg·N·ha^{-1}$ for corn. Pea received 6 $kg·N·ha^{-1}$ from monoammonium phosphate. Nitrogen fertilization rate to each crop was adjusted for soil NO_3-N content to a depth of 60 cm measured after harvest of previous crop in the autumn. Spring wheat (cv. Reeder) was planted from early April to mid-May and pea (cv. Majoret) and barley hay (cv. Haybet) from early to mid-April (**Table 1**) with a no-till drill at a spacing of 20.3 cm. Corn (cv. 39T67-RR) was planted in early May at a spacing of 76 cm in 2004 and 61 cm from 2005 to 2007. The narrow spacing for corn from 2005 to 2007 compared to 2004 was adapted to use soil water more efficiently and increase grain yield. Growing season weeds were controlled with selective post-emergence herbicides appropriate for each crop. Contact herbicides were applied at preplanting and post-harvest. In late June or early July, aboveground biomass of barley was harvested for hay with a self-propelled mower-conditioner and round baler after determining dry yield from two 0.5 m^2 plots at 60˚C for 3 d. In August, total aboveground biomass (stems, leaves, and grains) yield of spring wheat and pea was determined from two 0.5 m^2 areas outside yield rows after determining dry yields as above. At the same time, grain yield (oven-dried basis at 60˚C) was determined by harvesting grains from a swath of 1.5 m × 12.0 m using a combine harvester. In October, corn biomass and grain yields were determined from areas as described above. Biomass (stems and leaves) yield of spring wheat, pea, and corn was determined by deducting grain yield from total aboveground biomass. During harvest, grains were removed but biomass of spring wheat, pea, and corn were returned to the soil.

2.3. Surface Residue and Soil Sample Collection and Analysis

In November 2007, a month after crop harvest, soil surface residue samples were collected from five 30 cm × 30 cm areas randomly in the central rows of the plot, composited, washed with water to remove soil, and dried in the oven at 60˚C for 7 d to obtain dry matter weight. Samples were ground to pass a 1 mm screen to determine C concentration. After removing the residue, soil samples were collected with a hand probe (5 cm inside diameter) from the 0 - 20 cm depth from five places in the central rows of the plot. These were separated into 0 - 5, 5 - 10, and 10 - 20 cm depth increments, composited within a depth, air-dried, ground, and sieved to 2 mm for determining C fractions. The process was repeated again in April 2008 by collecting surface residue and soil samples and preparing exactly as above for C analysis. Because of the difficulty in getting intact soil cores due to low water content, only one additional core (5 cm inside diameter) containing undisturbed sample was collected from each depth in 2007 and 2008 to determine bulk

Table 1. Description of cultural practices (regular and ecological) used for crops in the rotation.

Crop	Cultural practice	Seeding rate (million seeds ha^{-1})	N fertilization at planting	Planting date	Stubble height (cm)
Spring wheat	Regular	2.23	Broadcast	Early to mid April	20
	Ecological	2.98	Banded	Early to mid May	30
Pea	Regular	0.60	Broadcast	Early to mid April	5
	Ecological	0.92	Banded	Early to mid April	5
Barley hay	Regular	2.23	Broadcast	Early to mid April	5
	Ecological	2.98	Banded	Early to mid April	5
Corn	Regular	0.04	Broadcast	Early May	5
	Ecological	0.05	Broadcast	Early May	5

density by dividing the mass of the oven-dried (105°C) soil by the volume of the core. Soil bulk density was not different among treatments and sampling dates averaged 1.34, 1.36, and 1.49 $Mg \cdot m^{-3}$ at 0 - 5, 5 - 10, and 10 - 20 cm, respectively.

Total C concentration ($g \cdot C \cdot kg^{-1}$) in the surface residue was determined by using a dry combustion C and N analyzer (LECO Corp., St Joseph, MI). Carbon content ($Mg \cdot C \cdot ha^{-1}$) in the residue was determined by multiplying dry matter weight by C concentration. The SOC concentration in soil samples was determined by using the C and N analyzer as above after grinding the samples to <0.5 mm and pretreating the soil with 5% H_2SO_3 to remove inorganic C [31]. For determining POC, 10 g soil sample was dispersed with 30 mL of 5 $g \cdot L^{-1}$ sodium hexametaphosphate by shaking for 16 h and the solution was poured through a 0.053 mm sieve [22]. The solution and particles that passed through the sieve and contained mineral associated and water soluble C were dried at 50°C for 3 to 4 d and SOC concentration was determined by using the analyzer as above after pretreating with 5% H_2SO_3. The POC concentration was determined by the difference between SOC in the whole-soil and that in the particles that passed through the sieve after correcting for the sand content.

The PCM in air-dried soils was determined by the method modified by Haney *et al.* [32]. Ten grams of soil sample was moistened with water at 50% field capacity (0.25 $m^3 \cdot m^{-3}$ [33]) and placed in a 1 L jar containing beakers with 2 mL of 0.5 M NaOH to trap evolved CO_2 and 20 mL of water to maintain high humidity. Soils were incubated in the jar at 21°C for 10 d. At 10 d, the beaker containing NaOH was removed from the jar and PCM concentration was determined by measuring CO_2 absorbed in NaOH, which was back-titrated with 1.5 M $BaCl_2$ and 0.1 M HCl. The moist soil used for determining PCM was subsequently used for determining MBC by the modified fumigation-incubation method for air-dried soils [34]. This container was incubated twice because MBC determination required moist-soil. The method also required mineralizable C to be flushed out during the first incubation [34]. The moist soil was fumigated with ethanol-free chloroform for 24 h and placed in a 1 L jar containing beakers with 2 mL of 0.5 M NaOH and 20 mL water. As with PCM, fumigated moist soil was incubated for 10 d and CO_2 absorbed in NaOH was back-titrated with $BaCl_2$ and HCl. The MBC concentration was calculated by dividing the amount of CO_2-C absorbed in NaOH by a factor of 0.41 [35] without subtracting the values from the non-fumigated control [34].

The SOC, POC, PCM, and MBC contents (Mg or $kg \cdot C \cdot ha^{-1}$) at 0 - 5, 5 - 10, and 10 - 20 cm depths were calculated by multiplying their concentrations ($g \cdot C \cdot kg^{-1}$) by bulk density at each depth and thickness of the soil layer. Total content at 0 - 20 cm was determined by summing the contents at 0 - 5, 5 - 10, and 10 - 20 cm.

2.4. Data Analysis

Data for crop biomass, surface residue, and soil C fractions at each depth were analyzed using the MIXED procedure of SAS, with year considered as the fixed effect [36]. Tillage, crop rotation, and cultural practice were considered as fixed effects and replication and tillage × replication as random effects. Since each phase of the crop rotation was present in every year, data for phases were averaged within a rotation and used for a crop rotation for the analysis. Means were separated by using the least square means test when treatments and interactions were significant [36]. Statistical significance was evaluated at $P \leq 0.05$, unless otherwise stated.

3. Results and Discussion

3.1. Crop Biomass Yield

Biomass (stems and leaves) yield varied between crop rotations from 2004 to 2007 (**Table 2**), as year and crop rotation × year interaction were significant ($P \leq 0.05$) but tillage, cultural practice, and their interactions with crop rotation and year were not (data not shown). Biomass, averaged across tillage and cultural practices, was greater in W-B-C-P than in CW in 2004 and 2005. Biomass was greater in 2004 and 2006 than in 2005 in CW and greater in 2004 than in other years in W-B-C-P. Averaged across tillage, crop rotations, and cultural practices, biomass was greater in 2004 than in other years.

The greater biomass in W-B-C-P than in CW in 2004 and 2005 was due to higher corn yield. During these years, biomass in corn was two to three times greater than that in spring wheat, barley hay, and pea. Biomass was lower in 2005 than in other years, probably due to less uniform distribution and/or below-average precipitation. This was especially true for spring wheat in CW. Precipitation amounting to 115 mm fell in June but only 2 mm in September in 2005 compared to 68-yr average (72 mm in June and 34 mm in September) and other years. The growing season (April-October) precipitation was 11 mm greater in 2006 but 10 mm lower in 2005 than the normal precipitation (311 mm). Since dryland crop yields depend on the amount and distribution of precipitation due to the absence of irrigation, large variations in crop yields can occur from year to year due to variations in growing season precipitation [3,5,12]. Tillage and cultural practice did not influence biomass yield. Several researchers [3,5,12] also found that tillage did not influence dryland crop biomass yield in the northern Great Plains. Since biomass, being a source of C input, was returned to the soil during grain harvest, the amount

Table 2. Effect of crop rotation on crop biomass (stems and leaves) returned to the soil from 2004 to 2007.

Year	Crop rotation[a]		
	CW ($Mg \cdot ha^{-1}$)	W-B-C-P ($Mg \cdot ha^{-1}$)	Mean ($Mg \cdot ha^{-1}$)
2004	5.40[ab]B[c]	6.71[a]A	6.06[a]
2005	3.83[b]B	5.32[b]A	4.58[c]
2006	5.61[a]A	4.77[b]A	5.19[b]
2007	5.10[ab]A	5.31[b]A	5.21[b]

[a]Crop rotations are CW, continuous spring wheat; and W-B-C-P, spring wheat-barley hay-corn-pea; [b]Numbers followed by different lowercase letters within a column (among years) are significantly different at $P \leq 0.05$ by the least square means test; [c]Numbers followed by different uppercase letters within a row (between crop rotations) are significantly different at $P \leq 0.05$ by the least square means test.

of biomass as influenced by crop rotation can influence surface residue and soil C fractions, as discussed below.

3.2. Soil Surface Residue Carbon

Residue incorporation into the soil and mineralization due to tillage, cultural practice, and time resulted in significant tillage, year, and tillage × cultural practice interaction in surface residue amount and C content (**Table 3**). Surface residue amount, averaged across crop rotations and years, was greater in NT with the regular and ecological practices than in CT with the regular practice. Surface residue C was greater in NT than in CT in the regular practice. Averaged across crop rotations, cultural practices, and years, surface residue amount and C content were 27% to 30% greater in NT than in CT. Averaged across treatments, surface residue amount and C content declined from 51% to 53% from autumn 2007 to spring 2008. Surface residue between crop rotations did not vary, possibly because mean biomass across rotations was not different (**Table 3**).

The increased surface residue amount and C content in NT with the regular practice could be a result of reduced mineralization due to surface placement of the residue. In contrast, incorporation of the residue into the soil to a greater depth could have increased mineralization, thereby reducing surface residue amount and C content in CT with the regular practice [37]. Although crop biomass yield was not influenced by tillage and cultural practice, biomass was somewhat lower in the regular than in the ecological practice in CT, which could have resulted in lower surface residue amount and C content in this treatment. Surface residue amount and C content declined from autumn 2007 to spring 2008 probably due to increased mineralization and/or loss of residue due to the actions of wind and water. During this period, more than half of surface residue amount and C were lost, indicating that substantial losses of residue C can occur even during the cold weather in the winter. Increased surface residue amount can reduce soil erosion, increase C sequestration, and sustain dryland crop yields due to increased soil water conservation [37,38]. Since C stored in surface residue form an important part of C cycle, management practices that increase surface residue C can also influence soil C fractions [37].

3.3. Soil Organic Carbon

Differences in crop residue quality and quantity and C mineralization rates among treatments also resulted in significant tillage × crop rotation × cultural practice interaction on SOC at 5 - 10, 10 - 20, and 0 - 20 cm (**Table 4**). At 10 - 20 and 0 - 20 cm, cultural practice, year, and tillage × crop rotation and tillage × cultural practice interactions were significant.

At 5 - 10 cm, SOC, averaged across years, was greater in NT with CW and the ecological practice or in NT with W-B-C-P and the regular practice than in CT with CW and the ecological practice (**Table 4**). At 10 - 20 cm, SOC was greater in NT with W-B-C-P and the regular practice than in NT with W-B-C-P and the ecological practice or in CT with CW and the regular and ecological practices. At 0 - 20 cm, SOC was greater in NT with CW and the ecological practice or in NT with W-B-C-P and the regular practice than in NT with W-B-C-P and the ecological practice or in CT with CW and the ecological practice. Averaged across tillage, crop rotations, and years, SOC at 10 - 20 and 0 - 20 cm was 2.4% to 4.3% greater with the regular than with the ecological practice. Averaged across treatments, SOC at 10 - 20 and 0 - 20 cm reduced by 3.7% to 4.8%, respectively from autumn 2007 to spring 2008.

Changes in SOC due to treatments occurred mostly at the subsurface than in the surface soil. This was kind of a surprise given the residue accumulation at the surface due to NT and incorporation into the soil due to CT occurred to a depth <10 cm. While the coefficient of variation for SOC ranged from 8.6% in 2007 to 9.1% in 2008 at 0 - 5 cm, it ranged from 8.5% in 2007 to 10.0% in 2008 at 5 - 10 and 10 - 20 cm. Since the variability in SOC at the surface and subsurface layers were similar, it could be possible that differences in the amount of residue incorporation into the soil due to CT and/or root growth among treatments resulted in SOC changes more in the subsurface than in the surface soil.

The greater SOC at 5 - 10, 10 - 20, and 0 - 20 cm in

Table 3. Effects of tillage and cultural practice on soil surface residue amount and C content in 2007 and 2008.

Tillage[a]	Cultural practice[b]	Year	Soil surface residue	
			Amount ($Mg \cdot ha^{-1}$)	C content ($Mg \cdot C \cdot ha^{-1}$)
NT	Regular		4.04	1.50
	Ecological		3.71	1.36
CT	Regular		2.72	1.03
	Ecological		3.21	1.22
LSD (0.05)			0.99	0.40
Means				
NT			3.87a[c]	1.43a
CT			2.97b	1.13b
		2007	4.64a	1.72a
		2008	2.26b	0.81b
Significance[d]				
Tillage (T)			*	*
Crop rotation (C)			NS	NS
T × C			NS	NS
Cultural practice (P)			NS	NS
T × P			*	*
C × P			NS	NS
T × C × P			NS	NS
Year (Y)			***	***
T × Y			NS	NS
C × Y			NS	NS
T × C × Y			NS	NS
P × Y			NS	NS
T × P × Y			NS	NS
C × P × Y			NS	NS
T × C × P ×Y			NS	NS

[a]Tillage practices are CT, conventional tillage; and NT, no-tillage; [b]See Table 2 for the description of cultural practices; [c]Numbers followed by different letters within a column in either tillage practice or year are significantly different at $P \leq 0.05$ by the least square means test; [d]Significance levels: $^{*}P \leq 0.05$; $^{**}P \leq 0.01$; $^{***}P \leq 0.001$; NS, Not significant.

NT with CW and the ecological practice or in NT with W-B-C-P and the regular practice than in CT with CW and the regular and ecological practices could be the results of increased crop residue returned to the soil, followed by decreased mineralization of residue and soil organic matter due to reduced soil disturbance. Crop biomass returned to the soil was greater in W-B-C-P than in CW in 2004 and 2005 (**Table 2**). While CT can reduce dryland soil C storage compared to NT [5,9,25], increased residue returned to the soil due to diversified crop rotation and increased cropping intensity can increase SOC levels [3-5]. The lower SOC at 10 - 20 and 0 - 20 cm in the ecological than in the regular practice was probably due to increased organic matter mineralization as a result of longer fallow period from delayed planting. The lower bulk density in the ecological practice also may have partly contributed to lower SOC in this treatment. During the fallow period in the winter, SOC did not change at 0 - 5 and 5 - 10 cm but declined at 10 - 20 cm, possibly a result of absence of crop residue input and/or increased mineralization of SOC and residue in the subsurface compared to the surface layers. Because of differences in SOC levels from autumn to spring at the subsurface soil layer, it may be a good idea to collect soil samples at the same time of the year for SOC analysis if deeper samples >10 cm are collected.

The SOC level at 0 - 20 cm at the initiation of the experiment in April 2004 was 30.5 $Mg \cdot ha^{-1}$. The SOC levels at 0 - 20 cm between 2007 and 2008 ranged from 28.0 $Mg \cdot C \cdot ha^{-1}$ in CT with CW and the ecological practice to 31.2 $Mg \cdot C \cdot ha^{-1}$ in NT with CW and the ecological practice (**Table 4**). This suggests that SOC decreased by

Table 4. Effects of tillage, crop rotation, and cultural practice on soil organic C (SOC) content at the 0 - 20 cm depth in 2007 and 2008.

Tillage[a]	Crop rotation[b]	Cultural practice[c]	Year	SOC content (Mg·C·ha^{-1})			
				0 - 5 cm	5 - 10 cm	10 - 20 cm	0 - 20 cm
NT	CW	Regular		8.1	7.2	14.8	30.1
		Ecological		8.7	7.9	14.6	31.2
	W-B-C-P	Regular		8.2	7.7	15.1	31.0
		Ecological		8.2	7.2	13.2	28.6
CT	CW	Regular		8.4	7.1	14.0	29.5
		Ecological		8.0	6.8	13.3	28.1
	W-B-C-P	Regular		8.1	7.4	14.3	29.8
		Ecological		8.0	7.3	14.6	29.9
LSD (0.05)				NS	0.7	1.0	2.1
Mean							
		Regular		8.2[ad]	7.3[a]	14.5[a]	30.0[a]
		Ecological		8.2[a]	7.3[a]	13.9[b]	29.4[b]
			2007	8.3[a]	7.4[a]	14.6[a]	30.3[a]
			2008	8.1[a]	7.1[a]	13.9[b]	29.1[b]
Significance[e]							
Tillage (T)				NS[b]	NS	NS	NS
Crop rotation (C)				NS	NS	NS	NS
T × C				NS	NS	**	*
Cultural practice (P)				NS	NS	***	**
T × P				NS	NS	***	*
C × P				NS	NS	NS	NS
T × C × P				NS	*	**	***
Year (Y)				NS	NS	***	**
T × Y				NS	NS	NS	NS
C × Y				NS	NS	NS	NS
T × C × Y				NS	NS	NS	NS
P × Y				NS	NS	NS	NS
T × P × Y				NS	NS	NS	NS
C × P × Y				NS	NS	NS	NS
T × C × P ×Y				NS	NS	NS	NS

[a]Tillage practices are CT, conventional tillage; and NT, no-tillage; [b]Crop rotations are CW, continuous spring wheat; and W-B-C-P, spring wheat-barley hay-corn-pea.; [c]See Table 2 for the description of cultural practices; [d]Numbers followed by different letters within a column in either cultural practice or year are significantly different at $P \leq 0.05$ by the least square means test; [e]Significance levels: $^{*}P \leq 0.05$; $^{**}P \leq 0.01$; $^{***}P \leq 0.001$; NS, Not significant.

500 kg·C·ha^{-1}·yr^{-1} in CT with CW and the ecological practice but increased by 140 kg·C·ha^{-1}·yr^{-1} in NT with CW and the ecological practice. The loss in SOC from 2004 to 2008 in CT with CW and the ecological practice is larger than the reported value 141 kg·C·ha^{-1}·yr^{-1} in CT with crop-fallow but the gain in NT with CW and the ecological practice is smaller than 233 kg·C·ha^{-1}·yr^{-1} in NT with continuous cropping under dryland farming system in the northern Great Plains [9]. Possibly, the experiment needs to be continued for more than five years to obtain a stable C sequestration rate, since several researchers [3,37,39] have concluded the need for long-term experiment on C sequestration to observe the effects of management practices on soil C storage under dryland cropping systems in the northern Great Plains. This is because lower crop biomass returned to the soil due to limited precipitation compared with humid regions and cold weather reduces turnover rate of plant C into SOC [3,5].

3.4. Soil Particulate Organic Carbon

Similar to SOC, POC content varied between cultural practices at 5 - 10 and 10 - 20 cm and between years at all depths (**Table 5**). Interactions were significant for tillage × year at 5 - 10 cm, crop rotation × year at 10 - 20 cm, and tillage × crop rotation × year at 0 - 5, 10 - 20, and 0 - 20 cm.

The POC at 0 - 5, 10 - 20, and 0 - 20 cm, averaged across cultural practices, was greater in NT with CW than in other tillage and crop rotations in 2007 (**Table 5**). In 2008, POC at 10 - 20 cm was greater in W-B-C-P than in CW in CT but at other depths, it was not different between tillage and crop rotations. From autumn 2007 to spring 2008, POC at 0 - 5 and 0 - 20 cm declined rapidly in NT with CW and CT with W-B-C-P compared to other treatments. Similarly, during this period, POC at 10 - 20 cm declined more in NT with CW or W-B-C-P than in other treatments. Averaged POC across treatments, however, declined from 16% to 18% at all depths from autumn 2007 to spring 2008. Averaged across tillage, crop rotations, and years, POC at 5 - 10 and 10 - 20 cm was greater in the regular than in ecological practice.

As with SOC, greater POC at 0 - 5, 10 - 20, and 0 - 20 cm in NT with CW than in other tillage and crop rotations in 2007 could be the results of an undisturbed soil

Table 5. Effects of tillage, crop rotation, and cultural practice on soil particulate organic C (POC) content at the 0 - 20 cm depth in 2007 and 2008.

Year	Tillage[a]	Crop rotation[b]	Cultural practice[c]	POC content (Mg C·ha^{-1})			
				0 - 5 cm	5 - 10 cm	10 - 20 cm	0 - 20 cm
2007	NT	CW		2.49	1.77	3.56	7.82
		W-B-C-P		2.04	1.57	2.96	6.57
	CT	CW		2.00	1.58	2.90	6.48
		W-B-C-P		2.14	1.69	2.91	6.74
2008	NT	CW		1.84	1.50	2.52	5.86
		W-B-C-P		1.86	1.52	2.55	5.93
	CT	CW		1.86	1.30	2.28	5.44
		W-B-C-P		1.64	1.26	2.76	5.66
LSD (0.05)				0.28	NS	0.48	0.82
Mean							
2007				2.16[ad]	1.65[a]	3.08[a]	6.89[a]
2008				1.80[b]	1.39[b]	2.53[b]	5.72[b]
			Regular	1.99[a]	1.55[a]	2.88[a]	6.42[a]
			Ecological	1.98[a]	1.49[b]	2.73[b]	6.20[a]
Significance[e]							
Tillage (T)				NS	NS	NS	NS
Crop rotation (C)				NS	NS	NS	NS
T × C				NS	NS	NS	NS
Cultural practice (P)				NS	*	*	NS
T × P				NS	NS	NS	NS
C × P				NS	NS	NS	NS
T × C × P				NS	NS	NS	NS
Year (Y)				***	***	***	***
T × Y				NS	*	NS	NS
C × Y				NS	NS	*	NS
T × C × Y				**	NS	*	*
P × Y				NS	NS	NS	NS
T × P × Y				NS	NS	NS	NS
C × P × Y				NS	NS	NS	NS
T × C × P ×Y				NS	NS	NS	NS

[a]Tillage practices are CT, conventional tillage; and NT, no-tillage; [b]Crop rotations are CW, continuous spring wheat; and W-B-C-P, spring wheat-barley hay-corn-pea; [c]See **Table 2** for the description of cultural practices; [d]Numbers followed by different letters within a column in either year or cultural practice are significantly different at $P \leq 0.05$ by the least square means test; [e]Significance levels: $^{*}P \leq 0.05$; $^{**}P \leq 0.01$; $^{***}P \leq 0.001$; NS, Not significant.

and surface placement of residue that reduced the mineralization of residue and coarse soil organic matter. The greater POC level with this treatment was, however, short lived because greater rate of decline in its level occurred in NT with CW than in other treatments from 2007 to 2008, thereby resulting in non-significant difference in POC between treatments at all depths, except at 10 - 20 cm, in 2008. This indicates that coarse fraction of soil organic matter can be enriched by using no-tillage with continuous spring wheat compared to conventional tillage with monocropping or crop rotation but the level was reduced from autumn to the following spring, probably by increased mineralization. The proportion of SOC in POC at 0 - 20 cm was also greater in NT with CW (226 $g \cdot POC \cdot kg^{-1}$ SOC) than in other tillage and crop rotations (207 to 208 $g \cdot POC \cdot kg^{-1}$ SOC) in 2007 but was not significantly different among treatments in 2008. This suggests that coarse organic matter was mineralized at a rapid rate compared to whole soil organic matter, a case similar to that observed by other researchers [5,18, 22,40].

The lower POC in the ecological than in the regular practice at the subsurface layers (5 - 10 and 10 - 20 cm) could be the results of increased mineralization of coarse organic matter due to longer fallow period and/or reduced residue input due to delayed planting, a case similar to that observed for SOC. As with SOC, changes in POC occurred more at the subsurface than in surface layers. Probably differences in crop root growth and/or various levels of residue incorporation to a greater depth due to tillage (e.g. greater spring wheat stubble height [**Table 2**] reduced the amount of residue incorporation into the soil in the ecological than in the regular practice) influenced POC levels between the two cultural practices more in the subsurface than in the surface soil.

3.5. Soil Potential Carbon Mineralization

The PCM content varied between years at all depths, except at 10 - 20 cm. (**Table 6**). Interactions were significant for tillage × year at all depths and tillage × cultural practice × year at 5 - 10, 10 - 20, and 0 - 20 cm. Crop rotation and its interaction with other treatments were not significant.

The PCM at 5 - 10 cm, averaged across crop rotations, was greater in NT with the regular practice than in CT with the ecological practice in 2007 (**Table 6**). At 10 - 20 and 0 - 20 cm, PCM was greater in NT with the regular practice than in CT with the regular and ecological practices. In 2008, PCM at 0 - 20 cm was greater in NT with the regular practice than in CT with the regular practice. From autumn 2007 to spring 2008, PCM at 5 - 10 and 10 - 20 cm declined rapidly in NT with the regular practice compared to other treatments, thereby resulting in non-significant difference among tillage and cultural practices in 2008. Averaged across crop rotations and cultural practices, PCM was greater in NT than in CT at all depths in 2007 and at 0 - 20 cm in 2008. In contrast to SOC and POC, PCM at all depths, averaged across treatments, declined from 17% to 34% from autumn 2007 to spring 2008.

The greater PCM in NT with the regular practice than in CT with the regular and ecological practices was probably a result of greater substrate availability, since SOC and POC were greater in this treatment (**Tables 4** and **5**). This is consistent with those reported by several researchers [18,21,40] who found that greater substrate availability increased C mineralization and microbial activity in NT than in CT. Increased C mineralization due to residue incorporation into the soil, followed by longer duration of fallow probably reduced PCM in CT with the regular and ecological practices. As with SOC and POC, greater level of PCM in NT with the regular practice was short lived, as substrate availability declined at a rapid rate with this treatment from autumn 2007 to spring 2008. Since PCM declined both at the surface and subsurface soil layers at higher rates during this period in contrast to slower rate of decline for SOC and POC at the subsurface layers, PCM could be a sensitive indicator of changes in soil organic matter with management practice and time, a case similar to those observed by various researchers [20,21]. Although variability was higher in PCM (CV 28% - 55%) than in SOC and POC (CV 7% - 25%), small changes in SOC and POC levels due to management practices and mineralization with time resulted in large changes in PCM level.

3.6. Soil Microbial Biomass Carbon

Like SOC and POC, changes in MBC occurred at the subsurface layers, with significant interactions for tillage × crop rotation × year at 0 - 20 cm, cultural practice × year at 5 - 10 cm, and tillage × crop rotation × cultural practice × year at 10 - 20 and 0 - 20 cm (**Table 7**). The MBC, however, varied significantly with crop rotation and years at all depths. The variability was lower for MBC (CV 15% - 40%) than for PCM but greater than for SOC and POC.

In 2007, MBC at 10 - 20 cm was greater in NT with CW and the regular practice than in NT with CW and the ecological practice, NT with W-B-C-P and the ecological practice, or in CT with CW and the regular practice (**Table 7**). At 0 - 20 cm, MBC was greater in NT with CW and the regular practice than in NT with CW and the ecological practice or in CT with CW and the regular practice. From 2007 to 2008, MBC at 10 - 20 and 0 - 20 cm declined rapidly in NT with CW and the regular practice and CT with CW or W-B-C-P with the ecological practice compared to other treatments, thereby resulting in non-significant difference between treatments in 2008.

Table 6. Effects of tillage and cultural practice on soil potential C mineralization (PCM) content at the 0 - 20 cm depth in 2007 and 2008.

Year	Tillage[a]	Cultural practice[b]	PCM content ($kg \cdot C \cdot ha^{-1}$)			
			0 - 5 cm	5 - 10 cm	10 - 20 cm	0 - 20 cm
2007	NT	Regular	39	35	78	152
		Ecological	42	32	60	134
	CT	Regular	33	25	47	105
		Ecological	28	23	50	101
2008	NT	Regular	24	18	43	85
		Ecological	24	19	52	95
	CT	Regular	31	20	52	103
		Ecological	30	19	50	99
LSD (0.05)			NS	12	28	48
Mean						
2007			36[ac]	29[a]	59[a]	124[a]
2008			27[b]	19[b]	49[a]	95[b]
Significance[d]						
Tillage (T)			NS	NS	NS	NS
Crop rotation (C)			NS	NS	NS	NS
T × C			NS	NS	NS	NS
Cultural practice (P)			NS	NS	NS	NS
T × P			NS	NS	NS	NS
C × P			NS	NS	NS	NS
T × C × P			NS	NS	NS	NS
Year (Y)			*	*	NS	*
T × Y			*	*	*	**
C × Y			NS	NS	NS	NS
T × C × Y			NS	NS	NS	NS
P × Y			NS	NS	NS	NS
T × P × Y			NS	*	*	*
C × P × Y			NS	NS	NS	NS
T × C × P ×Y			NS	NS	NS	NS

[a]Tillage practices are CT, conventional tillage; and NT, no-tillage; [b]See **Table 2** for the description of cultural practices; [c]Numbers followed by different letters within a column in a year are significantly different at $P \leq 0.05$ by the least square means test; [d]Significance levels: $^{*}P \leq 0.05$; $^{**}P \leq 0.01$; $^{***}P \leq 0.001$; NS, Not significant.

Averaged across tillage, cultural practices, and years, MBC at all depths was 11% to 14% greater in W-B-C-P than in CW. Similar to PCM, MBC at all depths, averaged across treatments, declined from 22% to 24% from autumn 2007 to spring 2008.

Greater substrate availability due to increased SOC and POC probably increased MBC in NT with CW and the regular practice at the subsurface soil in 2007. As the substrate availability decreased, MBC also declined from 2007 to 2008. Similar to PCM, MBC decline occurred at all depths, with a greater rate of decline than in SOC and POC. This verifies that MBC is also an active fraction of soil organic matter [20,21]. Diversified crop rotation clearly increased MBC in W-B-C-P than in CW, which was especially noted in CT in 2007. Increased amount of crop residue returned to the soil (**Table 2**) probably increased MBC in W-B-C-P compared to CW. Several researchers [4,26,40] also reported that MBC increased with diversified crop rotation compared to monocropping due to increased crop residue returned to the soil.

4. Conclusion

Differences in tillage, crop rotations, and cultural practices among treatments influenced crop biomass (stems and leaves) returned to the soil, surface residue, and soil C fractions. In contrast to our hypothesis, no-tillage with monocropping and the regular cultural practice increased

Table 7. Effects of tillage, crop rotation, and cultural practice on soil microbial biomass C (MBC) content at the 0 - 20 cm depth in 2007 and 2008.

Year	Tillage[a]	Cultural practice[b]	Crop rotation[c]	MBC content (kg·C·ha^{-1})			
				0 - 5 cm	5 - 10 cm	10 - 20 cm	0 - 20 cm
2007	NT	Regular	CW	99	79	196	374
			W-B-C-P	94	73	167	334
		Ecological	CW	101	72	115	288
			W-B-C-P	106	79	153	338
	CT	Regular	CW	82	54	120	256
			W-B-C-P	101	79	180	360
		Ecological	CW	84	61	165	310
			W-B-C-P	111	85	184	380
2008	NT	Regular	CW	59	58	114	231
			W-B-C-P	85	57	135	277
		Ecological	CW	93	52	108	253
			W-B-C-P	79	60	114	253
	CT	Regular	CW	67	61	129	257
			W-B-C-P	80	65	121	266
		Ecological	CW	67	52	122	241
			W-B-C-P	78	56	128	262
LSD (0.05)				NS	NS	38	59
Mean							
2007				97[ad]	73[a]	160[a]	330[a]
2008				76[b]	57[b]	122[b]	255[b]
			CW	81[b]	61[b]	134[b]	276[b]
			W-B-C-P	92[a]	68[a]	148[a]	308[a]
Significance[e]							
Tillage (T)				NS	NS	NS	NS
Crop rotation (C)				*	*	*	**
T × C				NS	NS	NS	NS
Cultural practice (P)				NS	NS	NS	NS
T × P				NS	NS	**	NS
C × P				NS	NS	NS	NS
T × C × P				NS	NS	NS	NS
Year (Y)				***	***	***	***
T × Y				NS	NS	NS	NS
C × Y				NS	NS	NS	NS
T × C × Y				NS	NS	NS	*
P × Y				NS	*	NS	NS
T × P × Y				NS	NS	NS	NS
C × P × Y				NS	NS	NS	NS
T × C × P ×Y				NS	NS	*	*

[a]Tillage practices are CT, conventional tillage; and NT, no-tillage; [b]See Table 2 for the description of cultural practices; [c]Crop rotations are CW, continuous spring wheat; and W-B-C-P, spring wheat-barley hay-corn-pea.; [d]Numbers followed by different letters within a column in either year or crop rotation are significantly different at $P \le 0.05$ by the least square means test; [d]Significance levels: $^{*}P \le 0.05$; $^{**}P \le 0.01$; $^{***}P \le 0.001$; NS, Not significant.

surface residue and soil C fractions but diversified crop rotation increased microbial biomass C. The increases occurred at subsurface soil layers for non-labile C fractions but occurred at all layers for labile C fractions. However, the increases were short lived as C fractions declined rapidly compared to other treatments from autumn 2007 to spring 2008, probably a result of increased mineralization during the fallow period in the winter. This resulted in similar levels of soil C fractions among treatments in 2008. The ecological cultural practice did not increase surface residue and soil C fractions compared to the regular practice probably due to longer fallow period and continued C and N mineralization. For increasing dryland C storage and soil quality, no-tillage with diversified crop rotation and/or the regular cultural practice can be used compared to conventional tillage with monocropping and/or the ecological practice in the northern Great Plains, USA. For obtaining stable C levels and sequestration rates, however, long-term experiment than the present four years of study may be needed because of declining levels of surface residue and C fractions from autumn to spring. As a result, surface residue and soil samples would be preferable to collect at the same time (e.g. either after crop harvest in the autumn or before planting in the spring) of the year for determinations of C fractions under dryland cropping systems.

5. Acknowledgements

We sincerely acknowledge the help provided by Joy Barsotti, Chris Russell, and Johnny Rieger for collection and analysis of surface residue and soil samples in the field and laboratory and Michael Johnson and Mark Gaffri for management of field plots for tilling, planting, herbicide and pesticide application, and harvest.

REFERENCES

[1] R. Lal, J. M. Kimble and B. A. Stewart, "World Soils as a Source or Sink for Radiatively-Active Gases," In: R. Lal, Ed., *Soil Management and Greenhouse Effect*, *Advances in Soil Science*, CRC Press, Boca Raton, 1995, pp. 1-8.

[2] K. Paustian, G. P. Robertson and E. T. Elliott, "Management Impacts on Carbon Storage and Gas Fluxes in Mid-Latitudes Cropland," In: R. Lal, Ed., *Soils and Global Climate Change, Advances in Soil Science*, CRC Press, Boca Raton, USA, 1995, pp. 69-83.

[3] A. D. Halvorson, G. A. Peterson and C. A., Reule, "Tillage System and Crop Rotation Effects on Dryland Crop Yields and Soil Carbon in the Central Great Plains," *Agronomy Journal*, Vol. 94, No. 6, 2002, pp. 1429-1436. doi:10.2134/agronj2002.1429

[4] L. A. Sherrod, G. A. Peterson, D. G. Westfall and L. R. Ahuja, "Cropping Intensity Enhances Soil Organic Carbon and Nitrogen in a No-Till Agroecosystem," *Soil Science Society of America Journal*, Vol. 67, No. 5, 2003, pp. 1533-1543. doi:10.2136/sssaj2003.1533

[5] U. M. Sainju, T. Caesar-TonThat, A. W. Lenssen, R. G. Evans and R. Kolberg, "Long-Term Tillage and Cropping Sequence Effects on Dryland Residue and Soil Carbon Fractions," *Soil Science Society of America Journal*, Vol. 71, No. 6, 2007, pp. 1730-1739. doi:10.2136/sssaj2006.0433

[6] P. E. Rasmussen, R. R. Allmaras, C. R. Rhoade and N. C. Roager Jr., "Crop Residue Influences on Soil Carbon and Nitrogen in a Wheat-Fallow System," *Soil Science Society of America Journal*, Vol. 44, No. 3, 1980, pp. 596-600. doi:10.2136/sssaj1980.03615995004400030033x

[7] G. A. Peterson, A. D. Halvorson, J. L. Havlin, O. R. Jones, D. G. Lyon and D. L. Tanaka, "Reduced Tillage and Increasing Cropping Intensity in the Great Plains Conserve Soil Carbon," *Soil and Tillage Research*, Vol. 47, No. 3-4, 1998, pp. 207-218. doi:10.1016/S0167-1987(98)00107-X

[8] A. Bauer and A. L. Black, "Quantification of the Effect of Soil Organic Matter Content on Soil Productivity," *Soil Science Society of America Journal*, Vol. 58, No. 1, 1994, pp. 185-193. doi:10.2136/sssaj1994.03615995005800010027x

[9] A. D. Halvorson, B. J. Wienhold and A. L. Black, "Tillage, Nitrogen, and Cropping System Effects on Soil Carbon Sequestration," *Soil Science Society of America Journal*, Vol. 66, No. 3, 2002, pp. 906-912. doi:10.2136/sssaj2002.0906

[10] T. O. West and W. M. Post, "Soil Organic Carbon Sequestration Rates by Tillage and Crop Rotation: A Global Data Analysis," *Soil Science Society of America Journal*, Vol. 66, No. 6, 2002, pp. 1930-1946. doi:10.2136/sssaj2002.1930

[11] A. W. Lenssen, G. D. Johnson and G. R. Carlson, "Cropping Sequence and Tillage System Influences Annual Crop Production and Water Use in Semiarid Montana," *Field Crops Research*, Vol. 100, No. 1, 2007, pp. 32-43. doi:10.1016/j.fcr.2006.05.004

[12] U. M. Sainju, A. W. Lenssen, T. Caesar-TonThat and R. G. Evans, "Dryland Crop Yields and Soil Organic Matter as Influenced by Long-Term Tillage and Cropping Sequence," *Agronomy Journal*, Vol. 101, No. 2, 2009, pp. 243-251. doi:10.2134/agronj2008.0080x

[13] M. F. Vigil, R. A. Anderson and W. E. Beard. "Base Temperature Growing-Degree-Hour Requirements for Emergence of Canola," *Crop Science*, Vol. 37, No. 3, 1997, pp. 844-849. doi:10.2135/cropsci1997.0011183X003700030025x

[14] P. R. Miller, B. McConkey, G. W. Clayton, S. A. Brandt, J. A. Staricka, A. M. Johnston, G. P. Lafond, B. G. Schatz, D. D. Baltensperger and K. E. Neill, "Pulse Crop Adaptation in the Northern Great Plains," *Agronomy Journal*, Vol. 94, No. 2, 2002, pp. 261-272. doi:10.2134/agronj2002.0261

[15] A. W. Lenssen, J. T. Waddell, G. D. Johnson and G. R. Carlson, "Diversified Cropping Systems in Semiarid Montana: Nitrogen Use during Drought," *Soil and Tillage Research*, Vol. 94, No. 2, 2007, pp. 362-375. doi:10.1016/j.still.2006.08.012

[16] J. K. Aase and G. M Schaefer, "Economics of Tillage Practices and Spring Wheat and Barley Crop Sequence in the Northern Great Plains," *Journal of Soil and Water Conservation*, Vol. 51, No. 2, 1996, pp. 167-170.

[17] P. J. Gregory, J. S. I. Ingram, R. Anderson, R. A. Betts, V. Brovkin, T. N. Chase, P. R. Grace, A. J. Gray, N. Hamilton, T. B. Hardy, S. M. Howden, A. Jenkins, M. Meybeck, M., Olsson, I. Ortiz-Montasterio, C. A. Palm, T. W. Payn, M. Rummukainen, R. E. Schulze, M. Thiem, C. Valentin and M. J. Wikinson, "Environmental Consequences of Alternative Practices for Intensifying Crop Production," *Agricultural Ecosystem and Environment*, Vol. 88, No. 3, 2002, pp. 279-290. doi:10.1016/S0167-8809(01)00263-8

[18] A. J. Franzluebbers, F. M. Hons and D. A. Zuberer, "Soil Organic Carbon, Microbial Biomass, and Mineralizable Carbon and Nitrogen in Sorghum," *Soil Science Society of America Journal*, Vol. 59, No. 2, 1995, pp. 460-466. doi:10.2136/sssaj1995.03615995005900020027x

[19] D. F. Bezdicek, D. F. Papendick and R. Lal, "Introduction: Importance of Soil Quality to Health and Sustainable Land Management," In: J. W. Doran and A. J. Jones, Eds., *Methods of Assessing Soil Quality*, Spec. Publ. 49, Soil Science Society of America, Madison, 1996, pp. 1-18.

[20] P. G. Saffigna, D. S. Powlson, P. C. Brookes and G. A. Thomas, "Influence of Sorghum Residues and Tillage on Soil Organic Matter and Soil Microbial Biomass in an Australian Vertisol," *Soil Biology and Biochemistry*, Vol. 21, No. 6, 1989, pp. 759-765. doi:10.1016/0038-0717(89)90167-3

[21] E. Bremner and C. Van Kissel, "Plant-Available Nitrogen from Lentil and Wheat Residues during a Subsequent Growing Season," *Soil Science Society of America Journal*, Vol. 56, No. 4, 1992, pp. 1155-1160. doi:10.2136/sssaj1992.03615995005600040025x

[22] C. A. Cambardella and E. T. Elliott, "Particulate Soil Organic Matter Changes across a Grassland Cultivation Sequence," *Soil Science Society of America Journal*, Vol. 56, No. 3, 1992, pp. 777-783. doi:10.2136/sssaj1992.03615995005600030017x

[23] J. Six, E. T. Elliott and K. Paustian, "Aggregate and Soil Organic Matter Dynamics under Conventional and No-Tillage Systems," *Soil Science Society of America Journal*, Vol. 63, No. 5, 1999, pp. 1350-1358. doi:10.2136/sssaj1999.6351350x

[24] H. J. Haas, W. O. Willis and J. J. Bond, "Summer Fallow in the Western United States," *USDA Conservation Research Report Number* 17, US Government Printing Office, Washington DC, 1974, pp. 2-35.

[25] R. A. Bowman, M. F. Vigil, D. C. Nielsen and R. L. Anderson, "Soil Organic Matter Changes in Intensively Cropped Dryland Systems," *Soil Science Society of America Journal*, Vol. 63, No. 1, 1999, pp. 186-191. doi:10.2136/sssaj1999.03615995006300010026x

[26] H. H. Schomberg and O. R. Jones, "Carbon and Nitrogen Conservation in Dryland Tillage and Cropping Systems," *Soil Science Society of America Journal*, Vol. 63, No. 5, 1999, pp. 1359-1366. doi:10.2136/sssaj1999.6351359x

[27] A. L. Black and D. L. Tanaka, "A Conservation Tillage Cropping System Study in the Northern Great Plains of the United States," In: E. A. Paul, Ed., *Soil Organic Matter in Temperate Agroecosystems: Long-term Experiments in North America*, CRC Press, Boca Raton, 1997, pp. 335-342.

[28] M. H. Entz, V. S. Baron, P. M. Carr, D. W. Meyer, S. R. Smith Jr. and W. P. McCaughey, "Potential of Forages to Diversity Cropping Systems in the Northern Great Plains," *Agronomy Journal*, Vol. 94, No. 2, 2002, pp. 240-250. doi:10.2134/agronj2002.0240

[29] S. M. Strydhorst, J. R. King, K. H. Lopetinsky and K. N. Harker, "Weed Interference, Pulse Species, and Plant Density Effects on Rotational Benefits," *Weed Science*, Vol. 56, No. 2, 2008, pp. 249-258. doi:10.1614/WS-07-118.1

[30] Eastern Agricultural Research Center, "Agricultural Research Update," Regional Rep. 2, Eastern Agricultural Research Center, Montana State University, Sidney, 1997.

[31] D. W. Nelson and L. E. Sommers, "Total Carbon, Organic Carbon, and Organic Matter," In: D. L. Sparks, Ed., *Methods of Soil Analysis, Part* 3, *Chemical Method,* SSSA Book Ser. 5, Soil Science Society of America, Madison, 1996, pp. 961-1010.

[32] R. L. Haney, A. J. Franzluebbers, E. B. Porter, F. M. Hons and D. A. Zuberer, "Soil Carbon and Nitrogen Mineralization: Influence of Drying Temperature," *Soil Science Society of America Journal*, Vol. 68, No. 2, 2004, pp. 489-492. doi:10.2136/sssaj2004.0489

[33] J. L. Pikul Jr. and J.K. Aase, J. K., "Water infiltration and Storage Affected by Subsoiling and Subsequent Tillage," *Soil Science Society of America Journal*, Vol. 67, No. 3, 2003, pp. 859-866. doi:10.2136/sssaj2003.0859

[34] A. J. Franzluebbers, R. L. Haney, F. M. Hons and D. A. Zuberer, "Determination of Microbial Biomass and Nitrogen Mineralization Following Rewetting of Dried Soil," *Soil Science Society of America Journal*, Vol. 60, No. 4, 1996, pp. 1133-1139. doi:10.2136/sssaj1996.03615995006000040025x

[35] R. P. Voroney and E. A. Paul, "Determination of k_C and k_N *in situ* for Calibration of the Chloroform Fumigation-Incubation Method," *Soil Biology and Biochemistry*, Vol. 16, No. 1, 1984, pp. 9-14. doi:10.1016/0038-0717(84)90117-2

[36] R. C. Littell, G. A. Milliken, W. W. Stroup and R. R. Wolfinger, "SAS System for Mixed Models," SAS Institute Inc., Cary, 1996.

[37] U. M. Sainju, A. W. Lenssen, T. Caesar-TonThat and J. T. Waddell, "Tillage and Crop Rotation Effects on Dryland Soil and Residue Carbon and Nitrogen," *Soil Science Society of America Journal*, Vol. 70, No. 2, 2006, pp. 668-678. doi:10.2136/sssaj2005.0089

[38] D. W. Fryrear, "Soil Cover and Wind Erosion," *Transaction American Society of Agricultural Engineers*, Vol. 28, No. 3, 1985, pp. 781-784.

[39] R. A. Ortega, G. A. Peterson and D. G. Westfall, "Residue Accumulation and Changes in Soil Organic Matter As Affected by Cropping Intensity in No-Till Dryland Agroecosystems," *Agronomy Journal*, Vol. 94, No. 4, 2002, pp. 944-954. doi:10.2134/agronj2002.0944

[40] U.M. Sainju, A. W. Lenssen, T. Caesar-TonThat and J. T. Waddell, "Dryland Plant Biomass and Soil Carbon and Nitrogen Fractions on Transient Land as Influenced by Tillage and Crop Rotation," *Soil and Tillage Research*, Vol. 93, No. 2, 2007, pp. 452-461. doi:10.1016/j.still.2006.06.003

32

Evaluation of Soil Organic Carbon and Soil Moisture Content from Agricultural Fields in Mississippi

Prem B. Parajuli*, Sarah Duffy

Department of Agricultural and Biological Engineering, Mississippi State University, Starkville, USA.
Email: *pparajuli@abe.msstate.edu

ABSTRACT

Independent observation of the effects of agricultural management practices on soil organic carbon (SOC) with soil moisture content (SMC) is essential to quantify their potential relationships for sustainable ecosystems. Soil water retention studies and soil carbon stocks have been mapped in some areas worldwide. However, few studies have been conducted in the southeastern US, particularly in Mississippi. The objectives of this research study were to collect soil samples from fields chosen to be representative of the watersheds they are contained within, analyze the soil samples for carbon content and soil moisture content, and evaluate the relationship between SOC and different parameters (land use, vertical distribution, temporal distribution, and soil moisture content). Field sites were chosen based on their compositional similarity shared with the watershed as a whole in the Town Creek watershed (TCW) and Upper Pearl River watershed (UPRW) in Mississippi. Monthly soil samples from different depths (6 inch, 12 inch, and 24 inch) were collected from crop, pasture, and forest field areas. Soil samples were analyzed using bench analysis, elemental analysis, and statistical analysis. This study was able to demonstrate the SOC distribution in the soil layers across all three land uses studied. It was also shown that there does seem to be an interactive effect of parameters such as land use type, vertical distribution, and time on carbon accretion within the soil. Results of this study also determined that the near surface (6-in) layer was found to contain significantly more carbon than either the 12 inch or 24 inch layers ($p < 0.01$) across all field types. There was found to be a high degree of variability within the soil moisture data and correlation between SOC and SMC. It was found that carbon amount is not influenced by SMC but SMC could be influenced by SOC.

Keywords: Soil Organic Carbon; Soil Moisture Content; Cropland; Land Uses

1. Introduction

In a modern world, teeming with an indifferent and expanding population, it has become imperative to study how human existence impacts the physical world in hopes of understanding how we might mitigate the strain on our planet. Carbon dioxide (CO_2) is only one of the many critical variables under scrutiny given its exponential increase in the atmosphere over the last two centuries [1]. Rising CO_2 concentrations are worrisome because of the serious threat it poses to global climate and ocean pH. However, it is anthropogenic activities that have easily overwhelmed the delicate balance that exists to keep the atmospheric CO_2 concentration at innocuous levels.

One way to control and reduce greenhouse gas emission is through terrestrial carbon sequestration [2]. A plant removes CO_2 from the atmosphere through photosynthesis whereby the CO_2 is then broken down into carbon and oxygen. Oxygen is released into the atmosphere as waste while carbon is used for food and incorporated into the plant. As plants die or are harvested, the carbon-based leaves, stems, and roots decay in the soil, and the carbon becomes soil organic carbon (SOC). The SOC constitutes more than twice as much stored carbon as that of the earth's vegetation and the atmosphere combined [3-5]. The SOC is estimated to be approximately 1500 Gigatonnes (Gt) globally [3,4,6]. SOC is strongly affected by human activity and its declining presence can be attributed to land conversion, soil disturbance, and increases in agricultural activity [6]. According to the US Environmental Protection Agency carbon sequestration rates vary by vegetative species, soil type, climate, topography and management practices [5]. A study esti-

*Corresponding author.

mated that anthropogenic pressures have caused the loss of anywhere between 42 and 78 Gt of soil's original carbon content [7]. As startling as this assessment is, it also represents hope. The introduction of more judicious land management practices could have the capacity to increase SOC storage by an additional 0.4 - 1.2 Gt C/yr [6, 8]. This additional storage capacity in SOC represents the clear possibility that soils globally could considerably offset greenhouse gas emissions, namely CO_2 [6,9,10]. However, "the present ability to forecast and improve the effects of climate and land cover change depends on SOC distributions, its control, source of inputs and outputs" [11].

An input of particular interest thought to have an interaction on SOC potential is soil moisture content (SMC). The SMC is an important parameter on its own, affecting runoff, land surface energy dynamics, and root zone productivity, and biomass yield [12,13]. Improved understanding and prediction of soil moisture can have far-reaching implications for irrigation planning, agricultural management, flooding and drought prediction, water quality assessment, and climate change [13]. A study compared several other studies done on the effects of SOC on SMC and found the results to be contradictory [14]. Soil carbon stocks and soil water retention studies have been mapped in some areas worldwide. However, few studies have been conducted in the southeastern US, particularly Mississippi. There have been some independent plot level studies done to observe the effects of agricultural management practices on carbon sequestration, but other variables interacting with the amount of SOC (such as SMC) have not been studied. The aims of this study were to: 1) examine the carbon content and soil moisture content of representative soil samples collected from two watersheds in central Mississippi; and 2) evaluate the relationship between SOC and different parameters (land use, vertical distribution, temporal distribution, and soil moisture content). By choosing sites that are compositionally similar to the watershed as a whole, the sites act as micro-watersheds that offer the potential to be scaled up and modeled. Long-term soil and land management datasets, in addition to model estimates, could provide valuable resource information to land managers and policy makers.

2. Materials and Methods

2.1. Study Area

The focus areas of this study were in two agricultural watersheds: Town Creek watershed (TCW) and Upper Pearl River watershed (UPRW). Watershed areas of the TCW is about 177,500 ha (438,612 ac) and UPRW is about 360,863 ha (891,711 ac). The TCW is located within Lee, Union, and Pontotoc counties with little areas in Chickasaw, Monroe and Itawamba counties (**Figure 1**). According to climate data the mean annual precipitation for the TCW is estimated as 154 cm, mean monthly minimum temperature of −19˚C and a maximum temperature of 42˚C [15]. Surface elevation of the TCW ranges from 48 m to 241 m with about 999 farms [16]. The Town Creek starts near Sherman and drains south of Nettleton [17].

The UPRW covers about ten counties (Attala, Choctaw, Kemper, Leake, Madison, Neshoba, Newton, Noxubee, Scott, and Winston; **Figure 2**). According to climate data the mean annual precipitation for the UPRW is estimated as 142 cm, mean monthly minimum temperature of −18˚C and a maximum temperature of 41˚C in August with a mean annual temperature of 17.7˚C

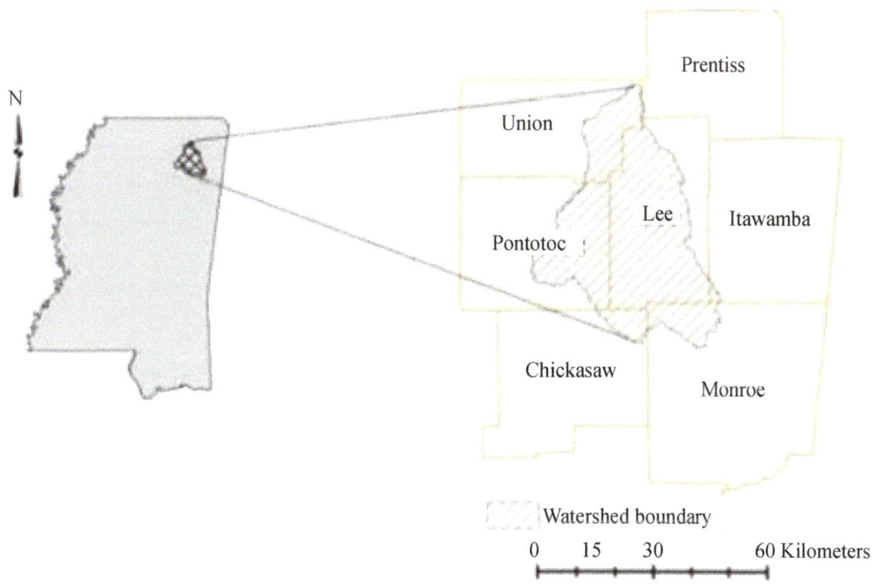

Figure 1. Location map of the Town Creek watershed in NE Mississippi.

[15]. Surface elevation of the UPRW ranges from 50 m to 221 m. The UPRW is classified as forested watershed with good livestock population [18]. The Pearl River starts near Nanih Waiya Indian mounds in Winston County and drains in the Ross Barnett Reservoir [19].

2.2. Site Selection

Study fields were chosen based on characteristics which are typical of each study watershed. Details of the locations and characteristics for each sampling point are summarized in **Table 1** and are pictured in **Figures 3** and **4**. TCW has six sampling points all in crop land use while the UPRW has six points split between forestland and pastureland. All study areas were located on privately owned property and sampling was conducted with the permission of the landowners. In the TCW two crop fields with historical corn-soybean plant rotations were chosen. Field 1 is 22 ha (54 ac) and has been harvested the last 30 years with a corn-soybean rotation. Field 2 is approximately 121 ha (300 ac) that has been planted with corn, soybeans and cotton for the last 20 years. Both properties are directly adjacent to Town Creek within the watershed.

In the UPRW a representative forest area (Field 3) and pastureland (Field 4) were chosen. Field 3 is a 2 ha (5 ac) pine plantation that was planted for the first time five years ago with 0.2 ha (0.5 ac) harvested in summer 2011. Previous to the pines, the field served as an open pasture. Field 4 is approximately 6 ha (15 ac) and currently has 10 head of cattle and one donkey that use the area as their primary grazing location.

Three points were chosen arbitrarily at each field location and identified as Point 1 through Point 12. In Fields 1, 2 and 3, the sampling area considered for the location of the points was limited to areas with close proximity to the field entrance due to the potential issues of accessing the points during peak crop growth.

2.3. Soil Physical and Hydraulic Properties

The physical and hydraulic properties of each field are shown in **Table 2**. Fields 1 and 2 in the TCW consist of Catalpa soil (MUID MS117). The Catalpa series, identified in Oktibbeha County, Mississippi in 1907, consists of deep, somewhat poorly drained to moderately well-drained silty clay loam and silty clay soils [21]. Slopes range from 0 to 3 percent, and the soil ranges from slightly acid to moderately alkaline. Most areas that consist of this soil series have been cleared and are used for growing pasture, hay, and row crops such as cotton, corn, and soybeans [21]. Determination of this soil type was based on data from the US General Soil Map (STATSGO2) [22].

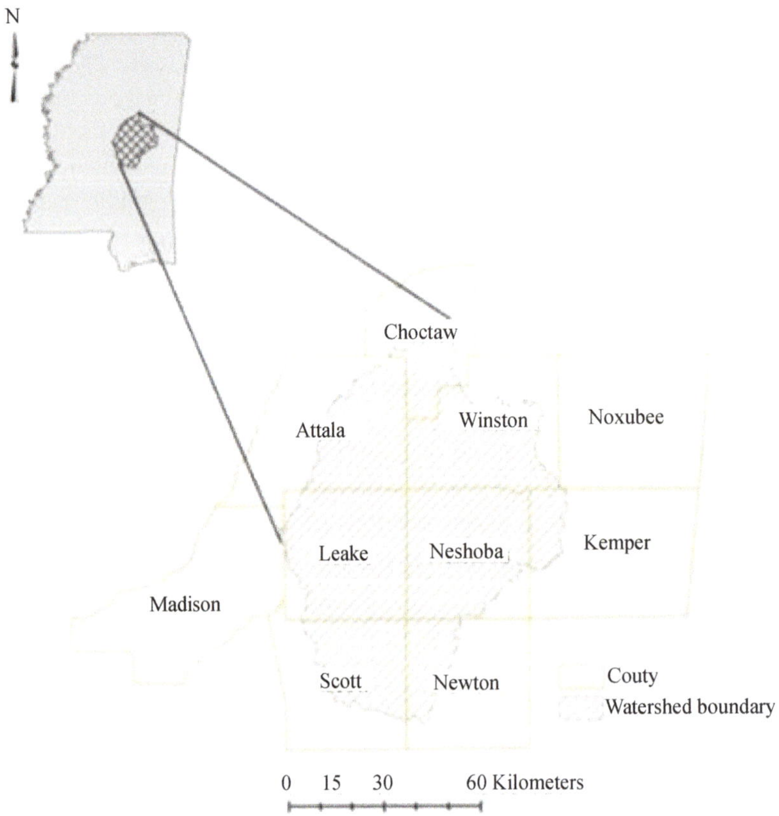

Figure 2. Location map of the Upper Pearl River watershed in central Mississippi.

Table 1. Locations of sampling points.

Point	Watershed	Latitude	Longitude	Land use
1	TCW	34°13'51.25"N	88°41'34.54"W	Crop
2	TCW	34°13'51.77"N	88°41'37.28"W	Crop
3	TCW	34°13'48.35"N	88°41'33.80"W	Crop
4	TCW	34°11'55.33"N	88°40'57.31"W	Crop
5	TCW	34°11'56.16"N	88°40'58.15"W	Crop
6	TCW	34°11'56.54"N	88°40'56.38"W	Crop
7	UPRW	32°43'2.41"N	89°33'27.02"W	Forest
8	UPRW	32°43'2.86"N	89°33'25.54"W	Forest
9	UPRW	32°43'0.25"N	89°33'26.28"W	Forest
10	UPRW	32°43'0.34"N	89°33'07.32"W	Pasture
11	UPRW	32°43'9.80"N	89°33'31.42"W	Pasture
12	UPRW	2°43'51.43"N	89°33'29.70"W	Pasture

Fields 1 Fields 2

Figure 3. Aerial imagery of Fields 1 and 2 located in TCW. Fields are outlined in orange and point locations 1 - 6 are noted [20].

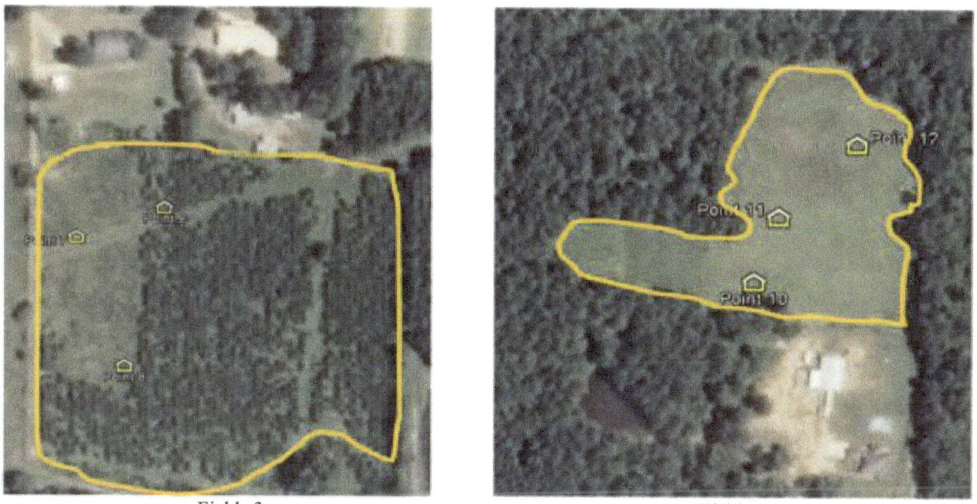

Fields 3 Fields4

Figure 4. Aerial imagery of Fields 3 and 4 located in the UPRW. Fields are outlined in orange and point locations 7 - 12 are identified [20].

Table 2. Physical and hydraulic properties of the soils.

Site	Depth (in)	ST	Bulk density ($g \cdot cm^{-3}$)	Sand (%)	Silt (%)	Clay (%)	HG
	6	SCL	1.37	7.97	51.53	40.5	
Field 1	12	SC	1.47	7.7	49.8	42.5	C
	24	SC	0.89	7.7	49.8	42.5	
	6	SCL	1.36	7.97	51.53	40.5	
Field 2	12	SC	1.42	7.7	49.8	42.5	C
	24	SC	0.97	7.7	49.8	42.5	
	6	SL	1.35	13.99	71.01	15	
Field 3	12	SL	1.55	13.67	69.33	17	C
	24	SL	1.02	13.67	69.33	17	
	6	SL	1.37	13.99	71.01	15	
Field 4	12	SL	1.73	13.67	69.33	17	C
	24	SL	0.94	13.67	69.33	17	

Acronymn: ST = soil texture, SCL = silty clay loam, SC = silty clay, SL = silt loam, HG = hydrologic group.

Fields 3 and 4 are composed of Ariel soil (MUID MS059 and MS067, respectively). The Ariel series consists of deep, well drained, nearly level silty loam soils on flood plains and low stream terraces with slopes ranging from 0 to 2 percent [23]. Most areas of Ariel soil are used for pasture and a small percentage is woodland [23]. Soil type determination was based on data from the US General Soil Map (STATSGO2) [22].

2.4. Soil Sampling

Earlier studies have shown that land use related change in soil properties is most strongly affected in the top surface layers [6]. As such, it was determined that soil cores would be taken at depths of 0 - 6 in, 6 - 12 in, and 12 - 24 in. Soil cores at the prescribed depths were collected once monthly at all twelve points. Samples were extracted using a thin-walled stainless steel T-handle soil probe (Forestry Suppliers Inc., item number 77654) of 11/16-in diameter with careful efforts made to minimize contamination between layers during extraction. Soil samples were collected in zipping plastic bags and were stored in cool (<21˚C) dark conditions until they could be processed. During the course of the study there were three months when data was not collected at Fields 1 and 2 (January 2011, April 2011, and August 2011) or at Fields 3 and 4 (November 2010, January 2011, and August 2011).

2.5. Soil Analysis

Bench Analysis. The bench analyses were performed in the Agricultural and Biological Engineering Water Quality Laboratory at Mississippi State University. To prepare the samples for dehydrating, the samples were transferred from plastic bags to standard paper bags. The samples were weighed ("wet weight") and then placed in a thermostatically controlled Grieve welded steel laboratory oven (Model LO-201C). Samples were dried at a temperature of 60˚C ± 5˚C for 24 hours. The samples were then weighed again ("dry weight") whereby their soil moisture percentage could be calculated (Equation 1).

$$\text{Percent moisture} = \frac{\text{wet weight(g)} - \text{dry weight (g)}}{\text{wet weight (g)}} \times 100 \quad (1)$$

Bulk density was determined on individual samples by dividing the oven-dry weight of the sample by the volume of the soil probe at the respective depth (see Equation (2)). The average bulk density based on depth was then taken for each field.

$$\text{Bulk density} = \frac{\text{dry weight}}{\pi r^2 h} \quad (2)$$

Soil densities vary over a wide range, and typically between 0.1 $g \cdot cm^{-3}$ for light peats and 1.8 $g \cdot cm^{-3}$ for very dense, compacted mineral soils with little pore space [24].

Elemental Analysis. Dry combustion using an elemental analyzer is the most accurate standard laboratory test for soil carbon, and is based on a solid-to-gas transformation by flash combustion of the sample material and measurement of concentrations via gas chromatography [24].

In this study samples from the same field were composited based on depth in order to achieve an average across the field at each respective depth. Once aggregated the samples were finely ground using a Dynacrush Soil Crusher (Custom Laboratory Equipment Inc., Orange City, FL). Pulverization was done until a majority of the sample could pass through a 60-mesh sieve. Sieving helped to ensure particles were 250 μm or less and also aided in the complete removal of any stray litter that may have been erroneously captured in the sample. A portion of the pulverized sample was then transferred to a 4 milliliter glass vial and dried at 105°C for one hour to remove any latent moisture acquired during pulverization. A subsample of approximately 30 milligrams was then measured out and sealed into an ultra-pure 5 × 9 mm tin capsule. The tin capsules were then loaded into a rotating autosample carousel and analyzed for carbon content using a Costech 4010 CHNS Dry Combustion Analyzer [25].

The Costech 4010 CHNS Dry Combustion Analyzer works by delivering one sample at a time into a 1050°C chromium oxide combustion reactor, pushed along by a helium carrier stream mixed with excess oxygen. Under these conditions, the tin capsule undergoes a flash combustion which raises the sample temperature to as much as 1800°C, forming water, CO_2, and nitrogen gas (N_2). Carbon content from the soil sample is changed to CO_2 during flash combustion. Water content from the soil sample is removed by a gas trap having magnesium perchlorate. Generally, clean gases from the sample are passed through a gas chromatograph column to separate the N_2 and CO_2. More chemical processes and analysis are described in the combustion analyzer document [25].

Calibration for the Costech 4010 CHNS system was performed using a standard procedure [25] using a LECO standard. In this particular study standards were measured in five increments (0.10, 0.20, 0.30, 0.45 and 0.60 micrograms) to generate the calibration curve; total C and total N content values were calculated using the standard chemical formulae and are entered before testing begins. In every ten samples an empty tin capsule blanks were used. Any detectable nitrogen or carbons in these blanks were subtracted from the sample to have a zero baseline. The correction for C traces were created from the tin capsules by blanks and for the small amount of N_2 gas introduced as an impurity in the oxygen pulse [25].

2.6. Statistical Analysis

Arithmetic means and standard deviations were calculated on the SOC data and SMC data collected from each field. Significance of the fixed effects of land use, depth of sample, sampling date, and SMC, and their interactions were analyzed using the generalized linear mixed model (GLMM) procedure (GLIMMIX) embedded in SAS software [26]. The GLMM was chosen due to its ability to model "data with correlations or nonconstant variability and where the response is not necessarily normally distributed" [26].

3. Results and Discussion

3.1. Carbon Analysis

Mean values and associated standard deviations of the SOC (percentage of carbon in the sample) are presented in **Tables 3** and **4**. Samples collected between September 2010 and February 2012 from four study sites at depths of 6 inch, 12 inch, and 24 inch were analyzed for percent carbon content using an elemental analyzer. Results were found to be consistent with the earlier studies in that the near surface (6-in) layer was found to contain significantly more carbon than either the 12- or 24-in layers ($p < 0.01$) across all field types [6,11]. Also as expected, the deeper layers show less variation with time. Furthermore, it was found that carbon amount was relatively consistent throughout the duration of the study period.

3.2. Soil Moisture

Mean values and associated standard deviations of the SMC (percentage of water in the sample) are presented in **Tables 5** and **6**. Soil moisture measurements were made on soil cores taken from the four field locations. There was found to be a high degree of variability within the soil moisture data. Across the entire dataset, there was

Table 3. Means and standard deviations of percent carbon based on field.

Depth		Number of samples	Mean	Standard deviation
6"	Field 1	15	0.841	0.184
	Field 2	15	0.926	0.173
	Field 3	15	1.194	0.285
	Field 4	15	1.116	0.242
12"	Field 1	15	0.537	0.217
	Field 2	15	0.668	0.161
	Field 3	14	0.683	0.244
	Field 4	15	0.659	0.230
24"	Field 1	15	0.385	0.164
	Field 2	15	0.497	0.192
	Field 3	13	0.485	0.268
	Field 4	14	0.427	0.162

Table 4. Overall means and standard deviations of percent carbon based on depth.

Depth	N	Mean	Standard deviation
6"	60	1.020	0.262
12"	59	0.636	0.217
24"	57	0.448	0.199

found to be a variance of 58.35 which is very high. This could be attributed to the inherent variability in collection dates after rain events or local variability of rain events (rain occurring at one site and not another). Because SMC was calculated as a percentage, sample size was not an attributing factor to the variability. Mean values and associated standard deviations of the SMC (percentage of water in the sample) are presented in **Tables 5** and **6**.

3.3. Interaction of Parameters

Secondary to the carbon content and SMC analyses which were the main justification for this portion of the study, fixed factors associated with each sampling location were analyzed to determine the degree to which they influenced carbon amount. Fixed factors included land use, depth of sample, time, and SMC. Although it is known that soil carbon content is the result of a myriad of factors, including soil physical and hydraulic properties, it was thought that there might be statistical evidence of interaction when carbon content was compared to the factors listed above. Already noted was a significant correlation of SOC and depth.

The statistical analysis procedures used for this portion of the study were the GLIMMIX and correlation procedures using SAS software. A univariate analysis was also performed on the carbon values to confirm that the distribution of the data was within acceptable range. The variable of interest is skewness and ideally it should be between −1 and 1. The carbon values for this dataset (n = 176) had a skew of 0.60 so no transformation of the data was necessary. Initially, a full model was run including all of the potential factors. The purpose of this was to identify the complexities of the responses (**Table 7**).

Several factors of the linear mixed model namely, field type ("Field"), sample depth ("Depth"), field by depth ("Field∗depth"), time ("Mon"), sampling date by depth ("Mon∗depth), and sampling date by sampling date ("Mon∗mon"), were found to be significant ($p < 0.05$).

Given the above, the next step was to simplify the analysis by isolating one variable and analyzing it at each level while everything else remained unchanged. A general statement about the four fields cannot be made because each field has its own characteristics. Therefore, the next model examined each depth (6, 12, and 24) at every field. At 6 in there were found to be linear effects as a result of field and time. LS Means ($\alpha = 0.05$) groupings of Fields 3 and 4 in A and Fields 2 and 1 in B. At 12 inches, responses were more complex; there was a linear effect of time and quadratic response over time. For this analysis, LS means ($\alpha = 0.05$) groupings were not sig-

Table 5. Means and standard deviations of SMC (percent) based on field.

Depth		N	Mean	Standard deviation
6"	Field 1	15	17.94	9.34
	Field 2	15	17.03	8.19
	Field 3	15	19.60	8.87
	Field 4	15	17.04	8.33
12"	Field 1	15	18.03	8.50
	Field 2	15	17.36	7.67
	Field 3	14	17.32	8.54
	Field 4	15	16.87	6.93
24"	Field 1	15	18.18	7.65
	Field 2	15	16.43	5.46
	Field 3	13	17.32	6.83
	Field 4	14	17.37	6.89

Table 6. Overall means and standard deviations of SMC (percent) based on depth.

Depth	N	Mean	Standard deviation
6"	60	17.90	1.209
12"	59	17.40	0.478
24"	57	17.33	0.715

Table 7. Full GLIMMIX model results showing factors and the complexity of their response.

Effect	Degrees of freedom	Den DF	F Value	Pr > F
Field	3	154	11.04	<0.0001
Depth	2	154	144.83	<0.0001
Field∗depth	6	154	3.22	0.0052
Mon	1	154	45.17	<0.0001
Mon∗field	3	154	1.27	0.2871
Mon∗depth	2	154	3.21	0.0432
Mon∗mon	1	154	12.27	0.0006
Mon∗mon∗field	13	154	1.38	0.2510
Mon∗mon∗depth	2	154	0.36	0.6981

nificantly different for any of the fields (all A grouping). At 24 inches across all fields a linear time effect was the only significant factor. Again, all LS means ($\alpha = 0.05$) groupings were not significantly different for any of the fields (all A grouping).

Then the reverse of the above was done first separating field type and then depth within every field. The effect of depth and time were highly significant at all four fields. Fields 2 and 4 also showed a quadratic response over time. LS means ($\alpha = 0.05$) groupings show significance of depth at each level: 6 inches = A, 12 inches = B, and 24 inches = C for Fields 1, 2 and 3. All four fields were found to have a difference of carbon concentration with depth but the means are different and concentration varies over time (three were linear and one was not).

Field type does have an effect at 6 inches but not at 12 inches or 24 inches as concluded by the depth analysis. The second analysis determined that two of the fields are the same but the other two are unique. Finding an effect of land use in the surface layer was consistent with what was reported by earlier study [6]. The results of both of these analyses meant further analysis to examine each depth in order to develop a mathematical model to predict carbon content. The coefficient of determination (R^2) was used to evaluate the efficacy of the equation to the observed data.

For the regression analysis the fields were recorded. Since the previous analysis determined there was not a statistical difference between Fields 1 and 2 (crop) or Fields 3 and 4 (non-crop) at the 6 inch depth, recoding reduced the data to two levels. At depths of 12 inches and 24 inches all fields were considered as one since there was no effect of field type on carbon content at those depths. The regression analysis consisted of fitting the best models to determine the regression equations. Regression equations derived from the models are as follows and are represented graphically in **Figure 5**:

Crop fields at 6 in: carbon concentration = 1.125 − 0.0281∗time.

Non-crop fields at 6 in: carbon concentration = 1.39 − 0.0281∗time.

All fields at 12 in: carbon concentration = 0.9567 − 0.07795∗time + 0.00368∗time^2.

All fields at 24 in: carbon concentration = 0.5717 − 0.01442∗time.

The R^2 values for each of the equations were 0.52, 0.22, and 0.12 respectively. All fields showed a carbon response over time. The linear response over time at the 24 in depth was not highly significant (significant at 0.05 level not 0.01). Therefore, this model may be statistically significant but biologically questionable. It is noteworthy to mention that R^2 is a function of how steep the line is (less slope means lower R^2). It is evident that Fields 3 and 4 contain higher carbon content than the combined crop fields and all fields at the 6 inch depth contain more carbon compared to all fields at 12 inch or 24 inch depths. These results are in accordance with other studies [11]. The fact that the carbon concentration is decreasing with time is unexpected. It is likely the decrease of carbon over time is a result of user error when analyzing for

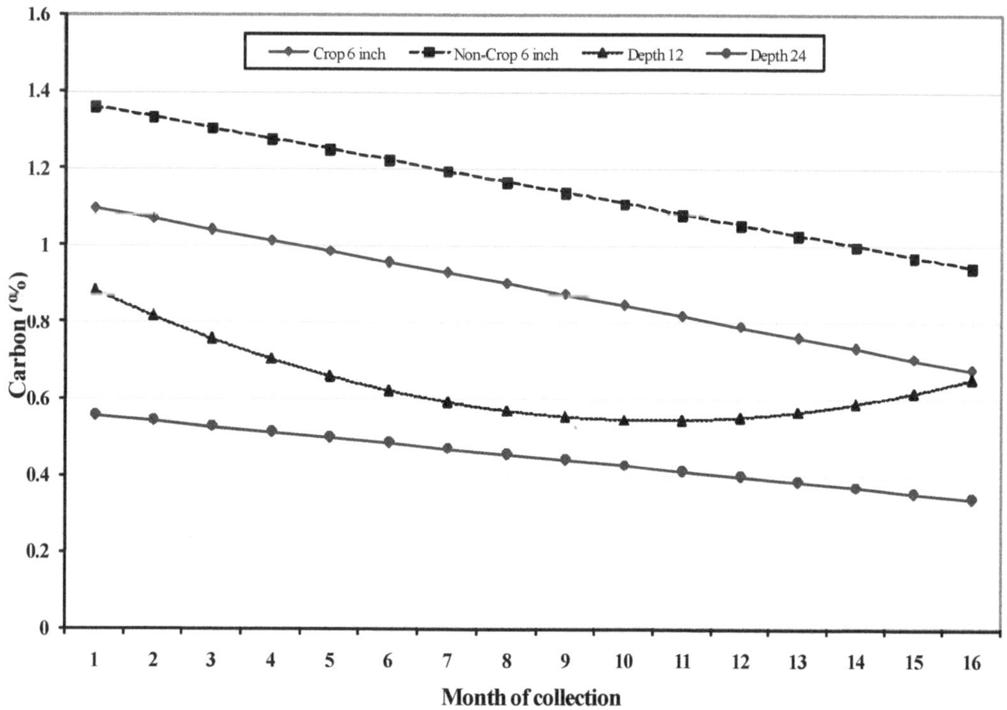

Figure 5. Regression plots using GLM model for all three depths.

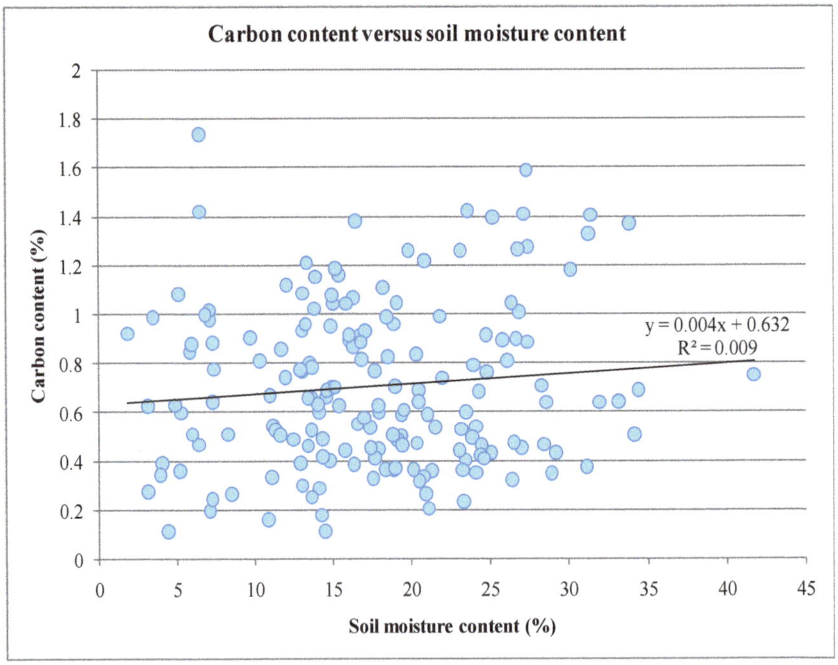

Figure 6. Correlation plot of carbon content (%) and SMC (%) across all fields at all depths (n = 176). No linear correlation exists.

carbon content and not a change that is occurring *in situ*.

A correlation procedure was performed on the carbon data and SMC data to determine if there was a relationship between the two variables. The reported correlation measures for the entire dataset (all fields and all depths) show a slight (but not significant according to the p-value) positive correlation between SMC and carbon amount (0.10). The correlation at Fields 3 and 4 at the 6 inch depth were 0.25 and 0.41 respectively. Correlation for Fields 1 and 2 (combined) was close to 0 (0.0041) and was not significant. **Figure 6** is a plot of the entire dataset (n = 176). This graph illustrates minimal correlation between the two variables of interest (SMC on the x-axis and carbon content on the y-axis). When graphed this way, it is apparent that there is no visual linear relationship between the two variables and the low R^2 confirms lack of correlation.

4. Conclusions

This study was able to demonstrate that the majority of SOC is contained within the near surface layer of soil across all three land uses studied. It was also shown that there does seem to be an interactive effect of parameters such as land use type, vertical distribution, and time on carbon accretion within the soil. There was found to be significantly more carbon in the forestland (Field 3) and pastureland (Field 4) at 6 inches than combined croplands possibly due to fewer disturbances in the soil column. A decrease in carbon amount was seen over time using regression models but this was likely a result of user error. There was no statistically significant correlation between carbon content and SMC although correlation (or lack thereof) does not imply causation.

These results are reasonable since the amount of SOC is dependent on the source input of carbon and well as management pressures. Since there were only 13 to 15 months of samples for each field, results may change as sample size increases. It was also found that carbon amount is not influenced by SMC but SMC could be influenced by SOC.

5. Acknowledgements

This material is based upon work performed through the Sustainable Energy Research Center at Mississippi State University and is supported by the Department of Energy under Award Number DE-FG3606GO86025; Micro CHP and Bio-fuel Center. We acknowledge the contributions of Jeffery Hatten in the Dept. of Forestry at Mississippi State University and landowners of the research fields in the watershed for this research.

REFERENCES

[1] D. Reay and M. Pidwirny, "Carbon Dioxide," In: C. J. Cleveland, Ed., *Encyclopedia of Earth*, 2011. http://www.eoearth.org/article/Carbon_dioxide

[2] V. Yadav, G. P. Malanson, E. Bekele and C. Lant, "Modeling Watershed-Scale Sequestration of Soil Organic Carbon for Carbon Credit Programs,". *Applied Geography*, Vol. 29, No. 4, 2009, pp. 488-500.

doi:10.1016/j.apgeog.2009.04.001

[3] N. H. Batjes, "Total Carbon and Nitrogen in the Soils of the World," *European Journal of Soil Science*, Vol. 47, No. 2, 1996, pp. 151-163. doi:10.1111/j.1365-2389.1996.tb01386.x

[4] R. A. Houghton, "The Contemporary Carbon Cycle," In: W. H Schlesinger, Ed., *Biogeochemistry*, Elsevier Science, 2005, pp. 473-513.

[5] US Environmental Protection Agency (USEPA), "Carbon Sequestration in Agriculture and Forestry," 2011. http://www.epa.gov/sequestration/tools_resources.html

[6] B. R. Wilson, T. B. Koen, P. Barnes, S. Ghosh and D. King, "Soil Carbon and Related Soil Properties along a Soil Type and Landuse Intensity Gradient, New South Wales, Australia," *Soil Use and Management*, Vol. 27, No. 4, 2011. pp. 437-447. doi:10.1111/j.1475-2743.2011.00357.x

[7] R. Lal and R. F. Follett, "Soil Carbon Sequestration and the Greenhouse Effect," 2nd Edition, Soil Science Society of America, Madison, 2009.

[8] R. Lal, "Soil Carbon Sequestration Impacts on Global Climate Change and Food Security," *Science*, Vol. 304, No. 5677, 2004. pp. 1623-1627. doi:10.1126/science.1097396

[9] S. M. Ogle, F. J. Breidt, M. D. Eve and K. Paustian, "Uncertainty in Estimating Land Use and Management Impacts on Soil Organic Carbon Storage for US Agricultural Lands between 1982 and 1997," *Global Change Biology*, Vol. 9, No. 11, 2003. pp. 1521-1542. doi:10.1046/j.1365-2486.2003.00683.x

[10] K. Y. Chan, M. K. Conyers, G. D. Li, K. R. Helyar, G. Poile, A. Oates and I. M. Barchia, "Soil Carbon Dynamics under Different Cropping and Pasture Management in Temperate Australia: Results of Three Long-Term Experiments," *Soil Research*, Vol. 49, No. 4, 2011. pp. 320-328. doi:10.1071/SR10185

[11] E. G. Jobbágy and R. B. Jackson, "The Vertical Distribution of Soil Organic Carbon and Its Relation to Climate and Vegetation," *Ecological Applications*, Vol. 10, No. 2, 2000, pp. 423-436. doi:10.1890/1051-0761(2000)010[0423:TVDOSO]2.0.CO;2

[12] M. S. Moran, C. D. Peters-Lidard, J. M. Watts and S. McElroy, "Estimating Soil Moisture at the Watershed Scale with Satellite-Based Radar and Land Surface Models," *Canadian Journal of Remote Sensing*, Vol. 30, No. 5, 2004, pp. 805-826. doi:10.5589/m04-043

[13] E. Han, "Soil Moisture Data assimilation at Multiple Scales and Estimation of Representative Field Scale Soil Moisture Characteristics," Ph.D. Thesis, Purdue University, West Lafayette, 2011.

[14] W. J. Rawls, Y. A. Pachepsky, J. C. Ritchie, T. M. Sobecki and H. Bloodworth, "Effect of Soil Organic Carbon on Soil Water Retention," *Geoderma*, Vol. 116, No. 1-2, 2003, pp. 61-76. doi:10.1016/S0016-7061(03)00094-6

[15] National Climatic Data Center (NCDC), "Locate Weather Observation Station Record," 2011. http://www.ncdc.noaa.gov/oa/climate/stationlocator.html

[16] Natural Resources Conservation Service (NRCS), "Mississippi Conservation Security Program (CSP)," 2011. http://www.ms.nrcs.usda.gov/programs/MissCSP.html

[17] US Environmental Protection Agency (USEPA), "Waterbody Report for Town Creek," 2006. http://oaspub.epa.gov/tmdl/attains_waterbody.control?p_list_id=MS013TE&p_cycle=2006&p_state=MS&p_report_type=T

[18] US Department of Agriculture, National Agricultural Statistics Service (USDA-NASS), "Mississippi County Data-Livestock.United States Department of Agriculture (USDA)," 2011. http://www.nass.usda.gov/Statistics_by_State/Mississippi/Publications/County_Estimates/index.asp

[19] Pearl River Basin Development District (PRBDD), "Pearl River Basin Development District: Topography and History," 2011. http://www.pearlriverbasin.com/topography_and_history.php

[20] Google Inc., "Google Earth (Version 6.2.2.6613) [Software]," 2012. http://www.google.com/earth/index.html

[21] M. C. Garber, "Soil Survey for Lee County Mississippi," United States Department of Agriculture, Soil Conservation Service, US Government Printing Office, 1973, p. 8.

[22] US Department of Agriculture, Natural Resources Conservation Service (USDA-NRCS), "US General Soil Map (STATSGO2) for Mississippi," 2006. http://soildatatmart.nrsc.usda

[23] F. T. Scott, "Soil Survey for Madison County, Mississippi," United States Department of Agriculture, Natural Resource Conservation Service, US Government Printing Office, 1984, pp. 2-13.

[24] P. Donovan, "Measuring Soil Carbon Change: A Flexible, Practical, Local Method," 2012. http://soilcarboncoalition.org/taxonomy/term/2

[25] Costech Instruments, "Elemental Combustion System CHNS-O," 2006. www.costechanalytical.com

[26] SAS Institute Inc., "The GLIMMIX procedure," 2006. http://support.sas.com/rnd/app/papers/glimmix.pdf

Permissions

All chapters in this book were first published in OJSS, by Scientific Research Publishing; hereby published with permission under the Creative Commons Attribution License or equivalent. Every chapter published in this book has been scrutinized by our experts. Their significance has been extensively debated. The topics covered herein carry significant findings which will fuel the growth of the discipline. They may even be implemented as practical applications or may be referred to as a beginning point for another development.

The contributors of this book come from diverse backgrounds, making this book a truly international effort. This book will bring forth new frontiers with its revolutionizing research information and detailed analysis of the nascent developments around the world.

We would like to thank all the contributing authors for lending their expertise to make the book truly unique. They have played a crucial role in the development of this book. Without their invaluable contributions this book wouldn't have been possible. They have made vital efforts to compile up to date information on the varied aspects of this subject to make this book a valuable addition to the collection of many professionals and students.

This book was conceptualized with the vision of imparting up-to-date information and advanced data in this field. To ensure the same, a matchless editorial board was set up. Every individual on the board went through rigorous rounds of assessment to prove their worth. After which they invested a large part of their time researching and compiling the most relevant data for our readers.

The editorial board has been involved in producing this book since its inception. They have spent rigorous hours researching and exploring the diverse topics which have resulted in the successful publishing of this book. They have passed on their knowledge of decades through this book. To expedite this challenging task, the publisher supported the team at every step. A small team of assistant editors was also appointed to further simplify the editing procedure and attain best results for the readers.

Apart from the editorial board, the designing team has also invested a significant amount of their time in understanding the subject and creating the most relevant covers. They scrutinized every image to scout for the most suitable representation of the subject and create an appropriate cover for the book.

The publishing team has been an ardent support to the editorial, designing and production team. Their endless efforts to recruit the best for this project, has resulted in the accomplishment of this book. They are a veteran in the field of academics and their pool of knowledge is as vast as their experience in printing. Their expertise and guidance has proved useful at every step. Their uncompromising quality standards have made this book an exceptional effort. Their encouragement from time to time has been an inspiration for everyone.

The publisher and the editorial board hope that this book will prove to be a valuable piece of knowledge for researchers, students, practitioners and scholars across the globe.

List of Contributors

Gaius D. Eudoxie, Dennison Phillips and Raymond Springer
Department of Food Production, University of the West Indies, Saint Augustine, Trinidad and Tobago

Cécile Villenave and Anne Jimenez
ELISOL Environnement and IRD UMR 210 ECO & SOLS, Montpellier, France

Muriel Guernion, Guénola Pérès and Daniel Cluzeau
UMR CNRS 6553 EcoBio, University of Rennes 1, Rennes, France

Thierry Mateille, Bernard Martiny, Mireille Fargette and Johannes Tavoillot
IRD UMR 1062 CBGP, Montferrier sur Lez, Monteferrier, France

Shunfeng Ge, Haigang Xu, Mengmeng Ji and Yuanmao Jiang
State Key Laboratory of Crop Biology, College of Horticulture Science and Engineering, Shandong Agricultural University, Taian, China

Victor H. M. Danelichen, Marcelo S. Biudes, Maísa C. Souza, Leone F. A. Curado and José S. Nogueira
Postgraduate Program in Environmental Physics, Physics Institute, Federal University of Mato Grosso (UFMT), Cuiaba, Brazil

Nadja G. Machado
Postgraduate Program in Environmental Physics, Physics Institute, Federal University of Mato Grosso (UFMT), Cuiaba, Brazil
Laboratory of Conservation Biology, Federal Institute of Mato Grosso, Cuiaba, Brazil

Marcela Quintero
Decision and Policy Analysis Research Area, International Center for Tropical Agriculture (CIAT), Cali, Colombia
University of Florida North Florida Research and Education Center, University of Florida (UF), Quincy, USA

Nicholas B. Comerford
University of Florida North Florida Research and Education Center, University of Florida (UF), Quincy, USA

Dominique Proust, Vivien Mathé and François Lévêque
LIENSs UMR 7266 CNRS-Université de La Rochelle, La Rochelle, France

Assize Touré
Centre de Suivi Ecologique (CSE), Dakar-Fann, Sénégal

Emile Temgoua
Faculty of Agronomy and Agricultural Sciences, University of Dschang, Dschang, Cameroon

Claire Guenat
Swiss Federal Institute of Technology EPFL, ENAC IIE ECOS, Lausanne, Switzerland

Bo Elberling
Institute of Geography, Copenhagen, Denmark

Kamala Haripal and Sunanda Sahoo
Ecology Section, School of Life Sciences, Sambalpur University, Burla, India

Kristine A. Nichols
USDA-ARS Northern Great Plains Research Laboratory, Mandan, USA

James Millar
Precision Soil Management, Redfield, USA

Paul Kristiansen
School of Environmental & Rural Science, University of New England, Armidale, Australia

Yadunath Bajgai
School of Environmental & Rural Science, University of New England, Armidale, Australia
RNR-Research and Development Centre, Bajo, Department of Agriculture, MoAF, Wangdue Phodrang, Bhutan

Nilantha Hulugalle
NSW Department of Primary Industries, Narrabri, Australia

Melinda McHenry
Centre for Plant and Water Sciences, Central Queensland University, Bundaberg, Australia

Lawrence H. Tanner, Morgan Nivison and David L. Smith
Department of Biological Sciences, Le Moyne College, Syracuse, USA

Ann E. Walker
Department of Earth & Planetary Sciences, and Science Alliance Center of Excellence, University of Tennessee, Knoxville, USA

Lakshmikanti Bhoi and Pramod Chandra Mishra
Department of Environmental Sciences, Sambalpur University, Orissa, India

Taoufik Saleh Ksiksi
Department of Biology, Faculty of Science, UAE University, Al-Ain, UAE

Everhard Christiaan Conradie and Valdon R. Smith
Department of Botany & Zoology, Stellenbosch University, Stellenbosch, South Africa

Asim K. Bhattacharyya
School of Environmental Sciences, Jawaharlal Nehru University, New Delhi, India

Sutapa Bose
School of Environmental Sciences, Jawaharlal Nehru University, New Delhi, India
Department of Earth Sciences, Indian Institute of Science Education and Research (IISER) Kolkata PO: BCKV Main Campus, Mohanpur Nadia, India

Farzana Rumi, Md. Abul Kashem and Khan Towhid Osman
Department of Soil Science, University of Chittagong, Chittagong, Bangladesh

Bekele Eshetu, Christel Baum and Peter Leinweber
Faculty of Agricultural and Environmental Sciences, Soil Science, University of Rostock, Rostock, Germany

Bola A. Senjobi
Federal University of Agriculture, Abeokuta, Nigeria

Olufunmilayo T. Ande
Institute of Agric Research and Training, Obafemi Awolowo University, Apata, Ibadan

A. E. Okulaja
Department of Crop Production, Olabisi Onabanjo University, Ago-Iwoye, Nigeria

Karolien Denef
Natural Resource Ecology Laboratory, Colorado State University, Fort Collins, Colorado, USA

Ilaria Del Galdo and Andrea Venturi
Environmental Science Department, Second University of Naples, Caserta, Italia

M. Francesca Cotrufo
Natural Resource Ecology Laboratory, Colorado State University, Fort Collins, Colorado, USA
Department of Soil and Crop Sciences, Colorado State University, Fort Collins, Colorado, USA

Maysoon M. Mikha, Joseph G. Benjamin and David C. Nielsen
US Department of Agricultural, Agricultural Research Service, Central Great Research Station, Akron, USA

Ardell D. Halvorson
US Department of Agricultural, Agricultural Research Service, Northern Plains Soil Plant Nutrient Research, Fort Collins, USA

Kofi Agyarko, Joseph Adu, Daniel Gyasi, Samuel Kumi and Lydia Mensah
College of Agriculture Education, University of Education, Winneba, Ghana

Orhan Dengiz, Mustafa Sağlam, F. Esra Sarioğlu, Fikret Saygin and Çağla Atasoy
Department of Soil Science and Plant Nutrition, Faculty of Agriculture, Ondokuz Mayıs University, Samsun, Turkey

Carles M. Rubio
Department Agri-Food Engineering and Biotechnology, Polytechnic University of Catalonia, Barcelona, Spain
Department Environmental Biophysics and Soils, Lab-Ferrer Soils and Environmental Consulting, Cervera, Spain

Francesc Ferrer
Department Envi-ronmental Biophysics and Soils, Lab-Ferrer Soils and Environmental Consulting, Cervera, Spain

Xavier Úbeda
Department Physical Geography, University of Barcelona, Barcelona, Spain

Amadou Keita and Hamma Yacouba
Joint Research Centre for Water and Climate, International Institute for Water and Environmental Engineering (2iE), Ouagadougou, Burkina Faso

Laszlo G. Hayde and Bart Schultz
Department of Water Science and Engineering, UNESCO-IHE-Institute for Water Education, Delft, The Netherlands

Wolińska Agnieszka, Stępniewska Zofia, Bułaś Aleksandra and Banach Artur
Department of Biochemistry and Environmental Chemistry, Institute of Biotechnology, The John Paul II Catholic University, Lublin, Poland

Lakshmy Kamala Sthanu
Centre for Environment and Development, Thozhuvancode, Vattiyoorkavu P.O., Thiruvananthapuram, India

Jaya Divakaran Sarasamma
Department of Envi- ronmental Sciences, University of Kerala, Kariavattom Campus, Thiruvananthapuram, India

Radhakrishnan Thanu Iyer
Division of Geo Informatics, Indian Institute of Information Technology and Management, Kerala, Techno Park Campus, Thiruvananthapuram, India

Mohandas Chellappan
Division of Crop Protection, Central Tuber Crops Research Institute, Sreekaryam, Thiruvananthapuram, India

Rudolf Schulz and Torsten Müller
University of Hohenheim, Institute of Crop Science, Fertilisation and Soil Matter Dynamics (340i), Stuttgart, Germany

Yonathan D. Redel
University of Hohenheim, Institute of Crop Science, Fertilisation and Soil Matter Dynamics (340i), Stuttgart, Germany Scientific and Technological Bioresource Nucleus (Bioren-UFRO), Universidad de La Frontera, Temuco, Chile

Richard L. Haney, Mike. J. White and Robert. D. Harmel
USDA-Agricultural Research Service (ARS), Grassland, Soil & Water Research Laboratory, Temple, USA

Alan. J. Franzluebbers
USDA-ARS, J. Phil Campbell, Sr. Natural Resource Conservation Center, Watkinsville, USA

Virginia. L. Jin
USDA-ARS, Agroecosystem Management Research Unit, Lincoln, USA

Mari-Vaughn. Johnson
4USDA-NRCS, Resources Inventory and Assessment Division, Temple, USA

Elizabeth. B. Haney
Texas AgriLife Research, Blackland Research & Extension Center, Temple, USA

Shao-Qiang Tao, Qiang Xia, Lin Zhu, Jing-Jing Chen, Ya-Nan Wang and Bing Qin
College of Resources and Environmental Sciences, Anhui Agricultural University, Hefei, China

Stella Asuming-Brempong
Department of Soil Science, School of Agriculture, University of Ghana, Legon, Ghana

Upendra M. Sainju, Thecan Caesar-TonThat, Jalal D. Jabro, Robert T. Lartey, Robert G. Evans and Brett L. Allen
US Department of Agriculture, Agricultural Research Service, Northern Plains Agricultural Research Laboratory, Sidney, USA

Andrew W. Lenssen
Agronomy Department, Iowa State University, Ames, USA

Prem B. Parajuli and Sarah Duffy
Department of Agricultural and Biological Engineering, Mississippi State University, Starkville, USA